Applications of Hybrid Nanofluids in Science and Engineering

Applications of Hybrid Nanofluids in Science and Engineering delves deep into the multifaceted realms in which these dynamic fluids are playing a pivotal role in various fields.

This comprehensive volume elucidates the diverse applications and promising potentials of hybrid nanofluids. It introduces hybrid nanofluids and their preparation methods, thermophysical properties, advantages, applications, and future scope. Models to compute the effective thermophysical properties of hybrid nanofluids are also discussed, along with their limitations. In the application section, mathematical models are formulated to contemplate the flow of hybrid nanofluids through different surfaces/geometries under different situations. Also, the entropy generation minimization in hybrid nanofluid flow is discussed with its application in refrigeration, power generation, and other processes.

The subject matter in this book will enable the reader to do the following:

- Learn the ins and outs of hybrid nanofluids—from how they are made to the special characteristics they embody.
- Explore hybrid nanofluids' potential in thermal management, energy systems, materials science, biomedical engineering, and more.
- Use advanced computational and analytical methods to analyse complex fluid dynamics models.
- Anticipate the impact of hybrid nanofluid research on upcoming sectors like renewable energy and innovative manufacturing.

This book is aimed at researchers and graduate students in mechanical and chemical engineering and materials science.

Emerging Materials and Technologies

Series Editor: Boris I. Kharissov

The *Emerging Materials and Technologies* series is devoted to highlighting publications centered on emerging advanced materials and novel technologies. Attention is paid to those newly discovered or applied materials with potential to solve pressing societal problems and improve quality of life, corresponding to environmental protection, medicine, communications, energy, transportation, advanced manufacturing, and related areas.

The series takes into account that, under present strong demands for energy, material, and cost savings, as well as heavy contamination problems and worldwide pandemic conditions, the area of emerging materials and related scalable technologies is a highly interdisciplinary field, with the need for researchers, professionals, and academics across the spectrum of engineering and technological disciplines. The main objective of this book series is to attract more attention to these materials and technologies and invite conversation among the international R&D community.

Metal Organic Framework Derived Materials: Design Strategies and Applications
Gomathi Nageswaran, Varsha M V, Arun Kumar Rajasekaran and M Shashank Rao

Hydrogen Production, Storage, and Utilization: Technologies and Applications
Abbas Tcharkhtchi, Hamidreza Vanaei, Albert Lucas and Sedigheh Farzaneh

MXenes for Energy Storage Applications: Emerging Characteristics, Compositions, and Synthesis Methods
Muhammad Rafique, M. Bilal Tahir, and Saira Anwar

Multi-scale and Multifunctional Coatings and Interfaces for Tribological Contacts
Ajit Behera, Kuldeep K Saxena, Dipen Kumar Rajak and Shankar Sehgal

Nanotechnology in Green Energy Generation
Ahmed Thabet Mohamed

Thermo-Acoustics of Nanofluids and Transfer Processes
Shriram S. Sonawane and Manjakuppam Malika

Hybrid Nanofluids: Heat and Mass Transfer Processes
Shriram S. Sonawane and Manjakuppam Malika

Applications of Hybrid Nanofluids in Science and Engineering
Edited by A. K. Pandey, H. Upreti, O. D. Makinde, and A. J. Chamkha

For more information about this series, please visit: www.routledge.com/Emerging-Materials-and-Technologies/book-series/CRCEMT

Applications of Hybrid Nanofluids in Science and Engineering

Edited by A. K. Pandey, H. Upreti, O. D. Makinde, and A. J. Chamkha

CRC Press
Taylor & Francis Group
Boca Raton London New York

CRC Press is an imprint of the
Taylor & Francis Group, an **informa** business

First edition published 2025
by CRC Press
2385 NW Executive Center Drive, Suite 320, Boca Raton FL 33431

and by CRC Press
4 Park Square, Milton Park, Abingdon, Oxon, OX14 4RN

CRC Press is an imprint of Taylor & Francis Group, LLC

ISBN: 978-1-032-53547-0 (hbk)
ISBN: 978-1-032-97854-3 (pbk)
ISBN: 978-1-003-59578-6 (ebk)

DOI: 10.1201/9781003595786

Typeset in Times
by Apex CoVantage, LLC

Contents

Preface

The fusion of nanotechnology and fluid mechanics has unleashed a realm of possibilities that transcend conventional boundaries in science and engineering. Nanofluids, colloidal suspensions of nanoparticles in a base fluid, have emerged as a promising class of materials with unique thermal, mechanical, and optical properties. Hybrid nanofluids, among the various types of nanofluids, have garnered significant attention for their enhanced performance and multifaceted applications.

This edited volume, *Applications of Hybrid Nanofluids in Science and Engineering*, presents a comprehensive exploration of hybrid nanofluids' advancements, challenges, and diverse applications across different domains. It brings together contributions from leading experts in the field, encompassing theoretical insights, experimental methodologies, and practical implementations.

The chapters within this volume traverse a wide spectrum of topics, spanning from fundamental principles to cutting-edge applications. Readers will delve into the synthesis of hybrid nanofluids and the techniques used for their characterization, unravelling the intricacies of nanoparticle dispersion, stability, and interaction mechanisms. Theoretical models and computer simulations help us understand how hybrid nanofluids behave in different flow conditions, thus allowing us to make better designs and predict how well they will work.

Furthermore, this book delves into the diverse array of applications wherein hybrid nanofluids demonstrate unparalleled efficacy. From heat transfer enhancement in thermal management systems to lubrication improvement in mechanical engineering, biomedical diagnostics to environmental remediation, the potential of hybrid nanofluids is boundless. Each chapter offers valuable insights, empirical data, and practical considerations that contribute to the burgeoning field of nanofluids.

As editors, we are immensely grateful to the contributing authors for their scholarly contributions and dedication to advancing knowledge in this field. We also extend our gratitude to the reviewers for their meticulous evaluation and constructive feedback, which have enriched the quality and rigour of this volume.

Editor Biographies

Alok Kumar Pandey is an Assistant Professor in the Department of Mathematics, Graphic Era (Deemed to be University), Dehradun, Uttarakhand, India. His area of research is Computational Fluid Dynamics. He is currently the Editor for the *Journal of Engineering Researcher and Lecturer*, Associate Editor for the *Journal of Advanced Research in Numerical Heat Transfer* and *Journal of Advanced Research in Micro and Nano Engineering*, and serving as a member of the editorial board for the journal *Teknomekanik.*

Himanshu Upreti is an Assistant Professor in the School of Engineering and Technology, BML Munjal University, Gurugram, Haryana, India. His research interest includes topics from fluid dynamics, heat and mass transfer analysis. He is member of the Indian Society of Theoretical and Applied Mechanics (ISTAM) and National Society of Fluid Mechanics and Fluid Power (NSFMFP).

Oluwole Daniel Makinde is a distinguished Professor of Applied Mathematics and Computation with over 26 years of experience in South African universities. He is affiliated with the Faculty of Military Science at Stellenbosch University, South Africa. Makinde has received several prestigious awards, including the 2011 African Union Kwame Nkrumah Continental Scientific Award, the 2009/2010 T.W. Kambule Senior Researcher Award, and the 2014 Nigerian National Honour Award (MFR). He is a fellow of multiple academies, including the African Academy of Sciences and the International Academy of Physical Sciences.

Ali J. Chamkha is a Distinguished Professor of Mechanical Engineering and Dean of Engineering at Kuwait College of Science and Technology. He is currently the Editor-in-Chief for the *Journal of Nanofluids* and has served as an Editor, Associate Editor or a member of the editorial board for many journals such as *ASME Journal of Thermal Science and Engineering Applications*, *International Journal of Numerical Method for Heat and Fluid Flow*, *Journal of Thermal Analysis and Calorimetry*, *Journal of Porous Media*, and others. Professor Chamkha was included in the World's Top 2% Scientists 2020, 2021, 2022, and 2023 (by Stanford University).

Editor Biographies

Contributors

Nilankush Acharya
NCP Umasashi Institution,
Department of Mathematics, Kolkata 700122, West Bengal, India.

Asgar Ali
Department of Mathematics,
Bajkul Milani Mahavidyalaya, Purba Medinipur 721655, West Bengal, India.

Taher Armaghani
Department of Mechanical Engineering,
Islamic Azad University, West Tehran Branch.

Amal T. Babu
Chemistry Division, School of Advanced Sciences,
Vellore Institute of Technology, Chennai Campus, Chennai 600127, Tamil Nadu, India.

Rupa Baithalu
Department of Mathematics,
Siksha 'O' Anusandhan Deemed to be University, Bhubaneswar 751030, Odisha, India.

Priya Bartwal
Department of Mathematics,
Graphic Era (Deemed to be University), Dehradun, Uttarakhand, India.

Ankita Bisht
Department of Mathematics,
Amity School of Physical Sciences, Amity University Punjab, India.

Arvind Singh Bisht
Centre for Energy,
NIT Hamirpur, Himachal Pradesh, India.

Ali J. Chamkha
Faculty of Engineering,
Kuwait College of Science and Technology, Doha District, Kuwait.

M. Chandrasekar
Department of Mechanical Engineering,
University College of Engineering BIT Campus, Tiruchirappalli 620024, Tamil Nadu, India.

Sanatan Das
Department of Mathematics,
University of Gour Banga, Malda 732103, West Bengal, India.

Sai Ganga
SoET, BML Munjal University, Gurgaon, Haryana, India.

Ramin Ghasemiasl
Department of Mechanical Engineering,
Islamic Azad University, West Tehran Branch.

Tanya Gupta
Department of Mathematics,
IAH, GLA University Mathura, India.

Sanjalee Maheshwari
School of Basic Sciences,
IIIT Una, Himachal Pradesh, 177209.

Oluwole Daniel Makinde
Faculty of Military Science,
Stellenbosch University, South Africa.

Ramakanta Meher
Department of Mathematics and Humanities,
Sardar Vallabhbhai National Institute of Technology Surat, Surat 395007, Gujarat, India.

S.R. Mishra
Department of Mathematics,
ITER, Siksha 'O' Anusandhan Deemed to be University, Bhubaneswar 751030, Odisha, India.

Aryan Moshiri
Department of Mechanical Engineering,
Islamic Azad University, West Tehran Branch.

Mahtab Nazarahari
Department of Mechanical Engineering,
Islamic Azad University, West Tehran Branch.

Subhajit Panda
Centre for Data Science,
Siksha 'O' Anusandhan Deemed to be University, Bhubaneswar 751030, Odisha, India.

Alok Kumar Pandey
Department of Mathematics,
Graphic Era (Deemed to be University), Dehradun, India.

Jayanta Parui
Chemistry Division, School of Advanced Sciences, Vellore Institute of Technology, Chennai Campus, Chennai 600127, Tamil Nadu, India.

J.R. Pattnaik
Department of Mathematics,
Odisha University of Technology and Research, Bhubaneswar 751029, Odisha, India.

P.K. Pattnaik
Department of Mathematics,
Odisha University of Technology and Research, Bhubaneswar 751029, Odisha, India.

Niraj Rathore
Department of Mathematics,
Central University of Karnataka, Kalaburagi 585367, Karnataka, India.

N. Sandeep
Department of Mathematics,
Central University of Karnataka, Kalaburagi 585367, Karnataka, India.

Soumitra Sarkar
Department of Mathematics,
Triveni Devi Bhalotia College, Paschim Bardhaman 713347, West Bengal, India.

Parthkumar P. Sartanpara
Department of Mathematics and Humanities,
Sardar Vallabhbhai National Institute of Technology Surat, Surat 395007, Gujarat, India.

R.S. Tripathy
Department of Mathematics,
Siksha 'O' Anusandhan Deemed to be University, Bhubaneswar 751030, Odisha, India.

Ziya Uddin
SoET, BML Munjal University, Gurgaon, Haryana, India.

Himanshu Upreti
SoET, BML Munjal University,
Gurgaon, Haryana, India.

Mohd Vaseem
SoET, BML Munjal University,
Gurgaon, Haryana, India.

Lalchand Verma
Department of Mathematics and
Humanities,
Sardar Vallabhbhai National
Institute of Technology
Surat, Surat 395007,
Gujarat, India.

1 Introduction to Hybrid Nanofluids

Ziya Uddin, Himanshu Upreti, and Mohd Vaseem

1.1 INTRODUCTION

The industrialization era has made humans realize the importance of optimal utilization of natural resources to meet the energy requirements of present and future generations. To achieve this, scientists explore various methods; among these, the intensification of heat transfer using heat transfer fluids, i.e., mixtures of base fluid with solid metal particles or their metallic oxides, is most significant.

Maxwell [1] conducted pioneering work in this area. In this work, he reported that dispersing millimetre-sized particles (metal or metal oxides) can enhance the heat transfer characteristics of conventional fluids such as water, ethylene glycol (EG), and oil. However, the yielded fluid was not stable and hence was not regarded as an efficient heat transfer fluid for the future. Thereafter, for more than a century, numerous researchers conducted several studies on the thermal conductivity enhancement of liquid–solid solutions (millimetre-sized solid particles), with the rapid settling of solid particles being the major hurdle. In 1993, Masuda et al. [2] extended the work of Maxwell [1] by reducing the size of solid particles in the range of 10^{-6} m. The solid particles of the prepared fluid still settled down, but not as rapidly as in the earlier case.

After a decade, researchers discovered that dispersion of ultrafine solid particles, i.e., particles with a size range of 1~100 nm in the conventional fluid, resulted in the development of a more stable coolant, named "nanofluid." Choi and Eastman [3] were the first to unturn this cornerstone. The solid particles present in nanofluid can be metal particles such as Ag, Au, and Cu; metal oxides such as TiO_2, Al_2O_3, CuO, and ZnO; carbon-based particles such as carbon nanotubes (CNTs); and many others. Nanofluids (mono) are considered the most suitable coolant or heat transfer fluid (HTF) due to their superior thermophysical properties. However, they still have several challenges, such as the long-term stability of nanoparticle dispersion, increased pressure drops, thermal performance in turbulent flow, higher viscosity, and lower specific heat. Upon further analysis, it has been shown that metal nanoparticles exhibit enhanced thermal conductivity and reduced chemical inertness and stability. On the other hand, metal oxide nanoparticles demonstrate higher stability but diminished thermal conductivity. Therefore, mono-nanofluid exhibits either high heat conductivity or enhanced stability.

Later, the concept of nanofluids comprising nanocomposites gained attention and became an intriguing subject of study. Nanocomposites involve synthesizing two or more nanoparticles into a unified structure. They represent materials that combine

DOI: 10.1201/9781003595786-1

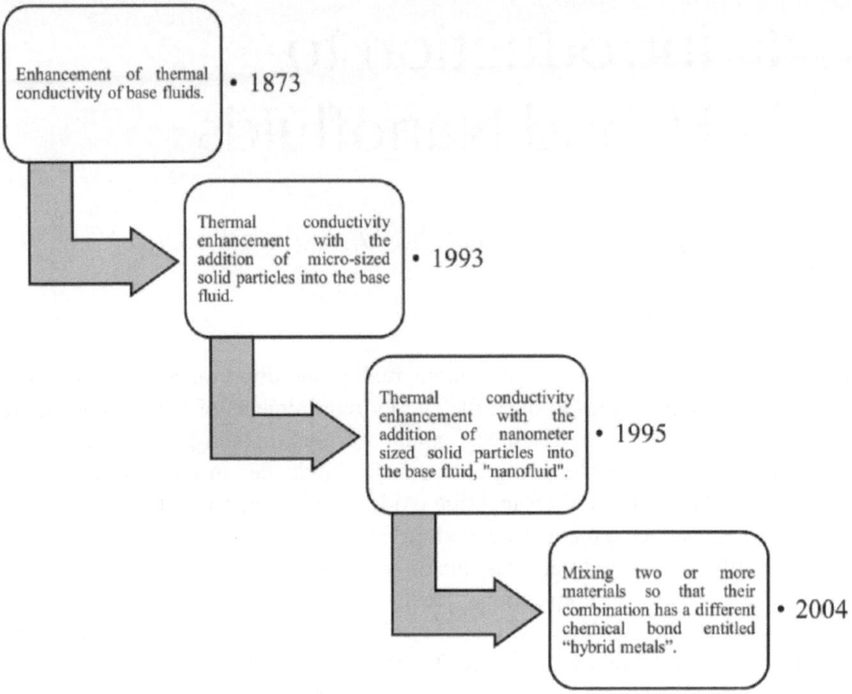

FIGURE 1.1 Journey of hybrid nanofluid.

the physical and chemical properties of multiple constituents, delivering these attributes in a homogeneous phase. "Hybrid nanofluids" is a buzzword used to refer to nanofluids formed by distributing nanocomposites into conventional fluids or by integrating various nanoparticles into the same base fluid. Hybrid nanofluids appear promising for improving heat transfer efficiency and thermo-physical characteristics when compared to traditional HTFs such as oil, water, pure water, propylene glycol, and EG, as well as nanofluids containing just one kind of nanoparticle. Researchers have discovered optimal properties in hybrid nanofluids, indicating their suitability for solar systems that demand good thermal, optical, and rheological working fluid characteristics. Over the last decade, the use of nanofluids has skyrocketed. Despite numerous discrepancies in the published data and a lack of knowledge of the heat transmission mechanism in nanofluids, it has emerged as a potential HTF. The use of hybrid nanofluids is meant to improve the transfer of heat and pressure drop properties by managing the pros and cons of each suspension well, especially the nanoparticles that have a high aspect ratio, a better thermal network, and an impact that works together [4–6].

The research landscape of nanofluid technology encompasses a variety of studies aimed at enhancing heat transfer and fluid flow characteristics. Esfahani et al. [7] made a ZnO–Ag/water nanofluid using ultrasonic and magnetic methods and studied how well it conducted heat at different temperatures and volume fractions.

Kakavandi et al. [8] studied the stability and thermal conductivity of a hybrid nanofluid made of water, EG, SiC, and multiwalled carbon nanotubes (MWCNTs). They found that higher concentrations of nanoparticles resulted in better thermal conductivity. Timofeeva et al. [9] studied how the shape of alumina nanoparticles in a fluid affected their thermal conductivity and viscosity and found that interfacial effects made the thermal conductivity improvements less noticeable. Esfe et al. [10] conducted experiments to study the thermal conductivity of an MWCNT–ZnO hybrid nanofluid and suggested using artificial neural networks (ANNs) to build a correlation model for understanding thermal conductivity. Suresh et al. [11] studied how heat moved and how much pressure dropped in a circular tube using a hybrid nanofluid of Al_2O_3–Cu and water. They found that this hybrid nanofluid had higher friction factors than a single nanofluid. In the same way, Esfe et al. [12] created Cu and TiO_2 hybrid nanofluids using a mixture of water–EG as the base fluid and suggested correlations for accurate measurements of thermal conductivity.

Huang et al. [13] analysed the heat transfer and pressure drop in Al_2O_3–water hybrid nanofluid flow through MWCNTs. Hayat et al. [14], on the other hand, investigated the three-dimensional flow of Ag-CuO–water hybrid nanofluid over a stretching surface, highlighting enhanced heat transfer compared to that in simple nanofluids.

The study by Esfe et al. [15] focused on making a hybrid nanofluid better at conducting heat by mixing magnesium oxide (MgO) nanoparticles with a base fluid made of single-walled carbon nanotubes and EG. They employed ANN techniques to accurately estimate the fluid's effective thermal conductivity under varying environmental conditions, thereby contributing to improved heat transfer performance in engineering applications. Bellos et al. [16] studied how well a parabolic solar collector worked with Syltherm 800, an oil-based fluid with nanoparticles of alumina (Al_2O_3) and titanium dioxide (TiO_2) added to it. By developing and validating a thermal model using Engineering Equation Solver (EES) software, they aimed to optimize the collector's heat transfer efficiency, offering potential benefits for solar energy utilization. Yıldız et al. [17] did a computer study on how heat moves through water-based nanofluids containing Al_2O_3 and SiO_2 nanoparticles. Their investigation included an analysis of different particle volume fractions and Rayleigh numbers, emphasizing the enhancement of heat transfer performance. They also looked into hybrid nanofluids containing both Al_2O_3 and SiO_2 nanoparticles. This helped us expand our knowledge on improving heat transfer in these kinds/categories of systems.

Sheikholeslami et al. [18] investigated the enhancement of convective heat transfer within a circular cavity containing a MWCNT–Fe_3O_4/H_2O hybrid nanofluid. They found that the Nusselt number and temperature boundary layer thickness increased. Tong et al. [19] looked into the thermal and optical features of the MWCNT–Fe_3O_4 hybrid nanofluid, focusing on its energy efficiency. Also, research by [20–22] looked into how to create entropy, control the boundary layer, and allow natural convection to happen in hybrid nanofluids. These investigations provide valuable insights into the potential applications and performance of hybrid nanofluids in various thermal systems.

Researchers [23–25] have further investigated heat transport and entropy formation in hybrid nanofluids using numerical simulations and predictive modelling research and demonstrated that algorithms based on artificial intelligence are capable of precisely predicting the behaviour of thermohydraulic systems. In addition, references [26–30] show that scientists studied how hybrid nanofluids acted in various situations, including how magnetic fields, bio-convection, and shape optimization affected the flow. The results of these experiments demonstrate the adaptability and helpfulness of hybrid nanofluids in a variety of engineering contexts.

Finally, Bhatti et al. [31] and Shoeibi et al. [32] investigated the movement of heat and mass across porous media and solar desalination systems using hybrid nanofluids. The researchers revealed that these fluids have the potential to improve energy efficiency and protect the environment.

The purpose of this work is to provide an in-depth review of hybrid nanofluids, including the techniques used for their synthesis, variables that influence their thermophysical characteristics, theoretical and experimental methods for correlating these properties, the problems encountered, and their limitations, applications, and future prospects. In addition, we have examined past evaluations and included significant final observations for the reader's convenience. Figure 1.1 displays the journey of hybrid nanofluids.

1.2 OVERVIEW OF HYBRID NANOFLUIDS

Hybrid nanofluids are a class of advanced HTFs that combine the unique properties of nanoparticles with a base fluid. These nanofluids exhibit enhanced thermal and physical characteristics compared to their base fluids alone, making them promising candidates for a wide range of applications. Hybrid nanofluids disperse nanoparticles in a base fluid, creating a stable suspension. The nanoparticles, typically with sizes ranging from 1 to 100 nanometres, provide high thermal conductivity due to their large surface area-to-volume ratio. The synergistic effects of mixing nanoparticles of different sizes and shapes in hybrid nanofluids can significantly influence their thermophysical properties. Incorporating nanoparticles can improve the thermal conductivity of the nanofluid, whereas adding another nanoparticle can enhance the stability of the nanofluid and prevent particle agglomeration. Moreover, hybrid nanofluids can exhibit altered rheological behaviour and unique flow characteristics compared to traditional HTFs. The properties of hybrid nanofluids can be tailored by carefully selecting the composition, concentration, size, and shape of the nanoparticles, along with the choice of the base fluid. Hybrid nanofluids are synthesized using a variety of techniques, including chemical and physical methods, with an emphasis on achieving homogeneous dispersion and particle stability in the base fluid.

1.2.1 Synthesis of Hybrid Nanofluids

The preparation of nanofluids significantly influences the enhancement of the thermo-physical properties of the base fluid and is regarded as the initial and crucial step in experimentation. Commonly used techniques for synthesizing nanofluids include the one-step method and the two-step method.

One-step method: This method involves the simultaneous synthesis and dispersion of nanoparticles (see Figure 1.2). Unlike multistep processes, this approach eliminates the need for time-consuming procedures such as

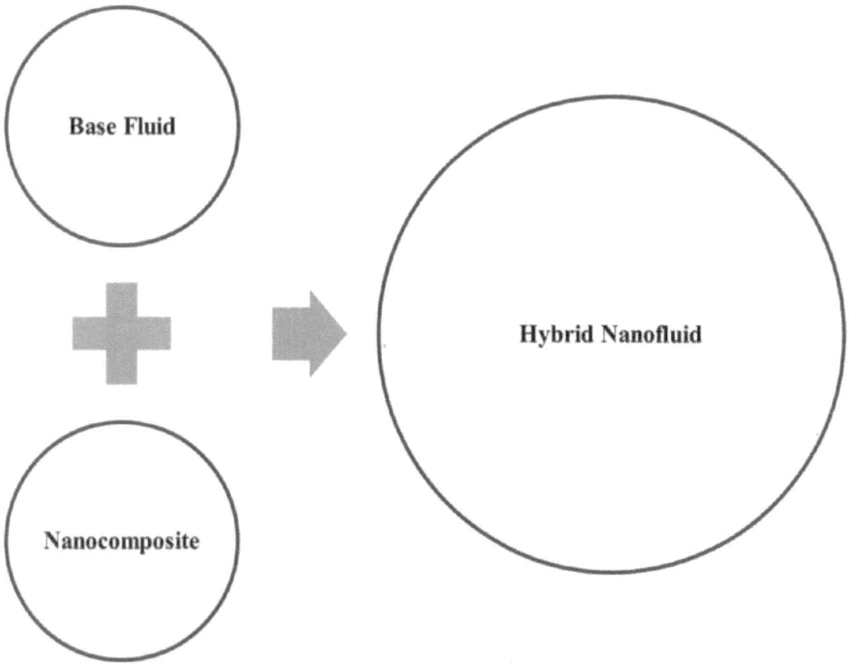

FIGURE 1.2 Illustration of the one-step method.

drying, storage, and particle mixing in the base fluid. As a result, nanofluids produced through this method exhibit enhanced stability and more uniform dispersion. However, it is important to note that the one-step method is associated with higher preparation costs and limitations in scalability for large-scale production.

Two-step method: This method involves the initial synthesis of nanocomposites, nanotubes, and nanoparticles in the form of a dry powder (see Figure 1.2). This is achieved through chemical or mechanical processes such as milling, grinding, vapour phase, and sol–gel methods. The nanoparticles are then evenly distributed in a dispersion medium, such as water, EG, or a combination of water, EG, or oils. However, during dispersion, the nanoparticles have a tendency to form agglomerates as a result of the attractive van der Waals and cohesive interactions between the particles. Different methods are used to avoid this clustering, which can in turn affect the thermophysical properties of the nanoparticles. These include high-shear mixing, ultrasonic disruption, ultrasonic bath, ultrasonic agitation, and magnetic force agitation. It is crucial to acknowledge that no one material contains all the necessary features and qualities, which means that there is a need to compromise between thermal and rheological capabilities. In such situations, hybrid nanofluids are the preferred choice because they provide enhanced thermal conductivity as a result of a synergistic impact.

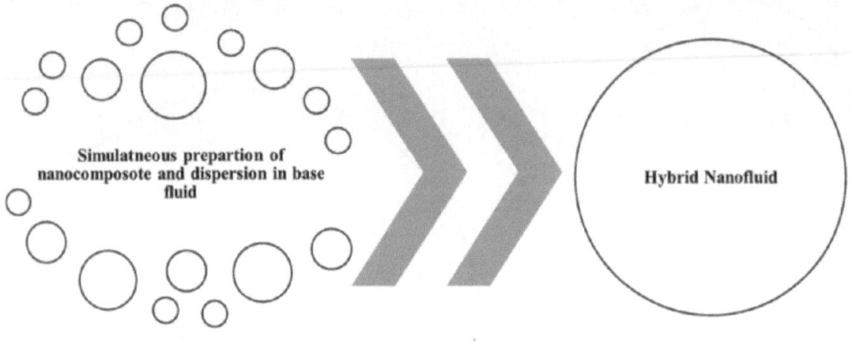

FIGURE 1.3 Illustration of the two-step method.

The advantages of employing the two-step method include the ability for large-scale production and affordability, while the drawbacks associated with this approach include nanoparticle aggregation and the complexity involved in nanofluid preparation.

1.2.2 Factors Affecting the Properties of Hybrid Nanofluids

The thermophysical properties of hybrid nanofluids mainly depend on the properties of the mixed nanoparticles and the properties of the base fluid. Several factors influence the thermal and physical properties of hybrid nanofluids. The major factors influencing the properties are as follows:

Nanoparticle material: The choice of nanoparticles used in the hybrid nanofluid has a substantial influence on its characteristics. The thermophysical parameters of a nanofluid are directly influenced by the unique thermal conductivities, specific heat capacities, and surface features of the different materials used.

Particle concentration: The concentration of different nanoparticles in the hybrid nanofluid plays a crucial role in determining its thermal and physical properties. As the concentration increases, the nanofluid's effective thermal conductivity improves, along with changes in other properties such as viscosity and specific heat capacity. However, we should wisely determine the optimum concentration of the different nanoparticles.

Particle size and shape: The size and shape of the nanoparticles and microparticles have a significant impact on the properties of hybrid nanofluids. Smaller particle sizes can improve thermal conductivity due to increased surface area, whereas particle shape can affect dispersion and interaction within the fluid.

Base fluid: In addition, the characteristics of the hybrid nanofluid are affected by the base fluid that is used for the dispersion of nanoparticles. Different base fluids have varied thermal conductivities, viscosities, and specific heat

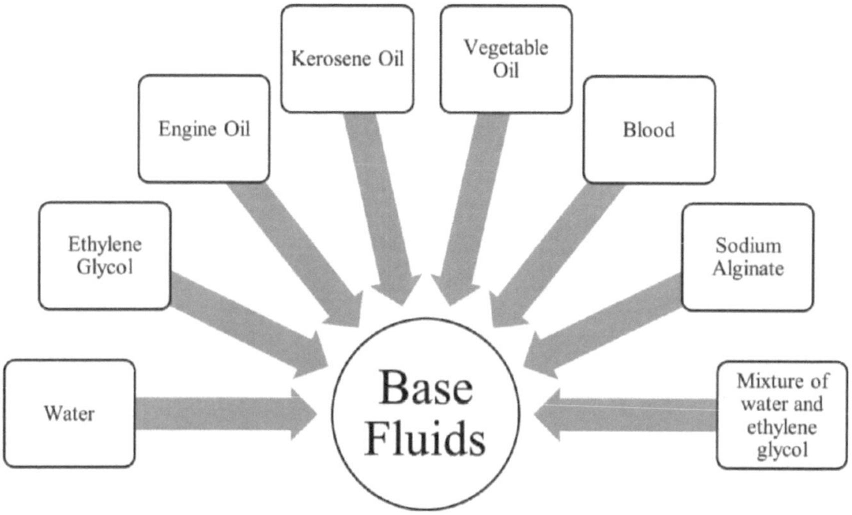

FIGURE 1.4 Base fluids used by researchers in their work.

capacities. These attributes interact with the particles and contribute to the overall properties of the nanofluid. Based on the available research, Figure 1.4 shows a selection of the base fluids that are most often used.

Surface modification or nanoparticle coating: The surface properties of nanoparticles may change as a result of surface modification. The thermophysical characteristics of hybrid nanofluid can be modified by coating or functionalization of the nanoparticle surface, which aids in mixing of the particles and maintaining their stability within the base fluid.

Temperature: The hybrid nanofluid's operating temperature can have a significant impact on its properties. Thermal conductivity, viscosity, and specific heat capacity can change with temperature, leading to variations in heat transfer performance and the overall behaviour of the nanofluid.

Agglomeration and sedimentation: The stability of hybrid nanofluids is crucial for maintaining their desired properties. Agglomeration and sedimentation of nanoparticles and microparticles can affect the dispersion and alter the thermal and physical characteristics of the nanofluid.

1.2.3 Thermophysical Properties of Hybrid Nanofluids

For examining the performance of any fluid (mono-nanofluid or hybrid nanofluid) with better accuracy, knowledge about its properties (both thermophysical and molecular) is important. Moreover, thermophysical properties, i.e., density, viscosity (dynamic), specific heat, expansion coefficient (thermal), and thermal conductivity, play a vital role in the heat transfer behaviour of hybrid nanofluid flow, and these features are affected by several factors (see Figure 1.5). The theoretical correlations

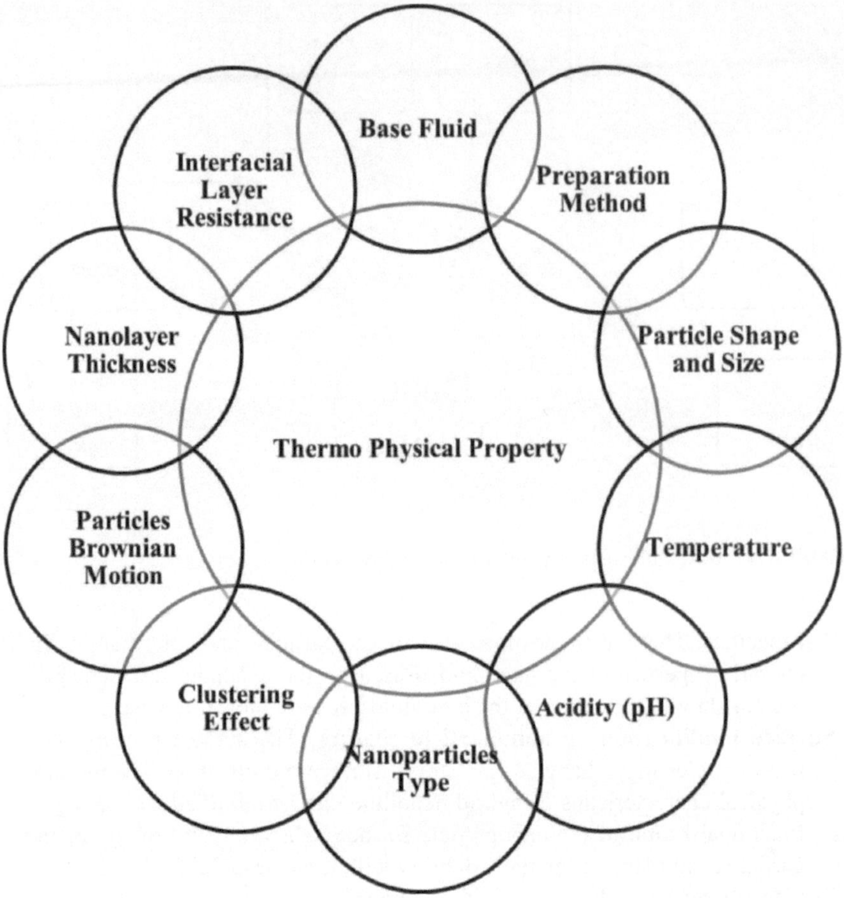

FIGURE 1.5 Factors affecting thermophysical properties.

commonly found in the literature for predicting the density, specific heat, and thermal expansion coefficient of nanofluids often rely on the fluid mixture rule. In this section, we examine the theoretical correlations used to calculate the effective thermophysical properties of hybrid nanofluids.

Density is defined as "the ratio of mass to volume of a substance," and it is measured in terms of kg/m^3. For hybrid nanofluid with nanoparticle volume fractions φ_{s1} and φ_{s2}, density is given as

$$\rho_{hnf} = (1 - \phi_1 - \phi_2)\rho_f + \phi_1\rho_{s1} + \phi_2\rho_{s2}, \tag{1.1}$$

where ρ_f, ρ_{s1}, and ρ_{s2} are the density of the base fluid and nanoparticles, respectively.

Specific heat capacity is "the amount of heat that should be given to one kilogram of a substance to increase its temperature by 1 degree Kelvin," and it is measured in units of J/kg·K. For hybrid nanofluid with nanoparticle volume fractions φ_{s1} and φ_{s2}, the specific heat capacity is given as

$$(\rho C_P)_{hnf} = (1-\phi_1-\phi_2)(\rho C_P)_f + \phi_1(\rho C_P)_{s1} + \phi_2(\rho C_P)_{s2}, \tag{1.2}$$

where $(C_P)_f$, $(C_P)_{s1}$, and $(C_P)_{s2}$ are the heat capacity of the base fluid and the nanoparticles, respectively.

The **volumetric thermal expansion** coefficient for a nanofluid is "the amount of change in nanofluid volume per one degree Kelvin increase in the temperature of the mixture," and it is measured in units of 1/K. For a hybrid nanofluid with nanoparticle volume fractions φ_{s1} and φ_{s2}, it is given as

$$(\rho\beta)_{hnf} = (1-\phi_1-\phi_2)(\rho\beta)_f + \phi_1(\rho\beta)_{s1} + \phi_2(\rho\beta)_{s2}, \tag{1.3}$$

where β_f, β_{s1}, and β_{s2} are the thermal expansion coefficients of the base fluid and the nanoparticles, respectively.

The **dynamic viscosity** of nanofluids is a crucial determinant of pumping power, flow resistance, and nanofluid availability. The viscosity of nanofluids is influenced by parameters such as temperature, particle morphology, particle size, volume concentration, presence of surfactants, and the properties of the base fluid. The Brinkman model, often used in previous numerical investigations on hybrid nanofluids, is used to evaluate the effective viscosity and is given as

$$\mu_{hnf} = \frac{\mu_f}{(1-\phi_1-\phi_2)^{2.5}}. \tag{1.4}$$

Thermal conductivity is defined as "heat transfer ability of material"; it is a cardinal property that decides the heat transfer characteristics of hybrid nanofluids. In case of hybrid nanofluids, thermal conductivity depends on both the dispersed nanoparticles or nanocomposites and the base fluid.

Modified Maxwell model is a widely employed model to compute the effective thermal conductivity of hybrid nanofluids, and it is expressed as

$$\frac{\kappa_{hnf}}{\kappa_f} = \frac{\dfrac{\varphi_{s1}\kappa_{s1}+\varphi_{s2}\kappa_{s2}}{\varphi_{s1}+\varphi_{s2}} + 2\kappa_f + 2(\varphi_{s1}\kappa_{s1}+\varphi_{s2}\kappa_{s2}) - 2(\varphi_{s1}+\varphi_{s2})\kappa_f}{\dfrac{\varphi_{s1}\kappa_{s1}+\varphi_{s2}\kappa_{s2}}{\varphi_{s1}+\varphi_{s2}} + 2\kappa_f - 2(\varphi_{s1}\kappa_{s1}+\varphi_{s2}\kappa_{s2}) + (\varphi_{s1}+\varphi_{s2})\kappa_f}. \tag{1.5}$$

Furthermore, the following are the recent models developed by researchers based on experimental findings:

Reference	Model
Esfahani et al. [7]	$\frac{\kappa_{hnf}}{\kappa_f} = \left(1 + 0.0008794(\varphi_{s1} + \varphi_{s2})^{0.5899} T^{1.345}\right)$
Kakavandi and Akbari [8]	$\frac{\kappa_{hnf}}{\kappa_f} = \left(0.981 + 0.0017(\varphi_{s1} + \varphi_{s2})^{0.698} T^{1.396}\right)$
Timofeeva et al. [9]	$\frac{\kappa_{hnf}}{\kappa_f} = 1 + 3(\varphi_{s1} + \varphi_{s2})$
Esfe et al. [10]	$\frac{\kappa_{hnf}}{\kappa_f} = 1.024 + 0.5988(\varphi_{s1} + \varphi_{s2})^{0.6029} \exp\left(\frac{\varphi_{s1} + \varphi_{s2}}{T}\right) - \left(\frac{8.059(\varphi_{s1} + \varphi_{s2})T^{0.2} + 2.24}{6.052(\varphi_{s1} + \varphi_{s2})^{0.2} + T}\right)$

1.3 HYBRID NANOFLUIDS AND ASSOCIATED CHALLENGES

Hybrid nanofluids are colloidal suspensions consisting of a mixture of two or more nanoparticles in the conventional base fluid. These new age fluids offer enhanced heat transfer and fluid properties compared to conventional fluids. However, they also present several challenges that need to be addressed. These challenges include dispersion stability of the nanofluid, particle settling issues, interparticle interactions, nanoparticle concentration ratios, particle agglomeration, stability with time, cost, and scalability. These challenges are explained further as follows.

Particle settling: In hybrid nanofluids, particle sedimentation is a significant challenge. Because different types of nanoparticles are present, settling rates may vary, which may result in non-uniform particle distribution. Devices using hybrid nanofluids may become clogged as a result of this settling. This may further reduce efficiency in heat transfer applications.

Interparticle interactions: Hybrid nanofluids involve interactions between different types of nanoparticles, which can influence the overall properties and behaviour of the hybrid nanofluids. The complex forces between different particle sizes, shapes, and surface chemistries can affect the dispersion stability and performance of the nanofluids.

Concentration ratio: Determining the optimal concentration ratio of different nanoparticles is challenging. Finding the right balance between two components is crucial for achieving the desired properties and maximizing the heat transfer performance of the nanofluid. However, determining the ideal concentration ratio requires extensive experimentation and optimization, and this may increase the synthesis cost of nanofluids.

Particle agglomeration: In hybrid nanofluids, aggregation or clustering of nanoparticles is a common problem. Agglomerates can form due to the attractive forces between particles, which reduce the effective surface area and hinder heat transfer. For improved performance, we must address the challenges involved in disrupting or preventing agglomeration.

Stability over time: Over time, hybrid nanofluids may experience changes in stability and properties due to particle settling, agglomeration, and interactions with the base fluid. Long-term stability is critical for practical applications, and addressing hybrid nanofluid ageing and stability issues is an ongoing challenge.

Cost and scalability: Hybrid nanofluid production and synthesis can be costly, especially when multiple particle types are involved. To make hybrid nanofluids commercially viable, we must address the challenges in achieving scalability and cost-effectiveness for large-scale industrial applications.

1.4 APPLICATIONS OF HYBRID NANOFLUIDS

Due to their enhanced thermophysical properties, hybrid nanofluids have a variety of industrial applications. The following list highlights some of the notable applications.

Heat transfer systems: Hybrid nanofluids have garnered considerable attention in research due to their ability to enhance heat transfer across a range of systems. They find use in a variety of applications, such as heat exchangers, cooling systems for electronic devices, and advanced engineering systems for thermal management, due to their superior thermal conductivity and convective heat transfer properties.

Solar thermal systems: Hybrid nanofluids are also useful in managing and improving the energy conversion efficiency of solar thermal systems. Hybrid nanofluids can improve solar collectors' performance by increasing sunlight absorption and facilitating heat transfer, resulting in increased overall energy output.

Electronics cooling: Hybrid nanofluids have the potential to address the cooling challenges associated with high-power electronic devices. Hybrid nanofluids, with their enhanced thermal conductivity, effectively dissipate heat from electronic components, thereby preventing overheating and enhancing device performance and reliability.

Automotive and aerospace cooling: The automotive and aerospace industries have investigated hybrid nanofluids for cooling applications. The use of hybrid nanofluids as coolants in engines, radiators, and heat exchangers improves overall thermal management and system efficiency.

Energy storage systems: Researchers have explored hybrid nanofluids to enhance the efficiency of energy storage systems such as phase-change materials (PCMs). Incorporation of nanoparticles and microparticles into PCMs enhances heat transfer during the charging and discharging processes, resulting in faster energy storage and release.

Biomedical applications: Hybrid nanofluids have potential applications in the field of biomedicine. Hybrid nanofluids can be used in targeted cancer therapy for hyperthermia treatment, wherein external energy sources generate heat with nanoparticles to selectively destroy cancer cells while minimizing damage to healthy tissues.

Oil and gas industry: The oil and gas industry has studied hybrid nanofluids for various applications. Hybrid nanofluids can be used in enhanced oil recovery (EOR) processes to enhance the fluid displacement and reservoir sweep efficiency. Additionally, hybrid nanofluids can aid in reducing frictional losses and improving the thermal conductivity of drilling fluids.

1.5 FUTURE ASPECTS OF HYBRID NANOFLUIDS

A number of promising directions for future research and development of hybrid nanofluids exist. The area is still in its early stages of development. Creating hybrid nanofluids with better dispersion properties and increased stability will be the primary goal of future studies. Surface functionalization, smart coatings, and customized surfactants are some of the creative ways that might be explored to provide long-term stability by preventing particle agglomeration and sedimentation. Concentrating on multicomponent hybrid nanofluids is another way to enhance the efficiency of hybrid nanofluids. Nanoparticles and a base fluid are the usual components of modern hybrid nanofluids. In the future, researchers may look at ways to improve the characteristics of hybrid nanofluids by adding new components, such as other kinds of nanoparticles or additives. Several researchers have already brought up the possibility that hybrid nanofluids might help make energy conversion systems more efficient. So, to maximize the efficiency of energy conversion, researchers will look into using hybrid nanofluids in thermoelectric devices, photovoltaics, and energy harvesting. Both mono- and hybrid nanofluids pose certain risks to the environment. Therefore, future improvements will undoubtedly consider the sustainability and environmental effects of hybrid nanofluids. Researchers will focus on using non-toxic and environmentally friendly materials, investigating recyclable nanofluids, and evaluating the environmental impacts of nanofluid use in the long run. Researchers in the domains of modelling and simulation are driving efforts to understand the behaviour of hybrid nanofluids and precisely forecast their performance. The complicated interactions between particles and the base fluid may be better understood by creating more accurate computer models, which can guide future studies into how these systems behave under varying operating circumstances.

1.6 CONCLUSION

This chapter provided a comprehensive overview of hybrid nanofluids and their applications. It discussed their synthesis methods, factors affecting their properties, limitations, and challenges. The chapter also emphasized the many uses of hybrid nanofluids, such as improving heat transfer, powering energy systems, advancing health sciences, and supporting developing technologies. The chapter ended by addressing potential areas for future study as well as the difficulties and possibilities that lie ahead in the realm of hybrid nanofluids. This chapter is a great resource for researchers, engineers, and scientists who are interested in comprehending and investigating the potential of hybrid nanofluids in many industrial and technical fields.

REFERENCES

1. Maxwell, J.C. (1873). *A treatise on electricity and magnetism* (Vol. 1). Clarendon Press.
2. Masuda, H., Ebata, A., & Teramae, K. (1993). Alteration of thermal conductivity and viscosity of liquid by dispersing ultra-fine particles. Dispersion of Al_2O_3, SiO_2 and TiO_2 ultra-fine particles. Netsu Bussei, 7, 227–233.
3. Choi, S.U., & Eastman, J.A. (1995). Enhancing thermal conductivity of fluids with nanoparticles (No. ANL/MSD/CP-84938; CONF-951135-29). Argonne National Lab. (ANL), Argonne, IL (United States).
4. Babu, J.R., Kumar, K.K., & Rao, S.S. (2017). State-of-art review on hybrid nanofluids. Renewable and Sustainable Energy Reviews, 77, 551–565.
5. Huminic, G., & Huminic, A. (2018). Hybrid nanofluids for heat transfer applications – a state-of-the-art review. International Journal of Heat and Mass Transfer, 125, 82–103.
6. Muneeshwaran, M., Srinivasan, G., Muthukumar, P., & Wang, C.C. (2021). Role of hybrid-nanofluid in heat transfer enhancement – A review. International Communications in Heat and Mass Transfer, 125, 105341.
7. Esfahani, N.N., Toghraie, D., & Afrand, M. (2018). A new correlation for predicting the thermal conductivity of ZnO–Ag (50%–50%)/water hybrid nanofluid: An experimental study. Powder Technology, 323, 367–373.
8. Kakavandi, A., & Akbari, M. (2018). Experimental investigation of thermal conductivity of nanofluids containing of hybrid nanoparticles suspended in binary base fluids and propose a new correlation. International Journal of Heat and Mass Transfer, 124, 742–751.
9. Timofeeva, E.V., Routbort, J.L., & Singh, D. (2009). Particle shape effects on thermophysical properties of alumina nanofluids. Journal of Applied Physics, 106(1), 014304.
10. Esfe, M.H., Esfandeh, S., Saedodin, S., & Rostamian, H. (2017). Experimental evaluation, sensitivity analyzation and ANN modelling of thermal conductivity of ZnO-MWCNT/EG-water hybrid nanofluid for engineering applications. Applied Thermal Engineering, 125, 673–685.
11. Suresh, S., Venkitaraj, K.P., Selvakumar, P., & Chandrasekar, M. (2012). Effect of Al_2O_3–Cu/water hybrid nanofluid in heat transfer. Experimental Thermal and Fluid Science, 38, 54–60.
12. Esfe, M.H., Wongwises, S., Naderi, A., Asadi, A., Safaei, M.R., Rostamian, H., Dahari, M., & Karimipour, A. (2015). Thermal conductivity of Cu/TiO_2–water/EG hybrid nanofluid: Experimental data and modelling using artificial neural network and correlation. International Communications in Heat and Mass Transfer, 66, 100–104.
13. Huang, D., Wu, Z., & Sunden, B. (2016). Effects of hybrid nanofluid mixture in plate heat exchangers. Experimental Thermal and Fluid Science, 72, 190–196.
14. Hayat, T., & Nadeem, S. (2017). Heat transfer enhancement with Ag-CuO/water hybrid nanofluid. Results in Physics, 7, 2317–2324.
15. Esfe, M.H., Alirezaie, A., & Rejvani, M. (2017). An applicable study on the thermal conductivity of SWCNT-MgO hybrid nanofluid and price-performance analysis for energy management. Applied Thermal Engineering, 111, 1202–1210.
16. Bellos, E., & Tzivanidis, C. (2018). Thermal analysis of parabolic trough collector operating with mono and hybrid nanofluids. Sustainable Energy Technologies and Assessments, 26, 105–115.
17. Yıldız, Ç., Arıcı, M., & Karabay, H. (2019). Comparison of a theoretical and experimental thermal conductivity model on the heat transfer performance of Al_2O_3-SiO_2/water hybrid-nanofluid. International Journal of Heat and Mass Transfer, 140, 598–605.
18. Sheikholeslami, M., Mehryan, S.A.M., Shafee, A., & Sheremet, M.A. (2019). Variable magnetic forces impact on magnetizable hybrid nanofluid heat transfer through a circular cavity. Journal of Molecular Liquids, 277, 388–396.

19. Tong, Y., Boldoo, T., Jeonggyun, H., & Cho, H. (2020). Improvement of photo-thermal energy conversion performance of MWCNT/Fe_3O_4 hybrid nanofluid compared to Fe_3O_4 nanofluid. Energy, 196, 117086.
20. Gul, T., Khan, A., Bilal, M., Alreshidi, N.A., Mukhtar, S., Shah, Z., & Kumam, P. (2020). Magnetic dipole impact on the hybrid nanofluid flow over an extending surface. Scientific Reports, 10(1), 1–13.
21. Huminic, G., & Huminic, A. (2020). Entropy generation of nanofluid and hybrid nanofluid flow in thermal systems: A review. Journal of Molecular Liquids, 302, 112533.
22. Ghalambaz, M., Doostani, A., Izadpanahi, E., & Chamkha, A.J. (2020). Conjugate natural convection flow of Ag–MgO/water hybrid nanofluid in a square cavity. Journal of Thermal Analysis and Calorimetry, 139(3), 2321–2336.
23. Muhammad, K., Hayat, T., Alsaedi, A., & Ahmad, B. (2021). Melting heat transfer in squeezing flow of base fluid (water), nanofluid (CNTs + water) and hybrid nanofluid (CNTs + CuO + water). Journal of Thermal Analysis and Calorimetry, 143(2), 1157–1174.
24. Muhammad, K., Hayat, T., Alsaedi, A., Ahmad, B., & Momani, S. (2021). Mixed convective slip flow of hybrid nanofluid (MWCNTs + Cu + water), nanofluid (MWCNTs + water) and base fluid (water): A comparative investigation. Journal of Thermal Analysis and Calorimetry, 143(2), 1523–1536.
25. Alizadeh, R., Abad, J.M.N., Fattahi, A., Mohebbi, M.R., Doranehgard, M.H., Li, L.K., Alhajri, E., & Karimi, N. (2021). A machine learning approach to predicting the heat convection and thermodynamics of an external flow of hybrid nanofluid. Journal of Energy Resources Technology, 143(7), 070908.
26. Hussain, S.M., & Jamshed, W. (2021). A comparative entropy-based analysis of tangent hyperbolic hybrid nanofluid flow: Implementing finite difference method. International Communications in Heat and Mass Transfer, 129, 105671.
27. Pandey, A.K., Upreti, H., Joshi, N., & Uddin, Z. (2022). Effect of natural convection on 3D MHD flow of MoS_2–GO/H_2O via porous surface due to multiple slip mechanisms. Journal of Taibah University for Science, 16 (1), 749–762.
28. Shah, S.A.A., Ahammad, N.A., Din, E.M.T.E., Gamaoun, F., Awan, A.U., & Ali, B. (2022). Bio-convection effects on Prandtl hybrid nanofluid flow with chemical reaction and motile microorganism over a stretching sheet. Nanomaterials, 12(13), 2174.
29. Chu, Y.M., Bashir, S., Ramzan, M., & Malik, M.Y. (2022). Model-based comparative study of magnetohydrodynamics unsteady hybrid nanofluid flow between two infinite parallel plates with particle shape effects. Mathematical Methods in the Applied Sciences, 46(10), 11568–11582.
30. Uddin, Z., Hassan, H., Harmand, S., & Ibrahim, W. (2022). Soft computing and statistical approach for sensitivity analysis of heat transfer through the hybrid nanoliquid film in rotating heat pipe. Scientific Reports, 12 (1), 14983.
31. Bhatti, M.M., Ellahi, R., & Doranehgard, M.H. (2022). Numerical study on the hybrid nanofluid (Co_3O_4-Go/H_2O) flow over a circular elastic surface with non-Darcy medium: Application in solar energy. Journal of Molecular Liquids, 361, 119655.
32. Shoeibi, S., Kargarsharifabad, H., Rahbar, N., Ahmadi, G., & Safaei, M.R. (2022). Performance evaluation of a solar still using hybrid nanofluid glass cooling-CFD simulation and environmental analysis. Sustainable Energy Technologies and Assessments, 49, 101728.

2 Empirical Correlations for the Estimation of Thermophysical, Friction Factor, and Heat Transfer Properties of Hybrid Nanofluids

M. Chandrasekar

2.1 INTRODUCTION

Thermal management systems for various applications rely upon newer, effective, and advanced coolants as conventional coolants like air and water have lower heat dissipation rates. During the early 1990s, nanotechnology opened the doors for the emergence of a novel class of coolants, namely, nanofluids [1]. Nanofluids could be viewed as a kind of composite coolant synthesized with the dispersion of nanometre-sized solid particles in conventional coolants. Initial research works related to nanofluids discussed the synthesis routes, evaluation of physical properties such as the density and viscosity, and thermal properties like specific heat and thermal conductivity. Although nanofluids exhibited enhanced thermal conductivity, the mechanism for its enhancement is not clear yet. However, a multimodal mechanism is thought to be the reason for the thermal conductivity enhancement [2].

Continued research efforts on nanofluids in the past 3 decades has resulted in the exploration of various categories of nanofluids [3–5]. A variety of nanofluids were proposed with respect to the nanoparticle dispersed and the base fluid used. The research progress on nanofluids resulted in the following classes of nanofluids:

- Mono nanofluid
- Hybrid nanofluid

Mono and hybrid nanofluids are developed with the dispersion of a single or more than one nanoparticle, respectively, in the base fluids. The current state of research on hybrid nanofluids indicates that hybrid nanofluids can be classified as follows:

- Binary nanofluid
- Ternary nanofluid

DOI: 10.1201/9781003595786-2

Hybrid nanofluids synthesized with two and three nanoparticles are termed as binary and Ternary nanofluids, respectively.

The thermo-hydraulic characteristics of heat transfer devices are very much influenced by the thermophysical characteristics of the coolants, which could be comprehended from non-dimensional parameters such as Reynolds number (Re) and Prandtl number (Pr) in the equations of friction factor (f) and Nusselt number (Nu). The fundamentals of convective heat transfer revealed that the heat transfer rates strongly depend on the specific heat and thermal conductivity, while the stability characteristics and pumping power depend on the density and viscosity, respectively. Therefore, nanofluids with higher specific heat and thermal conductivity and lower density and viscosity are preferable [6].

In this chapter, the various empirical correlations available in the literature for the estimation of the thermophysical characteristics of hybrid nanofluids like density, viscosity, specific heat, and thermal conductivity are presented. In addition, the correlations proposed for the prediction of the thermo-hydraulic performance of hybrid nanofluids are discussed. The applicability of these correlations with respect to the volume concentration range, mixing ratio, temperature, type of base fluid, and nanoparticle is also discussed.

2.2 DENSITY

Density of any material under consideration is evaluated based on the ratio of its mass and the volume it occupies. Hence, density is experimentally determined from the mass and volume of the material, which could be measured easily using a physical balance and a measuring jar, respectively. However, theoretical estimation of the density of mono- and hybrid nanofluids or a composite material is done with the application of the mixture rule.

The mixture rule formula is a straightforward one as given by Equation (2.1) [7, 8]. This is applicable for situations wherein the individual volume concentrations of the nanoparticles are known.

$$\rho_{hnf} = \phi_{np1}\rho_{np1} + \phi_{np2}\rho_{np2} + \left(1 - \phi_{np1} - \phi_{np2}\right)\rho_{bf}, \tag{2.1}$$

where the volume fraction of the individual particles could be determined using Equations (2.2) and (2.3).

$$\phi_{np1} = \frac{\left[\dfrac{w_{np1}}{\rho_{np1}}\right]}{\left[\dfrac{w_{np1}}{\rho_{np1}}\right] + \left[\dfrac{w_{np2}}{\rho_{np2}}\right] + \left[\dfrac{w_{bf}}{\rho_{bf}}\right]} \tag{2.2}$$

$$\phi_{np2} = \frac{\left[\dfrac{w_{np2}}{\rho_{np2}}\right]}{\left[\dfrac{w_{np1}}{\rho_{np1}}\right] + \left[\dfrac{w_{np2}}{\rho_{np2}}\right] + \left[\dfrac{w_{bf}}{\rho_{bf}}\right]} \tag{2.3}$$

Instead of considering the individual volume fractions of nanoparticles, it is possible to determine, for hybrid nanoparticles, the volume concentration and density using Equations (2.4) and (2.5), respectively. Finally, for hybrid nanofluids, density is determined with the application of the mixture rule as mentioned in Equation (2.6) [9].

$$\phi_{hp} = \frac{\left[\frac{w_{np1}}{\rho_{np1}}\right] + \left[\frac{w_{np2}}{\rho_{np2}}\right]}{\left[\frac{w_{np1}}{\rho_{np1}}\right] + \left[\frac{w_{np2}}{\rho_{np2}}\right] + \left[\frac{w_{bf}}{\rho_{bf}}\right]} \tag{2.4}$$

$$\rho_{hp} = \left[\rho_{np1+np2}\right] = \frac{\rho_{np1} w_{np1} + \rho_{np2} w_{np2}}{w_{np1+np2}} \tag{2.5}$$

$$\rho_{hnf} = \phi_{hp}\rho_{hp} + \left(1 - \phi_{hp}\right)\rho_{bf} \tag{2.6}$$

As the predicted density of the hybrid nanofluid varied marginally with the experimental values, a concept of application of correction factor was introduced by Vajjha et al. [10]. As the variation is not constant under different volume concentrations, the value for the correction factor Δρ was obtained in terms of volume concentration. Thus, in the case of nanofluids with ZnO nanoparticles, expressions for correction factor and corrected density were proposed, respectively, as

$$\Delta\rho = \frac{\left(0.9848\phi + 0.732\right)}{100}, \tag{2.7}$$

$$\rho_{nf} = \phi_p \rho_p + \left(1 - \phi_p\right)\rho_{bf} - \Delta\rho. \tag{2.8}$$

It is widely accepted that volume concentration and temperature have a direct and inverse relation, respectively, with density. Therefore, a polynomial equation for density of hybrid nanofluids with temperature and volume concentration was given by Sofiah et al. [11] for palm oil nanofluids containing CuO and polyaniline nanoparticles:

$$\rho_{hnf} = 0.92097 - 0.00064T + 0.03083\phi_{hp} - 0.0674\phi_{hp}^2 + 0.000645T\phi_{hp}^2. \tag{2.9}$$

2.3 SPECIFIC HEAT

The specific heat of a hybrid nanofluid can be theoretically estimated by two approaches:

- A model based on mixing theory
- A model based on thermal equilibrium.

As per the mixing theory model, the following equation is used based on the mixture rule [12]:

$$c_{p,hnf} = \phi_{hp} c_{p,hp} + \left(1 - \phi_{hp}\right) c_{p,bf}\,, \tag{2.10}$$

where $C_{p,hp} = \dfrac{C_{p,np1} w_{np1} + C_{p,np2} w_{np2}}{w_{np1+np2}}$ represents the specific heat for a hybrid particle.

As per the thermal equilibrium model [13, 14], the expression of specific heat is given by

$$C_{p,\,hnf} = \frac{\phi_{np1}\rho_{np1}c_{p,np1} + \phi_{np2}\rho_{np2}c_{p,np2} + (1 - \phi_{np1} - \phi_{np2})\rho_{bf}c_{p,bf}}{\rho_{hnf}}. \tag{2.11}$$

Although both the approaches are widely used by the research community, the thermal equilibrium model-based prediction of specific heat is found to be more accurate [15].

2.4 VISCOSITY

Viscosity of hybrid nanofluids is found to vary with the volume fraction, temperature, and shear stress. Numerous linear and non-linear correlations were proposed for predicting the viscosity of hybrid nanofluids made using a variety of nanoparticles, base fluids, volume concentrations (ϕ), temperatures (T), and shear rates (γ). Sepehrnia et al. [16] highlighted that the frequently proposed functional forms of the correlations for viscosity prediction are in the form of

$$\frac{\mu_{hnf}}{\mu_{bf}} = f(\phi)\,, \tag{2.12}$$

$$\frac{\mu_{hnf}}{\mu_{bf}} = f\left(\phi, T\right), \tag{2.13}$$

$$\frac{\mu_{hnf}}{\mu_{bf}} = f\left(\phi, T, \gamma\right). \tag{2.14}$$

It is widely accepted without any fear of contradiction by the research community that the increase in volume fraction enhances the viscosity of hybrid fluids. The correlations proposed in terms of volume concentration only are highlighted in Table 2.1 along with their validity [14, 17–22]. The studies revealed that the correlations involving volume concentration for the prediction of viscosity of hybrid nanofluids fell under any one of the following models:

- Linear model
- Polynomial model with higher orders
- Power law model.

TABLE 2.1
Viscosity Correlations for Hybrid Nanofluids in Terms of Volume Concentration

S.No.	Model type	Correlation	Validity: Hybrid nanoparticles	Validity: Base fluid	Validity: Volume concentration (%)	Validity: Temperature range (°C)	Reference
1	Linear	$\frac{\mu_{hnf}}{\mu_{bf}} = 0.312\phi + 1.3194$	SiC/TiO_2 (equal volumes)	Diathermic oil	≤0.8	20	Wei et al. [17]
2	Polynomial model with higher orders	$\frac{\mu_{hnf}}{\mu_{bf}} = a_0\phi + a_1\phi - a_2\phi^2 + a_3\phi^3$	$MWCNT/SiO_2$ (20:80)	SAE40	≤ 2	25–50	Efse et al. [18]
3		$\frac{\mu_{nf}}{\mu_{bf}} = a_0\phi + a_1\phi - a_2\phi^2 + a_3\phi^3 + a_4\phi^4$	$SiO_2/MWCNT$ (equal volumes)	SAE40	≤ 1	25–60	Afrand et al. [19]
4		$\frac{\mu_{hnf}}{\mu_{bf}} = 1 + 32.795\phi - 7214\phi^2 + 71400\phi^3 - 0.1941\times10^8\phi^4$	Ag/MgO (equal volumes)	Water	≤0.8	–	Efse et al. [20]
5	Power law	$\frac{\mu_{nf}}{\mu_{bf}} = 0.9595(1+\phi)^{2.399}$	Magnetic ND/Co_3O_4 (16:84)	Water and EG	<0.15	20–60	Sunder et al. [21]
6		$\frac{\mu_{haf}}{\mu_{bf}} = \frac{1}{(1-\phi_1)^{2.5}(1-\phi_2)^{2.5}}$	TiO_2/Cu	Water	$\phi_1 = 3$, $\phi_2 = 1,3,5$	–	Ghadikolaei et al. [22]
7	Exponential	$\frac{\mu_{nf}}{\mu_{bf}} = ae^{b\phi}$	Magnetic ND/Fe_3O_4 (72:28)	Water and EG	0.05–0.2	20–60	Sunder et al. [14]

The coefficients in the correlation of Efse et al. [18] and Afrand et al. [19] depend on the temperature.

The viscosity of hybrid nanofluids decreased with an increase in temperature for all volume concentrations. The reason for this behaviour is attributed to the gain in energy of the particles and the subsequent augmented mobility of the nanoparticles, reduced agglomeration, and molecular intermolecular forces [23]. A multivariable regression technique is being commonly adopted by researchers to propose correlations involving temperature and volume fraction for predicting viscosity. The multivariable correlations as provided in Tables 2.2 and 2.3 fall under any one of the following categories [24–37]:

- Linear model
- Quadratic and polynomial model
- Interactive model
- Exponential model
- Power model.

2.5 THERMAL CONDUCTIVITY

Thermal conductivity is found to vary directly with volume loading concentration and temperature. Similar to viscosity, a copious number of linear and non-linear correlations were proposed for predicting thermal conductivity too for a variety of nanoparticles, base fluids, volume concentrations (ϕ), and temperatures (T) [24–26, 38–55]. The frequently proposed linear and non-linear functional forms of the predictive correlations for thermal conductivity are as follows:

$$\frac{k_{hnf}}{k_{bf}} = f(\phi), \tag{2.15}$$

$$\frac{k_{hnf}}{k_{bf}} = f(\phi,T). \tag{2.16}$$

The reason for the direct dependency of correlations on volume loading concentration was due to the improved heat-carrying capacity due to a clustering effect, i.e., the presence of metallic particles with higher thermal conductivity. Brownian motion coupled with thermophoresis, i.e., the rapid movement of nanoparticles within the base fluids from the region of low temperature to the region of high temperature, enhances the net thermal conductivity. Furthermore, with the increase in temperature, the Brownian motion is also expected to increase. Therefore, the correlations describing thermal conductivity are in direct relationship with temperature. The correlations to compute the effective thermal conductivity of hybrid nanofluid are presented in Table 2.4.

2.6 NUSSELT NUMBER

In heat transfer calculations, Nusselt number (Nu) signifies the ratio of resistances to heat flow by conduction and convection modes. Higher Nu values indicate that

TABLE 2.2

Viscosity Correlations for Hybrid Nanofluids in Terms of Volume Concentration and Temperature

Sl. No.	Model type	Correlation	Validity: Hybrid nanoparticles	Validity: Base fluid	Validity: Volume concentration (%)	Validity: Temperature range (°C)	Reference
1	Power	$\mu_{nf} = 328201 \times T^{-2.053} \times \varphi^{0.09359}$	MWCNT/MgO (20:80)	Engine oil	0.25–2	25–50	Asadi et al [24]
2		$\frac{\mu_{hnf}}{\mu_{bf}} = 1.42(1+R)^{-0.1063}\left(\frac{T}{80}\right)^{0.2321}$	TiO_2-SiO_2 (20:80 to 80:20)	Water–EG (60:40)	1	30–80	Hamid et al. [25]
3		$\frac{\mu_{anf}}{\mu_{bf}} = 37\left(1+\frac{\phi}{100}\right)^{1.59}\left(0.1+\frac{T}{80}\right)^{0.31}$	TiO_2-SiO_2 50:50	Water–EG (60:40)	0.5–3	30–80	Nabil et al. [26]
4		$\mu_{hnf} = 0.9653 + 77.4567\left[\frac{\phi}{100}\right]^{1.1558}\left[\frac{T}{333}\right]^{0.6881}$	TiO_2-CuO/C (80:20)	EG	0–2	20–50	Akilu et al. [27]
5	Interactive	$\frac{\mu_{nf}}{\mu_{bf}} = 1.123 + 0.3251\phi - 0.08994T + 0.00255T^2 - 0.00002386T^3 + 0.9695\left(\frac{T}{\phi}\right)^{0.01719}$	Al_2O_3/MWCNT (75:25)	SAE40	0.0625–1	25–50	Dardan et al. [28]
6		$\mu_{hnf} = 796.8 + 76.26\varphi + 12.88T + 0.7695\varphi T + \frac{-196.6T - 16.53\varphi T}{\sqrt{T}}$	MWCNT/ZnO (20:80)	Engine oil	0.25–2	5–55	Asadi and Asadi [29]

(Continued)

TABLE 2.2
Continued

Sl. No.	Model type	Correlation	Validity: Hybrid nanoparticles	Validity: Base fluid	Validity: Volume concentration (%)	Validity: Temperature range (°C)	Reference
4	Exponential and interactive	$\frac{\mu_{hnf}}{\mu_{bf}}=\left[0.191\varphi+0.240\left(T^{-0.342}\varphi^{-0.473}\right)\right]e^{\left[1.45T^{0.120}\varphi^{0.158}\right]}$	MgO/MWCNT (50:50)	EG	0.1–1	30–60	Soltani and Akbari [30]
5		$\frac{\mu_{hnf}}{\mu_{bf}}=0.00337+e^{\left(0.07731\varphi^{1.452}T^{0.3387}\right)}$	$MWCNT/SiO_2$ (50:50)	SAE40	0–1	25–60	Afrand et al. [31]
6		$\frac{\mu_{hnf}}{\mu_{bf}}=0.09422-\left[\left(\frac{T}{\varphi}\right)^2+0.100556T^{0.8827}\varphi^{0.3148}\right]e^{\left(72474.75T\varphi^{37951}\right)}$	$MWCNT/SiO_2$ (20:80)	SAE 20W50	0.05–1	40–100	Motahari et al. [32]
7		$\frac{\mu_{hnf}}{\mu_{bf}}=1.035+\frac{\varphi e^{-1.023\varphi}\left(2.046\frac{\varphi}{T}+0.4015\varphi^2T\right)}{T^{0.8441}}$	ZnO/MWCNT (55:45)	10W40	0.05–1	5–55	Esfe et al. [33]
8		$\mu_{hnf}=3.371\times e^{\left(0.5782\varphi-.0000371T^2\right)}$	MgO-MWCNT (50:50)	Water–EG (60:40)	0.025–1	25–60	Zareie and Akbari [34]

TABLE 2.3
Correlations for Viscosity of Hybrid Nanofluids Involving Shear Rate

			Validity					
Sl. No.	Model type	Correlation	Hybrid nanoparticle	Base fluid	Volume concentration (%)	Temperature range (°C)	Shear rate (s^{-1})	Reference
1	$f(\phi,T,\gamma)$	$\mu_{hnf}=4\times10^{4}+145\varphi-240T-0.061\gamma+1.9\times10^{6}\varphi^{2}+0.36T^{2}$						
			MWCNT (COOH-functionalized)/MgO (20:80)	SAE50	0.0625–1	25–50	670–8700	Alirezaie et al. [35]
2		$\mu_{hnf}=15.88\phi^{0.8514}T^{-1.189\gamma^{-0.5639}}$						
			$MWCNT/SiO_2$	EG–water	0.0625–2	25–50	0.612–122.3	Eshgarf et al. [36, 37]
3	$f(\gamma)$	$\mu=m\dot{\gamma}^{n-1}$	$m=0.02048+2.189exp\left(-\frac{1.083}{\phi}-0.03327T\right)$					
			$n=0.6868\phi^{-0.0906-0.001474T}-0.006267T$					
		$MWCNT/SiO_2$	EG–water	0.0625–2	25–50	0.612–122.3		
4		$\mu=m\dot{\gamma}^{n-1}$	$m=0.01125+\left(\frac{38.19-0.3T}{7.655+0.6953T}\right)\left(0.01138\phi+0.5529\phi^{2}-0.3613\phi^{3}+0.07\phi^{4}\right)$					
			$n=0.8543+\left(\frac{-3.03+1.418T}{15.8+0.391T}\right)\left(-0.7366\phi+0.8519\phi^{2}-0.4552\phi^{3}+0.08871\phi^{4}\right)$					
			$MWCNT/SiO_2$	EG–water	0.0625–2	25–50		0.612–122.3

TABLE 2.4
Correlations for Thermal Conductivity of Hybrid Nanofluids

Sl. No.	Model type	Correlation	Validity				Reference
			Hybrid nanoparticles	Base fluid	Volume concentration (%)	Temperature range (°C) Shear rate (s^{-1})	
1	$f(\phi)$	$\frac{k_{nf}}{k_{bf}} = 1 + A\phi + B\varphi^2 + C\phi^3 + D\phi^4$	CNTs/Al_2O_3 (50:50)	Water	0.02–1	30–50	Efse et al. [38]
2		$k_{nf} = \frac{(0.1747\times10^5 + \phi)}{(0.1747\times10^5 - 0.1498\times10\phi^6 + 0.1117\times10^7\phi^2 + 0.1997\times10^8\phi^3)}$	Ag/MgO (50:50)	Water	0.5–2	–	Efse et al. [20]
3	$f(\phi, T)$	$\frac{k_{nf}}{k_{bf}} = 1 + 0.004503\phi^{0.8717}T^{0.7972}$	ZnO–TiO_2 (50:50)	EG	0–3.5	25–50	Toghraie et al. [39]
4		$\frac{k_{hnf}}{k_{bf}} = 0.8341 + 1.1\phi^{0.243}T^{-0.289}$	MgO/FMWCNTs	EG	0–0.6	25–50	Afrand [40]
5		$\frac{k_{nf}}{k_{bf}} = 1 + 0.0008794\phi^{0.5899}T^{1.345}$	ZnO/Ag (50:50)	Water	0.125–2	25–50	Esfahani et al. [41]

(Continued)

TABLE 2.4
(Continued)

Sl. No.	Model type	Correlation	Validity: Hybrid nanoparticles	Validity: Base fluid	Validity: Volume concentration (%)	Validity: Temperature range (°C) Shear rate (s^{-1})	Reference
6		$\frac{k_{nf}}{k_{bf}} = 1 + 0.0162\phi^{0.7038}T^{0.6009}$	FMWCNTs/Fe_3O_4 (50:50)	EG	0–2.3	25–50	Harandi et al. [42]
7		$\frac{k_{nf}}{k_{bf}} = 1 + 6.2299\left(\frac{\phi}{100}\right)^{0.9371}\left(\frac{T}{333}\right)^{10.2685}$	TiO_2/CuO/C (80:20)	EG	0.5–2	25–60	Akilu et al. [27]
8		$\frac{k_{nf}}{k_{bf}} = 0.963 + 0.008379 \times \left[\phi^{0.4439} \times T^{0.9246}\right]$	SWCNTs/Al_2O_3 (30:70)	EG	0.4–2.5	30–50	Efse et al. [43]
9		$k_{nf} = 0.1534 + (0.00026)T + (1.1193)\phi$	Al_2O_3/MWCNTs (85:15)	Thermal oil	0.125–1.5	25–50	Asadi et al. [44]
10		$K_{eff} = \frac{k_{hnf}}{k_{bf}} = 1.17(1+R)^{-0.1151}\left(\frac{T}{80}\right)^{0.0437}$	SiO_2/TiO_2 (20:80 to 80:20)	Water and EG (60:40)	1	30–80	Hamid et al. [25]

(Continued)

TABLE 2.4
(Continued)

Sl. No.	Model type	Correlation	Validity: Hybrid nanoparticles	Validity: Base fluid	Validity: Volume concentration (%)	Validity: Temperature range (°C) Shear rate (s^{-1})	Reference
11		$K_t = \frac{k_{nf}}{k_{bf}} = \left(1+\frac{\phi}{100}\right)^{5.25}\left(1+\frac{T}{70}\right)^{0.076}$	SiO_2/TiO_2 (60:40)	Water and EG (60:40)	0.5–3	30–70	Hamid et al. [45]
12		$\frac{k_{nf}}{k_{bf}} = 0.905 + 0.002069\varphi T + 0.04375\varphi^{0.09265}T^{0.3305} - 0.0063$	SiO_2/MWCNT (85:15)	EG	0.05–1.95	30–50	Efse et al. [46]
13		$\frac{k_{nf}}{k_{bf}} = 1.07 + 0.000589\times T + \frac{-0.000184}{T\times\varphi} + 4.44\times T\times\varphi\times\cos\left(6.11 + 0.00673T + 4.41\times T\times\varphi - 0.0414\sin(T)\right) - 32.5\varphi$	Cu/TiO_2	Water/EG (60:40)	0.1–2	30–60	Esfe et al. [47]
14		$\frac{k_{nf}}{k_{bf}} = 1.05 + 0.005A + 0.06B + 0.0099AB + 0.00317A^2 + 0.026B^2 + 0.0034A^2B + 0.00735AB^2$ A and B represent volume concentration and temperature, respectively	$CNTs/Al_2O_3$	Water	0.02–0.1	27–47	Esfe et al. [48]

(Continued)

TABLE 2.4
(Continued)

Sl. No.	Model type	Correlation	Validity: Hybrid nanoparticles	Validity: Base fluid	Validity: Volume concentration (%)	Validity: Temperature range (°C) Shear rate (s⁻¹)	Reference
15		$\frac{k_{nf}}{k_{bf}} = \frac{(9.6128+\phi)}{9.3885-0.00010759T^2} - \frac{0.0041099}{\phi}$	Al_2O_3/Cu (50:50)	EG	0.125–2	25–50	Parsian et al. [49]
		$\frac{k_{nf}}{k_{bf}} = \left(1+\frac{\phi}{100}\right)^{5.5}\left(\frac{T}{80}\right)^{0.01}$	SiO_2–TiO_2 50:50	Water and EG (60:40)	0.5–3	30–80	Nabil et al. [26]
16		$K_{eff} = \frac{k_{nf}}{k_{bf}} = 0.7054 + 0.009896T + 0.8717\phi - 6.479\times10^{-5}T^2 + 0.09749T\phi - 4.714\phi^2 - 0.0002718T^2\phi - 0.1174T\phi^2 + 10.09\phi^3$ $K_{eff} = \frac{k_{nf}}{k_{bf}} = 0.9842 - 0.0008376T - 2.121\phi + 2.677\times10^{-5}T^2 + 0.1497T\phi + 2.653\phi^2 - 0.0006927T^2\phi - 0.1386T\phi^2 + 2.1\phi^3$ $K_{eff} = \frac{k_{nf}}{k_{bf}} = 1.321 - 0.01661T - 4.723\phi + 0.000199T^2 + 0.2473T\phi + 3.689\phi^2 - 0.001766T^2\phi - 0.1222T\phi^2 - 0.1045\phi^3$	Cu/Zn (50:50) (75:25) (25:75)	Vegetable oil	0.1–0.5	30–60	Mechiri et al. [50]
17		$\frac{k_{nf}}{k_{bf}} = 0.01807 + 0.1593\phi - 0.00107T + 0.00001T^2 + 0.00707T\phi + 1.12568\phi^2$	ND/Co_3O_4 (40:60)	Water and EG	0.05–0.15	20–60	Esfe et al. [51]

(Continued)

TABLE 2.4
(Continued)

Sl. No.	Model type	Correlation	Validity: Hybrid nanoparticles	Validity: Base fluid	Validity: Volume concentration (%)	Validity: Temperature range (°C) Shear rate (s^{-1})	Reference
18		$\frac{k_{nf}}{k_{bf}} = 1.01 + 0.007685T\phi - 0.5136\phi^2 T^{-0.1578} + 11.5\phi^3 T^{-1.175}$	MWCNTs–SiO_2 (30:70)	EG	0.025–0.86	30–60	Efse et al. [52]
19		$\frac{k_{nf}}{k_{bf}} = 1.07 + 0.000589T - \frac{0.000184}{T \times \phi} + 4.44 \times T \times \phi \times \cos\left(6.11 + 0.00673T + 4.41 \times T \times \phi - 0.0414\sin(T)\right) - 32.5\phi$	Cu–TiO_2/	Water and EG (60:40)	0.025–1	25–60	Efse et al. [48]
20		$\frac{k_{nf}}{k_{bf}} = 0.0288 \times \ln(\phi) + 1.085e^{\left(0.001351T + 0.13\phi^2\right)}$	DWCNTs–ZnO (50:50)	Water and EG (60:40)	0.025–1	25–50	Efse et al. [53]
21		$\frac{k_{nf}}{k_{bf}} = 0.9787 + e^{\left(0.3081\phi^{0.3097} - 0.002T\right)}$	MgO/MWCNTs	EG	0.05–0.6	25–50	Vafaei et al. [54]
22		$\frac{k_{nf}}{k_{bf}} = \frac{A + T}{B + C\phi} + \frac{D}{T}$	CNTs–Al_2O_3	Water	0.02–1	30–49	Efse et al. [55]

resistance to convective heat transfer is lower, i.e., heat transfer occurs predominantly by convection mode and vice versa. For a plain tube, the functional form of Nu for hybrid nanofluids is

$$Nu = aRe^{b}Pr^{c}\left(1+\phi_{p}\right)^{d}. \tag{2.17}$$

The constants a, b, c, and d vary with the type of nanofluid and experimental conditions. For augmented heat transfer applications like inserts, magnetic fields, and axial distance, additional terms were incorporated in the plain tube correlation as represented in Equation (2.17) [56]. The available correlations for predicting Nu are presented in Table 2.5 [13, 57–65].

2.7 FRICTION FACTOR

The pumping power needed to circulate the coolant is directly proportional to the differential pressure between the entry and exit sections of a pipeline. This differential pressure is in direct correlation with the friction factor (f). Hence, it is better to have lower friction factors so that the pumping power will be minimum. For a plain tube, the functional form of f for hybrid nanofluids is

$$f = aRe^{b}\left(1+\phi_{p}\right)^{c}. \tag{2.18}$$

The constants a, b, and c vary with the type of nanofluid and the conditions of test. Again, for augmented heat transfer applications like inserts, magnetic fields, and axial distance, additional terms were incorporated in the plain tube correlation in Equation (2.18). The various correlations proposed for the prediction of f is presented in Table 2.5.

2.8 SUMMARY

Empirical correlations derived from the experimental results for the prediction of the thermophysical characteristics of hybrid nanofluids like density, viscosity, specific heat, and thermal conductivity were categorized and presented in this chapter. For this purpose, an exhaustive literature search was conducted with respect to the research articles that reported correlations for the prediction of the thermophysical and thermo-hydraulic characteristics of hybrid nanofluids. The consolidation and outlook of various available correlations resulted in the following outcomes:

- During the density prediction of hybrid nanofluids, the correlation that uses the mixture rule is found to be reasonably accurate and hence widely used.
- Thermal equilibrium-based correlation is accepted for the determination of the specific heat of nanofluids containing hybrid nanoparticles.
- For determining the viscosity of hybrid nanofluids, various functional forms of correlation were proposed to describe the Newtonian and non-Newtonian behaviour of hybrid nanofluids. Non-linear forms of correlations with shear

TABLE 2.5
Correlations for Thermophysical Characteristics of Hybrid Nanofluids

Sl. No.		Validity				Reference
		Hybrid nanoparticle	Base fluid	Volume concentration (%)	Flow regime	
1	$Nu = 0.031(RePr)^{0.68}(1+\phi_p)^{95.73}$ $f = 26.44Re^{-0.8737}(1+\phi_p)^{156.23}$	Al_2O_3/Cu	Water	0.1	$Re<2300$	Suresh et al [57]
2	$Nu = 0.125Re^{0.592}Pr^{0.5}(1+\phi_p)^{177.13}$ $f = 133.57Re^{-1.12}(1+\phi_p)^{112.45}$	Al_2O_3–Cu	Water	0.1		Moghadassi et al. [58]
3	$Nu = 0.012Re^{-0.8}Pr^{0.5}(1+\phi_p)^{0.78}$ $f = 0.3108Re^{-0.245}(1+\phi_p)^{0.42}$	MWCNTs–Fe_3O_4	Water	0–0.6	$Re=3000-22,000$ $Pr=3.72-6.37$	Sundar et al. [13]
4	$Nu = 0.0017066Re^{0.9253}Pr^{1.29001}$ $f = 0.567322Re^{-0.285869}\phi_p^{0.0271605}$	GNP–Ag	Water	0–0.1	$Re=5000-17,500$	Yarmand et al. [59]

(Continued)

TABLE 2.5
(Continued)

Sl. No.		Validity				
		Hybrid nanoparticle	Base fluid	Volume concentration (%)	Flow regime	Reference
5	$Nu = 0.5762Re^{-00.3572}Pr^{0.01181}(1+\phi)^{0.5606}\left(1+\frac{\overrightarrow{B_{\max}}}{\overrightarrow{B_{min}}}\right)^{0.5008}\left(1+\frac{x}{d}\right)^{-0.05199}$ $f = 30.22Re^{-0.8918}(1+\phi)^{0.3323}\left(1+\frac{\overrightarrow{B_{\max}}}{\overrightarrow{B_{min}}}\right)^{0.5437}\left(1+\frac{x}{d}\right)^{-0.06824}$	$CoFe_2O_4$–$BaTiO_3$	EG	1	$Re = 248.03 - 1995.43$	Sundar and Venkata Ramana [60]
6	$Nu = 0.02433Re^{0.8}Pr^{0.4}(1+\phi)^{1.193}(1+AR)^{0.0291}\left(1+\frac{D_h}{D_i}\right)^{-0.88}$ $f = 0.2689Re^{-0.2312}(1+\phi)^{0.3556}(1+AR)^{-0.0024}\left(1+\frac{D_h}{D_i}\right)^{-0.083}$ $AR < 12$	MWCNT/Fe_3O_4		0.3	$Re = 3{,}000 - 22{,}000$	Sundar et al. [61]
7	$Nu = 0.02433Re^{0.8}Pr^{0.4}(1+\phi)^{1.193}(1+AR)^{0.0291}\left(1+\frac{D_h}{D_i}\right)^{-0.88}$ $f = 0.2689Re^{-0.2312}(1+\phi)^{0.3556}(1+AR)^{-0.0024}\left(1+\frac{D_h}{D_i}\right)^{-0.083}$ $AR < 4$	ND/Ni		0.3	Re = 3,000–22,000	Sundar et al. [62]

(Continued)

TABLE 2.5
(Continued)

			Validity			
Sl. No.		**Hybrid nanoparticle**	**Base fluid**	**Volume concentration (%)**	**Flow regime**	**Reference**
8	$Nu = 0.4668Re^{0.7941}Pr^{-0.9025}(1+\phi)^{3.280}(1+AR)^{0.04553}\left(1+\frac{D_h}{D_i}\right)^{-0.9609}$ $f = 0.3761Re^{-0.2293}(1+\phi)^{0.7204}(1+AR)^{-0.06323}\left(1+\frac{D_h}{D_i}\right)^{-0.5892}$ AR < 4	rGO/Co_3O_4	Water	0.2	$Re = 2{,}000 - 21{,}000$	Sundar et al. [63]
9	$Nu = 0.0359Re^{0.776}Pr^{0.426}$ $f = 0.595Re^{-0.322}$	TiB_2/B_4C	Polyproylene glycol/water 20/80 wt%	2 wt%	$Re = 8{,}000 - 26{,}000$	Vallejo et al. [64]
10	$Nu = 0.09011Re^{0.7026}Pr^{0.1732}\left(1+\frac{\phi}{100}\right)^{28.69}$ $f = 0.3245Re^{-0.2549}\left(1+\frac{\phi}{100}\right)^{0.6410}$	Fly ash/Cu	Deionized water	0.5–2	6,800–45,200 $2.73 \le Pr \le 5.67$ $_{30} \le T \le 60°C$	Kanti et al. [65]

rates were generally used to describe the rheological characteristics of non-Newtonian behaviour. It was revealed that volume concentration-based viscosity correlation will suffice only at near ambient/room temperatures. For applications involving temperatures other than ambient/room conditions, correlations included both the volume loading concentration and temperature.

- As far as thermal conductivity is concerned, the number of correlations established in terms of both volume concentration and temperature was more compared with that of volume concentration alone. This is because of accounting of multi-mode mechanisms in explaining enhanced thermal conductivity.
- The correlations related to thermo-hydraulic characteristics of hybrid nanofluids were just an extended version of conventional heat transfer and pressure drop equations involving Reynolds number and Prandtl number.

Furthermore, it is a general opinion among the research community that the proposed correlations were applicable only to a particular hybrid type of nanofluid and that too for a narrow limit of volume concentration, temperature, shear rate, and Reynolds number. Hence future research works on hybrid nanofluids require the development of generalized predictive correlations for the estimation of the thermophysical and thermo-hydraulic characteristics of hybrid nanofluids [66].

REFERENCES

1. Masuda, H., Ebata, A., Teramae, K., & Hishinuma, N. (1993). Alteration of thermal conductivity and viscosity of liquid by dispersing ultra-fine particles. Dispersion of Al_2O_3, SiO_2 and TiO_2 ultra-fine particles. Netsu Bussei, 7(4), 227–233.
2. Choi, S. U. S., Singer, D. A., & Wang, H. P. (1995). Developments and applications of non-Newtonian flows. ASME Fed, 66, 99–105.
3. Sajid, M. U., & Ali, H. M. (2018). Thermal conductivity of hybrid nanofluids: A critical review. International Journal of Heat and Mass Transfer, 126, 211–234.
4. Jana, S., Salehi-Khojin, A., & Zhong, W. H. (2007). Enhancement of fluid thermal conductivity by the addition of single and hybrid nano-additives. Thermochimica Acta, 462(1–2), 45–55.
5. Babar, H., & Ali, H. M. (2019). Towards hybrid nanofluids: Preparation, thermophysical properties, applications, and challenges. Journal of Molecular Liquids, 281, 598–633.
6. Timofeeva, E. V., Routbort, J. L., & Singh, D. (2009). Particle shape effects on thermophysical properties of alumina nanofluids. Journal of Applied Physics, 106(1), 014304.
7. Sundar, L. S., Sharma, K. V., Singh, M. K., & Sousa, A. C. M. (2017). Hybrid nanofluids preparation, thermal properties, heat transfer and friction factor – a review. Renewable and Sustainable Energy Reviews, 68, 185–198.
8. Ho, C. J., Huang, J. B., Tsai, P. S., & Yang, Y. M. (2010). Preparation and properties of hybrid water-based suspension of Al_2O_3 nanoparticles and MEPCM particles as functional forced convection fluid. International Communications in Heat and Mass Transfer, 37(5), 490–494.
9. Kanti, P., Sharma, K. V., Khedkar, R. S., & Rehman, T. U. (2022). Synthesis, characterization, stability, and thermal properties of graphene oxide based hybrid nanofluids for thermal applications: Experimental approach. Diamond and Related Materials, 128, 109265.

10. Vajjha, R. S., Das, D. K., & Mahagaonkar, B. M. (2009). Density measurement of different nanofluids and their comparison with theory. Petroleum Science and Technology, 27(6), 612–624.
11. Sofiah, A. G. N., Samykano, M., Sudhakar, K., Said, Z., & Pandey, A. K. (2023). Copper oxide/polyaniline nanocomposites-blended in palm oil hybrid nanofluid: Thermophysical behavior evaluation. Journal of Molecular Liquids, 375, 121303.
12. Gao, Y., Xi, Y., Zhenzhong, Y., Sasmito, A. P., Mujumdar, A. S., & Wang, L. (2021). Experimental investigation of specific heat of aqueous graphene oxide Al_2O_3 hybrid nanofluid. Thermal Science, 25(1 Part B), 515–525.
13. Sundar, L. S., Singh, M. K., & Sousa, A. C. (2014). Enhanced heat transfer and friction factor of MWCNT–Fe_3O_4/water hybrid nanofluids. International Communications in Heat and Mass Transfer, 52, 73–83.
14. Sundar, L. S., Ramana, E. V., Graça, M. P. F., Singh, M. K., & Sousa, A. C. (2016). Nanodiamond-Fe_3O_4 nanofluids: Preparation and measurement of viscosity, electrical and thermal conductivities. International Communications in Heat and Mass Transfer, 73, 62–74.
15. Mercan, H. (2020). Thermophysical and rheological properties of hybrid nanofluids. In Hybrid nanofluids for convection heat transfer (pp. 101–142). Academic Press.
16. Sepehrnia, M., Mohammadzadeh, K., Veyseh, M. M., Agah, E., & Amani, M. (2022). Rheological behavior of engine oil-based hybrid nanofluid containing MWCNTs and ZnO nanopowders: Experimental analysis, developing a novel correlation, and neural network modeling. Powder Technology, 404, 117492.
17. Wei, B., Zou, C., Yuan, X., & Li, X. (2017). Thermo-physical property evaluation of diathermic oil-based hybrid nanofluids for heat transfer applications. International Journal of Heat and Mass Transfer, 107, 281–287.
18. Esfe, M. H., Afrand, M., Yan, W. M., Yarmand, H., Toghraie, D., & Dahari, M. (2016). Effects of temperature and concentration on rheological behavior of MWCNTs/SiO_2 (20–80)-SAE40 hybrid nano-lubricant. International Communications in Heat and Mass Transfer, 76, 133–138.
19. Afrand, M., Najafabadi, K. N., & Akbari, M. (2016). Effects of temperature and solid volume fraction on viscosity of SiO_2-MWCNTs/SAE40 hybrid nanofluid as a coolant and lubricant in heat engines. Applied Thermal Engineering, 102, 45–54.
20. Esfe, M. H., Arani, A. A. A., Rezaie, M., Yan, W. M., & Karimipour, A. (2015). Experimental determination of thermal conductivity and dynamic viscosity of Ag–MgO/water hybrid nanofluid. International Communications in Heat and Mass Transfer, 66, 189–195.
21. Sundar, L. S., Irurueta, G. O., Ramana, E. V., Singh, M. K., & Sousa, A. C. M. (2016). Thermal conductivity and viscosity of hybrid nanofluids prepared with magnetic nanodiamond-cobalt oxide (ND-Co_3O_4) nanocomposite. Case Studies in Thermal Engineering, 7, 66–77.
22. Ghadikolaei, S. S., Yassari, M., Sadeghi, H., Hosseinzadeh, K., & Ganji, D. D. (2017). Investigation on thermophysical properties of Tio_2–Cu/H_2O hybrid nanofluid transport dependent on shape factor in MHD stagnation point flow. Powder Technology, 322, 428–438.
23. Wanatasanappan, V. V., Kanti, P. K., Sharma, P., Husna, N., & Abdullah, M. Z. (2023). Viscosity and rheological behavior of Al_2O_3-Fe_2O_3/water-EG based hybrid nanofluid: A new correlation based on mixture ratio. Journal of Molecular Liquids, 375, 121365.
24. Asadi, A., Asadi, M., Rezaei, M., Siahmargoi, M., & Asadi, F. (2016). The effect of temperature and solid concentration on dynamic viscosity of MWCNT/MgO (20–80)–SAE50 hybrid nano-lubricant and proposing a new correlation: An experimental study. International Communications in Heat and Mass Transfer, 78, 48–53.

25. Hamid, K. A., Azmi, W. H., Nabil, M. F., Mamat, R., & Sharma, K. V. (2018). Experimental investigation of thermal conductivity and dynamic viscosity on nanoparticle mixture ratios of TiO_2-SiO_2 nanofluids. International Journal of Heat and Mass Transfer, 116, 1143–1152.
26. Nabil, M. F., Azmi, W. H., Hamid, K. A., Mamat, R., & Hagos, F. Y. (2017). An experimental study on the thermal conductivity and dynamic viscosity of TiO_2-SiO_2 nanofluids in water: Ethylene glycol mixture. International Communications in Heat and Mass Transfer, 86, 181–189.
27. Akilu, S., Baheta, A. T., & Sharma, K. V. (2017). Experimental measurements of thermal conductivity and viscosity of ethylene glycol-based hybrid nanofluid with TiO_2-CuO/C inclusions. Journal of Molecular Liquids, 246, 396–405.
28. Dardan, E., Afrand, M., & Isfahani, A. M. (2016). Effect of suspending hybrid nano-additives on rheological behavior of engine oil and pumping power. Applied Thermal Engineering, 109, 524–534.
29. Asadi, M., & Asadi, A. (2016). Dynamic viscosity of MWCNT/ZnO–engine oil hybrid nanofluid: An experimental investigation and new correlation in different temperatures and solid concentrations. International Communications in Heat and Mass Transfer, 76, 41–45.
30. Soltani, O., & Akbari, M. (2016). Effects of temperature and particles concentration on the dynamic viscosity of MgO-MWCNT/ethylene glycol hybrid nanofluid: Experimental study. Physica E: Low-Dimensional Systems and Nanostructures, 84, 564–570.
31. Afrand, M., Najafabadi, K. N., Sina, N., Safaei, M. R., Kherbeet, A. S., Wongwises, S., & Dahari, M. (2016). Prediction of dynamic viscosity of a hybrid nano-lubricant by an optimal artificial neural network. International Communications in Heat and Mass Transfer, 76, 209–214.
32. Motahari, K., Moghaddam, M. A., & Moradian, M. (2018). Experimental investigation and development of new correlation for influences of temperature and concentration on dynamic viscosity of MWCNT-SiO_2 (20–80)/20W50 hybrid nano-lubricant. Chinese Journal of Chemical Engineering, 26(1), 152–158.
33. Esfe, M. H., Rostamian, H., & Sarlak, M. R. (2018). A novel study on rheological behavior of ZnO-MWCNT/10w40 nanofluid for automotive engines. Journal of Molecular Liquids, 254, 406–413.
34. Zareie, A., & Akbari, M. (2017). Hybrid nanoparticles effects on rheological behavior of water-EG coolant under different temperatures: An experimental study. Journal of Molecular Liquids, 230, 408–414.
35. Alirezaie, A., Saedodin, S., Esfe, M. H., & Rostamian, S. H. (2017). Investigation of rheological behavior of MWCNT (COOH-functionalized)/MgO-engine oil hybrid nanofluids and modelling the results with artificial neural networks. Journal of Molecular Liquids, 241, 173–181.
36. Eshgarf, H., & Afrand, M. (2016). An experimental study on rheological behavior of non-Newtonian hybrid nano-coolant for application in cooling and heating systems. Experimental Thermal and Fluid Science, 76, 221–227.
37. Eshgarf, H., Sina, N., Esfe, M. H., Izadi, F., & Afrand, M. (2018). Prediction of rheological behavior of MWCNTs–SiO_2/EG–water non-Newtonian hybrid nanofluid by designing new correlations and optimal artificial neural networks. Journal of Thermal Analysis and Calorimetry, 132, 1029–1038.
38. Hemmat Esfe, M., Saedodin, S., Yan, W. M., Afrand, M., & Sina, N. (2016). Study on thermal conductivity of water-based nanofluids with hybrid suspensions of CNTs/Al 2 O 3 nanoparticles. Journal of Thermal Analysis and Calorimetry, 124, 455–460.

39. Toghraie, D., Chaharsoghi, V. A., & Afrand, M. (2016). Measurement of thermal conductivity of ZnO–TiO_2/EG hybrid nanofluid: Effects of temperature and nanoparticles concentration. Journal of Thermal Analysis and Calorimetry, 125, 527–535.
40. Afrand, M. (2017). Experimental study on thermal conductivity of ethylene glycol containing hybrid nano-additives and development of a new correlation. Applied Thermal Engineering, 110, 1111–1119.
41. Esfahani, N. N., Toghraie, D., & Afrand, M. (2018). A new correlation for predicting the thermal conductivity of ZnO–Ag (50%–50%)/water hybrid nanofluid: An experimental study. Powder Technology, 323, 367–373.
42. Harandi, S. S., Karimipour, A., Afrand, M., Akbari, M., & D'Orazio, A. (2016). An experimental study on thermal conductivity of F-MWCNTs–Fe_3O_4/EG hybrid nanofluid: Effects of temperature and concentration. International Communications in Heat and Mass Transfer, 76, 171–177.
43. Esfe, M. H., Rejvani, M., Karimpour, R., & Abbasian Arani, A. A. (2017). Estimation of thermal conductivity of ethylene glycol-based nanofluid with hybrid suspensions of SWCNT–Al_2O_3 nanoparticles by correlation and ANN methods using experimental data. Journal of Thermal Analysis and Calorimetry, 128, 1359–1371.
44. Asadi, A., Asadi, M., Rezaniakolaei, A., Rosendahl, L. A., Afrand, M., & Wongwises, S. (2018). Heat transfer efficiency of Al2O3-MWCNT/thermal oil hybrid nanofluid as a cooling fluid in thermal and energy management applications: An experimental and theoretical investigation. International Journal of Heat and Mass Transfer, 117, 474–486.
45. Hamid, K. A., Azmi, W. H., Nabil, M. F., & Mamat, R. (2017, October). Improved thermal conductivity of TiO_2–SiO_2 hybrid nanofluid in ethylene glycol and water mixture. In IOP Conference series: Materials science and engineering (Vol. 257, No. 1, p. 012067). IOP Publishing.
46. Esfe, M. H., Behbahani, P. M., Arani, A. A., & Sarlak, M. R. (2017). Thermal conductivity enhancement of SiO_2–MWCNT (85: 15%)–EG hybrid nanofluids. Journal of Thermal Analysis and Calorimetry, 128(1), 249–258.
47. Esfe, M. H., Wongwises, S., Naderi, A., Asadi, A., Safaei, M. R., Rostamian, H., & Karimipour, A. (2015). Thermal conductivity of Cu/TiO_2–water/EG hybrid nanofluid: Experimental data and modeling using artificial neural network and correlation. International Communications in Heat and Mass Transfer, 66, 100–104.
48. Esfe, M. H., Saedodin, S., Biglari, M., & Rostamian, H. (2015). Experimental investigation of thermal conductivity of CNTs-Al_2O_3/water: A statistical approach. International Communications in Heat and Mass Transfer, 69, 29–33.
49. Parsian, A., & Akbari, M. (2018). New experimental correlation for the thermal conductivity of ethylene glycol containing Al_2O_3–Cu hybrid nanoparticles. Journal of Thermal Analysis and Calorimetry, 131, 1605–1613.
50. Mechiri, S. K., Vasu, V., & Venu Gopal, A. (2017). Investigation of thermal conductivity and rheological properties of vegetable oil-based hybrid nanofluids containing Cu–Zn hybrid nanoparticles. Experimental Heat Transfer, 30(3), 205–217.
51. Esfe, M. H., & Hajmohammad, M. H. (2017). Thermal conductivity and viscosity optimization of nanodiamond-Co_3O_4/EG (40:60) aqueous nanofluid using NSGA-II coupled with RSM. Journal of Molecular Liquids, 238, 545–552.
52. Hemmat Esfe, M., Esfandeh, S., & Rejvani, M. (2018). Modeling of thermal conductivity of MWCNT-SiO_2 (30: 70%)/EG hybrid nanofluid, sensitivity analyzing and cost performance for industrial applications: An experimental based study. Journal of Thermal Analysis and Calorimetry, 131, 1437–1447.
53. Esfe, M. H., Yan, W. M., Akbari, M., Karimipour, A., & Hassani, M. (2015). Experimental study on thermal conductivity of DWCNT-ZnO/water-EG nanofluids. International Communications in Heat and Mass Transfer, 68, 248–251.

54. Vafaei, M., Afrand, M., Sina, N., Kalbasi, R., Sourani, F., & Teimouri, H. (2017). Evaluation of thermal conductivity of MgO-MWCNTs/EG hybrid nanofluids based on experimental data by selecting optimal artificial neural networks. Physica E: Low-Dimensional Systems and Nanostructures, 85, 90–96.
55. Hemmat Esfe, M., Saedodin, S., Yan, W. M., Afrand, M., & Sina, N. (2016). Study on thermal conductivity of water-based nanofluids with hybrid suspensions of CNTs/Al_2O_3 nanoparticles. Journal of Thermal Analysis and Calorimetry, 124, 455–460.
56. Hamzah, M. H., Sidik, N. A. C., Ken, T. L., Mamat, R., & Najafi, G. (2017). Factors affecting the performance of hybrid nanofluids: A comprehensive review. International Journal of Heat and Mass Transfer, 115, 630–646.
57. Suresh, S., Venkitaraj, K. P., Selvakumar, P., & Chandrasekar, M. (2012). Effect of Al2O3–Cu/water hybrid nanofluid in heat transfer. Experimental Thermal and Fluid Science, 38, 54–60.
58. Moghadassi, A., Ghomi, E., & Parvizian, F. (2015). A numerical study of water based Al_2O_3 and Al_2O_3–Cu hybrid nanofluid effect on forced convective heat transfer. International Journal of Thermal Sciences, 92, 50–57.
59. Yarmand, H., Gharehkhani, S., Ahmadi, G., Shirazi, S. F. S., Baradaran, S., Montazer, E.,. . . Dahari, M. (2015). Graphene nanoplatelets–silver hybrid nanofluids for enhanced heat transfer. Energy Conversion and Management, 100, 419–428.
60. Sundar, L. S., & Ramana, E. V. (2023). Influence of magnetic field location on the heat transfer and friction factor of $CoFe_2O_4$-$BaTiO_3$/EG hybrid nanofluids in laminar flow: An experimental study. Journal of Magnetism and Magnetic Materials, 579, 170837.
61. Sundar, L. S., Otero-Irurueta, G., Singh, M. K., & Sousa, A. C. (2016). Heat transfer and friction factor of multi-walled carbon nanotubes–Fe_3O_4 nanocomposite nanofluids flow in a tube with/without longitudinal strip inserts. International Journal of Heat and Mass Transfer, 100, 691–703.
62. Sundar, L. S., Singh, M. K., & Sousa, A. C. (2018). Heat transfer and friction factor of nanodiamond-nickel hybrid nanofluids flow in a tube with longitudinal strip inserts. International Journal of Heat and Mass Transfer, 121, 390–401.
63. Sundar, L. S., Said, Z., Saleh, B., Singh, M. K., & Sousa, A. C. (2020). Combination of Co_3O_4 deposited rGO hybrid nanofluids and longitudinal strip inserts: Thermal properties, heat transfer, friction factor, and thermal performance evaluations. Thermal Science and Engineering Progress, 20, 100695.
64. Vallejo, J. P., Ansia, L., Calviño, U., Marcos, M. A., Fernández-Seara, J., & Lugo, L. (2023). Convection behaviour of mono and hybrid nanofluids containing B4C and TiB2 nanoparticles. International Journal of Thermal Sciences, 189, 108268.
65. Kanti, P. K., Sharma, K. V., Said, Z., & Gupta, M. (2021). Experimental investigation on thermo-hydraulic performance of water-based fly ash–Cu hybrid nanofluid flow in a pipe at various inlet fluid temperatures. International Communications in Heat and Mass Transfer, 124, 105238.
66. Giwa, S. O., Adeleke, A. E., Sharifpur, M., & Meyer, J. P. (2023). From 2007 to 2021: Present state of research on hybrid nanofluids as advanced thermal fluids. In Materials for advanced heat transfer systems (pp. 63–174). Woodhead Publishing.

3 Synthetic Routes of Various Water-Based Hybrid Nanofluids and Their Thermal Conductivities

Amal T. Babu and Jayanta Parui

3.1 INTRODUCTION

Research and application of nanoscience are continuously contributing to the advancements in the fields of mechanics, optoelectronics, biopharmaceuticals, catalysis, energy harvesting, thermal engineering, and lightweight materials [1–4]. Not only heavy machineries, combustion engines, or nuclear power plants, but electronic devices also produce high amounts of heat when in operational mode [5, 6]. Hence irrespective of dimensionality, technological advancements are in need of thermal management to increase device reliability and efficiency. Foreseeing the wider applicability of nanotechnology, extensive research has been done on the utility of nanotechnology to tackle down problems associated with entropy and heat generation. During the past decade, cooling methods including a variety of heat sink designs, thermoelectric coolers, forced air systems, and fans and heat pipes have become the central attraction of researchers [5–8].

Inefficient thermal management can not only retard the quality of a product but can also increase the risks of devices' operation and lead to a financial loss. Therefore, to increase the efficiency of heavy machineries, automobiles, refrigerators, and cooling towers, in-depth studies on blending nanotechnology with conventional cooling technologies are being carried out [9, 10].

Conventional heat transfer fluids like water, glycerol, ethylene glycol, and engine oil have very low thermal conductivities, i.e., 0.615, 0.289, 0.252, and 0.145 W/mK, respectively, whereas solids like silver, copper, gold, aluminium, and alumina have comparatively very high thermal conductivities of 429, 401, 315, 237, and 40 W/mK, respectively [11]. Considering the difference in thermal conductivity in two different types of materials, researchers sought to enhance the thermal conductivity of conventional heat transfer fluids by suspending superior thermal conducting solids in them. In 1881, James Clerk Maxwell suspended micrometre-sized solids in the conventional fluids such as water and observed thermal conductivity enhancement

DOI: 10.1201/9781003595786-3

in them [11]. Despite the enhanced thermal conductivity, such fluids experienced sedimentation, clogging, fouling, erosion, and a decrease in pressure drop, which hindered their application as a cooling fluid [11]. Given the important characteristics of nanoparticles, nanofluids have been thought of as one of the most suitable and efficient options for heat transfer. Compared to micrometre- or bigger-sized particles, small particles in nanofluids ensure stability, and lesser sedimentation and clogging in the flow channels. One of the most significant factors driving researchers to use nanofluids in a variety of sectors and applications is their tremendous capacity to control the thermophysical characteristics of fluids.

In 1993, Masuda et al. [12] dispersed ultra-fine Al_2O_3, TiO_2, and SiO_2 particles of nanometre dimension in water to form electrostatically stabilized colloidal systems with enhanced thermal conductivity. In 1995, Stephen U.S. Choi dispersed nanosized metal oxides in conventional fluids, later known as base fluids, to prepare nanomaterial-enhanced fluids, i.e., nanofluids, that substantially overcame the drawbacks of solid particle-suspended fluids and showed enhanced thermal conductivity [11, 13]. Cho et al. [14] in 1998 reported thermal conductivity and viscosity enhancement of water as they prepared a stable aqueous colloidal solution of γ-alumina and TiO_2 nanoparticles with mean diameters of 13 and 27 nm, respectively.

With progress in the field, nanofluids are majorly defined as a type of engineered colloids that are meant to act as single-phase systems, which are generally prepared by the dispersion of nanostructures in a solvent, i.e., base fluid. The dispersed nanostructures retain their original identity as zero-dimensional [9, 15], one-dimensional [16], or two-dimensional [17] nanomaterials. Nanocomposites or nanohybrids are nowadays being studied for their possible technological advantages [18–20], and the field of nanofluids does not remain an exception. Although the dispersant mediums, i.e., the base fluids, are generally water [21], ethylene glycol [22], kerosene [23], petroleum oil [24], engine oil [25], or a fluid mixture like water/ethylene glycol [26] and propylene glycol/water [27], water is widely studied as the base fluid because of its relatively high thermal conductivity (~0.615 W/mK) and high dielectric constant (~78.3) at room temperature [28]. Also, its availability and economic feasibility compared to other conventional heat transfer fluids cannot be unnoticed.

Many studies on nanofluids have been conducted, such as on thermal conductivity [29–31], forced convection heat transfer [32], natural convection heat transfer [33], boiling heat transfer [34], heat exchangers [26, 35], and flat-plate solar collectors [36, 37], in search of simple thermal management. Recent studies have shown that hybrid nanofluids can take the place of single nanofluids in various thermal applications, including automotive, electromechanical, manufacturing, and solar energy systems. Hybrid nanofluids are expected not only to provide better heat transfer enhancement than plain nanofluids but also to lessen the pressure loss during their pumping with rheological control [15, 25, 38–40].

Hybrid nanofluids are created by dispersing two or more distinct nanoparticles or a hybrid nanostructure in a base fluid, resulting in thermal effects that are synergistically beneficial when compared to conventional fluids [40–43]. Each nanostructure of a nanohybrid will have its own contribution to the enhancement of the different physical and chemical properties of the nanohybrid structure. Hybrid nanofluids have outperformed single nanofluids in terms of thermal stability and other features, making them viable options for heat-related applications such as solar thermal systems, car cooling systems, heat sinks, and thermal energy storage. Suresh et al. [44] reported enhanced thermal conductivity (~12%) of aqueous Al_2O_3/Cu nanofluid compared to alumina-suspended single aqueous nanofluid. Additionally, they noticed that although the viscosity of the base fluid enhanced ~115%, Newtonian behaviour of water was not changed for 2% volume fraction of the hybrid nanofluid. They also reported that hybrid nanofluids exhibited higher viscosities than single nanofluids as the viscosity of the Al_2O_3/Cu hybrid nanofluid was ~1.5 mPas and for Al_2O_3 monodispersed nanofluid it was ~0.9 mPas at the loading of 2% volume fraction. It has been noticed that the synthesis of hybrid nanofluids not only has an impact on the thermal and rheological properties compared to single nanofluids but also has an influence on the stability of nanofluids. Although multiwalled carbon nanotubes (MWCNTs) have high thermal conductivity, they are hydrophobic by nature and disperse very little in polar base fluids like water. Gupta et al. [45] prepared aqueous hybrid nanofluids with MWCNT-Cu and observed high dispersion stability for up to 170 days along with thermal conductivity enhancement of ~ 76%. It is worthy to mention that the single nanofluid with MWCNT was stable only for 30 days and enhanced the thermal conductivity of water by ~25%.

Although hybrid nanofluids have various potential applications, their preparations are not straightforward. Turcu et al. [1], in 2006, successfully prepared a hybrid structure with MWCNTs and conducting polypyrrole (PPY). It was then dispersed into an aqueous suspension of Fe_3O_4 nanoparticles to obtain a hybrid nanofluid. Nanohybrid formation was confirmed by different characterization techniques like transmission electron microscopy (TEM), scanning electron microscopy (SEM), infrared (IR) spectroscopy, and X-ray diffraction (XRD) analysis. The research group of Hemmat Esfe, Imam Hossein University, Tehran, Iran, further extended the field of hybrid nanofluids as they worked on regulation of the viscosity of hybrid nanofluids in comparison to pure base fluids [46, 47]. The numerical analysis done by them suggested higher thermal conductivity for nanofluids with lower viscosities.

The selection of right combination of nanostructures and their relative proportion for the preparation of hybrid nanofluids play important roles in the stability of nanofluids and enhancement of their thermophysical properties. There are also reports of higher thermal conductivity enhancement in monodispersed nanofluids than hybrid nanofluids. Jana et al. [48], in 2007, observed that MWCNT/Cu and MWCNT/Au hybrid nanofluids prepared through the direct dispersion method had lower thermal conductivity enhancement than monodispersed nanofluids. For Cu-based single nanofluid, the enhanced thermal conductivity was ~38% higher than that of the MWCNT/Cu nanofluid. In the case of MWCNT/Au hybrid nanofluid, the observed values of thermal conductivity enhancement were similar to that of monodispersed nanofluid, and it was ~37% compared to that of the base fluid. The diminished thermal conductivity of the hybrid nanofluid was attributed to the weak interaction between the nanostructures in the hybrid nanofluid. Therefore, selected nanostructures for preparing hybrid nanofluids should be

i. compatible with each other to have necessary interaction for enhancing the thermophysical properties of the base fluid;
ii. compatible with the base fluid to form a stable nanofluid; and

iii. compared with appropriate size, as the size of the suspended nanostructures can significantly influence the thermal and rheological characters of the nanofluid.

Considering the above criteria to prepare an effective stable hybrid nanofluid, different combinations of nanostructures are used for their preparation, and they are considered to be of three types.

3.2 CLASSIFICATION OF HYBRID NANOFLUIDS

Hybrid nanofluids can be classified on the basis of nanostructures and type of interactions among the nanostructures as illustrated in Figure 3.1. Hybrid nanofluids are majorly classified into three types: core–shell, mutually functionalized, and co-dispersed hybrid nanofluids.

i. In nanohybrid structures of the core–shell type, one nanoparticle forms the shell, i.e., the outer layer material, over another core nanoparticle, i.e., the inner material (Figure 3.1(a)). It can be denoted as core@shell nanostructures or core–shell type nanostructures, and similar notations can also be used for hybrid nanofluids made of such nanomaterials. For example, in the ZnO@Ag nanohybrid structure [49], Ag nanoparticles form the shell and ZnO nanoparticles form the core, and the hybrid nanofluid made of this nanohybrid structure will be denoted as ZnO@Ag nanofluid.
ii. Mutually functionalized nanohybrids contain two or more nanoparticles that are covalently or non-covalently functionalized to have a synergic effect on their properties (Figure 3.1(b,c)). Covalently bonded nanohybrids

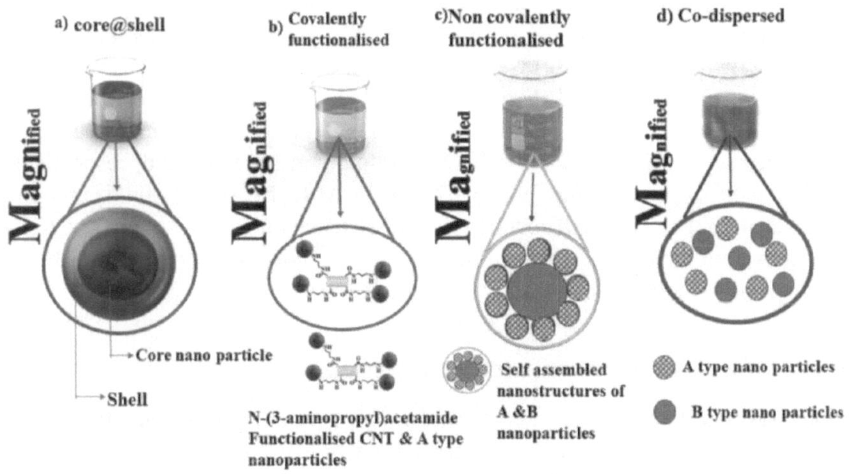

FIGURE 3.1 Illustrative presentation of hybrid nanofluids: a) core@shell type nanofluid, b) covalently functionalized nanofluid, c) non-covalently functionalized nanofluid, and d) co-dispersed nanofluid.

are represented as NPn-NPm, and non-covalently functionalized nanohybrids are represented as NPn∩NPm, where NPn and NPm are two different nanostructures of types n and m. For example, MWCNTs can be covalently [50] functionalized with 0D Fe_3O_4 (MWCNT-Fe_3O_4), and 0D Co_3O_4 can be coprecipitated on 0D nanodiamonds (ND) to get non-covalently functionalized (ND∩Co_3O_4) hybrid nanostructures [51]. The nanofluids formed of MWCNT-Fe_3O_4 in the base fluids are distinguished as MWCNT-Fe_3O_4 nanofluid, and the nanofluids formed of ND∩Co_3O_4 in the base fluids are recognized as ND∩Co_3O_4 nanofluids. Irrespective of their subtle differences, they are categorized as functionalized hybrid nanofluids.

iii. When two or more different nanostructures are dispersed in a base fluid to get a hybrid nanofluid (Figure 3.1(d)), and if they are not associated to co-precipitate, the hybrid nanofluid is termed as co-dispersed hybrid nanofluid and represented as NPn/NPm.

3.2.1 Core–Shell Hybrid Nanofluids

These hybrid nanostructures are widely studied as they can display unique features due to the combination of core and shell material, geometry, and design. For example, hollow Au@Cu_2O [52] can be tuned for their geometry-dependent optoelectrical properties, or eccentric SiO_2@Ag [53] nanostructures are studied due to their magneto-dielectric responses, and super paramagnetic properties of spherical and elliptical Au@Co [54] nanostructures can be utilized in magneto-hyperthermia. Hollow core@shell nanostructures have a hollow shell and a solid core (Figure 3.2(b)), while hollow eccentric core@shell nanostructures have a core that is not centred in the middle of the shell (Figure 3.2(d)). The core–shell structures are well engineered so that the shell material can improve the thermal and chemical stability of the core material. For example, the oxidative stability of VO_2 is improved by coating it with SnO_2 [55]. Core@shell type nanohybrids dispersed in base fluids can enhance the thermal conductivity of the base fluid. For instance, an increased thermal conductivity of 32% was found in ZnO@Ag water-based hybrid nanofluids, compared to maximum enhancements of 18% and 27% for ZnO nanofluid and Ag nanofluid, respectively, considering the same base fluid [34, 56, 57]. CuO@Al_2O_3 hybrid nanofluids prepared by Plant et al. [32] had thermal conductivity enhancement up to 34% compared to the base fluid where the enhancement of thermal conductivity was reported as 30% for monodispersed water-based alumina nanofluid [57].

3.2.2 Mutually Functionalized Nanohybrids

Nanohybrids can be mutually interacting through covalent bonds, ionic bonds, and other interactions like van der Waal's interactions or hydrogen bonds without forming into a core and shell. For example, a covalently bonded nanohybrid structure of 1D-0D nanostructures was produced by the chemical co-precipitation of Fe_3O_4 in

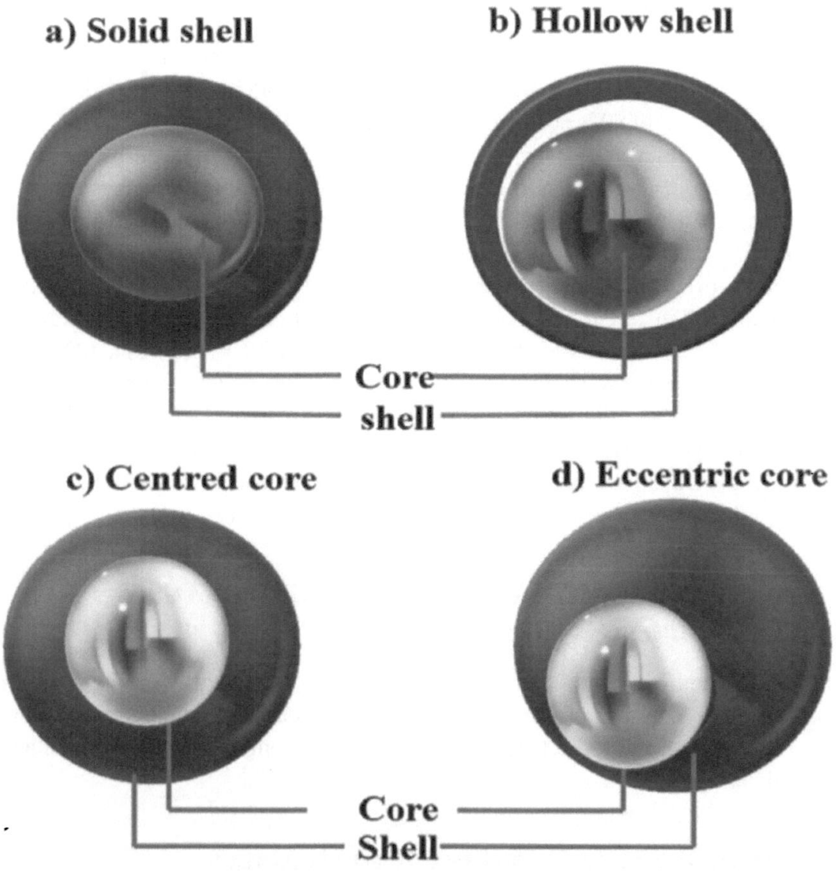

FIGURE 3.2 Pictorial representation of different types of core@shell nanostructures. a) Nanostructures with a solid core and a solid shell, b) nanostructures with a solid core and a hollow shell, c) nanostructures with the core at the centre of the shell, and d) nanostructures with the core away from the centre of the shell.

the presence of –COOH-functionalized MWCNTs [50]. The above-mentioned nanohybrid was then dispersed in water to obtain a stable hybrid nanofluid (zeta potential of –43mV) with thermal conductivity enhancement of 28.33%. When compared with base fluid water, the thermal conductivity of Fe_3O_4 monodispersed nanofluids was enhanced by ~3% [58]. Sundar et al. [59] prepared Co_3O_4-ND aqueous hybrid nanofluids that augmented the thermal conductivity and viscosity of the base fluid to 15.71% and 45.83%, respectively, whereas single nanofluids with Co_3O_4 provide a thermal conductivity enhancement of ~25% to water [60]. The zeta potential value of the synthesized hybrid nanofluid was –34 mV, indicating a stable nanofluid compared to the monodispersed Co_3O_4 nanofluid with a zeta potential of –19 mV [59, 61].

3.2.3 Co-Dispersed Nanostructures for Hybrid Nanofluids

Two or more different nanostructures are dispersed in a base fluid to produce hybrid nanofluids, and this method of dispersion is known as direct dispersion method [62]. Many 0D/0D co-dispersed nanohybrid fluids such as Cu/Al_2O_3 [62] and CuO/SiO_2 [63] and water-based hybrid nanofluids were produced by the direct dispersion method. On suspending independent nanostructures in a base fluid to form co-dispersed hybrid nanofluids, there is more freedom to customize the volume ratios of nanostructures to achieve the desired thermophysical properties. One of the common forms of co-dispersed hybrid nanofluids is ternary hybrid nanofluid, which contains 3 different nanostructures dispersed in a base fluid. For example, 2D-0D nanohybrids made of graphene nanoplatelet (GNP)-Fe_3O_4 and 0D Al_2O_3 were dispersed in water to achieve a heat transfer coefficient enhancement of ~92%. When GNP-Fe_3O_4 and 0D SiO_2 nanoparticles were dispersed in water, ~ 82% of enhancement in heat transfer coefficient was recoded [64]. When compared to water, the reported enhancements in heat transfer coefficients of single nanofluids of alumina [64] and SiO_2 [65] were 6.2% and 21.2%, respectively.

3.3 DIFFERENT COMBINATIONS OF NANOSTRUCTURES FOR HYBRID NANOFLUIDS

Superior thermophysical qualities of hybrid nanofluids over single nanofluids are the outcome of the synergic combination of nanostructures used to prepare hybrid nanofluids. Here some of the combinations are discussed to comprehend their advantages and disadvantages.

3.3.1 Metal and Metal Oxide-Based Hybrid Nanofluids

Metal and metal oxide nanoparticles are suspended in base fluids to produce various types of hybrid nanofluids as per the classification mentioned in Section 3.2.

For example, enhanced thermal conductivity of 32% was reported in ZnO@Ag water-based hybrid nanofluids, whereas a maximum enhancement of 18% was reported for ZnO nanofluids and 27% for Ag nanofluids considering the same base fluid [34, 56, 57]. Okonkwo et al. [66] reported 7.8% enhancement in thermal conductivity of water for Fe/Al_2O_3 hybrid nanofluids, whereas enhancement by monodispersed Al_2O_3 nanofluid was only 7.5%.

3.3.2 Metal Oxide and Metal Oxide-Based Hybrid Nanofluids

Metal oxide combinations such as ZrO_2/CeO_2 and $CuO \cap Al_2O_3$ were dispersed in water to obtain enhanced thermal conductivity of the base fluid of ~22% and 40%, respectively [68, 69], whereas water-based plain ZrO_2 and Al_2O_3 nanofluids were reported to have thermal conductivity enhancement of 2.01% [67] and 30 % [68], respectively. Such enhancements of thermal conductivity in hybrid nanofluids lead to more research activities in the field of hybrid nanofluids.

3.3.3 Carbon Nanostructured Hybrid Nanofluids

Metals like nickel or metal oxides such as Co_3O_4 or Fe_3O_4 can be co-precipitated on 0D ND to form mutually functionalized nanohybrids, which are proven to enhance thermal conductivity significantly [47, 59, 68–70]. For example, ND∩Ni aqueous hybrid nanofluids can enhance the thermal conductivity of water by ~30% [69, 71], whereas monodispersed ND nanofluids enhanced the thermal conductivity of water up to ~22.8% [72]. The high thermal conductivity of CNTs and graphene garnered the interests of researchers to prepare nanohybrids containing 1D or 2D nanostructures [45, 73–79]. A combination of Cu∩MWCNTs in water has shown thermal conductivity enhancement of ~78% [75]. Graphene-based hybrid nanostructures such as Fe_3O_4∩rGO have shown an 82% increase in convective heat transfer [80], while Fe_3O_4∩rGO-based nanofluids have shown an 83.44% increase in thermal conductivity of water [81]. The reported thermal conductivity and convective heat transfer enhancement by graphene oxide (GO)-based plain nanofluids are 28% and 40.3%, respectively [82]. Apart from metal-based nanostructures combined with carbon-based nanostructures, SiO_2 was also combined with graphene-based nanostructures to synthesize hybrid nanofluids that enhanced the thermal conductivity of water up to 26.93%, whereas plain SiO_2 nanofluids enhanced the thermal conductivity of water by ~3.4% [83, 84].

3.3.4 Metal- and Silicon-Based Hybrid Nanofluids

Fe and Si nanostructures were dispersed in water in the presence of carboxymethyl cellulose sodium salt to obtain stable Fe∩Si hybrid nanofluids [85] with 12.60% enhancement of thermal conductivity, which is higher than that of Si monodispersed nanofluids with ~–0.16% [86] thermal conductivity enhancement.

3.3.5 Metal or Metal Oxide- and SiO_2-Based Hybrid Nanofluids

A few metal-based nanostructures coupled with SiO_2 include Cu∩SiO_2 and TiO_2/SiO_2 nanostructures dispersed in conventional heat transfer fluids to enhance thermal conductivity [87, 88]. CuO/SiO_2 [63] dispersed in water and TiO_2/SiO_2 [89] in water and ethylene glycol mixture have shown enhanced thermal conductivity of up to 48% and 78%, respectively. As per the earlier reports, 40% thermal conductivity enhancement was recorded for CuO-loaded water-based nanofluids and 3.4% was recorded for water-based SiO_2 nanofluids [84, 90], whereas 33% thermal conductivity enhancement was reported for water-based TiO_2 nanofluids [91] and 10% enhancement for water–ethylene glycol-based TiO_2 nanofluids [92]. Cu∩SiO_2 [87] mutually functionalized nanostructures in water–ethylene glycol enhanced the thermal conductivity up to 96%, whereas the reported thermal conductivity enhancement of monodispersed SiO_2 in water–ethylene glycol was ~25% compared to the base fluid [93].

3.3.6 Metal- and Metal Sulphide-Based Hybrid Nanofluids

Theoretical studies on Ag∩MoS_2 water-based hybrid nanofluids were performed to estimate the thermophysical properties. Mathematical modelling of the proposed mutually functionalized 0D∩2D nanostructure of Ag∩MoS_2 dispersed in water predicted an enhancement of ~15% in heat transfer fluids compared with the base fluid [94], whereas the experimental studies on water-based MoS_2 plain nanofluids reported a thermal conductivity enhancement of ~0.8% [95].

The different types of hybrid nanofluids are tabulated in Table 3.1 to account for the enhancement of thermal conductivities along with the methods for the measurement of thermal conductivity, stability in terms if zeta potential, and viscosity. For comparison with individual nanomaterials, the properties of plain nanofluids are tabulated in Table 3.2, along with the same set of parameters as tabulated in Table 3.1. Many data are not available at present, which indicates the requirement of extensive research in this field and that some of the studies are yet to be published.

3.4 PREPARATION OF NANOFLUIDS

The methods used to prepare nanofluids can have a considerable impact on critical factors such as production cost and stability. For the simplicity of distinguishing the nanofluid preparation methods, they are typified as single-pot synthesis and multi-pot synthesis. Conventionally, single-pot synthesis of nanofluids refers to the simultaneous synthesis of nanoparticles and the formation of nanofluids. In this chapter, single-pot synthesis of hybrid nanofluids is defined as the simultaneous synthesis of a nanohybrid and the formation of a hybrid nanofluid. In multi-pot synthesis of hybrid nanofluids, nanohybrids or hybrid nanostructures are synthesized in a different reaction vessel before being dispersed in the base fluid.

3.4.1 Single-Pot Synthesis of Hybrid Nanofluids

As defined, in single-pot synthesis of hybrid nanofluids, the nanohybrid and nanofluid are produced in a single pot. This method's advantages include the nanofluid's stability, nanofluid homogeneity, and retarded particle aggregation. The disadvantages of this approach are limited yield and production of limited types of nanofluids. It has also been recorded that single-pot synthesis mostly produces low-pressure fluids with low volume fractions [46, 108, 109]. The various types of hybrid nanofluid synthesis using the single-pot method are described as follows.

3.4.1.1 Core–Shell Type Nanofluids

CuO@Al_2O_3 hybrid nanofluids were prepared by photochemical reduction of $CuSO_4$ in an aqueous dispersion of Al_2O_3 nanoparticles in a single pot [32]. In comparison to the base fluid water, an enhancement of thermal conductivity by 30% was observed in monodispersed Al_2O_3 water-based nanofluids [110], whereas a demonstration of thermal conductivity enhancement by 34% was made by Plant et al. [32] in CuO@Al_2O_3 hybrid nanofluids.

3.4.1.2 Mutually Functionalized Nanofluids

Thermal conductivity enhancement of 18.79% was observed in single-step synthesized Cu∩Pd bimetallic water-based hybrid nanofluids [111]. The experimental procedure followed for the single-step synthesis of the hybrid nanofluid is as follows.

TABLE 3.1
Different Types of Hybrid Nanofluids and Their Properties

Type of nanofluid	Combinations of nanostructures	Example	Thermal conductivity enhancement (%)	Measurement method for thermal conductivity	Stability (zetapotential/no. of days)	Viscosity (mPas)	Reference
Core–shell type							
	Metal–Metal oxide	*ZnO@Ag*	32	Transient hot wire (THW) method	>40mV		[34]
		ZnO@Ag	20.53	THW method	50.5 mV		[96]
		TiO_2@*Cu*	9.8	THW method			[97]
	Metal oxide–metal oxide	*CuO*@Al_2O_3	34	THW method	25 mV		[32]
Mutually Functionalized							
	Metal oxide–metal oxide	$TiO_2 \cap Al_2O_3$	13.5	THW method		0.91	[36]
		$Al_2O_3 \cap CuO$	40	THW method			[98]
	Carbon nanostructured	Cu-, Zn-, Fe-, Ag- COOH-MWCNT	Cu hybrid – 78, Fe hybrid – 25, Ag hybrid – 40	THW method	Cu hybrid–45days, Fe hybrid – 40 days, Ag hybrid–15 days, Zn hybrid–less than 1 day		[75]
	Carbon nanostructured	$ND \cap Fe_3O_4$	72.9	THW method			[99]
		$ND \cap Ni$	29.39	THW method		.970	[71]
		$ND \cap Co_3O_4$	16	THW method	−34 mV		[59]
		$ND \cap Co_3O_4$	15.71	THW method			[68]

(Continued)

TABLE 3.1
(Continued)

Type of nanofluid	Combinations of nanostructures	Example	Thermal conductivity enhancement (%)	Measurement method for thermal conductivity	Stability (zetapotential/no. of days)	Viscosity (mPas)	Reference
		ND∩Fe_3O_4	12.79	THW method		6	[99]
		MWCNT—Fe_3O_4	28.	THW method	–43 mv		[50]
		Ag∩rGO	18	C-Therm TCi thermal conductivity analyser			[100]
		Fe_3O_4∩rGO	83.4	THW method		2.09	[81]
		GO ∩Co_3O_4	19.14	THW method			[101]
		Fe_3O_4∩GO	18	THW method			[102]
		Fe_3O_4∩rGO	11	THW method			[80]
		GO∩Pd	17.77	THW method	22 days		[103]
		Cu –CNT	75.57	THW method	175 days	2.5	[45]
	Metal– Silicon	Fe∩Si	12.60	THW method	>–70 mV		[85]
		Fe∩Si		THW method	–73mV	1.2 mPas	[104]
	Metal–metal sulphide nanostructures	Ag∩MoS_2	14	Theoretical			[94]
Independent nano-structures							
	Metal–metal oxide	Fe/Al_2O_3	14	THW method		0.9 mPas	[66]
		Al_2O_3/Ag	17	THW method			[31]
	Metal oxide– metal oxide	ZrO_2/CeO_2	22	THW method			[105]

(Continued)

TABLE 3.1
(Continued)

Type of nanofluid	Combinations of nanostructures	Example	Thermal conductivity enhancement (%)	Measurement method for thermal conductivity	Stability (zetapotential/no. of days)	Viscosity (mPas)	Reference
		TiO_2/Al_2O_3	16.91	THW method		0.87 cp	[106]
	Carbon nanostructured	GO/SiO_2	26.93	THW method			[83]
		GO/Si	7.59	THW method			[79]
		$MWCNT/Fe_3O_4$	30.5	THW method		2 mPas	[74]
	Metal– silicon	Si/TiO_2	10.5	THW method	–43.33mV		[107]
	Metal based-SiO_2 nanostructures	SiO_2/CuO	58.95	THW method			[63]

TABLE 3.2
Individual Nanostructure-Based Nanofluids and Their Properties

Nanostructure	Thermal conductivity enhancement (%)	Measurement method	Stability (zetapotential/ no. of days)	Viscosity	Reference
ZnO	18	THW method		59% enhancement	[56]
Ag	27	THW method		~0.5 mPas	[57]
Al_2O_3	30	THW method	10 days		[57]
CuO	40	THW method	10 days		[90]
Fe_3O_4	3	THW method			[58]
ND	22.8	THW method	30 days	180% enhancement	[72]
Cu	14	THW method			[110]
Fe	18	THW method			[110]
GO	20–30	THW method		~10% enhancement	[111]
MWCNT	25	THW method	30 mV		[44]
Co_3O_4	25	THW method	−19 mV		[61,112]
SiO_2	3.4	THW method			[87]
ZrO_2	2.01	THW method			[70]
CeO_2	35.7	THW method		176% enhancement	[113]
TiO_2	33	THW method		1.83 mPaS	[94]
MoS_2	0.8	THW method			[114]
Si	–0.16	THW method			[89]

In brief, in a beaker an aqueous solution of trisodium citrate and polyvinyl pyrrolidone (PVP) was taken, and then to the beaker, various combinations of molar quantities of copper sulphate pentahydrate and palladium nitrate were added. To the homogeneous solution, formaldehyde and sodium hydroxide solution were gradually added. After stirring for 120 minutes at room temperature, the hybrid nanofluid was produced.

Another method of single-step nanofluid synthesis is the electrical explosion of wire method reported by Aberoumand et al. [112] to produce $WO_3 \cap Ag$ hybrid nanofluids in transformer oil as the base fluid, and 41% enhancement in thermal conductivity was recorded. In this method, the submerged thin metal wires of Ag and WO_3 are exploded into the base fluid on application of high voltage and current to form a stable colloidal suspension.

3.4.2 Multi-Pot Synthesis

The multiple pot or multi-pot synthesis begins with producing the nanostructures and then dispersing them in a suitable solvent with the help of ultrasonication, mechanical agitation, or magnetic stirring. As defined earlier in Section 3.4, the process is similar to hybrid nanofluid synthesis where hybrid nanostructures are synthesized in a different reaction pot before transferring them to another pot for fluid processing.

3.4.2.1 Core–Shell Type Nanofluids

Core@shell type nanostructure-based hybrid nanofluids can be prepared by multi-pot synthesis. Generally, the core of a nanostructure is prepared first, and the shell is coated over the core in a subsequent step. For example, in the preparation of ZnO@Ag, the core ZnO is prepared first through a sol–gel method and then it is coated with Ag as the shell by the reduction of silver nitrate in the next step [34]. After the preparation of the nanohybrid, they are dispersed in DI water to synthesize hybrid nanofluids of different volume fractions. It was recorded that 0.1% volume fraction of ZnO@Ag nanohybrid in water had an enhanced thermal conductivity of 32% compared to the base fluid. The reported enhancements by Ag and ZnO water-based plain nanofluids are 27% and 18%, respectively [56, 57].

3.4.2.2 Mutually Functionalized Nanofluids

Mutually functionalized hybrid nanofluids are also prepared by multi-pot synthesis. The nanostructures for the nanohybrid are individually prepared before their mutual functionalization, and the synthesized nanohybrid is then dispersed in a base fluid to obtain the hybrid nanofluid. The multi-pot synthesis for mutually functionalized hybrid nanofluids can be prepared through various methods, and some of the popular methods are mentioned as follows.

3.4.2.2.1 Laser Pyrolysis

In laser pyrolysis, a laser beam selectively pyrolyses a gas stream containing chemical precursors, decomposing them to nucleate for nanoparticles. For example, Fe∩Si nanohybrids have been prepared through laser pyrolysis and then dispersed in water in the presence of carboxymethyl cellulose sodium salt to obtain stable hybrid nanofluids [85, 104]. The precursors used for the preparation of Fe∩Si nanohybrids were iron pentacarbonyl and silane. A CO_2 laser of power 200 W was applied for this pyrolysis, and a flame temperature of 650 °C was recorded during the process. The Fe∩Si water-based hybrid nanofluids [85] had 12.60% enhancement of thermal conductivity, which is higher than that of Si monodispersed nanofluids with ~–0.16% [86] thermal conductivity enhancement, whereas Fe-based nanofluids are easily destabilized by aerial oxidation.

3.4.2.2.2 Chemical Co-Precipitation

Chemical co-precipitation is done in a monodispersed nanofluid to synthesize a mutually functionalized nanohybrid. In this approach, the precursors are typically inorganic salts, like nitrate, chloride, and sulphate, that are dissolved in a plain nanofluid to generate a homogenous dispersion of nanostructures surrounded by ion clusters. Similar clusters can also be formed if a nanostructure is added to a solution of salts to prepare a homogeneous sol. The pH of the sol is then adjusted or the sol is evaporated to precipitate as hydroxides, hydrous oxides, or oxalates. For example, chemical co-precipitation of iron salts in the presence of GO was facilitated by adjusting the pH, which eventually produced 2D-0D, graphene-based magnetic hybrid nanostructures [80]. Instead of iron hydroxides, the formation iron oxide 0D nanostructures was facilitated by the adjustment of high pH of, e.g., 11. The presence of tannic acid in a similar procedure resulted in the synthesis of $Fe_3O_4 \cap rGO$ nanohybrid [80], which was separated from the reaction mixture to disperse in water for the formation of a hybrid nanofluid. Thermal conductivity enhancement in the nanofluid was recorded as 11%, and the convective heat transfer enhancement was 82%. $Fe_3O_4 \cap rGO$ nanohybrids were also obtained by chemical co-precipitation and ultra-sonication of 500 W at 30 kHz, without using tannic acid. The synthesized ultrasound-assisted $Fe_3O_4 \cap rGO$ nanohybrids were separated and dispersed in water for obtaining hybrid nanofluids with an enhanced thermal conductivity of 83.44% [81]. It is to be noted that the reported thermal conductivity and convective heat transfer enhancement by GO-based plain nanofluids are 28% and 40.3%, respectively [82]. Two different modes of measurement of thermophysical property are mentioned here: thermal conductivity and convective heat transfer. Thermal conductivity measurement of a fluid is the measurement of the spatial heat transfer from a hot surface through the fluid, minimizing the mass transfer during the measurement. In convective heat transfer measurements, spatial heat transfer from a hot surface is associated with mass transfer in the fluid [113, 114]. A similar chemical co-precipitation method was employed for the formation Fe_3O_4 in the presence of –COOH-functionalized MWCNTs to produce a covalently bonded nanohybrid structure with a combination of 1D-0D nanostructures [50]. The above prepared nanostructure was then dispersed in water to obtain a stable hybrid nanofluid (zeta potential of −43 mV) with thermal conductivity enhancement of 28.33% when compared with its base fluid. For comparison, the enhanced thermal conductivity of water-based Fe_3O_4 plain nanofluids was recorded to be ~3% [58].

3.4.2.3 Nanostructures for Co-Dispersed Nanofluids

Various multi-pot methods are available to prepare co-dispersed nanofluids. The synthesis of an individual nanostructure is also part of it as individually prepared nanostructures are mixed to disperse in the multi-pot process for preparing co-dispersed hybrid nanofluids. Some commonly employed synthetic techniques are discussed below.

3.4.2.3.1 Chemical Exfoliation

2D nanostructures like GO sheets are prepared from bulk graphite using chemical exfoliation methods like modified Hummer's method [76, 81].

3.4.2.3.2 Sol–Gel Process

In sol–gel process [34, 49], the precursors for nanostructures are firstly synthesized as a "sol," which gradually progresses towards the formation of a gel-like diphasic system consisting of a liquid and solid phase. It is then dried to obtain the nanostructures.

3.4.2.3.3 Solvothermal Process

The synthesis of nanomaterials in the presence of a solvent at elevated temperatures ranging from 100 to 1,000 °C and pressure ranging from 1 to 10,000 atm is known as solvothermal synthesis [102]. The process is usually carried out in a sealed vessel at temperatures at or above the boiling point of the reaction solvent.

3.4.2.3.4 Chemical Reduction

Metal nanoparticles are chemically prepared by the reduction of metal salts. For instance, Ag nanoparticles can be synthesized by the reduction of $AgNO_3$ [115, 116]. The chemical reduction process consists of three key stages: reduction of metallic salts by reducing agents, stabilization of ionic complexes, and size control using a capping agent.

3.4.2.3.5 Chemical Co-Precipitation

Chemical co-precipitation or precipitation method is an efficient reaction route to form various nanostructures. For example, commercially available MWCNTs were coated with gum arabic (GA) to disperse in water, and tetramethylammonium hydroxide (TMAH)-coated Fe_3O_4 was prepared individually from iron salt by a precipitation method. The separated monodispersed nanofluids were mixed together to obtain co-dispersed hybrid nanofluids [74]. A thermal conductivity enhancement of 30.5% was recorded in the MWCNT/Fe_3O_4 hybrid nanofluid compared to the base fluid, whereas plain Fe_3O_4 nanofluids enhanced the thermal conductivity of water by ~3% [58].

3.4.2.3.6 Two-Step or Multi-Step Process

Two- or multi-step process for hybrid nanofluids are mainly in practice for their affordable and wide-scale manufacturing possibilities. Different types of nanofluids synthesized through different routes and their thermophysical properties are tabulated in Table 3.3. Irrespective of the processes and methods, agglomeration of nanostructures on the basis of cohesive strength and the Van der Waals force between them is the primary challenge for plain and hybrid nanofluids. Agglomeration is mostly reduced by surfactants or dispersants that are suitable at critical micelle concentrations [115, 117, 118]. Additionally, agglomeration is also reduced by employing dispersing tools, such as ultrasonic baths, magnetic stirrers, high-pressure homogenizers, and ultrasonic jammers [46, 11–35, 38–95, 98, 102, 104, 105, 108–122].

3.5 THERMAL CONDUCTIVITY EFFICACY OF HYBRID NANOFLUIDS

The primary purpose of hybrid nanofluids is to enhance the thermophysical properties of nanofluids, which could not be achieved with plain nanofluids. The combination of nanostructures in hybrid nanofluids has a vital role in the enhancement of thermal conductivity of the carrier fluid. Graphene-based hybrid nanostructures such as rGO$\cap Fe_3O_4$ [80] has shown ~82% enhancement in convective heat transfer, and $Fe_3O_4 \cap$rGO-based nanofluids have shown an enhanced thermal conductivity of water of 83.44% [81]. GO-Fe_3O_4/Si and GO-Fe_3O_4/Al_2O_3 ternary hybrid nanofluids [119] have shown ~82.7% and 92.1% enhancement in heat transfer coefficient. It is to be noted that GO-based simple nanofluids have been reported to improve convective heat transfer by 28% and 40% [82]. Hybrid nanofluids with MWCNTs have also shown significant enhancement in thermal conductivity up to 78% [75]. ZnO@Ag

TABLE 3.3
Different Types of Nanofluids Synthesized through Different Synthetic Routes

Type of synthesis	Synthetic routes for multi-pot synthesis	Type of nanofluid	Combinations of nanostructures	Example	Thermal conductivity enhancement (%)	Stability	Viscosity (mPas)	Number of publications (2016–2022)	Refer ence
Single-pot synthesis		Core@shell type							
			Metal oxide– metal oxide	$CuO@Al_2O_3$	34	25 mV			[32]
		Mutually functionalized	Metal oxide–metal	$WO_3 \cap Ag$	41				[112]
			Metal–metal	$Cu \cap Pd$	18.79				[111]
Multi-pot synthesis									
a	Laser pyrolysis	Core–shell type							
		Mutually functionalized							
			Metal–silicon	$Fe \cap Si$	12.6			3	[104]
		Independent nanostructures	Carbon nanostructured	GO/Si	7.59			1	[120]
b	Chemical co-precipitation	Mutually functionalized	Carbon nanostructured	$ND \cap Fe_3O_4$	17.76			6	[70]
				$ND \cap Co_3O_4$	16				[59]
				$ND \cap Co_3O_4$	15.71				[68]
				$ND \cap Fe_3O_4$	12.79		6		[99]

(Continued)

TABLE 3.3
(Continued)

Type of synthesis	Synthetic routes for multi-pot synthesis	Type of nanofluid	Combinations of nanostructures	Example	Thermal conductivity enhancement (%)	Stability	Viscosity (mPas)	Number of publications (2016–2022)	Refer ence
				Cu-, Zn-, Fe-, Ag- COOH-MWCNTs	Cu hybrid – 78, Fe hybrid – 25, Ag hybrid – 40	Cu hybrid – 45days, Fe hybrid – 40 days, Ag hybrid –15 days, Zn hybrid –less than 1 day	6		[75]
			Carbon nanostructured	MWCNT/Fe_3O_4	30.5		2	1	[50]
c	Chemical exfoliation	Mutually functionalized	Carbon nanostructured	GO∩Co_3O_4	19.14			6	[123]
				Fe_3O_4∩GO	18				[102]
				Fe_3O_4∩rGO	83.44		2.09		[81]
				Fe_3O_4∩GO	11				[80]
				GO∩Pd	17.77	22 days			[103]
				Ag∩rGO	18				[100]
		Independent nanostructures							
			Carbon nanostructured	GO/Si	7.59			2	[86]
d	Sol–gel process	Core@shell type	Metal–metal oxide	ZnO@Ag	20.53	50.5 mV		4	[96]
				TiO_2@Cu	9.8				[26]

(Continued)

TABLE 3.3
(Continued)

Type of synthesis	Synthetic routes for multi-pot synthesis	Type of nanofluid	Combinations of nanostructures	Example	Thermal conductivity enhancement (%)	Stability	Viscosity (mPas)	Number of publications (2016–2022)	Refer ence
Type of synthesis	**Synthetic routes for multi-pot synthesis**	**Type of nanofluid**	**Combinations of nanostructures**	**Example**	**Thermal conductivity enhancement (%)**	**Stability**	**Viscosity (mPas)**	**Number of publications (2016–2022)**	**Refer ence**
			Metal oxide– metal oxide	TiO_2@Al_2O_3	16.91		8.7	1	[124]
			Metal–metal oxide	ZnO@Ag	32	>40 mV		1	[34]
		Mutually functionalized	Metal oxide– metal oxide	$ZrO_2 \cap CeO_2$	22			1	[60]
e	Solvothermal process	Mutually functionalized	Carbon nanostructured	$Fe_3O_4 \cap GO$	18			1	[102]
f	Chemical reduction	Core@shell type	Metal–metal oxide	TiO_2@Cu	9.8			2	[26]
				Al_2O_3@Ag	17				[31]
		Mutually functionalized	Metal–metal oxide						
			Metal oxide– metal oxide	$CuO \cap Al_2O_3$		25 mV		1	[32]
			Carbon nanostructured	$Cu \cap CNT$	75.57	175 days	2.5	2	[45]
				$Ag \cap CNT$	58		0.9		[125]
				$ND \cap Ni$	29.39		0.97		[71]
			Metal– silicon						
				Al_2O_3/Fe	14		0.9		[66]

nanohybrids dispersed in water have shown thermal conductivity enhancement of up to 32% [34]. WO_3∩Ag hybrids in transformer oil enhanced the thermal conductivity of the base fluid by 41% [112]. The reported enhancements by Ag and ZnO water-based plain nanofluids are 27% and 18%, respectively [56, 57]. In general, hybrid nanofluids that contain only 0D-0D type nanostructures have relatively lesser enhancement of thermal conductivity than hybrid fluids containing 1D and 2D nanostructures.

3.6 PRESENT DEVELOPMENT AND FUTURE SCOPE

Current studies indicate that hybrid nanofluids have a significant impact on thermal conductivity enhancements of base fluids. Compared to monodispersed nanofluids, hybrid nanofluids have better thermal conductivity and flow characteristics. It is also realized that carbon-based hybrid nanofluids are widely preferred considering their high efficiency in thermal conductivity enhancement of base fluids. In syntheses considering cost-effectiveness, two-step or multi-pot synthesis of nanofluids is preferred.

Synthesis of hybrid nanofluids not only augments the thermophysical properties of monodispersed nanofluids but also opens wider research on the stability of otherwise unstable plain nanofluids. For example, Fe∩Si stable hybrid nanofluids (zeta potential > –70 mV) are produced by using laser pyrolysis [85], where it is challenging to produce stable Fe plain nanofluids due to the oxidative nature of iron.

The manufacture of nanofluids necessitates sophisticated processes, raising economic concerns about the use of nanofluids and preventing their commercial use. So, to expedite the commercial usage of nanofluids, economic analysis of such fluids must be addressed.

Furthermore, the fundamental problem with nanofluids is long-term stability, which is important for their application and a major constraint. Considering the potential applications of hybrid nanofluids, it is critical to analyse the life cycle of nanofluid-based thermal systems. It is also necessary to concentrate on sustainable materials suited for the use of nanofluids. To achieve stable conditions, nanofluid synthesis must be optimized using various parameters such as stirring duration, temperature, and speed; volume, density, and base fluid type; weight/volume concentration, density, nano-size, and type of mono or hybrid NPs used; type and quantity of surfactant used; sonication time; modes; frequency; and amplitude.

Although nanofluids were primarily developed as heat transfer fluids, they have a wide range of applications such as in the fields of mechanical engineering, biopharmaceuticals, optoelectronics, and energy harvesting. For example, CuO nanoparticles dispersed in water are good lubricants [126]. Silica-based nanofluids are widely studied for the enhancement of crude oil recovery from oil wells [127, 128]. Cysteine-capped ferrofluids have potential utility as MRI contrast agents [129]. TiO_2/Ag hybrid nanofluids are used as heat exchanger fluids [130], and Fe_3O_4 carbon quantum dot hybrid nanofluids are used as coolants in cooling towers [9].

3.7 SUMMARY

Nanofluids have wide applications in different fields of technology. Due to their enhanced thermophysical properties, hybrid nanofluids are the best choice as heat transfer fluids. Hybrid nanofluids considering mutual compatibility are prepared as core@shell type hybrid nanostructures, mutually functionalized nanohybrids, or nanofluids prepared by the direct dispersion of two or more nanostructures in the same base fluid. The combination of nanostructures used in the formation of hybrid nanofluids can be metal/metal oxide, metal oxide/metal oxide, carbon nanostructure based, metal/silicon, metal based/silica, and metal/metal sulphide. Hybrid nanofluids can be prepared by single-pot synthesis or multi-pot synthesis. Nanofluids are generally prepared by a single-step or a two-step method. Carbon-based nanostructures, especially 1D and 2D nanostructures like MWCNTs and graphene-based structures, show significant thermal property enhancement by enhancing convective heat transfer up to 82% and thermal conductivity up to 83.4%. In case of Ag@ZnO aqueous hybrid nanofluids, thermal conductivity enhancement was reported as 32%, which is higher than that of their individual monodispersed nanofluids. The synthesis of hybrid nanofluids not only augments the thermophysical properties of the base fluid compared to the base fluid but also proffers to produce otherwise challenging monodispersed nanofluids. For instance, stable hybrid nanofluids of $Fe \cap Si$ are produced where it is difficult to produce stable Fe plain nanofluids. Although hybrid nanofluids are potential heat transfer fluids, the economic feasibility, long-term stability, and recyclability of nanofluids are still being investigated.

REFERENCES

1. Turcu R, Darabont A, Nan A, et al. New polypyrrole-multiwall carbon nanotubes hybrid materials. J Optoelectron Adv Mater. 2006;8(2):643–647.
2. Barrera CC, Groot H, Vargas WL, Narváez DM. Efficacy and molecular effects of a reduced graphene oxide/Fe_3O_4 nanocomposite in photothermal therapy against cancer. Int J Nanomed. 2020;15:6421–6432.
3. Iskandar F, Dwinanto E, Abdullah M, Khairurrijal, Muraza O. Viscosity reduction of heavy oil using nanocatalyst in aquathermolysis reaction. KONA Powder Part J. 2016;2016(33):3–16. doi:10.14356/kona.2016005
4. Okonkwo EC, Wole-Osho I, Almanassra IW, Abdullatif YM, Al-Ansari T. An updated review of nanofluids in various heat transfer devices. Vol 145. Springer International Publishing; 2021. doi:10.1007/s10973-020-09760-2
5. Skorek AW, Blé SV, Gryko-Nikitin A, Nazarko J. Nanothermal management in nanoelectronics systems. Can Conf Electr Comput Eng. 2007;1(1):1298–1301. doi:10.1109/CCECE.2007.330
6. Cavin RK, Zhirnov VV, Herr DJC, Avila A, Hutchby J. Research directions and challenges in nanoelectronics. J Nanoparticle Res. 2006;8(6):841–858. doi:10.1007/s11051-006-9123-4
7. Muneeshwaran M, Srinivasan G, Muthukumar P, Wang CC. Role of hybrid-nanofluid in heat transfer enhancement—A review. Int Commun Heat Mass Transf. 2021;125(May):105341. doi:10.1016/j.icheatmasstransfer.2021.105341
8. Khalili Z, Sheikholeslami M. Analyzing the effect of confined jet impingement on efficiency of photovoltaic thermal solar unit equipped with thermoelectric generator in existence of hybrid nanofluid. J Clean Prod. 2023;406(March):137063. doi:10.1016/j.jclepro.2023.137063
9. Mousavi H, Mostafa S, Ghomshe T, Rashidi A. Hybrids carbon quantum dots as new nanofluids for heat transfer enhancement in wet cooling towers. Heat Mass Transf. 2022;58(2):1–12. doi:10.1007/s00231-021-03077-y

10. Aberoumand S, Jafarimoghaddam A, Moravej M, Aberoumand H, Javaherdeh K. Experimental study on the rheological behavior of silver-heat transfer oil nanofluid and suggesting two empirical based correlations for thermal conductivity and viscosity of oil based nanofluids. Appl Therm Eng. 2016;101:362–372. doi:10.1016/j.applthermaleng.2016.01.148
11. Das SK, Choi SUS, Argonne WY, Pradeep T. Nanofluids: Science and technology. John Wiley & Sons; 2007.
12. Masuda H, Ebata A, Teramae K, Hishinuma N. Alteration of thermal conductivity and viscosity of liquid by dispersing ultra-fine particles. dispersion of Al_2O_3, SiO_2 and TiO_2 ultra-fine particles. Netsu Bussei. 1993;7(4):227–233. doi:10.2963/jjtp.7.227
13. Choi SUS, Eastman JA. Enhancing thermal conductivity of fluids with nanoparticles. Proceedings of the ASME International Mechanical Engineering Congress and Exposition. 1995;66:99–105.
14. Pak BC, Cho Y. Hydrodynamic and heat transfer study of dispersed fluids with submicron metallic oxide. Exp Heat Transf J Therm Energ Transp Stor Convers. 1998;11(2):151–170. doi:10.1080/08916159808946559
15. Izadi N, Nasernejad B. Newly engineered alumina quantum dot-based nanofluid in enhanced oil recovery at reservoir conditions. Sci Rep. 2022;12(1):1–19. doi:10.1038/s41598-022-12387-y
16. Soleimani H, Baig MK, Yahya N, et al. Impact of carbon nanotubes based nanofluid on oil recovery efficiency using core flooding. Results Phys. 2018;9:39–48. doi:10.1016/j.rinp.2018.01.072
17. Yarmand H, Zulkifli NWBM, Gharehkhani S, et al. Convective heat transfer enhancement with graphene nanoplatelet/platinum hybrid nanofluid. Int Commun Heat Mass Transf. 2017;88(September):120–125. doi:10.1016/j.icheatmasstransfer.2017.08.010
18. Kumar V, Sarkar J. Research and development on composite nanofluids as next-generation heat transfer medium. J Therm Anal Calorim. 2019;137(4):1133–1154. doi:10.1007/s10973-019-08025-x
19. Devendiran DK, Amirtham VA. A review on preparation, characterization, properties and applications of nanofluids. Renew Sustain Energy Rev. 2016;60:21–40. doi:10.1016/j.rser.2016.01.055
20. Nabil MF, Azmi WH, AbdulHamid K, Mamat R, Hagos FY. An experimental study on the thermal conductivity and dynamic viscosity of TiO_2-SiO_2 nanofluids in water: Ethylene glycol mixture. Int Commun Heat Mass Transf. 2017;86(June):181–189. doi:10.1016/j.icheatmasstransfer.2017.05.024
21. Verma SK, Tiwari AK, Tiwari S, Chauhan DS. Performance analysis of hybrid nanofluids in flat plate solar collector as an advanced working fluid. Sol Energy. 2018;167(April):231–241. doi:10.1016/j.solener.2018.04.017
22. Toghraie D, Chaharsoghi VA, Afrand M. Measurement of thermal conductivity of ZnO–TiO_2/EG hybrid nanofluid: Effects of temperature and nanoparticles concentration. J Therm Anal Calorim. 2016;125(1):527–535. doi:10.1007/s10973-016-5436-4
23. Yu W, Xie H, Chen L, Li Y. Enhancement of thermal conductivity of kerosene-based Fe_3O_4 nanofluids prepared via phase-transfer method. Colloids Surfaces A Physicochem Eng Asp. 2010;355(1–3):109–113. doi:10.1016/j.colsurfa.2009.11.044
24. Wei B, Zou C, Yuan X, Li X. Thermo-physical property evaluation of diathermic oil based hybrid nanofluids for heat transfer applications. Int J Heat Mass Transf. 2017;107:281–287. doi:10.1016/j.ijheatmasstransfer.2016.11.044
25. Asadi M, Asadi A. Dynamic viscosity of MWCNT/ZnO-engine oil hybrid nanofluid: An experimental investigation and new correlation in different temperatures and solid concentrations. Int Commun Heat Mass Transf. 2016;76:41–45. doi:10.1016/j.icheatmasstransfer.2016.05.019

26. Leong KY, Razali I, KuAhmad KZ, Ong HC, Ghazali MJ, Abdul Rahman MR. Thermal conductivity of an ethylene glycol/water-based nanofluid with copper-titanium dioxide nanoparticles: An experimental approach. Int Commun Heat Mass Transf. 2018;90:23–28. doi:10.1016/j.icheatmasstransfer.2017.10.005
27. Leena M, Srinivasan S. A comparative study on thermal conductivity of TiO_2/ethylene glycol–water and TiO_2/propylene glycol–water nanofluids. J Therm Anal Calorim. 2018;131(2):1987–1998. doi:10.1007/s10973-017-6616-6
28. Malmberg CG, Maryott AA. Dielectric constant of water from 0 to 100 °C. J Res Natl Bur Stand (1934). 1956;56(1):1. doi:10.6028/jres.056.001
29. Azwadi N, Sidik C, Adamu IM, Jamil MM. Preparation methods and thermal performance of hybrid nanofluids. J Adv Res Appl Mech. 2020;66(1): 7–16. https://doi.org/10.37934/aram.66.1.716.
30. Ranjbarzadeh R, Akhgar A, Musivand S, Afrand M. Effects of graphene oxide-silicon oxide hybrid nanomaterials on rheological behavior of water at various time durations and temperatures: Synthesis, preparation and stability. Powder Technol. 2018;335:375–387. doi:10.1016/j.powtec.2018.05.036
31. Aparna Z, Michael M, Pabi SK, Ghosh S. Thermal conductivity of aqueous Al_2O_3/Ag hybrid nanofluid at different temperatures and volume concentrations: An experimental investigation and development of new correlation function. Powder Technol. 2019;343:714–722. doi:10.1016/j.powtec.2018.11.096
32. Plant RD, Hodgson GK, Impellizzeri S, Saghir MZ. Experimental and numerical investigation of heat enhancement using a hybrid nanofluid of copper oxide/alumina nanoparticles in water. J Therm Anal Calorim. 2020;141(5):1951–1968. doi:10.1007/s10973-020-09639-2
33. Pourpasha H, Heris SZ, Mohammadpourfard M. The effect of TiO_2 doped multi-walled carbon nanotubes synthesis on the thermophysical and heat transfer properties of transformer oil: A comprehensive experimental study. Case Stud Therm Eng. 2023;41:102607. doi:10.1016/j.csite.2022.102607
34. Sharma PO, Barewar SD, Chougule SS. Experimental investigation of heat transfer enhancement in pool boiling using novel Ag/ZnO hybrid nanofluids. J Therm Anal Calorim. 2021;143(2):1051–1061. doi:10.1007/s10973-020-09922-2
35. Allahyar HR, Hormozi F, ZareNezhad B. Experimental investigation on the thermal performance of a coiled heat exchanger using a new hybrid nanofluid. Exp Therm Fluid Sci. 2016;76:324–329. doi:10.1016/j.expthermflusci.2016.03.027
36. Nasirzadehroshenin F, Maddah H, Sakhaeinia H, Pourmozafari A. Investigation of exergy of double-pipe heat exchanger using synthesized hybrid nanofluid developed by modeling. Int J Thermophys. 2019;40(9):1–24. doi:10.1007/s10765-019-2551-z
37. Syam Sundar L, Mesfin S, Tefera SintieY, PunnaiahV, ChamkhaAJ, Sousa ACM. A review on the use of hybrid nanofluid in a solar flat plate and parabolic trough collectors and its enhanced collector thermal efficiency. J Nanofluids. 2021;10(2):147–171. doi:10.1166/jon.2021.1783
38. Hamzah MH, Sidik NAC, Ken TL, Mamat R, Najafi G. Factors affecting the performance of hybrid nanofluids: A comprehensive review. Int J Heat Mass Transf. 2017;115:630–646. doi:10.1016/j.ijheatmasstransfer.2017.07.021
39. Bahrami M, Akbari M, Karimipour A, Afrand M. An experimental study on rheological behavior of hybrid nanofluids made of iron and copper oxide in a binary mixture of water and ethylene glycol: Non-Newtonian behavior. Exp Therm Fluid Sci. 2016;79:231–237. doi:10.1016/j.expthermflusci.2016.07.015
40. Dharmakkan N, Manikandan P, Muthusamy S, Jomde A, Shamkuwar S, Sonawane C. Case studies in thermal engineering a case study on analyzing the performance of micro-plate heat exchanger using nanofluids at different flow rates and temperatures. Case Stud Therm Eng. 2023;44:102805. doi:10.1016/j.csite.2023.102805

41. Sidik NAC, Adamu IM, Jamil MM, Kefayati GHR, Mamat R, Najafi G. Recent progress on hybrid nanofluids in heat transfer applications: A comprehensive review. Int Commun Heat Mass Transf. 2016;78:68–79. doi:10.1016/j.icheatmasstransfer.2016.08.019
42. Esfe MH, Afrand M. An updated review on the nanofluids characteristics: Preparation and measurement methods of nanofluids thermal conductivity. J Therm Anal Calorim. 2019;138(6):4091–4101. doi:10.1007/s10973-019-08406-2
43. Soltani O, Akbari M. Effects of temperature and particles concentration on the dynamic viscosity of MgO-MWCNT/ethylene glycol hybrid nanofluid: Experimental study. Phys E: Low Dimens Syst Nanostruct. 2016;84:564–570. doi:10.1016/j.physe.2016.06.015
44. Suresh S, Venkitaraj KP, Selvakumar P, Chandrasekar M. Synthesis of Al_2O_3-Cu/water hybrid nanofluids using two step method and its thermo physical properties. Colloids Surfaces A: Physicochem Eng Asp. 2011;388(1–3):41–48. doi:10.1016/j.colsurfa.2011.08.005
45. Gupta N, Gupta SM, Sharma SK. Synthesis, characterization and dispersion stability of water-based Cu–CNT hybrid nanofluid without surfactant. Microfluid Nanofluidics. 2021;25(2):1–14. doi:10.1007/s10404-021-02421-2
46. Ali HM, ed. Hybrid nanofluids for convection heat transfer. Elsevier; 2020. doi:10.1016/C2018-0-04602-2
47. Hemmat Esfe M, Hajmohammad MH, Razi P, Ahangar MRH, Arani AAA. The optimization of viscosity and thermal conductivity in hybrid nanofluids prepared with magnetic nanocomposite of nanodiamond cobalt-oxide (ND-Co_3O_4) using NSGA-II and RSM. Int Commun Heat Mass Transf. 2016;79:128–134. doi:10.1016/j.icheatmasstransfer.2016.09.015
48. Jana S, Salehi-Khojin A, Zhong WH. Enhancement of fluid thermal conductivity by the addition of single and hybrid nano-additives. Thermochim Acta. 2007;462(1–2):45–55. doi:10.1016/j.tca.2007.06.009
49. Barewar SD, Tawri S, Chougule SS. Experimental investigation of thermal conductivity and its ANN modeling for glycol-based Ag/ZnO hybrid nanofluids with low concentration. J Therm Anal Calorim. 2020;139(3):1779–1790. doi:10.1007/s10973-019-08618-6
50. Said Z, Sharma P, Syam Sundar L, Afzal A, Li C. Synthesis, stability, thermophysical properties and AI approach for predictive modelling of Fe_3O_4 coated MWCNT hybrid nanofluids. J Mol Liq. 2021;340:117291. doi:10.1016/j.molliq.2021.117291
51. Sousa ACM. Case studies in thermal engineering thermal conductivity and viscosity of hybrid nanofluids prepared with magnetic nanodiamond-cobalt oxide (ND-Co_3O_4) nanocomposite. Case Stud Therm Eng. 2016;7:66–77. doi:10.1016/j.csite.2016.03.001
52. Lu B, Liu A, Wu H, Shen Q, Zhao T, Wang J-S. Hollow Au-Cu_2O core-shell nanoparticles with geometry-dependent optical properties as efficient plasmonic photocatalysts under visible light. Published online 2016. doi:10.1021/acs.langmuir.6b00331
53. Barreda ÁI, Gutiérrez Y, Sanz JM, González F. Light guiding and switching using eccentric core-shell geometries. 2017;(August):1–10. doi:10.1038/s41598-017-11401-y
54. Dhawan U, Tseng C, Wang H, Hsu S, Tsai M. Assessing suitability of Co@ Au core/shell nanoparticle geometry for improved theranostics in colon carcinoma. Nanomaterials (Basel). 2021;11(8):2048. doi:10.3390/nano11082048.
55. Li D, Deng S, Zhao Z, et al. VO_2(M)@SnO_2 core–shell nanoparticles: Improved chemical stability and thermochromic property rendered by SnO_2 shell. Appl Surf Sci. 2022;598:153741. doi:10.1016/J.APSUSC.2022.153741
56. Jeong J, Li C, Kwon Y, Lee J, Kim SH, Yun R. Particle shape effect on the viscosity and thermal conductivity of ZnO nanofluids. Int J Refrig. 2013;36(8):2233–2241. doi:10.1016/j.ijrefrig.2013.07.024
57. Godson L, Raja B, Lal DM, Wongwises S. Experimental investigation on the thermal conductivity and viscosity of silver-deionized water nanofluid. Exp Heat Transf. 2010;23(4):317–332. doi:10.1080/08916150903564796

58. Ebrahimi S, Saghravani SF. Experimental study of the thermal conductivity features of the water based Fe_3O_4/CuO nanofluid. Heat Mass Transf und Stoffuebertragung. 2018;54(4):999–1008. doi:10.1007/s00231-017-2188-z
59. Sundar LS, Irurueta GO, Venkata Ramana E, Singh MK, Sousa ACM. Thermal conductivity and viscosity of hybrid nanfluids prepared with magnetic nanodiamond-cobalt oxide (ND-Co_3O_4) nanocomposite. Case Stud Therm Eng. 2016;7:66–77. doi:10.1016/j.csite.2016.03.001
60. Sekhar T, Nandan G, Prakash R, Muthuraman M. Investigations on viscosity and thermal conductivity of cobalt oxide-water nano fluid. Mater Today Proc. 2018;5(2):6176–6182. doi:10.1016/j.matpr.2017.12.224
61. Savi M, Bocchi L, Cacciani F, et al. Cobalt oxide nanoparticles induce oxidative stress and alter electromechanical function in rat ventricular myocytes. Part Fibre Toxicol. 2021;18(1):1. doi: 10.1186/s12989-020-00396-6
62. Siddiqui FR, Tso CY, Chan KC, Fu SC, Chao CYH. On trade-off for dispersion stability and thermal transport of Cu-Al_2O_3 hybrid nanofluid for various mixing ratios. Int J Heat Mass Transf. 2019;132:1200–1216. doi:10.1016/j.ijheatmasstransfer.2018.12.094
63. Castro F, Sathish N, Kannaiyan K, Boobalan C. A Proficient approach to enhance heat transfer using cupric oxide/silica hybrid nanoliquids. J Therm Anal Calorim. 2022;147(10):5589–5598. doi:10.1007/s10973-021-10956-3
64. Sandhu H, Gangacharyulu D, Singh MK. Experimental investigations on the cooling performance of microchannels using alumina nanofluids with different base fluids. J Enhanc Heat Transf. 2018;25(3):283–291. doi:10.1615/JEnhHeatTransf.2018026139
65. Lahari MLRC, Sai PHVST, Sharma KV, Narayanaswamy KS, Bee PH, Devaraj S. Analysis of parallel flow heat exchanger using SiO_2 nanofluids in the laminar flow region. J Phys Conf Ser. 2021;2070(1). doi:10.1088/1742-6596/2070/1/012230
66. Okonkwo EC, Wole-Osho I, Kavaz D, Abid M. Comparison of experimental and theoretical methods of obtaining the thermal properties of alumina/iron mono and hybrid nanofluids. J Mol Liq. 2019;292:111377. doi:10.1016/j.molliq.2019.111377
67. Çolak AB. Experimental study for thermal conductivity of water-based zirconium oxide nanofluid: Developing optimal artificial neural network and proposing new correlation. Int J Energy Res. 2021;45(2):2912–2930. doi:10.1002/er.5988
68. Syam Sundar L, MisganawAH, Singh MK, Sousa ACM, Ali HM. Efficiency analysis of thermosyphon solar flat plate collector with low mass concentrations of ND–Co_3O_4 hybrid nanofluids: An experimental study. J Therm Anal Calorim. 2021;143(2):959–972. doi:10.1007/s10973-020-10176-1
69. Sundar LS, Singh MK, Sousa ACM. Turbulent heat transfer and friction factor of nanodiamond-nickel hybrid nanofluids flow in a tube: An experimental study. Int J Heat Mass Transf. 2018;117:223–234. doi:10.1016/j.ijheatmasstransfer.2017.09.109
70. Said Z, Jamei M, Sundar LS, PandeyAK, Allouhi A, Li C. Thermophysical properties of water, water and ethylene glycol mixture- based nanodiamond + Fe_3O_4 hybrid nanofluids: An experimental assessment and application of data-driven approaches. J Mol Liq. 2022;347:117944. doi:10.1016/j.molliq.2021.117944
71. Syam Sundar L, Singh MK, Sousa ACM. Heat transfer and friction factor of nanodiamond-nickel hybrid nanofluids flow in a tube with longitudinal strip inserts. Int J Heat Mass Transf. 2018;121:390–401. doi:10.1016/j.ijheatmasstransfer.2017.12.096
72. Sundar LS, Hortiguela MJ, Singh MK, Sousa ACM. Thermal conductivity and viscosity of water based nanodiamond (ND) nanofluids: An experimental study. Int Commun Heat Mass Transf. 2016;76:245–255. doi:10.1016/j.icheatmasstransfer.2016.05.025
73. Megatif L, Ghozatloo A, Arimi A, Shariati-Niasar M. Investigation of laminar convective heat transfer of a novel TiO_2-carbon nanotube hybrid water-based nanofluid. Exp Heat Transf. 2016;29(1):124–138. doi:10.1080/08916152.2014.973974

74. Shahsavar A, Bahiraei M. Experimental investigation and modeling of thermal conductivity and viscosity for non-Newtonian hybrid nanofluid containing coated CNT/Fe_3O_4 nanoparticles. Powder Technol. 2017;318:441–450. doi:10.1016/j.powtec.2017.06.023
75. Gupta N, Mital S, Sharma SK. Materials today: Proceedings preparation of stable metal/COOH-MWCNT hybrid nanofluid. Mater Today Proc. 2020. doi:10.1016/j.matpr.2020.04.492
76. Cakmak NK, Said Z, Sundar LS, Ali ZM, Tiwari AK. Preparation, characterization, stability, and thermal conductivity of rGO-Fe_3O_4-TiO_2 hybrid nanofluid: An experimental study. Powder Technol. 2020;372:235–245. doi:10.1016/j.powtec.2020.06.012
77. Borode A, Tshephe T, Olubambi P, Sharifpur M. Stability and thermophysical properties of GNP-Fe_2O_3 hybrid nanofluid: Effect of volume fraction and temperature. Nanomaterials. 2023;13(7):1238. doi:10.3390/nano13071238
78. Yang L, Ji W, Zhang Z, Jin X. Thermal conductivity enhancement of water by adding graphene nano-sheets: Consideration of particle loading and temperature effects. Int Commun Heat Mass Transf. 2019;109(October):104353. doi:10.1016/j.icheatmasstransfer.2019.104353
79. Huminic G, Alexandru V, Huminic A, Fleac C, Dumitrache F. Water-based graphene oxide—silicon hybrid nanofluids—experimental and theoretical approach. Int. J. Mol. Sci. 2022;23:3056. doi:10.3390/ijms23063056
80. Sadeghinezhad E, Mehrali M, Akhiani AR, et al. Experimental study on heat transfer augmentation of graphene based ferrofluids in presence of magnetic field. Appl Therm Eng. 2017;114:415–427. doi:10.1016/j.applthermaleng.2016.11.199
81. Barai DP, Bhanvase BA, Saharan VK. Reduced graphene oxide-Fe_3O_4 nanocomposite based nanofluids: Study on ultrasonic assisted synthesis, thermal conductivity, rheology, and convective heat transfer. Ind Eng Chem Res. 2019;58(19):8349–8369. doi:10.1021/acs.iecr.8b05733
82. Ranjbarzadeh R, Karimipour A, Afrand M, Isfahani AHM, Shirneshan A. Empirical analysis of heat transfer and friction factor of water/graphene oxide nanofluid flow in turbulent regime through an isothermal pipe. Appl Therm Eng. 2017;126:538–547. doi:10.1016/j.applthermaleng.2017.07.189
83. Nguyen Q, Rizvandi R, Karimipour A, Malekahmadi O, Bach QV. A novel correlation to calculate thermal conductivity of aqueous hybrid graphene oxide/silicon dioxide nanofluid: Synthesis, characterizations, preparation, and artificial neural network modeling. Arab J Sci Eng. 2020;45(11):9747–9758. doi:10.1007/s13369-020-04885-w
84. Guo W, Li G, Zheng Y, Dong C. Measurement of the thermal conductivity of SiO_2 nanofluids with an optimized transient hot wire method. Thermochim Acta. 2018;661(August 2017):84–97. doi:10.1016/j.tca.2018.01.008
85. Huminic G, Huminic A, Fleacă C, Dumitrache F, Morjan I. Experimental study on viscosity of water based Fe–Si hybrid nanofluids. J Mol Liq. 2021;321:114938. doi:10.1016/j.molliq.2020.114938
86. Chun SY, Bang IC, Choo YJ, Song CH. Heat transfer characteristics of Si and SiC nanofluids during a rapid quenching and nanoparticles deposition effects. Int J Heat Mass Transf. 2011;54(5–6):1217–1223. doi:10.1016/j.ijheatmasstransfer.2010.10.029
87. Lahari MLRC, Talpa PHVS, Sharma KV, Narayanaswamy KS. Materials today: Proceedings thermal conductivity and viscosity of glycerine-water based Cu-SiO_2 hybrid nanofluids. Mater Today Proc. 2022. doi:10.1016/j.matpr.2022.05.284
88. Bergmann CP, Sharma V, Nor K, Hamid HB. Topics in mining, metallurgy and materials engineering series editor: Engineering applications of nanotechnology from energy to drug delivery; 2017. http://www.springer.com/series/11054
89. Hamid KA, Azmi WH, Nabil MF, Mamat R. Heat transfer augmentation of mixture ratio TiO_2 to SiO_2 in hybrid nanofluid. J Mech Eng. 2017;SI 4(3):51–63.

90. Agarwal R, Verma K, Agrawal NK, Duchaniya RK, Singh R. Synthesis, characterization, thermal conductivity and sensitivity of CuO nanofluids. Appl Therm Eng. 2016;102:1024–1036. doi:10.1016/j.applthermaleng.2016.04.051
91. Murshed SMS, Leong KC, Yang C. Enhanced thermal conductivity of TiO_2—water based nanofluids. Int J Therm Sci. 2005;44(4):367–373. doi:10.1016/j.ijthermalsci.2004.12.005
92. Islam MR, Shabani B, Rosengarten G. Electrical and thermal conductivities of 50/50 water-ethylene glycol based TiO_2 nanofluids to be used as coolants in PEM fuel cells. Energy Procedia. 2017;110:101–108. doi:10.1016/j.egypro.2017.03.113
93. Manikandan ASP, Nivetha T, Uma R, Saranya L, Pradeep T. Effect of SiO_2 nanoparticle suspension on thermal conductivity and viscosity of ethylene glycol/water base fluid. Int J Sci Technol Res. 2020;9(3):3067–3071.
94. Sani FH, Pourfallah M, Gholinia M. The effect of MoS_2–Ag/H_2O hybrid nanofluid on improving the performance of a solar collector by placing wavy strips in the absorber tube. Case Stud Therm Eng. 2022;30(January):101760. doi:10.1016/j.csite.2022.101760
95. Parametthanuwat T, Bhuwakietkumjohn N, Rittidech S, Ding Y. Experimental investigation on thermal properties of silver nanofluids. Int J Heat Fluid Flow. 2015;56(Mebe):80–90. doi:10.1016/j.ijheatfluidflow.2015.07.005
96. Barewar SD, Chougule SS, Jadhav J, Biswas S. Synthesis and thermo-physical properties of water-based novel Ag/ZnO hybrid nanofluids. J Therm Anal Calorim. 2018;134(3):1493–1504. doi:10.1007/s10973-018-7883-6
97. Leong KY, KuAhmad KZ, Ong HC, Ghazali MJ, Baharum A. Synthesis and thermal conductivity characteristic of hybrid nanofluids—a review. Renew Sustain Energy Rev. 2017;75(November 2016):868–878. doi:10.1016/j.rser.2016.11.068
98. Kannaiyan S, Boobalan C, Umasankaran A, Ravirajan A, Sathyan S, Thomas T. Comparison of experimental and calculated thermophysical properties of alumina/cupric oxide hybrid nanofluids. J Mol Liq. 2017;244:469–477. doi:10.1016/j.molliq.2017.09.035
99. Syam Sundar L, Mesfin S, Venkata Ramana E, Said Z, Sousa ACM. Experimental investigation of thermo-physical properties, heat transfer, pumping power, entropy generation, and exergy efficiency of nanodiamond + Fe_3O_4/60:40% water-ethylene glycol hybrid nanofluid flow in a tube. Therm Sci Eng Prog. 2021;21(August 2020). doi:10.1016/j.tsep.2020.100799
100. Mehrali M, Ghatkesar MK, Pecnik R. Full-spectrum volumetric solar thermal conversion via graphene/silver hybrid plasmonic nanofluids. Appl Energy. 2018;224(April):103–115. doi:10.1016/j.apenergy.2018.04.065
101. Syam Sundar L, Singh MK, Ferro MC, Sousa ACM. Experimental investigation of the thermal transport properties of graphene oxide/Co_3O_4 hybrid nanofluids. Int Commun Heat Mass Transf. 2017;84:1–10. doi:10.1016/j.icheatmasstransfer.2017.03.001
102. Damodaran SP. Novel nanohybrid containing magnetite nanocluster-decorated reduced graphene oxide nanosheets for heat transfer applications. ChemistrySelect. 2021;6(26):6698–6706. doi:10.1002/slct.202101692
103. Yarmand H, Gharehkhani S, Shirazi SFS, et al. Nanofluid based on activated hybrid of biomass carbon/graphene oxide: Synthesis, thermo-physical and electrical properties. Int Commun Heat Mass Transf. 2016;72:10–15. doi:10.1016/j.icheatmasstransfer.2016.01.004
104. Huminic G, Huminic A, Dumitrache F, Fleacă C, MorjanI. Study of the thermal conductivity of hybrid nanofluids: Recent research and experimental study. Powder Technol. 2020;367:347–357. doi:10.1016/j.powtec.2020.03.052
105. Vidhya R, Balakrishnan T, Kumar BS, et al. An experimental study of ZrO_2–CeO_2 hybrid nanofluid and response surface methodology for the prediction of heat transfer performance: The new correlations. J Nanomater. 2022(1) doi:10.1155/2022/6596028.

106. Bhattad A, Sarkar J, Ghosh P. Heat transfer characteristics of plate heat exchanger using hybrid nanofluids: Effect of nanoparticle mixture ratio. Heat Mass Transf und Stoffuebertragung. 2020;56(8):2457–2472. doi:10.1007/s00231-020-02877-y
107. LeBa T, Várady ZI, Lukács IE, et al. Experimental investigation of rheological properties and thermal conductivity of SiO_2–P_{25} TiO_2 hybrid nanofluids. J Therm Anal Calorim. 2021;146(1):493–507. doi:10.1007/s10973-020-10022-4
108. Urmi WT, Shafiqah AS, Rahman MM, Kadirgama K, Maleque MA. Preparation methods and challenges of hybrid nanofluids: A review. J Adv Res Fluid Mech Therm Sci. 2021;78(2):56–66. doi:10.37934/ARFMTS.78.2.5666
109. Safiei W, Rahman MM, Yusoff AR, Radin MR. Preparation, stability and wettability of nanofluid: A review. J Mech Eng Sci. 2020;14(3):7244–7257. doi:10.15282/jmes.14.3.2020.24.0569
110. Agarwal R, Verma K, Agrawal NK, Singh R. Sensitivity of thermal conductivity for Al_2O_3 nanofluids. Exp Therm Fluid Sci. 2017;80:19–26. doi:10.1016/j.expthermflusci.2016.08.007
111. Jaiswal AK, Wan M, Singh S, et al. Experimental investigation of thermal conduction in copper-palladium nanofluids. J Nanofluids. 2016;5(4):496–501. doi:10.1166/jon.2016.1243
112. Aberoumand S, Jafarimoghaddam A. Tungsten (III) oxide (WO_3)–silver/transformer oil hybrid nanofluid: Preparation, stability, thermal conductivity and dielectric strength. Alexandria Eng J. 2018;57(1):169–174. doi:10.1016/j.aej.2016.11.003
113. WhatIstheDifferencebetweenThermalConductivityandHeatTransferCoefficient|Compare theDifferencebetweenSimilarTerms.AccessedJuly2,2023.https://www.differencebetween.com/what-is-the-difference-between-thermal-conductivity-and-heat-transfer-coefficient/
114. ConvectiveHeatTransfer—LouisC. Burmeister—Google Books. Accessed July 2, 2023. https://books.google.co.in/books?hl=en&lr=&id=pJaiReRZvHMC&oi=fnd&pg=PA1&dq=what+is+convective+heat+transfer&ots=wauU0q_F-J&sig=ThtH4blBoGW9kyTxAH_jxmOpLkY&redir_esc=y#v=onepage&q&f=false
115. Pavithra KS, Gurumurthy SC, Yashoda MP, et al. Polymer-dispersant-stabilized Ag nanofluids for heat transfer applications. J Therm Anal Calorim. 2021;146(2):601–610. doi:10.1007/s10973-020-10064-8
116. Parwin S, Parui J. Ag nanofluids synthesis in presence of citrate at different stirring rotation and their post reaction stability. J Dispers Sci Technol. 2020;0(0):1–12. doi:10.1080/01932691.2020.1789469
117. Tiwari AK, Pandya NS, Said Z, Öztop HF, Abu-Hamdeh N. 4S consideration (synthesis, sonication, surfactant, stability) for the thermal conductivity of CeO_2 with MWCNT and water based hybrid nanofluid: An experimental assessment. Colloids Surfaces A Physicochem Eng Asp. 2021;610(September 2020). doi:10.1016/j.colsurfa.2020.125918
118. Akhgar A, Toghraie D. An experimental study on the stability and thermal conductivity of water-ethylene glycol/TiO_2-MWCNTs hybrid nanofluid: Developing a new correlation. Powder Technol. 2018;338:806–818. doi:10.1016/j.powtec.2018.07.086
119. Ma X, Song Y, Wang Y, et al. Experimental study of boiling heat transfer for a novel type of GNP-Fe_3O_4 hybrid nano fluids blended with different nanoparticles. 2022;396:92–112. doi:10.1016/j.powtec.2021.10.029
120. Huminic G, Huminic A, Fleaca C. Synthesis, characterization and thermal conductivity of water based graphene oxide–silicon hybrid nanofluids: An experimental approach. Published online 2022:12111–12122. doi:10.1016/j.aej.2022.06.012
121. Silva-Yumi J, MorenoRomero T, ChangoLescano G. Nanofluids, synthesis and stability—brief review. ESPOCH Congr Ecuadorian J STEAM. 2021;1(2):998–1006. doi:10.18502/espoch.v1i2.9520

122. Babar H, Ali HM. Towards hybrid nanofluids: Preparation, thermophysical properties, applications, and challenges. J Mol Liq. 2019;281:598–633. doi:10.1016/j.molliq.2019.02.102
123. Said Z, Sundar LS, Rezk H, Nassef AM, Ali HM, Sheikholeslami M. Optimizing density, dynamic viscosity, thermal conductivity and specific heat of a hybrid nanofluid obtained experimentally via ANFIS-based model and modern optimization. J Mol Liq. 2021;321:114287. doi:10.1016/j.molliq.2020.114287
124. Nasirzadehroshenin F, Pourmozafari A, Maddah H, Sakhaeinia H. Experimental and theoretical investigation of thermophysical properties of synthesized hybrid nanofluid developed by modeling approaches. Arab J Sci Eng. 2020;45(9):7205–7218. doi:10.1007/s13369-020-04352-6
125. Jin C, Wu Q, Yang G, Zhang H, Zhong Y. Investigation on hybrid nanofluids based on carbon nanotubes filled with metal nanoparticles: Stability, thermal conductivity, and viscosity. Powder Technol. 2021;389:1–10. doi:10.1016/j.powtec.2021.05.007
126. He J, Sun J, Yang F, Meng Y, Jiang W. Water-based Cu nanofluid as lubricant for steel hot rolling: Experimental investigation and molecular dynamics simulation. Lubr Sci. 2022;34(1):30–41. doi:10.1002/LS.1569
127. Nowrouzi I, Mohammadi AH, Manshad AK. Water-oil interfacial tension (IFT) reduction and wettability alteration in surfactant flooding process using extracted saponin from *Anabasissetifera* plant. J Pet Sci Eng. 2020;189(January):106901. doi:10.1016/j.petrol.2019.106901
128. Sagala F, Hethnawi A, Nassar NN. Hydroxyl-functionalized silicate-based nanofluids for enhanced oil recovery. Fuel. 2020;269(January). doi:10.1016/j.fuel.2020.117462
129. Alaghmandfard A, Madaah Hosseini HR. A facile, two-step synthesis and characterization of Fe_3O_4–LCysteine–graphene quantum dots as a multifunctional nanocomposite. Appl Nanosci. 2021;11(3):849–860. doi:10.1007/s13204-020-01642-1
130. Benkhedda M, Boufendi T, Tayebi T, Chamkha AJ. Convective heat transfer performance of hybrid nanofluid in a horizontal pipe considering nanoparticles shapes effect. J Therm Anal Calorim. 2020;140(1):411–425. doi:10.1007/s10973-019-08836-y

4 Hybrid Nanofluid Heat Transfer in an Inclined Saturated Porous Cavity

An Experimental Study

Mahtab Nazarahari, Ramin Ghasemiasl, Taher Armaghani, and Aryan Moshiri

Nomenclature

SSA	Surface area of the solid particles divided by the mass of the solid particles (m^2 kg^{-2})
Nu	Nusselt number
h	Convection heat transfer coefficient (W m^{-2} K^{-1})
H	Characteristic length (m)
k	Conduction heat transfer coefficient (W m^{-1} K^{-1})
k_{pl}	Copper plate conduction
q"	Heat flux (w m^{-2})
t	Copper plate thickness
$T_{h\text{-}corr}$	Corrected temperatures of the hot surface (K)
$T_{c\text{-}corr}$	Corrected temperatures of the cold surface (K)
V	Voltage (v)
I	Electric current intensity (A)
T_h	Average temperature of the hot surface (K)
T_c	Average temperature of the cold surface (K)
ΔT	Temperature difference (K)
k_{p1}	Conduction of the first particle of a nanofluid
k_{p2}	Conduction of the second particle of a hybrid nanofluid
k_f	Base fluid conduction
k_{nf}	Nanofluid conductivity
k_{hnf}	Hybrid nanofluid conductivity
k_{eff}	Effective conduction
Ra	Rayleigh number
G	Gravitational acceleration (m s^{-2})

DOI: 10.1201/9781003595786-4

Nu*	Effect of the porous medium and nanofluids on the increase in Nusselt number
Greek symbol	
	Porosity ratio
α	Thermal diffusivity ($m^2\ s^{-1}$)
β	Coefficient of thermal expansion (K^{-1})
ν	Kinematic viscosity ($m^2\ s^{-1}$)
Θ	Angle of inclination of the cavity
Φ	Nanoparticle volume concentration
Subscript	
H	Hot
C	Cold
Pl	Plate

4.1 INTRODUCTION

Hybrid nanofluid (NFL) is a promising novel type of NFL with vast potential for various applications in the industry. However, further research is needed to optimize their preparation process, improve stability, and ensure their safety for use at high concentrations.

With continued research and development, hybrid NFLs will likely play a significant role in future industrial applications.

Understanding the free convection of hybrid NFLs is vital for many practical applications, including heat transfer (HTR) enhancement and studying the effects of various factors such as cavity inclination angle, cavity shape, and NFL volume concentrations (VCs) on the natural convection of hybrid NFLs. Overall, exploring natural convection in hybrid NFLs is currently at an early stage. Furthermore, comprehensive research is necessary to enhance our understanding of the characteristics and behaviour of these fluids.

Various HTR systems such as radiators, heat exchangers, electronic devices, and automobiles necessitate implementing suitable techniques to enhance HTR for optimizing energy devices. NFL technology is an approach that has emerged relatively recently with the potential to achieve this objective [1–3]. NFLs are artificially synthesized colloidal suspensions composed of nanoparticles with sizes ranging from 10 to 100 nm dispersed within a base fluid. These nanoparticles can be made of carbon-based materials, metals, or metallic oxides [4]. By incorporating these nanoparticles into base fluids like water, oils, or ethylene glycol (EG), the thermal conductivity of the fluids can be improved as nanoparticles generally exhibit higher heat conductivity than traditional thermal transfer fluids. Consequently, NFLs hold considerable potential as a category of thermal transfer fluids with superior thermal properties [5–7].

Numerical and experimental investigations have been devoted to understanding the consequence of the mentioned factors on the natural convective thermal transfer of non-hybrid and hybrid NFLs.

Multiple studies have demonstrated the potential of NFLs to enhance HTR compared to traditional fluids [8]. Ghodsinezhad et al. [9] empirically studied the thermal convection of Al_2O_3–water NFLs and stated a 15% rise in HTR coefficient at a VC of 0.1%. Sharifpur et al. [10] examined the HTR augmentation of TiO_2–water NFLs in a cavity flow, finding that a VC of 0.05% led to an 8.2% increase in HTR. Choudhary and Subudhi [11] empirically analysed the natural convection performance of Al_2O_3 NFLs in a rectangular cavity and perceived a maximum of 29.5% augmentation at 0.01 vol%. Giwa et al. [12] scrutinized the natural convection of Al_2O_3-MWCNT/water hybrid NFLs and found that the most significant improvement in HTR occurred at a percentage weight of 60% Al_2O_3 and 40% MWCNT nanoparticles. However, Rashad et al. [13] found that a hybrid NFL containing equal amounts of copper (Cu) and aluminium oxide (Al_2O_3) nanoparticles in water did not yield a remarkable impact on HTR when compared to a mono-NFL. The HTR augmentation of hybrid NFLs is influenced by various factors, such as preparation method, base fluid, nanoparticle shape and size, nanoparticle purity and stability, the fluid's temperature, and the nanoparticles' thermophysical properties [8]. Torki and Etesami [14] experimentally studied SiO_2/water NFLs' HTR through natural convection in a rectangular cavity at different VCs and inclination angles. Results showed that low concentrations had a negligible impact on the HTR coefficient (h), but concentrations above 0.005 resulted in a reduction of the coefficient. Nusselt (Nu) number was more affected by the inclination angle at low concentrations but less involved as the cavity inclination angle increased or approached a vertical state. The HTR rate was highest at a zero or horizontal inclination angle, decreasing as the inclination angle increased.

Simulation results presented by Sheikhzadeh et al. [15] suggested that the highest Nu number occurred in the middle–middle and top–bottom case of the active portion cases of a squared cavity with partially activated side walls, containing a NFL composed of Cu–water. The effect of the inclination angle is more pronounced in Rayleigh (Ra) numbers, which are more than 10^4. The orientation angle from 0° to 30° enhances the HTR rate and lowers it from 30° to 90°.

Armaghani et al. [16] investigated the analysis of the production of thermodynamic irreversibility and natural convection within a cavity that features partial porosity, inclined and filled with a NFL composed of water and Cu particles. If the Ra is below 10^5, adding more nanoparticles increases the amount of wasted heat and enhances the overall efficiency of HTR. In addition, for low Ra values, increasing the thickness of the porous layer and orienting the cavities in a particular way improve thermal efficiency. Conversely, at high Ra values, these same adjustments lead to decreased thermal performance.

Cho et al. [17] performed a numerical analysis to scrutinize the natural heat convection and entropy production of Al_2O_3–water NFL in a sloped wavy-wall cavity. The study's findings indicated that the cavity's inclination angle plays a pivotal role in influencing the streamlines of flow, isotherms, and local entropy production. Moreover, using an increased nanoparticle VC increased the average Nu number and decreased the total entropy production. Eventually, a wavy-surface geometry was observed to outperform a regular (flat-walled) cavity in terms of thermal transfer performance at a specific Ra number and cavity inclination angle.

Chamkha et al. [18] examined natural convection in a semi-circular chamber filled with Al_2O_3/Cu–water NFL through a numerical method. They measured the influence of nanoparticle VC and thermal conductivity ratio on flow patterns and HTR. It was found that HTR is improved by adding a low concentration of hybrid nanoparticles to the chamber.

Mehryan et al. [19] investigated the natural convection of hybrid Al_2O_3-Cu/water NFLs in a porous chamber subjected to differential heating. Compared with Al_2O_3 nanoparticles, the nano-velocity reduction from Cu-Al_2O_3 hybrid nanoparticles is much more significant. The strength of recirculating cells increased using a glass-ball porous medium (PM), compared to an aluminium foam. In addition, the thermal conductivity of each solid matrix will increase when the porosity increases.

Alsabery et al. [20] investigated the natural convection flow of a NFL within a sloped square enclosure. Using numerical methods, the enclosure features a partially saturated porous layer and two opposing sidewalls with variable sinusoidal temperature profiles. Results showed that a lower inclination angle significantly enhances the flow structure, resulting in a clockwise rotating cell in the NFL layer. The streamlines form a singular rotating cell that resembles constant temperature distribution. Moreover, as the thickness of the porous layer increases, the Al_2O_3 nanoparticles tend to transfer more heat within the cavity due to their lower thermal expansion.

Ghalambaz et al. [21] analysed local thermal non-equilibrium in conjugate and free convection within a porous enclosure containing Ag-MgO hybrid NFLs. The conclusions show that utilizing hybrid nanoparticles reduced the flow strength and HTR rate. The power of the flow will increase as Ra grows stronger. At low Ra values, the hybrid nanoparticles have a more pronounced effect on the thermal fields. The size and strength of the vortexes created in the PM are also significantly increased by increasing ε.

Selimefendigil et al. [22] performed numerical investigations of free convection and entropy production in a slanted cavity containing a conductive partition with a curved shape. The cavity was filled with a NFL exposed to an inclined magnetic field. Results showed that the growth of the Ra number and nanoparticles increased the average Nu number, while increasing the Hartmann number decreased it. The cavity's inclination angle significantly affected the convective HTR characteristics, and improved HTR occurred with the augmented conductivity of the partition.

Hakim Kadhim et al. [23] studied Cu-Al_2O_3 hybrid NFLs' natural convection in an inclined, wavy cavity partly covered with thin PM layers. The Galerkin finite element method and the Darcy–Brinkman model were used. A Cu- Al_2O_3 hybrid NFL with Al_2O_3-water as a single NFL indicates augmented HTR performance.

Izadi et al. [24] scrutinized the natural convection HTR and fluid flow of a hybrid nanoliquid consisting of Ag-MgO nanoparticles in water in a porous square chamber under a periodically tilted magnetic field using a local thermal non-equilibrium approach. The study investigated the impact of diverse control factors on the HTR performance and flow structures, showing that the inclination angle and wavelength of the periodic magnetic field can significantly alter the HTR performance in both the liquid and solid phases.

Selimefendigil and Chamkha [25] numerically scrutinized the natural convection of a hybrid NFL consisting of Al_2O_3/Cu and water in a triangular annular cavity that features an aperture located on the inclined side of the outer triangle. The study explored the impact of diverse parameters such as Ra ranging from 10^4 to 5×10^5, the VC of NFLs ranging from 0 to 0.02, and the opening ratio ranging from 0 to 0.625 on fluid flow and HTR. Results showed that the effect of the opening balance on HTR augmentation was more pronounced for higher Ra, and the Nu number increased with an increase in Ra and opening ratio.

Asmadi et al. [26] investigated the thermal performance of an Al_2O_3-CuO/water hybrid NFL in buoyancy-driven HTR of a U-shaped cavity with a heated wavy wall. The method used is a three-node triangular finite element method. If the length of a hot wall is longer than that of a cold wall, including hybrid NFLs may hinder HTR. Compared to pure water, using hybrid NFLs results in as low as 4% and up to 16% increase in thermal performance in the curved U-shaped enclosure.

A hybrid NFL flow from Cu-Al_2O_3/water was carried out within a porosity chamber in the study by Algehyne et al. [27]. Results have shown that two different types of nanoparticles accelerate HTR through a porous structure. The percentage analysis showed that hybrid NFLs (Cu-Al_2O_3/water) significantly improve the thermal distribution of traditional fluids.

Utilizing the successive under-relaxation (SUR) technique by Chamkha et al. [28], the influence of VC, Ra number, heat generation, and heat source length and location on the magneto-free convective flow of a hybrid NFL inside a slant porous cavity was investigated. The results demonstrated that higher VCs lead to a more significant reduction in thermal performance than lower concentrations. Furthermore, adding nanoparticles at various Ra numbers results in a decline in thermal performance.

The study conducted by Armaghani et al. [29] focuses on the numerical analysis of magnetohydrodynamic (MHD) natural convection and entropy generation in an Al_2O_3-water NFL within a T-shaped baffled cavity subjected to a magnetic field. The study revealed that the effectiveness of the NFL in enhancing the Nu number is more pronounced with a rising aspect ratio of the enclosure. The results also indicate that the Nu number improves as the baffle length within the cavity increases. Furthermore, the entropy production was found to decrease with a higher Hartmann number and a lower Nu number.

Armaghani et al. [30] investigated the impacts of discrete heat source location on HTR and entropy generation in an open inclined L-shaped enclosure filled with an Ag-water NFL. The results of the study demonstrate that an increase in the inclination angle results in augmented HTR. Furthermore, it is figured out that as the VC of nanoparticles increases, the thermal performance decreases. However, the thermal performance improves as the inclination angle increases.

Chamkha et al. [31] investigated free convection and entropy production in an L-shaped porous cavity containing NFLs obeying the Buongiorno two-phase model. The study findings indicate that the thermophoresis diffusion outcome has a dominant influence compared to the Brownian diffusion outcome. The Nu number, which represents the HTR performance and the total entropy production, is higher for a slender cavity. Based on the findings, a cavity with an aspect ratio of 0.5 and utilizing Al_2O_3 and CuO nanoparticles is suggested to achieve the best thermal performance.

A review article by Nabwey et al. [32] presented the recent breakthroughs in utilizing NFLs for HTR applications in porous materials. Moreover, Sadeghi et al. [33] provided a comprehensive summary and discussion of the latest research literature regarding NFLs' natural convection in different enclosures. The review focuses on five specific enclosure geometries, namely, square, circular, triangular, trapezoidal, and unconventional shapes.

4.2 EXPERIMENTAL SETUP

4.2.1 Methodology

4.2.1.1 Experiment Description

The current research outlines a device that comprises a square cube-shaped box made of Plexiglas, with a thickness of 10 mm and dimensions of 100 × 100 × 100 mm. Inside this box, a small compartment on the right wall enables the circulation of cold water. This compartment is connected to a cold-water bath through reciprocating pipes, regulated by a pump, thus allowing temperature control. To establish a connection between the cold-water chamber and the main chamber, a copper plate measuring 2 × 100 × 100 mm is utilized.

On the right wall, an additional copper plate matching the thickness of the previous one is linked to an electric heater, which operates on direct current (DC) voltage. By modifying the voltage, it is possible to adjust the current consumption of the heater, a parameter that can be monitored using a multimeter. To measure temperature variations on different points of the copper plates, three K-type thermocouples are connected to the back of the copper plates on both sides of the walls. These thermocouples employ two distinct alloys as sensors.

The temperatures on the hot and cold sides are obtained by determining the average readings from these three thermocouples on each respective side. However, in order to accurately measure the temperature of the fluid in proximity to the hot and cold walls, adjustments must be made to the average temperature values using appropriate calculations. Furthermore, the temperature of the hot plate can be controlled and modified by adjusting the voltage and current intensity of the power supply.

Incorporating an inclined surface whose angle can be adjusted, the HTR chamber is positioned within the setup. An overview of the entire test setup is presented in Figure 4.1, while Figure 4.2 illustrates the changes in cavity inclination angles ($\theta = \{0°, 30°, 60°\}$), providing valuable insights.

For the HTR material, stainless steel with a thickness of 1 mm is employed. To maintain separation between the thermocouples and the stainless-steel material at specific distances, a thin layer of silicone paste is applied around the sides of the heater. Additionally, narrow grooves are created on the copper plates to allow proper contact between the heater and the copper plate, ensuring accurate temperature readings.

Figure 4.3 presents the appropriate arrangement of the thermocouples and the cold-water return pipes, which are attached to the cold wall. Similarly, the placement of the heater on the hot wall is also indicated in the figure.

4.2.1.2 Hybrid NFLs

The present experimental investigation utilized a hybrid NFL comprising titanium oxide and aluminium oxide in conjunction with a water-based fluid, with both NFLs mixed in equal proportions. The VCs of the NFLs employed in the experiment were 1000 ppm (1%) and 500 ppm (0.5%), respectively.

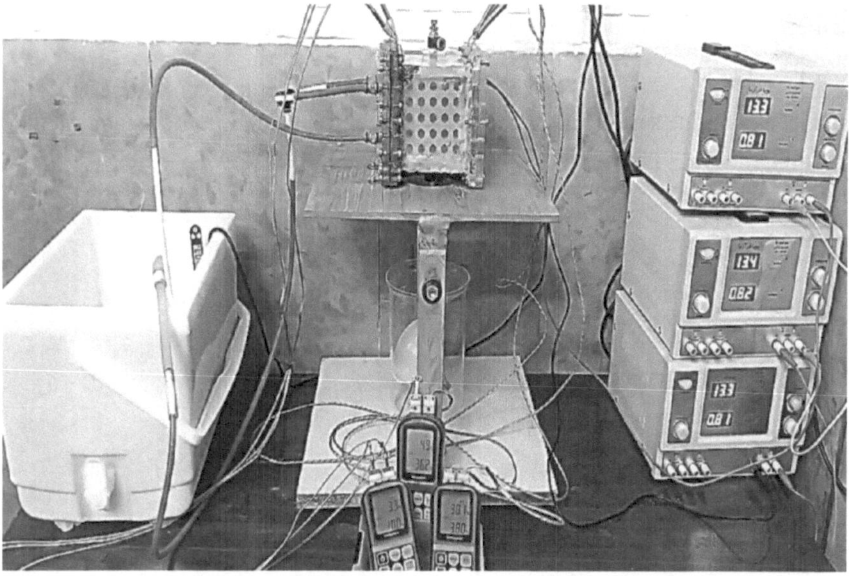

FIGURE 4.1 Overview of the laboratory equipment (WTIAU Lab).

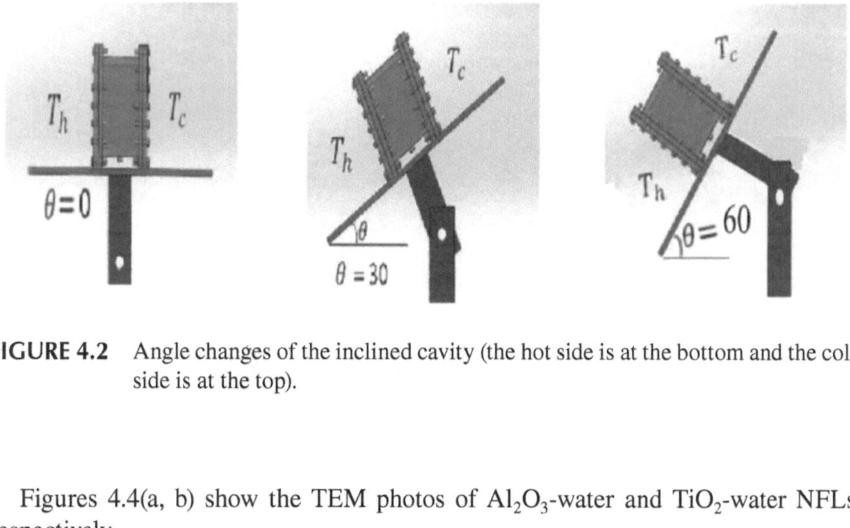

FIGURE 4.2 Angle changes of the inclined cavity (the hot side is at the bottom and the cold side is at the top).

Figures 4.4(a, b) show the TEM photos of Al_2O_3-water and TiO_2-water NFLs, respectively.

The specifications of the Al_2O_3 and TiO_2 NFLs are shown in Table 4.1.

Figure 4.5 displays the NFLs employed in the present study. The figure also illustrates the temporal stability of the NFLs.

4.2.1.3 Thermocouple Calibration

There are two main approaches for the calibration of temperature sensors. The initial method involves comparing the temperature sensor being evaluated with highly precise thermometers. The second method involves using fixed temperature points to calibrate the temperature sensor. In this specific investigation, the calibration of thermocouples was carried out by measuring the equilibrium temperature of water and ice, as well as the equilibrium temperature of water at room temperature. To ensure accurate measurements, a thermometer was immersed in the water, allowing for a comparison with a reference thermometer known for its high accuracy. This reference thermometer validates the precision of these temperature measurements. The calibration process of the experiment's thermocouples is presented in Figure 4.6.

FIGURE 4.3 Cold wall (left) and hot wall (right).

FIGURE 4.4(A) TEM photo of aluminium oxide NFL.

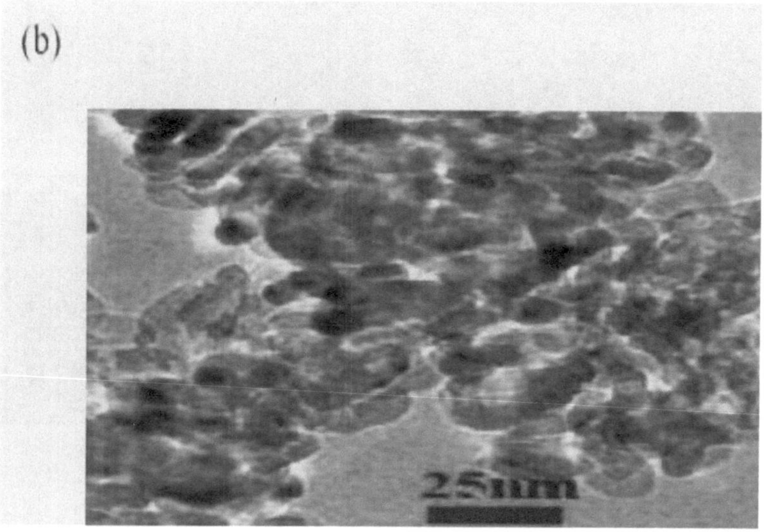

FIGURE 4.4(B) TEM photo of titanium oxide NFL.

TABLE 4.1
Specifications of the Al_2O_3 and TiO2 NFLs

Al_2O_3 Details:	TiO_2 Details:
Nanopowder (gamma) – hydrophilic	Nanopowder (TiO_2, anatase)
Purity: 99%	Purity: >99%
SSA: >188 m^2/g	SSA: 200–240 m^2/g
Morphology: nearly spherical	Morphology: nearly spherical
Color: white	Color: white
Specific heat capacity: 850 J/(Kg.K)	Bulk density: 0.24 g/cm^3
Density: 3990 Kg/m^3	True density: 3.9 g/cm^3

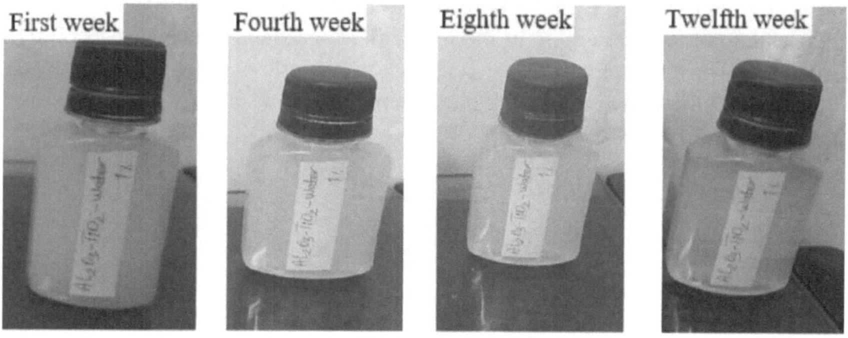

FIGURE 4.5 NFLs' stability over time.

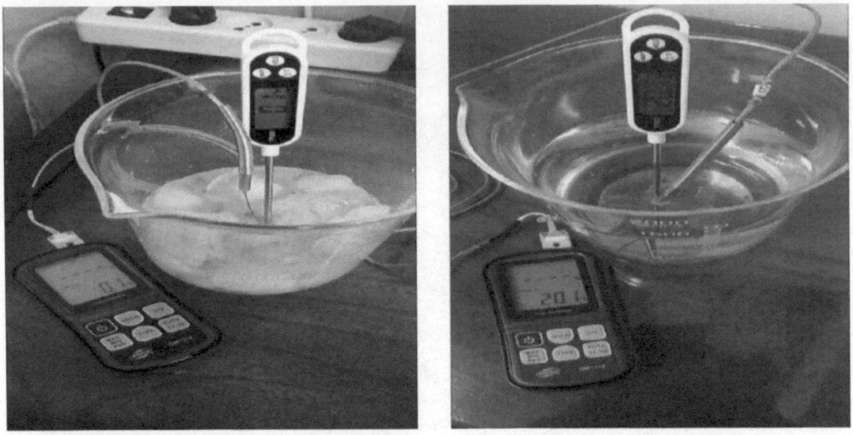

FIGURE 4.6 Thermocouple calibration.

4.2.1.4 Pore Shape and Porosity

To examine the impact of PMs' porous shapes on HTR, three PM models were constructed, each with identical material composition and dimensions but differing pore shapes. Specifically, the pore shapes utilized were circles, hexagons, and squares. The porosity ratio (ε), defined as the volume of hollow space to the total volume of an object, was set at 0.7 for each porous cube. The cubes were manufactured from polylactic acid (PLA), a material characterized by low conductivity ($k_s \approx 2$ W m^{-1} K^{-1}), and were produced via 3D printing using fused deposition modelling (FDM) technology. Figure 4.7 shows the PM with different pore shapes (circular, hexagonal, and squared).

4.2.2 Mathematical Formulation

Determining the heat convection coefficient and the Nu number is vital for analysing the process of increasing or decreasing HTR. Various experimental conditions, such as adjustments in the inclination angle of the setup, changes in the VC of nanoparticles, and variations in the pore shapes of the PM with consistent porosity and material, lead to fluctuations in the temperature gradient between the hot and cold surfaces. These temperature differences consequently affect the heat convection coefficient, ultimately influencing the Nu number.

$$h = \frac{I \times V}{\left(T_{h-corr} - T_{c-corr}(x)\right) * A} \tag{4.1}$$

$$Nu = \frac{hH}{K} \tag{4.2}$$

$$q'' = \frac{I \times V}{A} \tag{4.3}$$

FIGURE 4.7 3D design and 3D printed porous cube-shaped parts.

$$T_h = \frac{T_{h1} + T_{h2} + T_{h3}}{3} \tag{4.4}$$

$$T_c = \frac{T_{c1} + T_{c2} + T_{c3}}{3} \tag{4.5}$$

$$T_{h-corr} = T_h - \frac{q''t}{k_{pl}} \tag{4.6}$$

$$T_{c-corr} = T_c + \frac{q''t}{k_{pl}} \tag{4.7}$$

$$Ra = \frac{g\beta\left(T_{h-corr} - T_{c-corr}\right)H^3}{\alpha\vartheta} \tag{4.8}$$

$$Nu = 0.082\ Ra_h^{0.329}\left(for\ 10^6 \le Ra \le 10^{12}\right) \tag{4.9}$$

$$k_{nf} = \left[\frac{k_{p1} + 2k_f - 2\phi_1\left(k_f - k_{p1}\right)}{k_{p1} + 2k_f + \phi_1\left(k_f - k_{p1}\right)}\right]k_f \tag{4.10}$$

$$k_{hnf} = \left[\frac{k_{p2} + 2k_{nf} - 2\phi_2\left(k_{nf} - k_{p2}\right)}{k_{p2} + 2k_{nf} + \phi_2\left(k_{nf} - k_{p2}\right)}\right]k_{nf} \tag{4.11}$$

$$k_{eff} = \varepsilon * k_{nf} + \left(1 - \varepsilon\right) * k_s \tag{4.12}$$

$$Nu_1 = hl / k_{eff} \tag{4.13}$$

$$Nu_2 = hl / k_f \tag{4.14}$$

$$Nu^* = Nu_1 / Nu_2 \tag{4.15}$$

The above equations express the correlations among the heat flux *q*"; the corrected temperatures of the hot and cold surfaces $T_{c\text{-}corr}$ and $T_{h\text{-}corr}$, respectively; the thickness of the copper plate *t*, and the characteristic length *H*. The heat convection coefficient (*h*) and the heat conduction coefficient (*k*) are also included in the equations, as well as the voltage *V* and the electric current intensity *I*. Moreover, T_h and T_c denote the average temperature of the hot and cold surfaces, respectively. The Ra number formula is presented in Equation (4.8), where *g* represents gravitational acceleration, *v* denotes kinematic viscosity, α stands for thermal diffusivity, and β represents the coefficient of thermal expansion. Equation (4.9), also known as the Markatos [34] correlation, presents the Nu number formula. Correlations (4.10) and (4.11) are Maxwell formulas [35] to calculate the thermal conductivities of an NFL and hybrid NFL. To determine the conductivity of both the hybrid NFL and PM, Equation (4.12) is utilized. The parameter k_s represents the PM conduction coefficient, while ε stands for porosity. Equation (4.13) elucidates the application of the Nu_1 correlation for assessing the Nu number pertaining to the PM and NFL. Conversely, Equation (4.14) delineates the Nu_2 correlation for estimating the Nu number of the base fluid, which, in this investigation, is water. Equation (4.13) also introduces the Nu* correlation, which appraises the collective effect of the PM and NFL on HTR.

4.3 VALIDATION PROCESS

4.3.1 Validation

A preliminary experiment was conducted using pure water without any PM to ensure the precision of the setup and validate the test. The experimental conclusions were obtained at a zero-degree angle and compared with the Markatos [34] correlation (Equation [4.9]) to verify the accuracy of the setup.

The thermodynamic properties of water used in the experiment were obtained from thermodynamic tables by averaging the temperature data. These properties are summarized in Table 4.2.

TABLE 4.2
Thermodynamic Characteristics of Water

$\beta(k^{-1})$	$\vartheta\left(\frac{m^2}{s}\right)$	$\alpha\left(\frac{m^2}{s}\right)$	$k\left(\frac{W}{m\cdot K}\right)$	T(k)
0.000142	11.625×10^{-7}	1.4045×10^{-7}	0.5881	288

Table 4.3 compares the Nu number values obtained from the experimental relations with those calculated from the Markatos [34] correlation. Furthermore, the percentage error between these two values is also calculated and presented in the table.

4.3.2 Uncertainty

Table 4.4 presents a comprehensive list of the equipment utilized in the experimental setup and the uncertainty associated with each instrument. It is essential to consider the uncertainty associated with each measuring device as it can affect the accuracy and precision of the results acquired from the experiment. The uncertainty associated with each measuring instrument was calculated using the manufacturer's specifications, and appropriate methods were used to ensure that the devices were calibrated correctly.

In this experimental investigation, an uncertainty analysis was conducted for various essential parameters such as Ra, heat flux, heat convection coefficient, and Nu. The combined uncertainty was specified utilizing the Beckwith method [36], a commonly employed approach in uncertainty assessment. The results of the uncertainty analysis for the aforementioned parameters are detailed in Table 4.5, offering precious information regarding the precision and dependability of the experimental measurements.

4.3.3 Repeatability

Before implementing any PM in the cavity, the temperature difference was measured using the heater and cold-water circulation, with pure water fluid at zero-degree

TABLE 4.3
Comparison between the Nu Results Obtained Experimentally and Theoretically

Error percentage	Nu of current experiment	Nu [34]	Ra
3.2%	20.3	19.6	17,176,060

TABLE 4.4
Measurement Equipment, Characteristics, and Uncertainty

No.	Tool	Interval	Variable	Minimum measured value	Measurement interval	Uncertainty percentage
1	Thermocouple	−5–1260	T_h	0.1	20–50	0.21
2	Thermocouple	−5–1260	T_c	0.1	0–10	0.42
3	Voltmeter	0–220	V	0.01	10–60	0.01
4	Ampere meter	0–15	I	0.001	0–0.8	0.125
5	Height	0–30	W-H	0.00005 m	0.1	0.05

TABLE 4.5
Uncertainty

Percentage of uncertainty	Variable
0.73	Heat flux
1.08	*h* (non-porous)
1.2	Nu (non-porous)
0.99	*h* (circular pore PM)
1.12	Nu (circular pore PM)
1.07	*h* (hexagonal pore PM)
1.18	Nu (hexagonal pore PM)
1.14	*h* (squared pore PM)
1.24	Nu (squared pore PM)

TABLE 4.6
Results and Date of Repeatability

No.	Repeatability date	Empirical repeatability	Empirical validation Nu 2022/09/01
1	2022/09/03	21.93	20.32
2	2022/09/06	22.58	
3	2022/09/10	21.26	

cavity inclination angle. After the thermocouples were calibrated, the HTR coefficient and Nu number values were computed. The results were documented to assess repeatability and comparison with the validation outcome and are shown in Table 4.6. The repeatability value was close to the validation value.

It has been demonstrated that the laboratory's environmental conditions, encompassing relative humidity, ambient temperature, and consistent airflow, were maintained at a stable level within acceptable limits.

4.4 RESULT AND DISCUSSION

The present experimental study investigates different factors, including cavity inclination angles (0°, 30°, and 60°), VCs (0%, 0.5%, and 1%), type of hybrid NFL (Al_2O_3-TiO_2-Water), and pore shapes (circular, hexagonal, and square) while maintaining a constant porosity of 0.7.

The resulting *h*, Nu, and Nu* values were graphed with respect to the cavity inclination angles for PM with circular, hexagonal, and square porosities in the presence of various NFLs.

The following sections (4.4.1 and 4.4.2) present the influence of each NFL's angle and VC on the *h*, Nu, and Nu* parameters for every PM.

4.4.1 Cavity Inclination Angle Effects

In this section, the diagrams of h–θ (Figure 4.8), Nu–θ (Figure 4.9), and Nu*–θ (Figure 4.10) are presented.

The highest and lowest values of the heat convection coefficient, Nu number, and Nu* are described in Table 4.7.

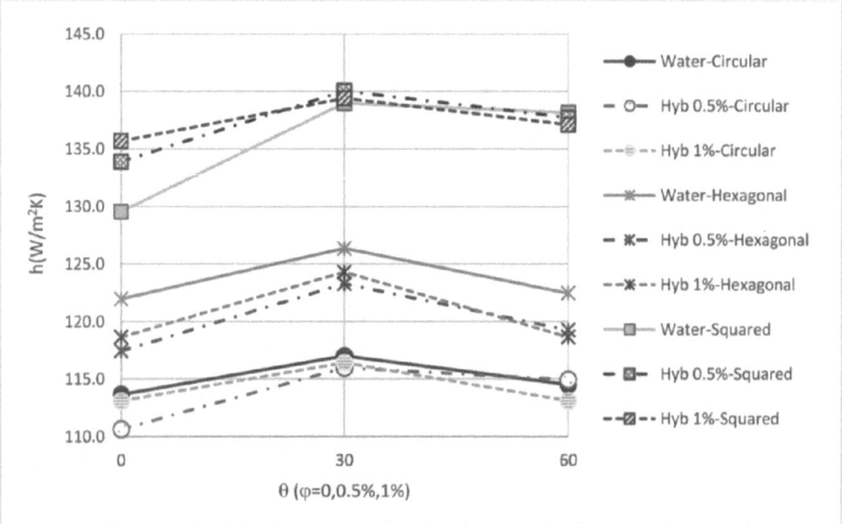

FIGURE 4.8 h diagram of hybrid NFLs at different inclination angles (θ) for three types of the PM.

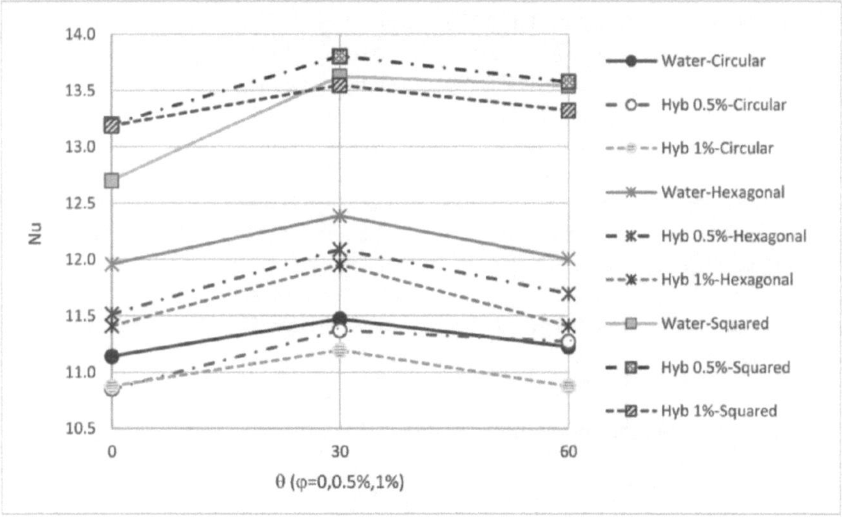

FIGURE 4.9 Nu diagram of hybrid NFLs at different inclination angles (θ) for three types of the PM.

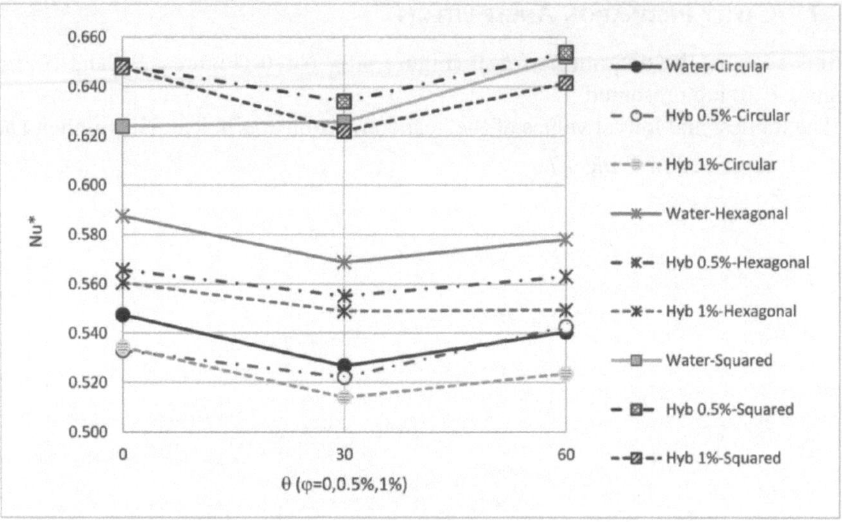

FIGURE 4.10 Nu* diagram of hybrid NFLs at different inclination angles (θ) for three types of PM.

TABLE 4.7
Highest and Lowest Values of *h*, Nu, and Nu* of Water and Hybrid NFLs for Angles of 0, 30, and 60 Degrees for PMs

60°	30°	0°	Cavity inclination angle
Min h = 113.1 (circular hybrid 1%)	Min h = 115.96 (circular hybrid 0.5%)	Min h = 110.65 (circular hybrid 0.5%)	h (W m^{-2} K^{-1})
Max h = 138.145 (squared water)	Max h = 140.04 (squared hybrid 0.5%)	Max h = 135.7 (squared hybrid 1%)	
Min Nu = 10.87 (circular hybrid 1%)	Min Nu = 11.19 (circular hybrid 1%)	Min Nu = 10.85 (circular hybrid 0.5%)	Nu
Max Nu = 13.57 (squared hybrid 0.5%)	Max Nu = 13.8 (squared hybrid 0.5%)	Max Nu = 13.2 (squared hybrid 0.5%)	
Min Nu* = 0.524 (circular hybrid 1%)	Min Nu* = 0.514 (circular hybrid 1%)	Min Nu* = 0.533 (circular hybrid 0.5%)	Nu*
Max Nu* = 0.6536 (squared hybrid 0.5%)	Max Nu* = 0.634 (squared hybrid 0.5%)	Max Nu* = 0.6484 (squared hybrid 0.5%)	

The table presents the following:

- The lowest value of the *h* parameter corresponds to 0-degree angle of the chamber and the PM with circular pores and hybrid NFL with a VC of 0.5%.
- The highest value of the *h* parameter is related to the 30-degree angle of the chamber and the PM with square pores and hybrid NFL with a VC of 0.5%.

- The lowest values of the Nu parameter are related to the 0-degree angle of the chamber and the PM with circular pores and hybrid NFL with a VC of 0.5%.
- The highest values of the Nu parameter are related to the 30 degree angle of the chamber and the PM with square pores and hybrid NFL with a VC of 0.5%.
- The lowest values of the Nu* parameter are related to the 30-degree angle of the chamber and the PM with circular pores and hybrid NFL with a VC of 1%.
- The highest values of the Nu* parameter are related to the 60-degree angle of the chamber and the PM with square pores and hybrid NFL with a VC of 0.5%.

4.4.2 Volume Concentration Effects

In this section, the diagrams of h–ϕ (Figure 4.11), Nu–ϕ (Figure 4.12), and Nu*–ϕ (Figure 4.13) are presented.

The highest and lowest values of the heat convection coefficient, Nu, and Nu*, at different VC percentages, are described in Table 4.8.

Table 4.8 presents the following:

- The lowest value of the h parameter corresponds to hybrid 0.5% and the PM with circular pores at 0° angle of the enclosure.
- The highest value of the h parameter is related to hybrid 0.5% and the PM with square pores at 30° angle of the enclosure.
- The lowest value of the Nu parameter is related to hybrid 0.5% and the PM with circular pores at 0° angle of the enclosure.
- The highest value of the Nu parameter is related to hybrid 0.5% and the PM with square pores at 30° angle of the enclosure.

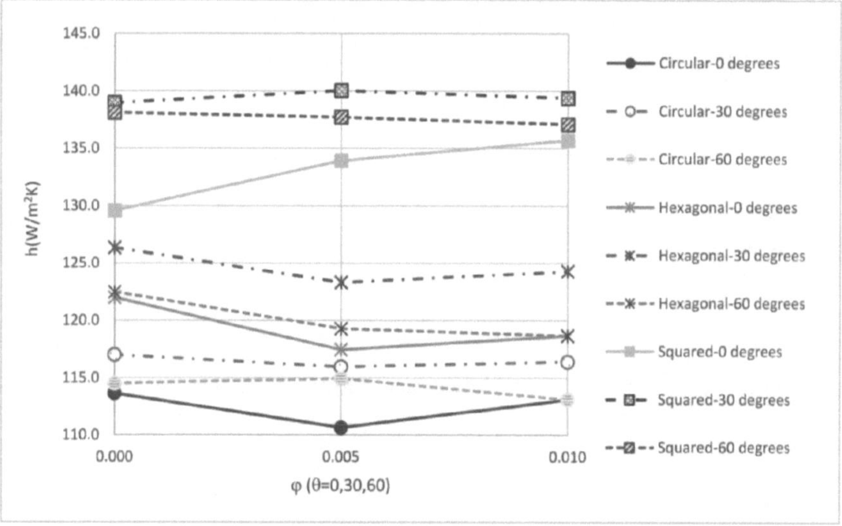

FIGURE 4.11 Diagram of convective HTR coefficient by volume percentage of hybrid NFLs at three inclination angles.

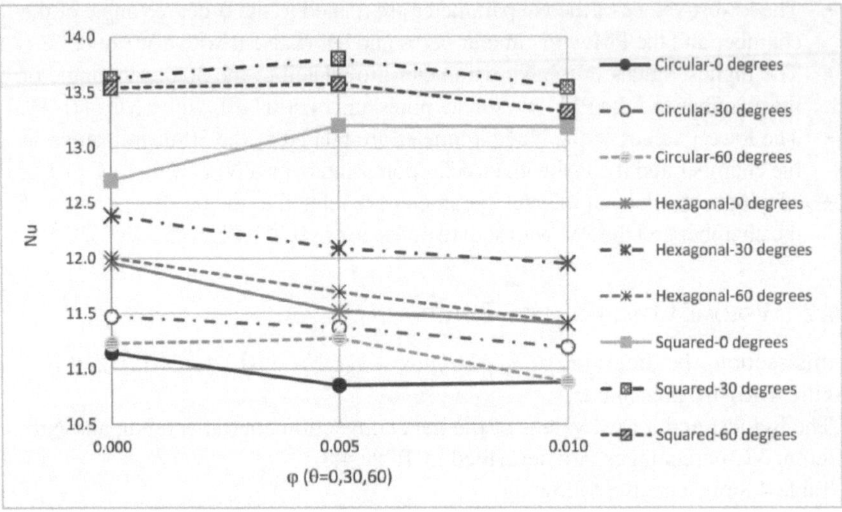

FIGURE 4.12 Diagram of Nu by VC percentage of hybrid NFLs at three different inclination angles.

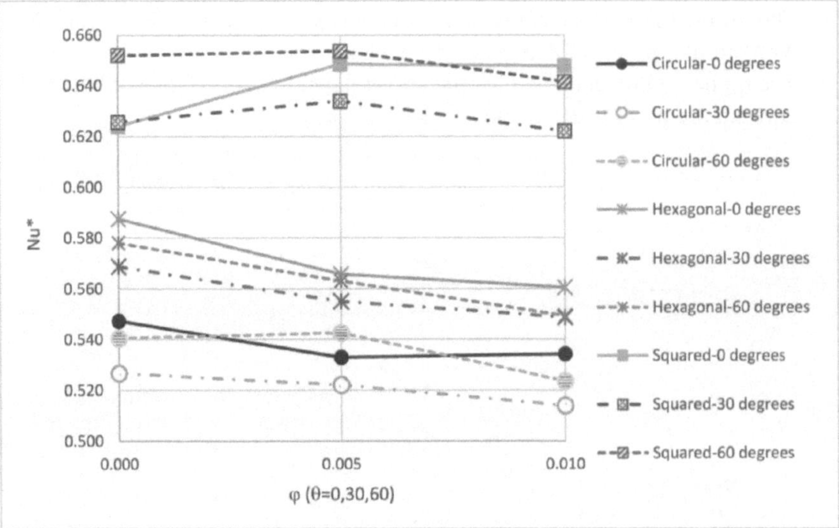

FIGURE 4.13 Diagram of Nu* by VC percentage of hybrid NFLs at three different inclination angles.

- The lowest value of the Nu* parameter is related to hybrid 1% and the PM with circular pores at 30° angle of the enclosure.
- The highest values of the Nu* parameter are related to hybrid 0.5% and the PM with square pores at 60° angle of the enclosure.

TABLE 4.8
Highest and Lowest Values of *h*, Nu, and Nu* of Hybrid NFLs for Different VC Percentages

1%	0.5%	0% (water)	VC percentage
Min *h* = 113.105 (circular 0° and 60°)	Min *h* = 110.65 (circular 0°)	Min *h* = 113.665 (circular 0°)	*h* ($W\ m^{-2}\ K^{-1}$)
Max *h* = 139.4 (squared 30°)	Max *h* = 140.04 (squared 30°)	Max *h* = 138.98 (squared 30°)	
Min Nu = 10.877 (circular 60°)	Min Nu = 10.85 (circular 0°)	Min Nu = 11.227 (circular 60°)	Nu
Max Nu = 13.5473 (squared 30°)	Max Nu = 13.8 (squared 30°)	Max Nu = 13.6257 (squared 30°)	
Min Nu* = 0.514 (circular 30°)	Min Nu* = 0.522 (circular 30°)	Min Nu* = 0.527 (circular 30°)	Nu*
Max Nu* = 0.6478 (squared 0°)	Max Nu* = 0.6536 (squared 60°)	Max Nu* = 0.652 (squared 60°)	

4.5 CONCLUSIONS

A system of HTR between hot and cold sources was simulated in this study. The system can rotate at various angles, and integrating porous materials can alter the shape of its internal chamber. In this research, three cubic porous solid bodies (made of PLA) with different pore geometries of circular, squared, and hexagonal pores were used to change the shape of the cavity. Also, testing under different inclination angles (0, 30, and 60 degrees) with the horizon and towards the hot source was repeated.

The observed results are as follows:

- The maximum value of *h* and Nu number for all VCs and in all three types of PMs occurred at an angle of 30 degrees, and the minimum value of *h* and Nu number in all three types of PMs occurred at an angle of 0 degrees.
- With the increase in the cavity inclination angle from 0 to 30 degrees, the HTR rate and the Nu also increase. From an angle of 30 to 60 degrees, the heat convection coefficient and Nu decrease. However, the values of *h* and Nu resulting from this decrease remain higher than that at the 0-degree state for all concentrations and NFLs.
- For all concentrations, the natural convective HTR coefficient and Nu value at all angles were higher for the PM with square pores than hexagonal and circular.
- The maximum value of *h* and Nu was obtained for the angle of 30 degrees, in the presence of a squared PM and for a hybrid NFL with a VC of 0.5%. The lowest values of these parameters were obtained for an angle of 0 degrees for a hybrid NFL with a VC of 0.5% and in the presence of the circular PM.

- The parameter *h* for the VC of 0 to 0.5% for hexagonal pores and hybrid NFLs decreases at all three angles of 0, 30, and 60 degrees.
- At all three angles of 0, 30, and 60 degrees, a PM with square pores in all VCs causes the Nu* parameter to be greater than that of other pore shapes.
- The highest value of Nu* in the squared PM belongs to the angle of 60 degrees and a volume percentage of 0.5% and is equal to 0.6536. The lowest value of Nu* belongs to the circular PM at the angle of 30 degrees and the hybrid NFL with a VC of 1% and is equal to 0.514.
- At an inclination angle of 60 degrees, the simultaneous presence of a PM with square pores and hybrid NFL with a VC of 0.5% causes the Nu* parameter to have the highest value. As a result, this shows that the combination of the mentioned conditions has the least decreasing effect on natural convection HTR compared to other situations.

The recommendations from the study are as follows:

- The least significant decrease of Nu from the state without a PM and hybrid NFL to the state with PM presence and NFL (Nu* parameter) is dedicated to the presence of the hybrid NFL with a VC of 0.5% plus PM with square pores at an inclination angle of 60 degrees.
- The most appropriate state of this study to reduce the Nu from the state without PM and hybrid NFL to the state with PM presence and NFL (Nu* parameter) is the state of hybrid NFL with a VC of 1% plus PM with circular pores at an inclination angle of 30 degrees.
- The maximum HTR coefficient is observed in a cavity containing hybrid NFL 0.5% in the presence of square pores at an inclination angle of 30 degrees.
- Generally, the PM with squared pores augmented the HTR rate.

This work was done at an advanced heat transfer laboratory in West Tehran Branch, Islamic Azad University, Tehran, Iran.

REFERENCES

1. Dey, D., & Sahu, D. S. (2021). Experimental study in a natural convection cavity using nanofluids. Materials Today: Proceedings, 41, 403–412.
2. Suresh, S., Venkitaraj, K. P., Selvakumar, P., & Chandrasekar, M. (2011). Synthesis of Al_2O_3–Cu/water hybrid nanofluids using two step method and its thermo physical properties. Colloids and Surfaces A: Physicochemical and Engineering Aspects, 388(1–3), 41–48.
3. Adio, S. A., Olalere, A. E., Olagoke, R. O., Alo, T. A., Veeredhi, V. R., Ewim, D. R., & Olakoyejo, O. T. (2021). Thermal and entropy analysis of a manifold microchannel heat sink operating on CuO–water nanofluid. Journal of the Brazilian Society of Mechanical Sciences and Engineering, 43, 1–15.
4. Okonkwo, E. C., Wole-Osho, I., Almanassra, I. W., Abdullatif, Y. M., & Al-Ansari, T. (2021). An updated review of nanofluids in various heat transfer devices. Journal of Thermal Analysis and Calorimetry, 145, 2817–2872.

5. Kumar, D. D., & Arasu, A. V. (2018). A comprehensive review of preparation, characterization, properties and stability of hybrid nanofluids. Renewable and Sustainable Energy Reviews, 81, 1669–1689.
6. Onyiriuka, E. J., Ighodaro, O. O., Adelaja, A. O., Ewim, D. R. E., & Bhattacharyya, S. (2019). A numerical investigation of the heat transfer characteristics of water-based mango bark nanofluid flowing in a double-pipe heat exchanger. Heliyon, 5(9), e02416.
7. Adio, S. A., Alo, T. A., Olagoke, R. O., Olalere, A. E., Veeredhi, V. R., & Ewim, D. R. (2021). Thermohydraulic and entropy characteristics of Al_2O_3-water nanofluid in a ribbed interrupted microchannel heat exchanger. Heat Transfer, 50(3), 1951–1984.
8. Sidik, N. A. C., Jamil, M. M., Japar, W. M. A. A., & Adamu, I. M. (2017). A review on preparation methods, stability and applications of hybrid nanofluids. Renewable and Sustainable Energy Reviews, 80, 1112–1122.
9. Ghodsinezhad, H., Sharifpur, M., & Meyer, J. P. (2016). Experimental investigation on cavity flow natural convection of Al_2O_3–water nanofluids. International Communications in Heat and Mass Transfer, 76, 316–324.
10. Sharifpur, M., Solomon, A. B., Ottermann, T. L., & Meyer, J. P. (2018). Optimum concentration of nanofluids for heat transfer enhancement under cavity flow natural convection with TiO_2–water. International Communications in Heat and Mass Transfer, 98, 297–303.
11. Choudhary, R., & Subudhi, S. (2016). Aspect ratio dependence of turbulent natural convection in Al_2O_3/water nanofluids. Applied Thermal Engineering, 108, 1095–1104.
12. Rashad, A. M., Chamkha, A. J., Ismael, M. A., & Salah, T. (2018). Magnetohydrodynamics natural convection in a triangular cavity filled with a Cu-Al_2O_3/water hybrid nanofluid with localized heating from below and internal heat generation. Journal of Heat Transfer, 140(7).
13. Giwa, S. O., Sharifpur, M., & Meyer, J. P. (2020). Experimental study of thermo-convection performance of hybrid nanofluids of Al_2O_3-MWCNT/water in a differentially heated square cavity. International Journal of Heat and Mass Transfer, 148, 119072.
14. Torki, M., & Etesami, N. (2020). Experimental investigation of natural convection heat transfer of SiO_2/water nanofluid inside inclined enclosure. Journal of Thermal Analysis and Calorimetry, 139, 1565–1574.
15. Kheirkhah, M. H., Arefmanesh, A., Sheikhzadeh, G. A., & Abdollahi, R. (2011). Numerical study of natural convection in an inclined cavity with partially active sidewalls filled with cu-water nanofluid. International Journal of Engineering, 24(3), 279–292.
16. Armaghani, T., Ismael, M. A., & Chamkha, A. J. (2017). Analysis of entropy generation and natural convection in an inclined partially porous layered cavity filled with a nanofluid. Canadian Journal of Physics, 95(3), 238–252.
17. Cho, C. C., Chiu, C. H., & Lai, C. Y. (2016). Natural convection and entropy generation of Al_2O_3–water nanofluid in an inclined wavy-wall cavity. International Journal of Heat and Mass Transfer, 97, 511–520.
18. Chamkha, A. J., Miroshnichenko, I. V., & Sheremet, M. A. (2017). Numerical analysis of unsteady conjugate natural convection of hybrid water-based nanofluid in a semicircular cavity. Journal of Thermal Science and Engineering Applications, 9(4), 041004.
19. Mehryan, S. A., Kashkooli, F. M., Ghalambaz, M., & Chamkha, A. J. (2017). Free convection of hybrid Al_2O_3-Cu water nanofluid in a differentially heated porous cavity. Advanced Powder Technology, 28(9), 2295–2305.
20. Alsabery, A. I., Chamkha, A. J., Saleh, H., & Hashim, I. (2017). Natural convection flow of a nanofluid in an inclined square enclosure partially filled with a porous medium. Scientific Reports, 7(1), 2357.

21. Ghalambaz, M., Sheremet, M. A., Mehryan, S. A. M., Kashkooli, F. M., & Pop, I. (2019). Local thermal non-equilibrium analysis of conjugate free convection within a porous enclosure occupied with Ag–MgO hybrid nanofluid. Journal of Thermal Analysis and Calorimetry, 135, 1381–1398.
22. Selimefendigil, F., & Öztop, H. F. (2020). Effects of conductive curved partition and magnetic field on natural convection and entropy generation in an inclined cavity filled with nanofluid. Physica A: Statistical Mechanics and Its Applications, 540, 123004.
23. Kadhim, H. T., Jabbar, F. A., & Rona, A. (2020). Cu-Al_2O_3 hybrid nanofluid natural convection in an inclined enclosure with wavy walls partially layered by porous medium. International Journal of Mechanical Sciences, 186, 105889.
24. Izadi, M., Sheremet, M. A., & Mehryan, S. A. M. (2020). Natural convection of a hybrid nanofluid affected by an inclined periodic magnetic field within a porous medium. Chinese Journal of Physics, 65, 447–458.
25. Selimefendigil, F., & Chamkha, A. J. (2016). Natural convection of a hybrid nanofluid-filled triangular annulus with an opening. Computational Thermal Sciences: An International Journal, 8(6).
26. Asmadi, M. S., Kasmani, R. M., Siri, Z., Saleh, H., & Ghani, N. C. (2023). Buoyancy-driven heat transfer performance, vorticity and fluid flow analysis of hybrid nanofluid within a U-shaped lid with heated corrugated wall. Alexandria Engineering Journal, 71, 21–38.
27. Algehyne, E. A., Raizah, Z., Gul, T., Saeed, A., Eldin, S. M., & Galal, A. M. (2023). Cu and Al_2O_3-based hybrid nanofluid flow through a porous cavity. Nanotechnology Reviews, 12(1), 20220526.
28. Chamkha, A. J., Armaghani, T., Mansour, M. A., Rashad, A. M., & Kargarsharifabad, H. (2022). MHD convection of an Al_2O_3–Cu/water hybrid nanofluid in an inclined porous cavity with internal heat generation/absorption. Iranian Journal of Chemistry and Chemical Engineering, 41(3), 936–956.
29. Armaghani, T., Kasaeipoor, A., Izadi, M., & Pop, I. (2018). MHD natural convection and entropy analysis of a nanofluid inside T-shaped baffled enclosure. International Journal of Numerical Methods for Heat & Fluid Flow, 28(8), 1930–1955.
30. Armaghani, T., Rashad, A. M., Vahidifar, O., Mishra, S. R., & Chamkha, A. J. (2019). Effects of discrete heat source location on heat transfer and entropy generation of nanofluid in an open inclined L-shaped cavity. International Journal of Numerical Methods for Heat & Fluid Flow, 29(4), 1363–1377.
31. Chamkha, A. J., Jomardiani, G., Ismael, M. A., Ghasemiasl, R., & Armaghani, T. (2020). Thermal and entropy analysis in L-shaped non-Darcian porous cavity saturated with nanofluids using Buongiorno model: Comparative study. Mathematical Methods in the Applied Sciences, 3(10), 5827–5846.
32. Nabwey, H. A., Armaghani, T., Azizimehr, B., Rashad, A. M., & Chamkha, A. J. (2023). A comprehensive review of nanofluid heat transfer in porous media. Nanomaterials, 13(5), 937.
33. Sadeghi, M. S., Anadalibkhah, N., Ghasemiasl, R., Armaghani, T., Dogonchi, A. S., Chamkha, A. J., & Asadi, A. (2020). On the natural convection of nanofluids in diverse shapes of enclosures: An exhaustive review. Journal of Thermal Analysis and Calorimetry, 1–22.
34. Markatos, N. C., & Pericleous, K. A. (1984). Laminar and turbulent natural convection in an enclosed cavity. International Journal of Heat and Mass Transfer, 27(5), 755–772.
35. Maxwell, J. C. (1873). A treatise on electricity and magnetism (Vol. 1). Oxford: Clarendon Press.
36. Beckwith, T. G., Marangoni, R. D., & Lienhard, J. H. (2007). Mechanical measurements (Vol. 768). Upper Saddle River, NJ: Pearson Prentice Hall.

5 Effects of Different Shapes of the Porous Cavity on Natural Convective Heat Transfer of Hybrid Nanofluids

Experimental Study

Aryan Moshiri, Ramin Ghasemiasl, Taher Armaghani, and Mahtab Nazarahari

Nomenclature

SSA	Surface area of the solid particles divided by the mass of the solid particles ($m^2\ kg^{-2}$)
Nu	Nusselt number
h	Convection heat transfer coefficient ($W\ m^{-2}\ K^{-1}$)
H	Characteristic length (m)
k	Conduction heat transfer coefficient ($W\ m^{-1}\ K^{-1}$)
k_{pl}	Copper plate conduction
q''	Heat flux ($w\ m^{-2}$)
t	Copper plate thickness
T_{h_corr}	Corrected temperature of the hot surface (K)
T_{c_corr}	Corrected temperature of the cold surface (K)
V	Voltage (v)
I	Electric current intensity (A)
T_h	Average temperature of the hot surface (K)
T_c	Average temperature of the cold surface (K)
ΔT	Temperature difference (K)
k_{pl}	Conduction of the first particle of a nanofluid

DOI: 10.1201/9781003595786-5

k_{p2}	Conduction of the second particle of a hybrid nanofluid
k_f	Base fluid conduction
k_{nf}	Nanofluid conduction
k_{hnf}	Hybrid nanofluid conduction
k_{eff}	Effective conduction
Ra	Rayleigh number
G	Gravitational acceleration (m s^{-2})
Nu*	Effect of the porous medium and nanofluids on the increase in Nusselt number
Greek symbols	
$\mathcal{E}$	Porosity ratio
A	Thermal diffusivity (m^2 s^{-1})
B	Coefficient of thermal expansion (K^{-1})
N	Kinematic viscosity (m^2 s^{-1})
θ	Angle of inclination of the cavity
ϕ	Nanoparticle volume fraction
Subscripts	
h	Hot
c	Cold
pl	Plate

5.1 INTRODUCTION

An innovative new type of nanofluids, with great potential for applications in the industrial sector, is hybrid nanofluids. However, to optimize the product preparation process, improve stability, and provide safety for use at larger concentrations, it is necessary to carry out additional research. Future industrial uses of hybrid nanomaterials are probably going to be very significant because of ongoing research and development into these materials. In many practical applications, such as heat transfer enhancement and studies of the effects of several factors like cavity angle, shape, or nanofluid volume density on natural convection in the hybrid nanofluids, an understanding of free convection is essential. In general, investigation on natural convection in hybrid nanofluids is still in its early stages, and extensive research is needed to understand the behaviour of these fluids better.

Thermal conductivity of fluids can be enhanced by introducing small solid particles into the fluids, leveraging high thermal conductivity of solids. Previous studies by researchers highlighted certain limitations when utilizing dispersions of solid particles of sizes ranging from 2 mm to μm [1]. But when Choi and Eastman from Argonne National Laboratory investigated nanoscale metallic particles and carbon nanotube suspensions, breakthroughs were achieved in this area [2]. Since then, a great deal of research has been carried out on the suspension of various metal and metal oxide nanoparticles in different fluids [3–6], with encouraging outcomes. However, a number of unanswered questions about these suspensions of nanostructured materials, or "nanofluids" as Choi and Eastman referred to them, still remain [1–4, 7]. Extensive experimental and numerical research has been conducted to determine how the aforementioned variables affect the natural thermal convection of hybrid and non-hybrid nanofluids. A multitude of scholarly investigations have documented improvements in heat transfer when nanofluids are employed in contrast to conventional fluids [8]. To illustrate, Ghodsinezhad et al. [9] observed an empirical increase of 15% in heat transfer coefficient when a volume fraction of 0.1% Al_2O_3-water nanofluids was utilized. Similarly, Sharifpur et al. [10] showed that a maximum boost of 8.2% at a volume concentration of 0.05% was achieved in heat transfer when TiO_2 nanoparticles were added to water. With deionized water serving as the base fluid, Choudhary and Subudhi [11] carefully examined the natural convection heat transfer of Al_2O_3 nanofluids and discovered a maximum increase of 29.5% at a volume concentration of 0.01%. In contrast to mono-nanofluids in a triangular cavity, Rashad et al. [12] found that a hybrid nanofluid with equal numbers of Cu and Al_2O_3 nanoparticles dispersed in the water had no appreciable impact on heat transmission. According to these research, a number of parameters, including the size and shape of the nanoparticles; the base fluid; the manufacturing technique; and the purity, stability, temperature, and thermophysical characteristics of the nanoparticles, might affect the increase in heat transfer in hybrid nanofluids [8]. In a rectangular enclosure, Torki and Etesami [13] conducted experiments to examine the inherent heat convection properties of SiO_2/water nanofluids at different concentrations and inclination angles. The results of the investigation demonstrated that although the heat transfer coefficient did not vary considerably at low nanofluid concentrations, it did decrease with concentration for nanoparticle volume fractions greater than 0.005. The influence of inclination angle on Nusselt number was found to be more pronounced when the nanofluid concentration was low. With increasing cavity inclination angle or vertical approach to the heated wall, the effect of nanofluid concentration on Nusselt number exhibited a diminishing effect. At a tilt angle of zero from the horizontal plane, the heat transfer rate was greatest; conversely, as the inclination angle increased, heat transfer rate decreased. In view of heat generation effects, Mansour et al.'s research [14] deals with natural heat convection in an inclined triangular enclosure containing nanosaturated porous media. An increase in the volume fraction of nanoparticles substantially enhances the average Nusselt number. Furthermore, the average Nusselt number diminished as a consequence of elevated heat generation parameters. Srinivasacharya and Kumar [15] analysed the heat and mass transfer phenomena of free convection over a slanted, undulating surface submerged in a nanofluid-saturated porous medium. The heat flux and nanoparticle volume fraction fluxes were both uniform at the surface and significant influence was exerted by the Brownian motion parameter Nb on flow behaviour. The mass transfer rate of nanoparticles increased with an increase in Nb, but the fluid's heat transfer rate decreased. Higher levels of wave amplitude were observed to enhance the mass transfer rates of nanoparticles and the local heat of the fluid. Furthermore, the fluid's heat and nanoparticle mass transfer rates were increased by the wavy surface's tilted angle. Adriana [16] assessed heat transfer characteristics of various hybrid nanofluids using a thermometric estimation method and numerical comparison using a 3D tube model. The following conclusions were obtained from computational fluid dynamics (CFD) analysis: the heat convection coefficient of nanofluids increased with the increase in Reynolds number and hybrid nanoparticle concentration. When the total volume concentration reached up to 4%, the hybrid nanofluid MHS3's heat convection coefficient increased by 257%. As the Reynolds number increased, so did the Nusselt number. The Nusselt number for MHS3 nanofluid increased up to 241% when examined numerically in a hybrid nanofluid with turbulent flow. Buongiorno's non-homogeneous two-phase model was used by Motlagh and Soltanipour [17] to study the Al_2O_3-water nanofluid's spontaneous convection in a tilted square container. They varied the slope angle ($0^\circ \leq \theta \leq 60^\circ$) and Rayleigh number ($10^2 \leq Ra \leq 10^6$). The study found that particle volume fraction had little effect on the average Nusselt number, and heat transfer augmentation percentage for low Rayleigh numbers with increased inclination angle. At high Rayleigh numbers, the average Nusselt number initially increased and then decreased with slope angle, while the percentage of heat transfer augmentation consistently increased. Moreover, particle distribution was non-uniform at low Rayleigh numbers and nearly uniform at high Rayleigh numbers at all inclination angles.

Esfandiary et al. [18] conducted a numerical investigation to examine the impact of inclination angle (between 0 and 60 degrees) on fluid flow and natural convection heat transfer in a chamber containing Al_2O_3-water nanofluid with different volume fractions (0%–3%) and Rayleigh numbers (10^5–10^7). The results of the investigation showed that the slip velocity mechanism is responsible for the Nusselt number decreasing with increasing volumetric concentration. Furthermore, at a 30-degree inclination angle, the Nusselt number reaches its maximum value. Mansour et al. [19] scrutinized the problem of MHD natural convection flow within a square cavity filled with CuO-Al_2O_3/water hybrids, which were heated and cooled using heat sources or sinks. The study results showed that if the heat source location changes, the Nusselt number will be reduced significantly for hybrid suspension. In comparison with other nanoparticles, addition of Cu was found to provide the highest thermal transfer rate when testing various hybrid nanofluids.

2D numerical simulations in a C-shaped open cavity with hybrid nanofluid were carried out by Kalidasan et al. [20]. The main vortex's intensity decreased with higher Rayleigh numbers due to buoyancy factors suppressing wave conduction. The heat transmission increased as the proportion of hybrid nanocomposites increased at $Ra = 10^4$ and 10^5. The bottom half of the hot wall experienced more heat transmission than the top part due to the fusing of hybrid nanocomposites. The investigation of natural heat convection in a sloping porous cavity containing Cu-water nanofluid was conducted by Rajarathinam and Nithyadevi [21]. The impact of the nanoparticles on fluid flow and temperature fields was negligible. Without heat generation, the optimal inclination angles for a greater heat transfer rate at all volumetric concentrations were 135° and 45°. The addition of nanoparticles decreased the heat transfer rate at a 90° inclined angle, suggesting that the utilization of a nanofluid was superfluous at that particular angle. Xiong et al. [22] used the control volume finite element method (CVFEM) method to study hybrid nanofluid free flow through a porous medium, including magnetic effects. Consideration was given to the hybrid nanoparticle Fe_3O_4 + multiwalled carbon nanotubes (MWCNTs) with water as the base fluid. The results showed that average Nusselt values were significantly increased by Rayleigh and Darcy parameters and thermal radiation.

Cimpean and Pop [23] investigated free convection in a square inclined porous cavity containing a nanofluid and subjected to sinusoidal temperature profiles on both sidewalls through numerical investigations. The results indicate that the Nusselt number increases in conjunction with the inclination angle of the cavity and decreases in tandem with the amplitude ratio of the sinusoidal temperature. The Rayleigh–Darcy convection is enhanced in its heat transport due to the inclination of the cavity. Furthermore, with a low Lewis number, a high thermophoresis parameter, and a moderate cavity inclination, the nanoparticles are significantly dispersed in an inhomogeneous manner within the cavity. In Shehzad et al. [24] analysed the thermal transferability and irreversibility of the hybrid nanomaterial in a porous medium. At the point Nhs* = 1, Ha = 1, an increase in Ra yields values of Nu that are 3.86 times smaller. When Ha (Hartman number) increases by Ra* = 5, Nhs = 0.01, approximately 70.1% less Nu is produced. By incorporating Nhs*, the detrimental impact of MHD on Nu is diminished. When Ra* = 5, Ha = 1, the addition of Nhs results in approximately 72.5% reduction in Nu. Abu Libdeh et al. [25] investigated a recently devised cavity containing porous medium and Ag/MgO/H_2O nanofluids. A constant magnetic field was applied to the cavity, and the behaviour of natural convection and total entropy within this system were investigated. Nanofluid fluxes are regarded as laminar and incompressible. The results indicate that the heat transfer rate becomes increasingly limited as the Hartmann number increases. This fact suggests that a magnetic field can serve as an effective control mechanism in heating applications.

In their study, Reddy et al. [26] utilized numerical simulations to examine heat transfer and buoyancy-driven convection within an inclined porous annulus that was filled with a hybrid nanofluid. The findings demonstrated that as Darcy's number increased, the average Nusselt number improved, suggesting enhanced heat transfer. Furthermore, an inclination angle of 30 degrees dissipated the maximum thermal energy. The study further revealed that thin annular enclosures showed higher heat transport than other aspect ratios. However, an increase in internal heat generation led to a reduction in the heat transfer rate, with a steeper decline observed at higher levels of internal heat production.

Multiple numerical investigations [27–30] were undertaken by Armaghani et al. with the objective of examining natural convection in inclined cavities. The aforementioned studies present comprehensive numerical analyses of the fluid flow and heat transmission properties within inclined cavities subjected to natural convection. Furthermore, comprehensive review articles by Armaghani et al. [31, 32] discuss the recent headways in using nanofluids for heat transfer applications in porous materials; extensive scrutiny of five distinct enclosure geometries: square, circular, triangular, trapezoidal, and unconventional shapes; the relationship among thermophysical properties; and their mutual influence within each geometry.

5.2 EXPERIMENTAL SETUP

5.2.1 Methodology

5.2.1.1 Description of the experiment

The setup comprises a square-shaped box device made of Plexiglas with a thickness of 10 mm, measuring 100 × 100 × 100 mm. A compartment for circulating cold water is designed on the right wall of the setup. The temperature of the compartment is regulated via reciprocating pipelines, and the compartment is connected to a cold-water bath via a circulator. In order to connect the cold-water chamber to the main chamber, a 2 × 100 × 100-mm copper plate is employed. An additional copper plate of equivalent thickness is affixed to the right wall, where it is powered by a DC-voltage electric heater. After establishing the intended voltage, the current consumption of the heater can be observed using a multimeter. In addition to affixing three K-type thermocouples to the reverse side of each copper plate, a thermometer is employed to ascertain the temperature at various locations across the plates. The sensors comprising these thermocouples are constructed from two distinct alloys. The temperatures of the hot and cold sides are calculated as the average of three thermocouple readings for each side. These resulting temperatures represent the temperature behind the copper plates. In order to ascertain the temperature of the fluid in proximity to the hot and cold walls, specific equations must be employed to rectify the average temperatures acquired for each side. Moreover, by adjusting the current and voltage intensity of the power supply, it is possible to regulate and optimize the temperature of the heated plate. The heat transfer chamber is positioned on a sloping surface that is capable of being modified in relation to the horizon. The

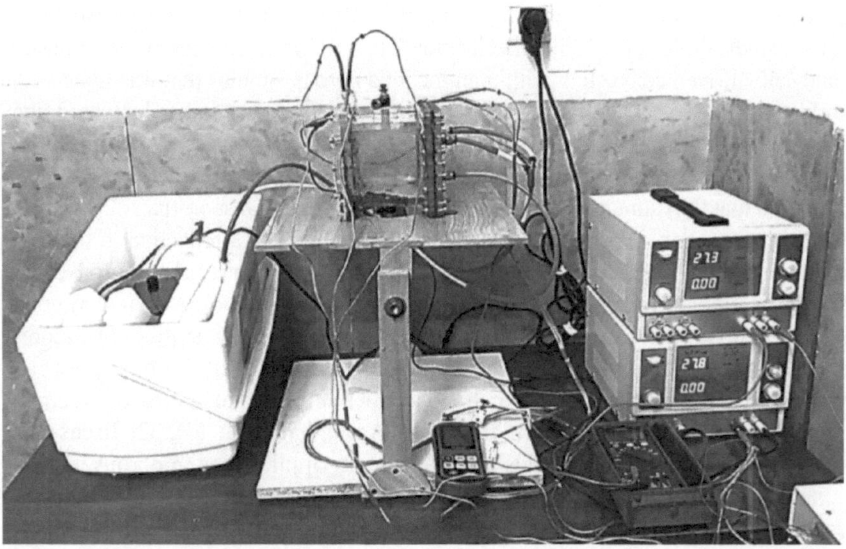

FIGURE 5.1 Overview of laboratory equipment.

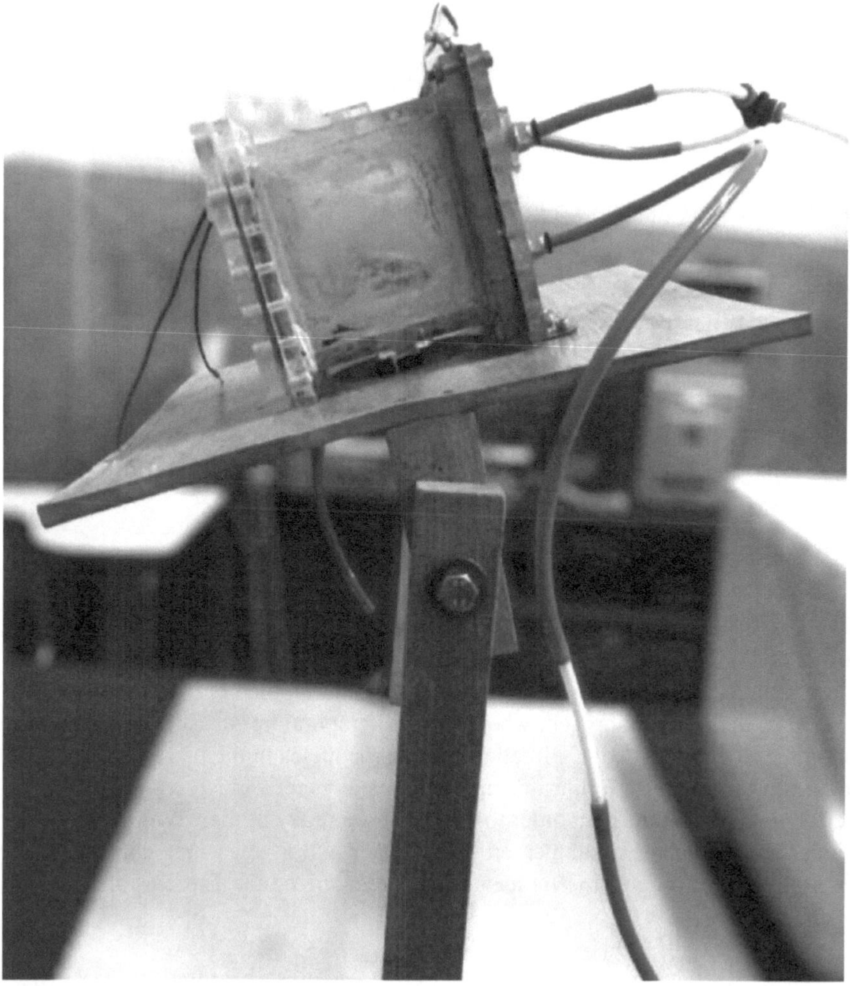

FIGURE 5.2 Alterations in the cavity's angle of inclination (the hot side is at the bottom and the cold side is at the top).

examination configuration is illustrated in Figure 5.1. The proper configuration of the thermocouples and the return pipes for cold water attached to the cold wall is shown in Figure 5.1. The illustration also shows where the heater should be placed in relation to the hot wall. The alterations in the cavity's angle of inclination are shown in Figure 5.2. The heat transfer material utilized in this investigation is stainless steel, which has a thickness of 1 mm. To ensure that the thermocouples remain at a precise distance from the stainless steel material, a thin layer of silicone paste is applied along the sides of the furnace. To achieve an ideal interface between the furnace and the copper plates, the thermocouples are meticulously inserted into shallow channels that have been etched into the copper plates.

5.2.1.2 Hybrid Nanofluids

This study utilized a hybrid nanofluid consisting of titanium oxide and aluminium oxide. The base fluid used was water, and an equal amount of each nanofluid was mixed. The hybrid nanofluids were prepared at volume fractions of 1000 ppm (1%) and 500 ppm (0.5%). The specifications of the aluminium oxide and titanium oxide nanofluids are shown in Table 5.1.

Al_2O_3 Details:	TiO_2 Details:
Nanopowder (gamma) – hydrophilic	Nanopowder (TiO_2, anatase)
Purity: 99%	Purity: >99%
SSA: >188 $m^2\ g^{-1}$	SSA: 200–240 $m^2\ g^{-1}$
Morphology: nearly spherical	Morphology: nearly spherical
Color: white	Color: white
Specific heat capacity: 850 J/(Kg-K)	Bulk density: 0.24 g cm^{-3}
Density: 3990 Kg m^{-3}	True density: 3.9 g cm^{-3}

5.2.1.3 Thermocouple Calibration

Two methodologies are accessible for the purpose of calibrating the temperature sensors. Comparative analysis is performed using thermometers renowned for their precision in the initial approach, whereas fixed temperature measurements are utilized in the second method. Calibration of thermocouples in the present experiment entails ascertaining the equilibrium temperatures of water and ice, as well as that of water at room temperature. In order to verify the accuracy of these temperatures, an additional thermometer is submerged in the water and its measurements are compared to those of a reliable thermometer that is recognized for its precision.

5.2.1.4 Porous Shapes

The influence of the porosity geometry of porous media on heat transfer was investigated using four distinct models. Despite having identical material compositions

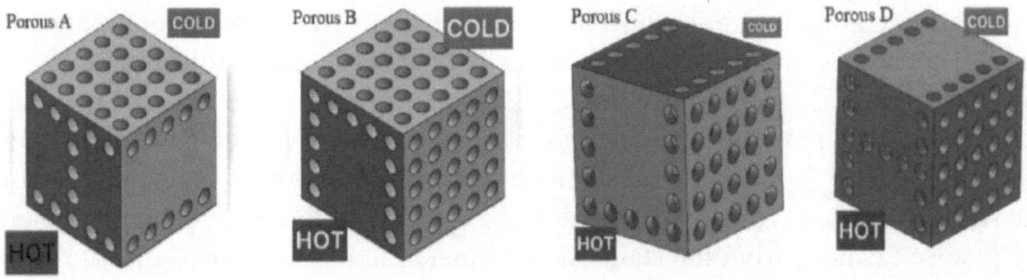

FIGURE 5.3 Designed porous cubes in four different pore shapes.

FIGURE 5.4 3D design and 3D-printed porous cube-shaped parts.

and dimensions, the pores of these models varied. More precisely, as illustrated in Figure 5.3, the utilized pore configurations are designated A, B, C, and D. For each porous cube, the porosity ratio (ε), which represents the proportion of vacant space volume to total volume of the object, was established at 0.4. The cubes in question were produced utilizing polylactic acid (PLA), a substance renowned for its exceptionally low heat transfer coefficient ($ks \approx 2$ W m^{-1} K^{-1}). As illustrated in Figure 5.4, the manufacturing procedure entailed 3D printing by means of fused deposition modelling (FDM) technology. As depicted in the schematic figures provided in Figure 5.3, upon insertion of the porous media A, B, C, and D into the heat transfer chamber, the walls designated as "hot" and "cold" established direct contact with corresponding hot and cold walls of the chamber.

5.2.2 Mathematical Formulation

In order to evaluate the consequences of increasing or decreasing heat transfer, it is critical to calculate the heat convection coefficient and the Nusselt number. While maintaining the same porosity and material, the temperature gradient between the hot and cold surfaces fluctuates under various experimental conditions, including modifications to the inclination angle of the apparatus, changes in the volumetric concentration, and variations in the shape of the porous media. The modifications in the heat convection coefficient that result from these changes in temperature difference have an effect on the Nusselt number.

$$h = \frac{I \times V}{\left(T_{h-corr} - T_{c-corr}(x)\right) * A} \tag{5.1}$$

$$Nu = \frac{hH}{K} \tag{5.2}$$

$$q'' = \frac{I \times V}{A} \tag{5.3}$$

$$T_h = \frac{T_{h1} + T_{h2} + T_{h3}}{3} \tag{5.4}$$

$$T_c = \frac{T_{c1} + T_{c2} + T_{c3}}{3} \tag{5.5}$$

$$T_{h-corr} = T_h - \frac{q''t}{k_{pl}} \tag{5.6}$$

$$T_{c-corr} = T_c + \frac{q''t}{k_{pl}} \tag{5.7}$$

$$Ra = \frac{g\beta\left(T_{h-corr} - T_{c-corr}\right)H^3}{\alpha\vartheta} \tag{5.8}$$

$$Nu = 0.082\ Ra_h^{0.329}\left(for\ 10^6 \le Ra \le 10^{12}\right) \tag{5.9}$$

$$k_{nf} = \left[\frac{k_{p1} + 2k_f - 2\phi_1\left(k_f - k_{p1}\right)}{k_{p1} + 2k_f + \phi_1\left(k_f - k_{p1}\right)}\right]k_f \tag{5.10}$$

$$k_{hnf} = \left[\frac{k_{p2} + 2k_{nf} - 2\phi_2 \left(k_{nf} - k_{p2} \right)}{k_{p2} + 2k_{nf} + \phi_2 \left(k_{nf} - k_{p2} \right)} \right] k_{nf} \tag{5.11}$$

$$k_{eff} = \varepsilon * k_{nf} + (1 - \varepsilon) * k_s \tag{5.12}$$

$$Nu_1 = hl / k_{eff} \tag{5.13}$$

$$Nu_2 = hl / k_f \tag{5.14}$$

$$Nu^* = Nu_1 / Nu_2 \tag{5.15}$$

The aforementioned equations delineate the correlations among the characteristic length (H), the heat flux (q''), the corrected temperatures of the hot and cold surfaces ($T_{c\text{-}corr}$ and $T_{h\text{-}corr}$), and the thickness of the copper plate (t). The heat convection and heat conduction coefficients (h and k), in addition to voltage (V) and electric current intensity (I), are included in these equations. Furthermore, the average temperatures of the heated and frigid surfaces are denoted by T_h and T_c, respectively. The formula for the Rayleigh number is given in Equation (5.8), in which γ denotes the coefficient of thermal expansion, g represents gravitational acceleration, and v signifies kinematic viscosity. Equation (5.9) concludes with the formulation of the Nusselt number. Equation (5.9), which represents the Markatos [33] correlation, provides the mathematical expression for the numerical Nusselt number. Conductivities of nanofluids and hybrid nanofluids are stipulated by Equations (5.10) and (5.11), respectively. The conductivity of porous and nanofluid media is determined using Equation (5.12), and Nusselt number is computed using this formula. Porosity is denoted by ε, and the conduction coefficient of the porous media is denoted by k_s. Equation (5.13) delineates the correlation between Nu_1 and the Nusselt value of the nanofluid and porous medium. Equation (5.14) delineates the correlation between Nu_2 and the Nusselt number of the base fluid, i.e., water in the present investigation. Equation (5.15) describes the Nusselt Star or Nu* correlation, which investigates the combined effect of porous medium and nanofluid on heat transfer.

5.3 VALIDATION PROCESS

5.3.1 Validation

A preliminary experiment was conducted using purified water devoid of any porous media in order to verify the accuracy of the setup and the test. In order to evaluate the setup's dependability, experimental results were obtained at a zero-degree cavity angle and compared to the Markatos [33] correlation (Equation [5.9]).

By calculating the mean temperature values, the thermodynamic properties of the water used in the experiment were derived from the thermodynamic tables.

The Nusselt number values obtained from the laboratory data and experimental equations are compared in Table 5.2 to those obtained through the application of the Markatos [33] relation. Furthermore, the table provides the percentage error that exists between the aforementioned value sets.

5.3.2 Uncertainty

The equipment used in the experiment and the uncertainty of each of them are shown in Table 5.3.

The uncertainties pertaining to the Rayleigh number, heat flux, heat convection coefficient, and Nusselt number were ascertained through an experimental investigation. Utilizing the Beckwith method [34], the combined uncertainty was computed. The results of the examination of uncertainty pertaining to these parameters are displayed in Table 5.4.

5.3.3 Repeatability

Before introducing any porous medium into the cavity, the temperature gradient between heater and cold-water circulation was measured using pure water fluid at a zero-degree cavity angle. The Nusselt number and heat convection coefficient were calculated after thermocouple calibration. These findings, which are shown in Table 5.5, were noted in order to evaluate repeatability and contrast them with the preliminary validation results. Following confirmation that the empirical validation value and the empirical repeatability value matched within a suitable tolerance, more analysis was performed, and the further processes were executed.

TABLE 5.2
Comparison of Experimental and Theoretical Nusselt Number Results

Ra	Nu [33]	Nu of current work	Error (%)
102,375,411.2	35.41	35.15	**0.7%**

TABLE 5.3
Measurement Equipment Used, Their Characteristics, and Uncertainty

No.	Tool	Interval	Variable	Lowest measured value	Measurement interval	Uncertainty (%)
1	Thermocouple	−5–1260	T_h	0.1	20–50	0.21
2	Thermocouple	−5–1260	T_c	0.1	0–10	0.42
3	Voltmeter	0–220	V	0.01	10–60	0.01
4	Ampere meter	0–15	I	0.001	0–1	0.125
5	Height	0–30	W-H	0.00005 m	0.1	0.05

TABLE 5.4
Uncertainty

Variable	Uncertainty (%)
Heat flux	0.72
h (non-porous)	2.1
Nu (non-porous)	2.16
h (porous A)	1.05
Nu (porous A)	1.17
h (porous B)	1.09
Nu (porous B)	1.2
h (porous C)	0.97
Nu (porous C)	1.1
h (porous D)	1.3
Nu (porous D)	1.4

TABLE 5.5
Results and Date of Repeatability

No.	Repeatability date before each transaction	Experimental repeatability	Validation Nusselt 02/10/2022
1	05/10/2022	34.1	35.15
2	08/10/2022	33.5	35.15
3	11/10/2022	35.05	35.15
4	15/10/2022	33.2	35.15

5.4 RESULT AND DISCUSSION

This study aimed to model a heat transfer system that can function with different fluids, such as nanofluids and water, by modifying the system setup. The purpose was to enhance or reduce heat transfer using hybrid nanofluids with various porous materials and by adjusting the cavity angle. This section presents the findings and data acquired from the heat transfer experiments conducted under different conditions, including the presence or lack of a porous medium and changes in shapes of the porous medium at angles ranging from 0 to 60 degrees from the horizontal plane when hybrid nanofluids are present.

This experimental study examined the heat transfer properties of different porous shapes placed inside a cavity at different angles. Once the experiment's repeatability was confirmed, four types of porous cubes were inserted into the cavity one after another, filling the system with pure water. Tests of heat transmission were carried out

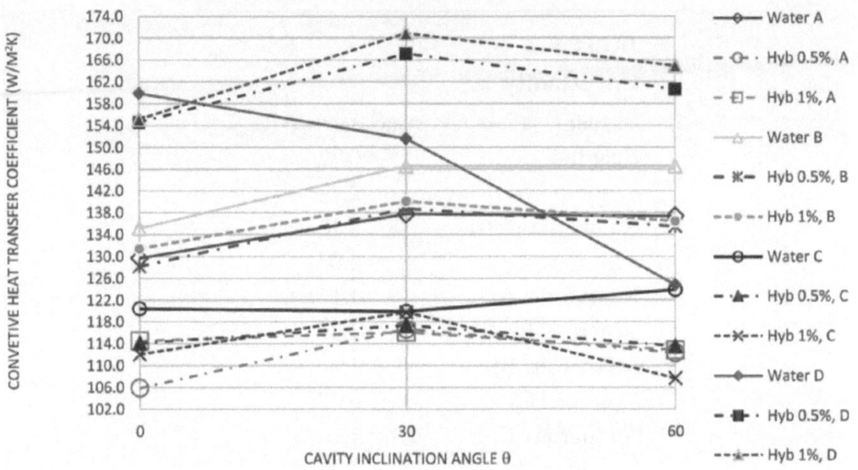

FIGURE 5.5 *h* diagram of nanofluids at different inclination angles for the four types of porous medium.

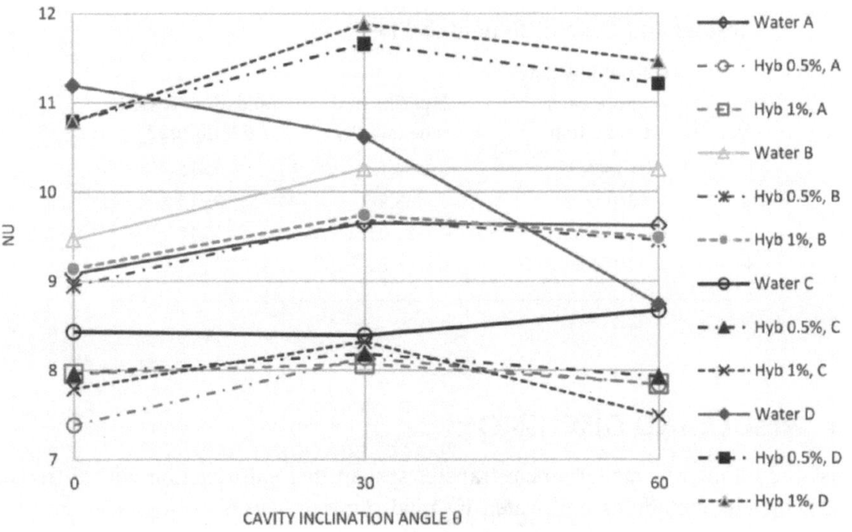

FIGURE 5.6 Nu diagram of nanofluids at different inclination angles for the four types of porous medium ($\phi = 0$, 0.5%, 1%).

at horizon-facing angles of 0, 30, and 60 degrees. Furthermore, hybrid nanofluids with water base fluid and aluminium oxide–titanium oxide volume percentages of 0.5% and 1% were examined at the same angles. Measurement of the steady-state temperature differential (ΔT) was used to compute the Nusselt number and the heat convection coefficient (*h*). The heat transfer coefficient of nanofluids was determined using Maxwell's equation [35]. Furthermore, Nu* was computed using Equations (5.13) and (5.14). The obtained values of *h*, Nu, and Nu* were plotted against the angles for each porous cube.

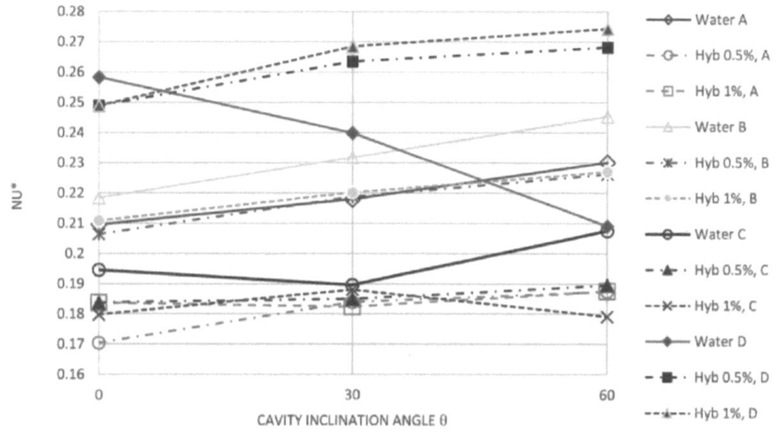

FIGURE 5.7 Nu* diagram of nanofluids at different inclination angles for the four types of porous medium ($\phi = 0, 0.5\%, 1\%$).

TABLE 5.6
Highest and Lowest Values of *h*, Nu, and Nu*of Hybrid Nanofluids for Angles of 0, 30, and 60 Degrees for Porous Media

Cavity inclination angle	0°	30°	60°
h (W m^{-2} K^{-1})	Min h = 105.9	Min h = 116.1	Min h = 107.7
	(shape A, hybrid 0.5%)	(shape A, hybrid 1%)	(shape C, hybrid 1%)
	Max h = 159.9	Max h = 170.9	Max h = 164.9
	(shape D, water)	(shape D, hybrid 1%)	(shape D, hybrid 1%)
Nu	Min Nu = 7.38	Min Nu = 8.07	Min Nu = 7.49
	(shape A, hybrid 0.5%)	(shape A, hybrid 1%)	(shape C, hybrid 1%)
	Max Nu = 11.19	Max Nu = 11.88	Max Nu = 11.468
	(shape D, water)	(shape D, hybrid 1%)	(shape D, hybrid 1%)
Nu*	Min Nu* = 0.17	Min Nu* = 0.182	Min Nu* = 0.179
	(shape A, hybrid 0.5%)	(shape A, hybrid 1%)	(shape C, hybrid 1%)
	Max Nu* = 0.258	Max Nu* = 0.268	Max Nu* = 0.274
	(shape D, water)	(shape D, hybrid1%)	(shape D, hybrid 1%)

To investigate the impact of the hybrid nanofluid and porous media on the Nusselt number, Nu* was presented (Equation [5.15]).

Section 5.4.1 provides information on the impacts of cavity angle, volumetric concentration, and various shapes of the porous media on h, Nu, and Nu*.

5.4.1 Effects of the Cavity Angle

In this section, the diagrams of h–θ (Figure 5.5), Nu–θ (Figure 5.6), and Nu*–θ (Figure 5.7) are presented.

The Nusselt number and Nu* as well as the maximum and lowest values of the convective heat transfer coefficient are shown in Table 5.6.

The following can be inferred from Table 5.6:

- Comparison of the lowest h values according to cavity angles:

  $h_{0^\circ\_A\_0.5\%} < h_{60^\circ\_C\_1\%} < h_{30^\circ\_A\_1\%}$

- Comparison of the highest h values according to cavity angles:

  $h_{0^\circ\_D\_Water} < h_{60^\circ\_D\_1\%} < h_{30^\circ\_D\_1\%}$

- Comparison of the lowest Nu values according to cavity angles:

  $\text{Nu}_{0^\circ\_A\_0.5\%} < \text{Nu}_{60^\circ\_C\_1\%} < \text{Nu}_{30^\circ\_A\_1\%}$

- Comparison of the highest Nu values according to cavity angles:

  $\text{Nu}_{0^\circ\_D\_Water} < \text{Nu}_{60^\circ\_D\_1\%} < \text{Nu}_{30^\circ\_D\_1\%}$

- Comparison of the lowest Nu* values according to cavity angles:

  $\text{Nu}^*_{0^\circ\_A\_0.5\%} < \text{Nu}^*_{60^\circ\_C\_1\%} < \text{Nu}^*_{30^\circ\_A\_1\%}$

- Comparison of the highest Nu* values according to cavity angles:

  $\text{Nu}^*_{0^\circ\_D\_Water} < \text{Nu}^*_{30^\circ\_D\_1\%} < \text{Nu}^*_{60^\circ\_D\_1\%}$

5.4.2 Effects of Volume Fraction

In this section, diagrams of h–ϕ (Figure 5.8), Nu–ϕ (Figure 5.9), and Nu*–ϕ (Figure 5.10) are presented.

The heat convection coefficient's greatest and lowest values, as well as the Nusselt number and Nu* at different volume concentration percentages, are shown in Table 5.7.

Table 5.7 shows the following:

- Comparison of the lowest h values according to volume concentration:

  $\text{h}_{0.5\%\_A\_0^\circ} < \text{h}_{1\%\_C\_60^\circ} < \text{h}_{0\%\_C\_30^\circ}$

- Comparison of the highest h values according to volume concentration:

  $\text{h}_{0\%\_D\_0^\circ} < \text{h}_{0.5\%\_D\_30^\circ} < \text{h}_{1\%\_D\_30^\circ}$

- Comparison of the lowest Nu values according to volume concentration:

  $\text{Nu}_{0.5\%\_A\_0^\circ} < \text{Nu}_{1\%\_C\_60^\circ} < \text{Nu}_{0\%\_C\_30^\circ}$

- Comparison of the highest Nu values according to volume concentration:

  $\text{Nu}_{0\%\_D\_0^\circ} < \text{Nu}_{0.5\%\_D\_30^\circ} < \text{Nu}_{1\%\_D\_30^\circ}$

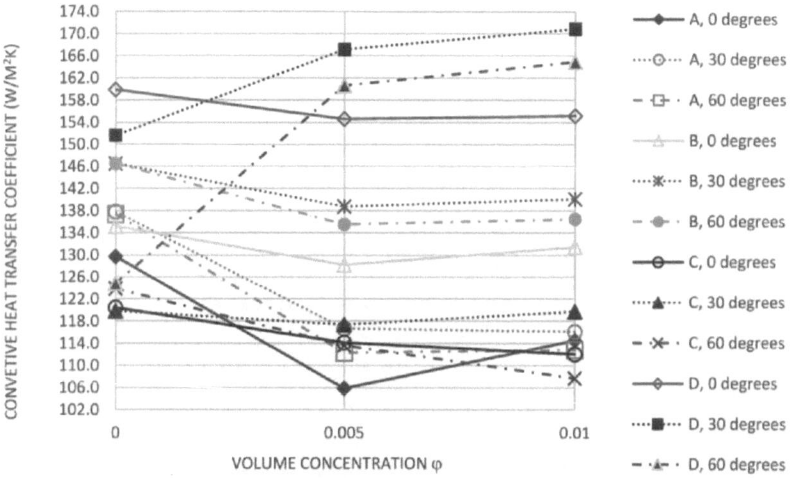

FIGURE 5.8 Diagram of convective heat transfer coefficient by volume percentage of nanofluids for the four types of porous medium (θ = 0, 30, 60).

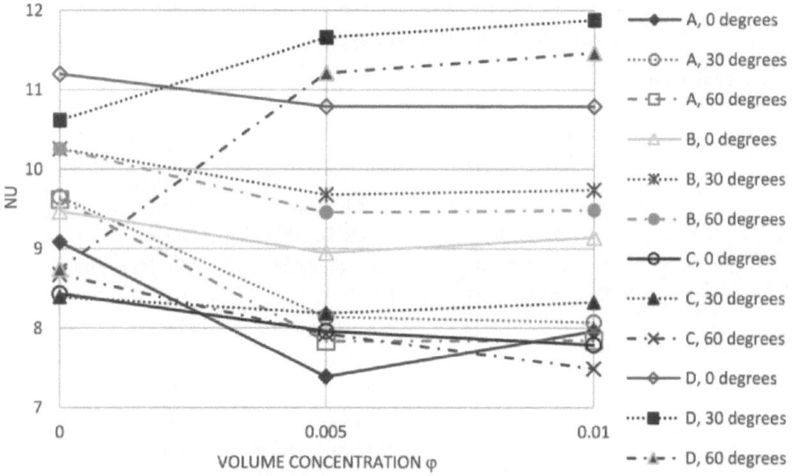

FIGURE 5.9 Diagram of Nu by volume concentration percentage of nanofluids for the four types of porous medium (θ = 0, 30, 60).

- Comparison of the lowest Nu* values according to volume concentration:

$$Nu^*_{0.5\%\_A\_0^\circ} < Nu^*_{1\%\_C\_60^\circ} < Nu^*_{0\%\_C\_30^\circ}$$

- Comparison of the highest Nu* values according to volume concentration:

$$Nu^*_{0\%\_D\_0^\circ} < Nu^*_{0.5\%\_D\_60^\circ} < Nu^*_{1\%\_D\_60^\circ}$$

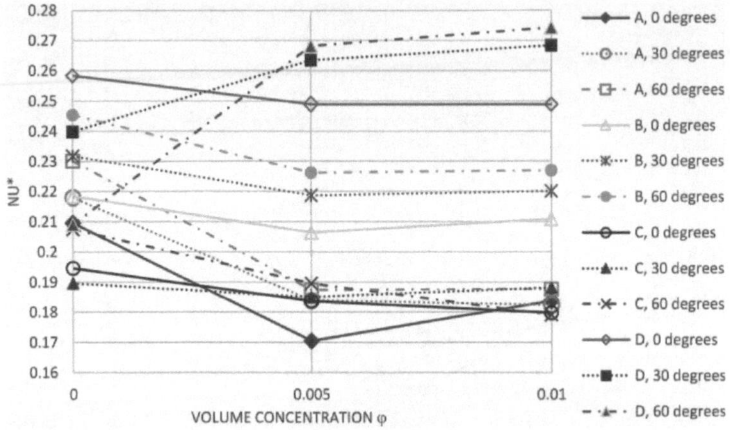

FIGURE 5.10 Diagram of Nu* by volume concentration percentage of nanofluids for the four types of porous medium (θ = 0, 30, 60).

TABLE 5.7
Highest and Lowest Values of *h*, Nu, and Nu* of Hybrid Nanofluids for Different Volume Concentration Percentages

Volume concentration percentage	0%	0.5%	1%
h (W m^{-2} K^{-1})	Min h = 119.9 (shape C, 30°)	Min h = 105.9 (shape A, 0°)	Min h = 107.7 (shape C, 60°)
	Max h = 159.9 (shape D, 0°)	Max h = 167.1 (shape D, 30°)	Max h = 170.9 (shape D, 30°)
Nu	Min Nu = 8.39 (shape C, 30°)	Min Nu = 7.38 (shape A, 0°)	Min Nu = 7.49 (shape C, 60°)
	Max Nu = 11.19 (shape D, 0°)	Max Nu = 11.66 (shape D, 30°)	Max Nu = 11.88 (shape D, 30°)
Nu*	Min Nu* = 0.189 (shape C, 30°)	Min Nu* = 0.17 (shape A, 0°)	Min Nu* = 0.179 (shape C, 60°)
	Max Nu* = 0.258 (shape D, 0°)	Max Nu* = 0.268 (shape D, 60°)	Max Nu* = 0.274 (shape D, 60°)

5.5 CONCLUSIONS

The purpose of this experimental work was to explore a naturally occurring convective heat transfer system that can be adjusted in terms of the heat transfer chamber's geometry and the presence or absence of nanofluid. The main objective was to investigate the properties of heat transmission between two plates, which correspond to hot and cold surfaces, at 0-, 30-, and 60-degree cavity angles. The experimental setup involves rotating the cavity to position the hot wall at the bottom and the cold wall at the top. The cavity shape was modified four times to explore the effects of different porous media on the system. Distilled water and hybrid nanofluid with 0.5% and 1% volume concentration were utilized as the working fluids for each specific angle and shape configuration.

The findings observed from this empirical study are summarized as follows:

1. Heat convection coefficient and Nusselt number trends:
   - The heat convection coefficient and Nusselt number increased (except for water, shape C) when the angle was changed from 0 to 30 degrees.
   - When the angle was changed from 30 to 60 degrees relative to the horizon, the heat convection coefficient and Nusselt number decreased (except for water, shape C).
2. Influence of cavity angle and nanofluid concentration:
   - The highest heat transfer coefficient was observed for shape D at an angle of 30 degrees for a hybrid nanofluid with 1% volume concentration, while the lowest *h* was observed for shape A with a 0.5% hybrid nanofluid at a 0-degree cavity angle.
   - The highest Nusselt number was recorded for shape D at an angle of 30 degrees and 1% hybrid nanofluid, while the lowest Nusselt number was observed for shape A with a 0.5% hybrid nanofluid at a 0-degree cavity angle.
3. Nu* parameter:
   - The highest Nu* value was obtained for shape D using a 1% hybrid nanofluid at an angle of 60 degrees, while the lowest Nu* value was recorded for shape A with a 0.5% hybrid nanofluid at 0 degrees.

Recommendations:

- The most significant increase in Nusselt number from the condition without porous medium and nanofluid to the condition with the presence of porous medium and nanofluid (Nu* parameter) was observed for shape D at a cavity angle of 60 degrees for the 1% hybrid nanofluid.
- The most effective configuration in this study for reducing the Nusselt number based on Nu* parameter, from the condition without a porous medium and nanofluid to the condition with both a porous medium and a nanofluid, was the configuration of shape A with 0.5% volume concentration at an angle of 0 degrees.
- The maximum heat transfer coefficient was observed for the cavity filled with 1% hybrid nanofluid in the presence of porous D at an inclination angle of 30 degrees.
- In general, porous D enhanced the heat transfer rate.

REFERENCES

1. Das, S. K., Choi, S. U., & Patel, H. E. (2006). Heat transfer in nanofluids—a review. Heat Transfer Engineering, 27(10), 3–19.
2. Eastman, J. A., Choi, U. S., Li, S., Thompson, L. J., & Lee, S. (1996). Enhanced thermal conductivity through the development of nanofluids. MRS Online Proceedings Library (OPL), 457, 3.

3. Choi, S. U. S. (1998). Nanofluid technology: Current status and future research (No. ANL/ET/CP-97466). Argonne, IL: Argonne National Lab (ANL).
4. Choi, S. U. S., Zhang, Z. G., Yu, W., Lockwood, F. E., & Grulke, E. A. (2001). Anomalous thermal conductivity enhancement in nanotube suspensions. Applied Physics Letters, 79(14), 2252–2254.
5. Chon, C. H., Paik, S. W., Tipton Jr, J. B., & Kihm, K. D. (2006). Evaporation and dryout of nanofluid droplets on a microheater array. Journal of Heat Transfer, 128(7), 753–762.
6. Eastman, J. A., Choi, S. U. S., Li, S., Yu, W., & Thompson, L. J. (2001). Anomalously increased effective thermal conductivities of ethylene glycol-based nanofluids containing copper nanoparticles. Applied Physics Letters, 78(6), 718–720.
7. Sivashanmugam, P. (2012). Application of nanofluids in heat transfer. In S. N. Kazi (Ed.), An overview of heat transfer phenomena (pp. 411–440). IntechOpen.
8. Sidik, N. A. C., Jamil, M. M., Japar, W. M. A. A., & Adamu, I. M. (2017). A review on preparation methods, stability and applications of hybrid nanofluids. Renewable and Sustainable Energy Reviews, 80, 1112–1122.
9. Ghodsinezhad, H., Sharifpur, M., & Meyer, J. P. (2016). Experimental investigation on cavity flow natural convection of Al_2O_3–water nanofluids. International Communications in Heat and Mass Transfer, 76, 316–324.
10. Sharifpur, M., Solomon, A. B., Ottermann, T. L., & Meyer, J. P. (2018). Optimum concentration of nanofluids for heat transfer enhancement under cavity flow natural convection with TiO_2–water. International Communications in Heat and Mass Transfer, 98, 297–303.
11. Choudhary, R., & Subudhi, S. (2016). Aspect ratio dependence of turbulent natural convection in Al_2O_3/water nanofluids. Applied Thermal Engineering, 108, 1095–1104.
12. Rashad, A. M., Chamkha, A. J., Ismael, M. A., & Salah, T. (2018). Magnetohydrodynamics natural convection in a triangular cavity filled with a Cu-Al_2O_3/water hybrid nanofluid with localized heating from below and internal heat generation. Journal of Heat Transfer, 140(7).
13. Torki, M., & Etesami, N. (2020). Experimental investigation of natural convection heat transfer of SiO_2/water nanofluid inside inclined enclosure. Journal of Thermal Analysis and Calorimetry, 139, 1565–1574.
14. Mansour, M. A., & Ahmed, S. E. (2015). A numerical study on natural convection in porous media-filled an inclined triangular enclosure with heat sources using nanofluid in the presence of heat generation effect. Engineering Science and Technology, an International Journal, 18(3), 485–495.
15. Srinivasacharya, D., & Kumar, P. V. (2015). Free convection of a nanofluid over an inclined wavy surface embedded in a porous medium with wall heat flux. Procedia Engineering, 127, 40–47.
16. Minea, A. A. (2017). Hybrid nanofluids based on Al_2O_3, TiO_2 and SiO_2: Numerical evaluation of different approaches. International Journal of Heat and Mass Transfer, 104, 852–860.
17. Motlagh, S. Y., & Soltanipour, H. (2017). Natural convection of Al_2O_3-water nanofluid in an inclined cavity using Buongiorno's two-phase model. International Journal of Thermal Sciences, 111, 310–320.
18. Esfandiary, M., Mehmandoust, B., Karimipour, A., & Pakravan, H. A. (2016). Natural convection of Al_2O_3–water nanofluid in an inclined enclosure with the effects of slip velocity mechanisms: Brownian motion and thermophoresis phenomenon. International Journal of Thermal Sciences, 105, 137–158.
19. Mansour, M. A., Siddiqa, S., Gorla, R. S. R., & Rashad, A. M. (2018). Effects of heat source and sink on entropy generation and MHD natural convection of Al_2O_3-Cu/water

hybrid nanofluid filled with square porous cavity. Thermal Science and Engineering Progress, 6, 57–71.

20. Kalidasan, K., Velkennedy, R., & Kanna, P. R. (2017). Laminar natural convection of copper-titania/water hybrid nanofluid in an open ended C-shaped enclosure with an isothermal block. Journal of Molecular Liquids, 246, 251–258.
21. Rajarathinam, M., & Nithyadevi, N. (2017). Heat transfer enhancement of Cu-water nanofluid in an inclined porous cavity with internal heat generation. Thermal Science and Engineering Progress, 4, 35–44.
22. Manh, T. D., Tlili, I., Shafee, A., Nguyen-Thoi, T., & Hamouda, H. (2020). Modeling of hybrid nanofluid behavior within a permeable media involving buoyancy effect. Physica A: Statistical Mechanics and its Applications, 554, 123940.
23. Cimpean, D. S., & Pop, I. (2019). Free convection in an inclined cavity filled with a nanofluid and with sinusoidal temperature on the walls: Buongiorno's mathematical model. International Journal of Numerical Methods for Heat & Fluid Flow, 29(12), 4549–4568.
24. Shehzad, S. A., Sheikholeslami, M., Ambreen, T., & Shafee, A. (2020). Convective MHD flow of hybrid-nanofluid within an elliptic porous enclosure. Physics Letters A, 384(28), 126727.
25. Abu-Libdeh, N., Redouane, F., Aissa, A., Mebarek-Oudina, F., Almuhtady, A., Jamshed, W., & Al-Kouz, W. (2021). Hydrothermal and entropy investigation of Ag/MgO/H_2O hybrid nanofluid natural convection in a novel shape of porous cavity. Applied Sciences, 11(4), 1722.
26. Reddy, N. K., Swamy, H. K., Sankar, M., & Jang, B. (2023). MHD convective flow of Ag-TiO_2 hybrid nanofluid in an inclined porous annulus with internal heat generation. Case Studies in Thermal Engineering, 102719.
27. Chamkha, A. J., Rashad, A. M., Armaghani, T., & Mansour, M. A. (2018). Effects of partial slip on entropy generation and MHD combined convection in a lid-driven porous enclosure saturated with a Cu–water nanofluid. Journal of Thermal Analysis and Calorimetry, 132, 1291–1306.
28. Rashad, A. M., Armaghani, T., Chamkha, A. J., & Mansour, M. A. (2018). Entropy generation and MHD natural convection of a nanofluid in an inclined square porous cavity: Effects of a heat sink and source size and location. Chinese Journal of Physics, 56(1), 193–211.
29. Dogonchi, A. S., Armaghani, T., Chamkha, A. J., & Ganji, D. D. (2019). Natural convection analysis in a cavity with an inclined elliptical heater subject to shape factor of nanoparticles and magnetic field. Arabian Journal for Science and Engineering, 44, 7919–7931.
30. Chamkha, A., Abdelrahman, Z., Mansour, M., Armaghani, T., & Rashad, A. (2021). Effects of magnetic field inclination and internal heat sources on nanofluid heat transfer and entropy generation in a double lid driven L-shaped cavity. Thermal Science, 25(2 Part A), 1033–1046.
31. Nabwey, H. A., Armaghani, T., Azizimehr, B., Rashad, A. M., & Chamkha, A. J. (2023). A comprehensive review of nanofluid heat transfer in porous media. Nanomaterials, 13(5), 937.
32. Sadeghi, M. S., Anadalibkhah, N., Ghasemiasl, R., Armaghani, T., Dogonchi, A. S., Chamkha, A. J., & Asadi, A. (2020). On the natural convection of nanofluids in diverse shapes of enclosures: An exhaustive review. Journal of Thermal Analysis and Calorimetry, 1–22.

33. Markatos, N. C., & Pericleous, K. A. (1984). Laminar and turbulent natural convection in an enclosed cavity. International Journal of Heat and Mass Transfer, 27(5), 755–772.
34. Beckwith, A. (2011). Detailing coherent, minimum uncertainty states of gravitons, as semi classical components of gravity waves, and how squeezed states affect upper limits to graviton mass. Journal of Modern Physics, 2011.
35. Maxwell, J. C. (1873). A treatise on electricity and magnetism (Vol. 1). Oxford: Clarendon Press.

6 Hydrothermal Performance of Magnetized Al_2O_3-TiO_2-Water Hybrid Nanofluid Flow within a Triangular Enclosure with a Circular Heater

A Multiple Linear Regression Analysis

Nilankush Acharya

Nomenclature

(u, v)	Velocity components (m s^{-1})
g	Gravitation (m s^{-2})
L	Cavity length (m)
T_h	Temperature of heated wall (K)
T_c	Temperature of cold wall (K)
T	Hybrid nanofluid temperature (K)
ρ	Density (kg m^{-3})
μ	Dynamic viscosity (kg m^{-1} s^{-1})
κ	Thermal conductivity (W $m_{-1\ K-1)}$
ρC_p	Heat capacitance (J m^{-3} K^{-1})
β	Thermal expansion coefficient (K^{-1})
α	Thermal diffusivity (m^2s^{-1})

DOI: 10.1201/9781003595786-6

σ	Electrical conductivity ($\Omega_{-1\ m-1)}$
B_0	Magnetic field ($\Omega_{1/2\ m-1\ s-1/2\ kg1/2)}$
ϕ	Nanoparticle volume fraction
$Ra\left(=\dfrac{g\beta_f(T_h-T_c)L^3}{\nu_f\alpha_f}\right)$	Rayleigh number
$Pr\left(=\dfrac{\mu_f(\rho C_p)_f}{\rho_f\kappa_f}\right)$	Prandtl number
$Ha\left(=B_0L\sqrt{\dfrac{\sigma_f}{\mu_f}}\right)$	Hartmann number
Nu_{loc}	Local Nusselt number
Nu_{avg}	Average Nusselt number

Subscripts

f	Base fluid
nf	Mono nanofluid
hnf	Hybrid nanofluid
s	Nanoparticle
1	First particle
2	Second particle

6.1 INTRODUCTION

Buoyancy-induced convection or natural convective motion describes a heat transfer mechanism that initiates fluid motion owing to the density discrepancies caused by thermal gradients. When the fluid is warmed, the nearest zone of the heated object becomes less dense because of the enhanced kinetic energy of the fluid particles. Then the less dense fluid rises and the cold highly dense fluid sinks. This procedure repeats itself as the less dense fluid moves away from the heated source, cools down and sinks again, whereas the fluid near the heated source makes it rise. This produces convection currents that are not reliant on any external source like a fan, pump, or suction device to let the fluid be in motion. The buoyancy-induced motion within a heated or discretely heated closed or partitioned chamber finds significant application in cooling devices [1], solar collectors [2], electronic industries [3], food sterilization [4], solidification and melting [5], medical applications [6], geothermal systems [7], etc. Depending on the geometrical configuration and flow structure, buoyancy-induced motion can be grouped into two families: internal and external convection. During internal convective motion, fluid motion is confined within some specific geometry, whereas during external convection liquids go through some solid objects. Internal convection is typically more complex compared to the earlier mentioned external convection, as the latter can be demonstrated by classic boundary layer assumption. Generally, buoyancy-driven motion within square-shaped enclosures attracts the most for testing and validating computational fluid dynamics (CFD) codes in numerical simulations and experimental and theoretical studies owing to their simple configuration. But few inaccuracies still exist in the case of natural convective motion within square-shaped chambers [8]. McQuain et al. [9] remarked that the numerous fallouts collected from square-shaped chambers may be unreliable for flows through triangular compartments. In practice, the latter geometric configurations are mostly introduced because of their noteworthy applications in solar power, attic designing, electronic cooling, antenna designing, coatings, etc. [10–13]. Li and Tang [14] illustrated the viscous steady flow through a triangular chamber following the finite difference technique. Using the finite element method, Gaskell [15] investigated the viscous steady Newtonian fluid motion within a solid triangular compartment whose upper wall was set to motion. They noted that enhancement in the corner angle exceeding 40° will decline the secondary recirculation size. Nazeer et al. [16] discussed forced convective transport of microfluids through a porous medium within a lid-driven right-angled triangular chamber and noted average heat transfer enhancement of microfluid parameters and reduction in Hartmann numbers. Ali et al. [17] opted a numerical system to decode the radiative mixed convective bi-viscous transport through a lid-driven triangular enclosure. They claimed the positive influence of the Grashof number on heat transmission. Shah et al. [18] numerically scrutinized the impacts of cold cylinder walls on a power-law fluid's behaviour within a non-uniformly heated isosceles triangular section and observed both heat transfer and kinetic energy enhancement for Rayleigh numbers, but the power-law index shows the reverse. Turkyilmazoglu and Duraihem [19] described a fully developed magnetized flow through a long triangular channel and spotted recirculating regimes within the triangular path. More literature on this can be found in [20–22].

Heat exchange is the most serious challenging issue for engineers working in various technological sectors. Thus, the use of heat sinks or heat exchangers is foreseeable in several sectors. Irregular heat generation causes obstruction and malfunction in many devices [23] and thereby leads to serious obstructions in the smooth functioning of solar panels, microtechnological equipment, heat exchangers, electronic devices, etc. Attachment of various geometric-shaped fins and their optimization have been introduced by scientists to augment heat transfer in those devices. Another innovative technique that can improve heat transfer is the use of working fluid within complex devices. Water is the most low-cost and available fluid, while different types of coolants and oils are employed in industries, but low thermal conductivity prevents them from being an effective medium that can accelerate heat transmission. Initially, the idea of dispersing high thermal conductive solid metal particles into the base liquid was thought to upsurge the resulting fluid's thermal conductivity, but pumping power and suspension instability were some major drawbacks of such practice. A stable mixture of nanosized (1–100 nm) metallic, oxide, carbon nanotubes, carbides, etc., in base liquids introduced a novel concept, i.e., nanofluids [24]. The noteworthy applications of nanofluids include in heat exchangers, solar receivers, energy storage, electronic devices, and biomedical applications [25–28]. Hybrid nanofluids [29, 30] are the advanced version of classical mono-nanofluids formed by the blending of two dissimilar nanosized particles within the same host liquid to have improved rheological, morphological, optical, and thermophysical features. The latter is originated to replace the classical mono-nanofluids due to their small frictional losses, low-pressure drop, pumping power, and elevated thermal conductivity compared to traditional nanofluids. They have been successfully tested for several applications like electronic cooling, photovoltaic thermal managing, automotive cooling, heat exchangers, solar receivers, etc. [31–33]. Zahan et al. [34] discussed Cu-Al_2O_3-water transport through a lid-driven wavy-bottom-based triangular section. They observed that hybrid nanofluids enhance heat transfer compared to mono-nanofluids. Izadi [35] examined the unsteady Al_2O_3-Cu-water nanofluidic motion within a porous triangular chamber and noted an adverse effect on heat transfer at increased volume concentrations. Chabani et al. [36] scrutinized the behaviour of Cu-TiO_2-ethylene glycol hybrid nanofluids within a triangular zigzag chamber with an elliptical obstacle. They noted flow deceleration due to magnetic effect. Redouane et al. [37] numerically observed the flow of magnetized Ag-MgO-H_2O hybrid nano-suspended stream through a porous wavy circular compartment and noted a drop in heat transmission for $Ra = 10^5$, $Ha = 50$. Ahmed and Raizah [38] simulated Casson hybrid nanofluid motion within a heat-generated porous triangular chamber and claimed that an engine-oil hybrid nanofluid medium extracts higher heat transmission. Islam et al. [39] illustrated the flow of carbon nanotubes through a non-uniformly warm triangular compartment. They perceived periodic flow patterns were due to the non-uniformly heated wall. Abdelhak et al. [40] depicted the CuO-water flow through various triangular regions. The outcomes revealed heat transport deterioration for magnetic strength but enhancement for enhanced concentration. More outcomes are presented in [41–44].

An impressive hydrothermal variation is spotted when a heated object or heat source is fitted in a geometric enclosure [45, 46]. Normally, heat-generating objects do exist within electronic and technical devices. The heated object's position (central, corner, topmost, or lower), shape (square, arc, triangular, wavy, spiral, etc.), and size (small, big, or medium) affect the interrelated transition [47]. Some useful arc-shaped heater applications are traced in solar heaters, reactors, fuel oil, electronic devices, storage devices, etc. [48–50]. Roy [51] examined unsteady magnetized Al_2O_3-Cu-water flow through a square duct with multiple heat sources fitted at the lower boundary. Results ensured enhancement in heat transmission with the number of heat sources used. Gul et al. [52] deliberated Al_2O_3-Cu-water flow within a porous elliptical chamber containing a circular heated source and perceived heat transport improvement for both Rayleigh and Darcy numbers. Manna et al. [53] worked on irreversibility variations of hybrid nanofluids for the discrete heating–cooling strategy of a heated circular chamber. They observed the best heat transmission for the heater–cooler's third and second quadrant arrangements. Mahalakshmi [54] investigated the variations in Ag-CuO-water, Ag-TiO_2-water, and Ag-MgO-water nanofluids within lid-driven square compartments with heat sources. Maximum heat transport was observed for the Ag-CuO-water combination. Thirumalaisamy and Ramachandran [55] analysed the influence of heat sink/source on Fe_3O_4-H_2O and Fe_3O_4-Cu-H_2O nanofluids confined within a tilted square chamber. The study claims heat transfer amplification for heat sources. Mansour and Bakier [56] numerically scrutinized mixed convective magnetically driven Cu-TiO_2-water flow through a wavy tilted double lid-driven chamber with heat sources. Results confirmed suppression in the flow for magnetic strength. Similar reports are presented in [57–60].

Motivated by previous investigations, this numeric attempt investigated the buoyancy-driven motion of Al_2O_3-TiO_2-water nanofluids inside a triangular chamber fitted with a circular heater below. The boundary surface, except the circular heater, was set as adiabatic, whereas the other triangular sections were made cold. The Galerkin finite element method was used to run the simulation. Both experimental and numeric simulation-based studies are presented to compare the current investigation's accuracy. Various plots are rendered to exhibit the hydrothermal transition of hybrid nanofluids within the confined geometry. Al_2O_3-based nanofluids are significantly used in solar receivers, heat exchangers, electronic cooling, biomedical applications, heat pipes, etc. [61, 62], whereas TiO_2-based nanofluids are widely applicable in heat exchangers, photovoltaic systems, automobile cooling, etc. [63, 64]. Thus, the key queries that will be extracted through this attempt are as follows:

- What are the changes in velocities, isotherms, and streamlines for the variation of magnetic strength, Rayleigh numbers, and nanoparticle concentrations?
- How does the size of circular heaters affect the hydrothermal scenario within a triangular chamber?
- How do the heater's size and other flow parameters affect heat transport?

6.2 MATHEMATICAL DESCRIPTION

6.2.1 Problem Statement and Foremost Equations

This investigation assumes a steady, incompressible, laminar, and buoyancy-driven hybrid nanofluidic transportation through a triangular chamber whose vertices are named as (A, B, C) in Figure 6.1 and whose specific geometric locations are listed in Table 6.1. A circular heater with radius $R = 0.20\,L$ is fitted at the triangular bottom wall. Three different sizes of circular heaters with radii $R = 0.20\,L$, $0.25\,L$, $0.30\,L$ are set to inspect the hydrothermal transitions that occurred within the chamber. The thermal setup along the boundary is restricted as if the inclined triangular sides are made cold at temperature $T = T_c$, the circular heater is heated at temperature $T = T_h (> T_c)$, and the remaining bottom walls remain insulated. A uniform magnetic strength B_0 acts horizontally, and the well-known gravitational force g acts downwards. The surrounding chamber's walls and the circular heater are supposed to be impermeable. To override the induced magnetic impact, the classic magnetic Reynolds number is thought to be insignificantly small. The interrelated joule heating, viscous dissipation, and any slips are disregarded during the model simulation. Moreover, the dispersed tiny nanoparticles maintain a thermal balance with the base liquid and remain to refrain from any chemical reactions. After integrating the classic Boussinesq approximation owing to the buoyancy effect, the leading two-dimensional flow equations transpire as [11, 16, 17, 34–36] follows:

Continuity equation

$$\frac{\partial u}{\partial x} + \frac{\partial v}{\partial y} = 0 \tag{6.1}$$

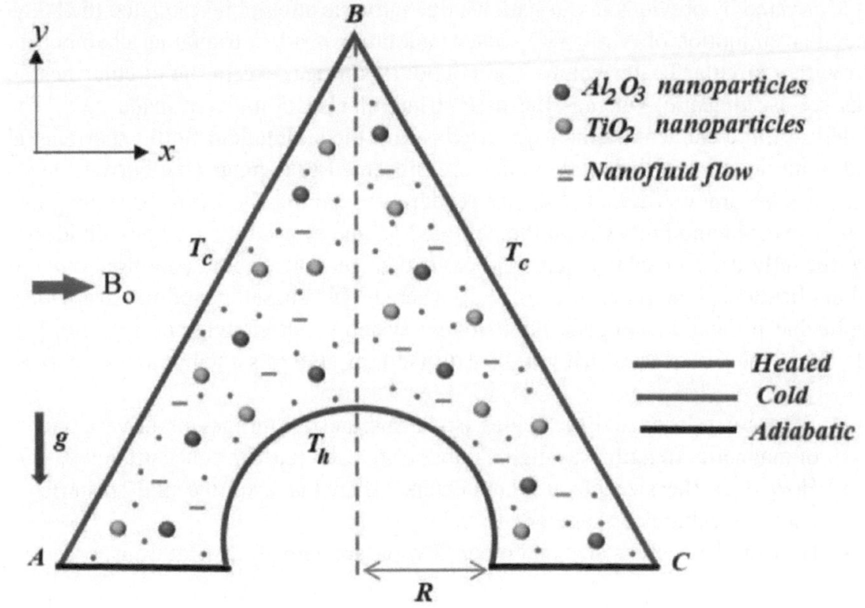

FIGURE 6.1 Schematic of the hybrid nanofluid flow.

Momentum equations

$$u\frac{\partial u}{\partial x}+v\frac{\partial u}{\partial y}=-\frac{1}{\rho_{hnf}}\frac{\partial p}{\partial x}+\frac{\mu_{hnf}}{\rho_{hnf}}\left(\frac{\partial^2 u}{\partial x^2}+\frac{\partial^2 u}{\partial y^2}\right) \tag{6.2}$$

$$u\frac{\partial v}{\partial x}+v\frac{\partial v}{\partial y}=-\frac{1}{\rho_{hnf}}\frac{\partial p}{\partial y}+\frac{\mu_{hnf}}{\rho_{hnf}}\left(\frac{\partial^2 v}{\partial x^2}+\frac{\partial^2 v}{\partial y^2}\right)+g\frac{(\rho\beta)_{hnf}}{\rho_{hnf}}(T-T_c)-\frac{\sigma_{hnf}}{\rho_{hnf}}B_0^2 v \tag{6.3}$$

Energy equation

$$u\frac{\partial T}{\partial x}+v\frac{\partial T}{\partial y}=\frac{\kappa_{hnf}}{(\rho C_p)_{hnf}}\left(\frac{\partial^2 T}{\partial x^2}+\frac{\partial^2 T}{\partial y^2}\right) \tag{6.4}$$

TABLE 6.1
Vertices Location of the Triangular Chamber

Vertices name	A	B	C
Geometric location	$(0,0)$	$(0.5\,L, L)$	$(L,0)$

The predetermined boundary restrictions are mathematically noted as follows:

Along the two inclined triangular walls, $u=0,\ v=0,\ T=T_c$. (6.5)
Along the circular designed heater, $u=0,\ v=0,\ T=T_h$. (6.6)
Along the remaining parts of the lowermost walls, $u=0,\ v=0,\ \dfrac{\partial T}{\partial n^*}=0$. (6.7)

6.2.2 Thermophysical Properties

During simulation, working hybrid nanofluids consist of water, 30-nm TiO_2, and 43-nm Al_2O_3 tiny nanoparticles. Thermophysical data of the hybrid nanofluid's ingredients are enumerated in Table 6.2. Noteworthy thermophysical correlations of both mono- and hybrid nanofluids at the 20°C–30°C reference temperature are listed in Table 6.3. Experimentally verified thermal conductivity and dynamic viscosity correlations prescribed by Moldoveanu et al. [29, 30] are integrated.

6.2.3 Similarity Transformation

To implement the non-dimensionalization procedure, the below similarity variables [11, 16, 17, 34–36] are introduced within Equations (6.1)–(6.4):

$$\left.X=\frac{x}{L},Y=\frac{y}{L},AR=\frac{R}{L},U=\frac{uL}{\alpha_f},V=\frac{vL}{\alpha_f},P=\frac{pL^2}{\rho_f\alpha_f^2},\theta=\frac{T-T_c}{T_h-T_c}\right\}. \tag{6.8}$$

Consequently, the foremost dimensional equations as in Equations (6.1)–(6.4) are translated as

$$\frac{\partial U}{\partial X}+\frac{\partial V}{\partial Y}=0, \tag{6.9}$$

$$U\frac{\partial U}{\partial X}+V\frac{\partial U}{\partial Y}=-\frac{1}{\Sigma_1}\frac{\partial P}{\partial X}+\frac{\Sigma_2}{\Sigma_1}\Pr\left(\frac{\partial^2 U}{\partial X^2}+\frac{\partial^2 U}{\partial Y^2}\right), \tag{6.10}$$

TABLE 6.2
Thermophysical Properties of the Base Liquid and Al_2O_3 and TiO_2 Nanoparticles [29, 30]

Physical properties	H_2O	Al_2O_3	TiO_2
C_p (J Kg^{-1} K)	4179	765	686.2
ρ (Kg m^{-3})	997.1	3970	4250
κ (W mK^{-1})	0.613	40	8.9538
$\sigma(\Omega m)^{-1}$	0.05	35×10^{-6}	2.38×10^{6}
β (1 K^{-1})	21×10^{-5}	0.846×10^{-5}	0.9×10^{-10}

TABLE 6.3
Thermophysical Models of Mono-Nanofluid and Hybrid Nanofluid [29, 30, 34, 36]

Properties	Mono-nanofluid (Al_2O_3-H_2O)
Density	$\rho_{nf}=(1-\phi)\rho_f+\phi\rho_s$
Heat capacity	$(\rho C_p)_{nf}=(1-\phi)(\rho C_p)_f+\phi(\rho C_p)_s$
Viscosity	$\mu_{nf}=(1-\phi)^{-2.5}\mu_f$
Thermal conductivity	$\dfrac{\kappa_{nf}}{\kappa_f}=\dfrac{\kappa_s+2\kappa_f-2\phi(\kappa_f-\kappa_s)}{\kappa_s+2\kappa_f+\phi(\kappa_f-\kappa_s)}$
Thermal expansion coefficient	$(\rho\beta)_{nf}=(1-\phi)(\rho\beta)_f+\phi(\rho\beta)_s$
Electrical conductivity	$\dfrac{\sigma_{nf}}{\sigma_f}=1+\dfrac{3(\sigma-1)\phi}{(\sigma+2)-(\sigma-1)\phi}$, *where* $\sigma=\dfrac{\sigma_s}{\sigma_f}$
Properties	**Hybrid nanofluid (Al_2O_3-TiO_2-H_2O)**
Density	$\rho_{hnf}=(1-\phi_{hnf})\rho_f+\phi_1\rho_1+\phi_2\rho_2$
Heat capacity	$(\rho C_p)_{hnf}=(1-\phi_{hnf})(\rho C_p)_f+\phi_1(\rho C_p)_1+\phi_2(\rho C_p)_2$
Viscosity	$\mu_{hnf}=\mu_f(2.06-1.32\phi_1-0.96\phi_2+0.58\phi_1^2+0.39\phi_2^2+1.89\phi_1\phi_2)$, where $0.00\le\phi_1,\phi_2\le 0.04$
Thermal conductivity	$\kappa_{hnf}=\kappa_f\begin{pmatrix}0.995+10.097\phi_1-120.835\phi_1^2+23.227\phi_2\\-43.648\phi_2^2+22380.350\phi_2^3\end{pmatrix}$, where $0.00\le\phi_1,\phi_2\le 0.04$
Thermal expansion coefficient	$(\rho\beta)_{hnf}=(1-\phi_{hnf})(\rho\beta)_f+\phi_1(\rho\beta)_1+\phi_2(\rho\beta)_2$
Electrical conductivity	$\dfrac{\sigma_{hnf}}{\sigma_f}=1+\dfrac{3\left(\dfrac{\phi_1\sigma_1+\phi_2\sigma_2}{\sigma_f}\right)-3\phi_{hnf}}{2+\left\{\dfrac{\phi_1\sigma_1+\phi_2\sigma_2}{\phi_{hnf}\sigma_f}\right\}-\left\{\dfrac{\phi_1\sigma_1+\phi_2\sigma_2}{\sigma_f}-\phi_{hnf}\right\}}$

$$U\frac{\partial V}{\partial X}+V\frac{\partial V}{\partial Y}=-\frac{1}{\Sigma_1}\frac{\partial P}{\partial Y}+\frac{\Sigma_2}{\Sigma_1}\Pr\left(\frac{\partial^2 V}{\partial X^2}+\frac{\partial^2 V}{\partial Y^2}\right)+\frac{\Sigma_3}{\Sigma_1}Ra\Pr\theta-\frac{\Sigma_4}{\Sigma_1}\mathrm{Ha}^2\Pr V, \quad (6.11)$$

$$U\frac{\partial\theta}{\partial X}+V\frac{\partial\theta}{\partial Y}=\frac{\Sigma_5}{\Sigma_6}\left(\frac{\partial^2\theta}{\partial X^2}+\frac{\partial^2\theta}{\partial Y^2}\right), \quad (6.12)$$

where $\Sigma_1 = \frac{\rho_{hnf}}{\rho_f}, \Sigma_2 = \frac{\mu_{hnf}}{\mu_f}, \Sigma_3 = \frac{(\rho\beta)_{hnf}}{(\rho\beta)_f}, \Sigma_4 = \frac{\sigma_{hnf}}{\sigma_f}, \Sigma_5 = \frac{\kappa_{hnf}}{\kappa_f}, \Sigma_6 = \frac{(\rho C_p)_{hnf}}{(\rho C_p)_f}$,

$\phi_{hnf} = \phi_1 + \phi_2$.

The fixed boundary constraints have been translated to non-dimensional forms as follows:

Along the two inclined triangular walls, $U = 0,, V = 0, \theta = 0$. (6.13)

Along the circular heater, $U = 0, V = 0, \theta = 1$. (6.14)

Along the remaining parts of the lowermost walls, $U = 0, V = 0, \frac{\partial \theta}{\partial n^*} = 0$. (6.15)

6.2.4 Physical Quantities

Heat transfer outcomes are extremely imperative for designing thermal devices or equipment, and Nusselt number plots or numeric outcomes help engineers visualize or predict the heat transport scenario and how to control the flow parameters to avail better heat transfer. Along the semi-circular heater, the Nu_{loc} is defined as

$$Nu_{loc} = -\frac{\kappa_{hnf}}{\kappa_f}\frac{\partial \theta}{\partial n}, \tag{6.16}$$

where n indicates the outward-directed normal. Thus, Nu_{avg} is expressed as

$$Nu_{avg} = \frac{1}{\pi}\int_0^{\pi} Nu_{loc} d\varsigma . \tag{6.17}$$

6.3 SOLUTION METHODOLOGY AND VALIDATION

The classic renowned Galerkin finite element method [65] has been integrated with the dimensionless flow Equations (6.9)–(6.15). The flow province is firstly discretized into several small triangular sections, and high mesh density subregions are used to avail the solution at desired accuracy. The domain discretization or mesh formation is depicted in Figure 6.2.

The penalty factor $\lambda\left(\approx 10^7\right)$ is used to deal with the pressure term with the equation $P = -\lambda\left(\frac{\partial U}{\partial X} + \frac{\partial V}{\partial Y}\right)$. At every node of the discretized triangular element, the Galerkin method is introduced and thus one gets the non-linear residuals. Consequently, the interconnected element equations for respective nodes are combined, and thereby a global set of equations is obtained in which the Newton–Raphson iterative method is merged with those boundary predefined restrictions. The convergence limit of the iterative method is set to 10^{-6}. The detailed formulation and methodological mechanism of finite element method (FEM) are well demonstrated in [65], so further recurrence is not entertained here for the conciseness of the chapter demonstration. Rather, a flowchart related to FEM is provided in Figure 6.3.

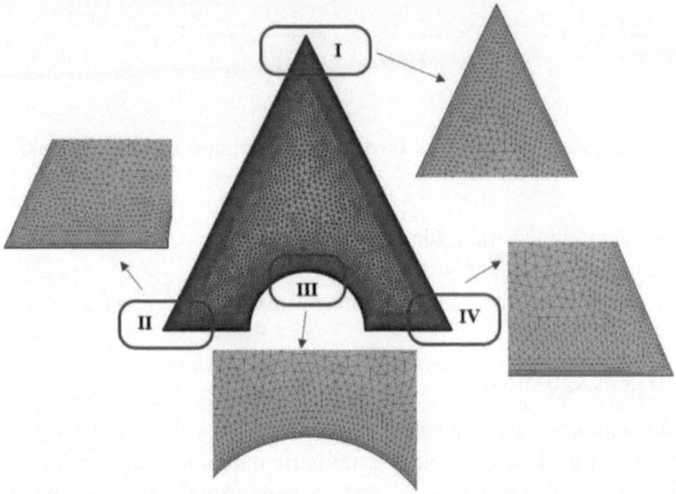

FIGURE 6.2 Grid formation and large view of some subregions.

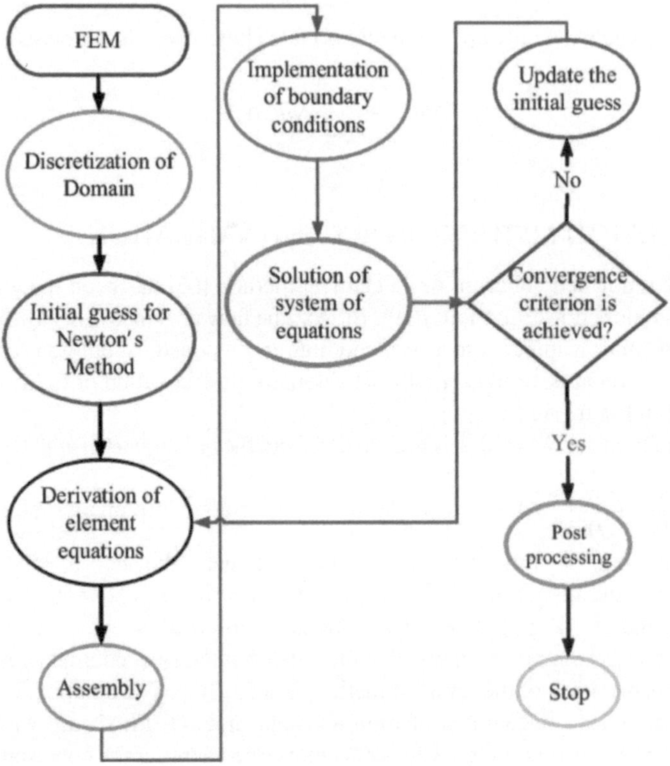

FIGURE 6.3 Flowchart of FEM.

One most vital requirement of such cavity flow simulation lies in the grid independence test, i.e., the numeric outcomes must be grid-independent. One can choose several grids to present the outcomes, but the most fitted and accurate approach is to verify which grid maintains the numeric consistency, and that grid would be the most fitted grid or mesh to serve the purpose. Table 6.4 enlightens that the numeric consistency in Nu_{avg} for hybrid Al_2O_3-TiO_2-water nanofluids is spotted after grid G2, but the major problem of selecting a highly dense grid G3 is that a highly dense grid will slow down the computational procedure; thus, such a grid will be time-consuming. Hence, the most suitable and timely fitted grid for the entire simulation is G2. Both numerical and experimental validations are clutched here to validate the model setup. The experimental validation executed by Yesiloz and Aydin [11] is depicted and compared with this numeric model-generated streamline plot in Figure 6.4 for $Ra = 10^5$, $Ra = 5\times10^5$, $Ra = 10^6$, $Ra = 5\times10^6$, $Ra = 10^7$, and $\phi = 0.0$. Also, isotherms are portrayed in Figure 6.5 for the same aforementioned Rayleigh numbers, and in both cases, outcomes are completely in agreement with the study by Yesiloz and Aydin [11].

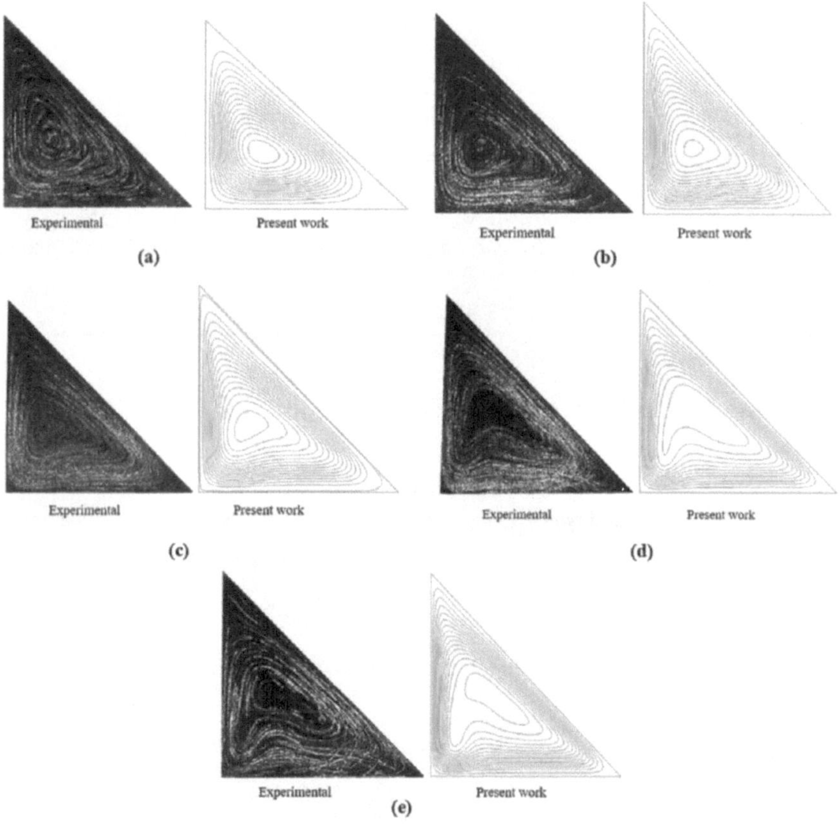

FIGURE 6.4 Comparison of streamlines with the experimental [11] and present work for (a) $Ra = 10^5$, (b) $Ra = 5\times10^5$, (c) $Ra = 10^6$, (d) $Ra = 5\times10^6$, and (e) $Ra = 10^7$.

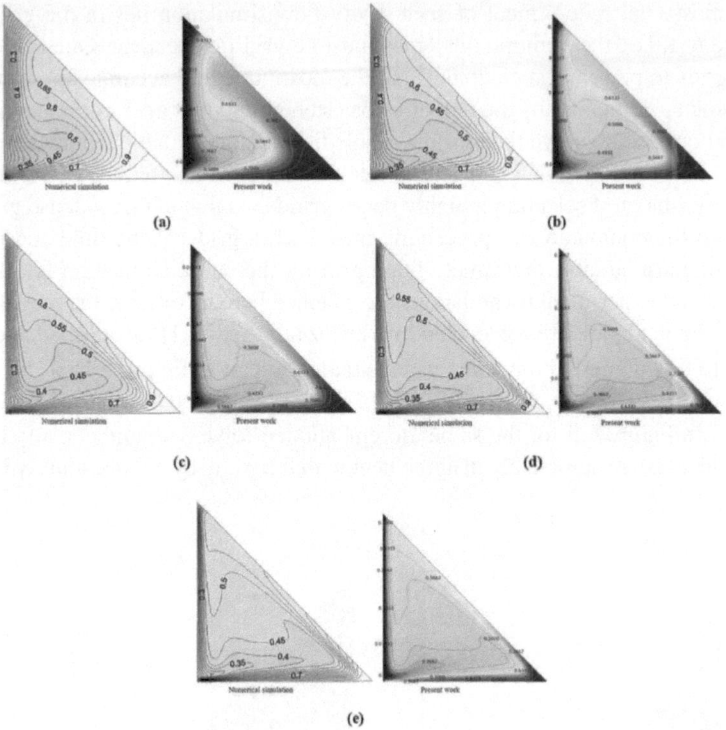

FIGURE 6.5 Comparison of isotherms with numerical simulation [11] and the present work for (a) $Ra = 10^5$, (b) $Ra = 5\times10^5$, (c) $Ra = 10^6$, (d) $Ra = 5\times10^6$, and (e) $Ra = 10^7$.

TABLE 6.4
Effect of Mesh Density on Nu_{avg} for $Ra = 10^5$, $Ha = 40$ for Al_2O_3-TiO_2/Water Nanofluid

Configuration	Grid type	Number of elements	Nu_{avg}	Time (sec)
$AR = 0.20$	G1	4536	4.4401	11
$AR = 0.20$	G2	12462	4.4529	12
$AR = 0.20$	G3	15070	4.4525	14
$AR = 0.25$	G1	4530	4.5455	15
$AR = 0.25$	G2	12478	4.4565	17
$AR = 0.25$	G3	14590	4.4562	19
$AR = 0.30$	G1	4298	5.1619	12
$AR = 0.30$	G2	12034	5.1725	15
$AR = 0.30$	G3	13834	5.1722	18

6.4 RESULTS AND DISCUSSION

This section describes the detailed hydrothermal orientation of the Al_2O_3-TiO_2/water hybrid nanofluids when the liquid is running through the triangular chamber in which a semi-circular heater is fitted below. The sizes of the fitted circular heater are varied to perceive the hydrothermal changes. The requisite number of velocities, isotherms, streamlines, and heat transfer outcomes are illustrated through various charts and graphs. The default choice of flow constraints is set as $10^4 \leq Ra \leq 10^6$, $0 \leq Ha \leq 100, 0.00 \leq \phi_2 \leq 0.02$. Also, the fixed value of specific parameters during simulation is set as $Ra = 10^5$, $Ha = 40$, $\phi_1 = 0.01$, $\phi_2 = 0.01$, and $Pr = 6.2$.

6.4.1 Impact of Rayleigh Number

Figure 6.6 discloses the streamlines and U-velocity variation for several inputs of Ra and various semi-circular heater sizes. For $AR = 0.20$ and $Ra = 10^4$, two intense circular regimes are spotted on either lower side of the triangular chamber. Initially, high circular intensity at the lower walls is detected for $Ra = 10^4$, while the upper portion exhibits lower circular intensity. But enhancing Rayleigh numbers from $Ra = 10^4 \rightarrow Ra = 10^5$, the lower intense circular regimes tend to migrate towards the upper corner; i.e., the circular intensity seems to stretch along the upper conic section. Again for $Ra = 10^6$, the streamlines exhibit high circulation throughout the triangular compartment. The uniform circular high-intense regimes at the lower portion completely move towards the upper conic region. The conductive heat transmission seems to dominate mostly at low Rayleigh numbers ($Ra = 10^4$); thus, less intensified streamlines are detected. But buoyancy-driven convection current dominates for higher Rayleigh numbers ($Ra = 10^6$); hence, more concentrated and highly intense circulations are perceived. The liquid close to the semi-circular heater gets warmed, becomes less dense. and migrates upwards, whereas the upgoing liquid becomes highly dense when it comes to the cold slant sides of the triangular chamber and gravitational acceleration makes it to fall. Such a scenario gets enhanced or intensified for high Rayleigh numbers. But when the circular heater's size is fitted to $AR = 0.25$, then the circulation intensity drops off. For $AR = 0.25$ and $Ra = 10^4$, the circulating zones are traced at a similar place as it was for $AR = 0.20$, but the circulation intensity undoubtedly breaks down. For $AR = 0.25$ and $Ra = 10^5$, the circulation tends to go upwards, and a slight augmentation in circulations is received. Highly concentrated streamlines are spotted for $Ra = 10^6$, but the circulation intensity seems still low compared to that for $AR = 0.20$. Thus, increasing the heater's size reduces the buoyancy-assisted convection within the compartment. The circulation intensity gradually declines with the circular heater size. Hence, a slightly lower intensity is revealed for $AR = 0.30$ and $Ra = 10^4$ compared to that for $AR = 0.25$. One interesting scenario is detected when the heater's size is again enhanced and fitted to $AR = 0.30$; two small secondary circulating regimes are spotted near the bottom wall, and they are primarily connected with larger circulating zones. When the parametric shift $Ra = 10^4 \rightarrow Ra = 10^5$ is made, then the intensity of both smaller and larger circulation upsurges, and finally for $Ra = 10^6$, such small secondary

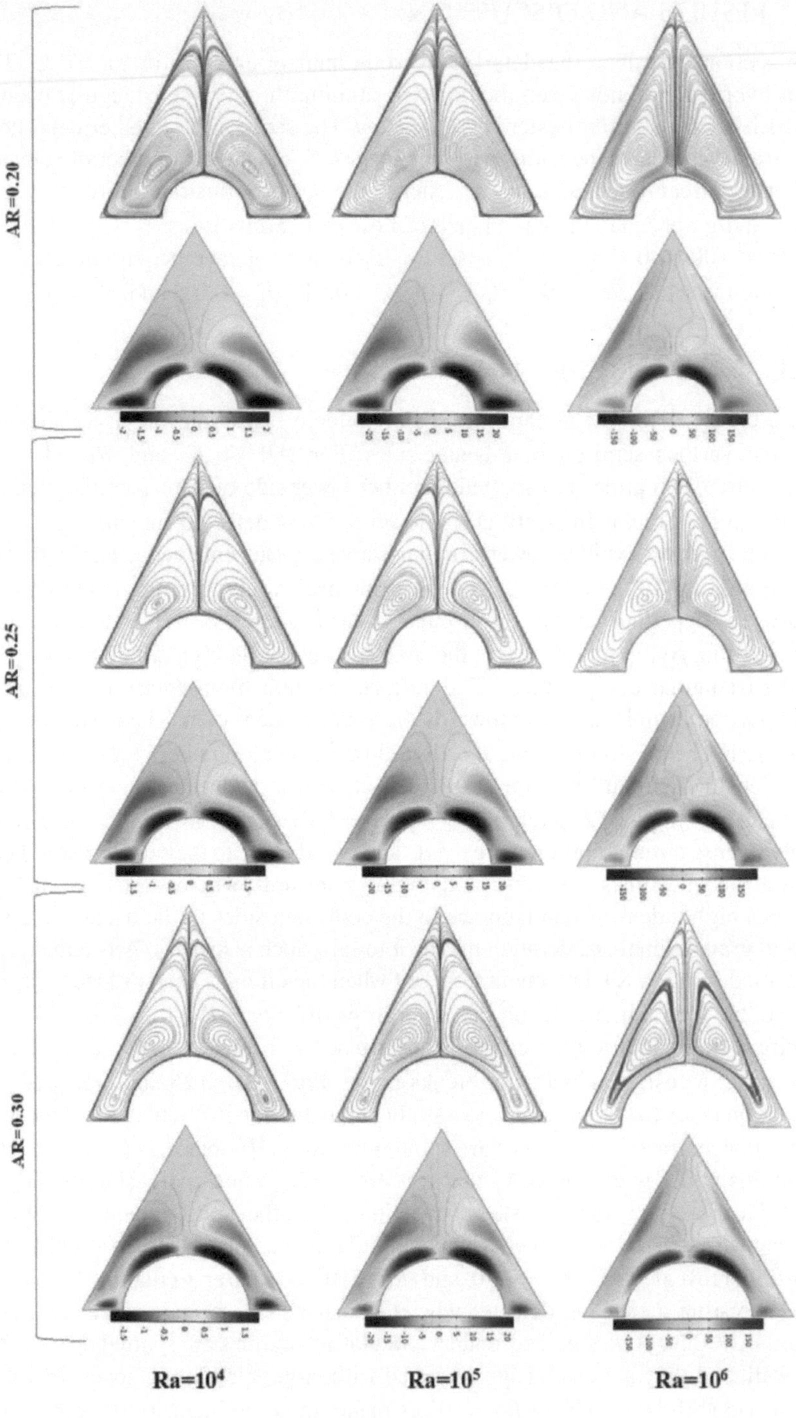

FIGURE 6.6 Streamline variation (1st, 3rd, and 5th rows) and velocity U variation (2nd, 4th, and 6th rows) for varied AR when Ra varies and $Ha = 40$, $\phi_1 = 0.01$, and $\phi_2 = 0.01$.

circulation vanishes and the big circulating regimes' intensity becomes high for the higher buoyancy-driven convection. But those circulation seems to be squeezed a little bit compared to those for $AR = 0.25$ and $AR = 0.20$. The x-directional velocity profiles, i.e., U, seem to increase with Ra. Initially, when $AR = 0.20$ and $Ra = 10^4$, the maximum magnitude of U velocity stretches over the left lower wall and left circular periphery, including the lower half of the right inclined wall. Similarly, the minimum magnitude of U velocity regimes resides along the opposite face of the maximum zones. The U velocity is amplified with increasing Rayleigh numbers followed by the enhanced buoyancy-driven current. The minimum velocity regimes drop their strength and are mostly covered by relatively high-velocity regions. Also, due to buoyancy-induced convection, the high U velocity stretches within the inner chamber and can be spotted through the right inclined triangular wall plus along the left circular periphery for $Ra = 10^6$. The U-profile behaves on average a similar trend for $AR = 0.25$ and $AR = 0.30$, but the lowest U-magnitude is perceived for $AR = 0.30$.

The velocity V and velocity magnitude distribution profiles are presented in Figure 6.7. Due to a like cause as mentioned earlier, the V-profile is also amplified. Maximum V-velocity magnitude regimes for $AR = 0.20$ and $Ra = 10^4$ are spotted along the heated circular periphery, whereas the minimum zones remain along the two cold inclined walls. But with improved buoyancy-assisted convection current followed by Ra, the higher velocity zones started to migrate upwards and minimum zones too. A red-shaded plume-shaped high-velocity precinct is noted near the circular heater's top surface. For other circular heaters' size, i.e., for $AR = 0.25$ and $AR = 0.30$, the V-profiles act more or less similarly, but the significant point is that velocity magnitude or intensity declines gradually with the heater's size. Thus, lower magnitude is perceived for $AR = 0.30$. The distribution of velocity magnitude in Figure 6.7 follows a similar trajectory. The maximum magnitude is observed near the circular arc of the heater, lower wall, and lower portion of the inclined walls, while within the other area of the chamber, comparatively lower profiles are spotted. But buoyancy-driven convection enhances the magnitude distribution and pushes the maximum magnitude towards the upper section. Additionally, the plume-shaped maximum magnitude region on the circular heater tip seems to appear gradually with Rayleigh number enhancement, but the maximum is detected for $Ra = 10^6$ and $AR = 0.20$.

Figure 6.8 explores the isotherms within the triangular compartment. The isotherms are spotted to be aligned along the heated circular periphery at $Ra = 10^4$. No significant thermal distortion is perceived, but slight shifting from $Ra = 10^4 \rightarrow Ra = 10^5$ causes little distortion in isotherms. A little plume-shaped distortion is traced out near the circular heater tip. Also, for $Ra = 10^6$, those aligned isotherms near the circular periphery are scattered completely, and maximum thermal distortion is witnessed near the circular heater's tip. The little plume-shaped zone that was initially noted for $Ra = 10^5$ is now clearly traceable and gets maximized. Hence, a higher thermal profile can be anticipated within the chamber. Such massive distortion is the effect of high convective motion within the chamber for increasing Ra. The smaller size circular heater $(AR = 0.20)$ causes a slightly higher thermal distortion. The Nu_{loc} in Figure 6.8 is positively connected with Ra. Although initially,

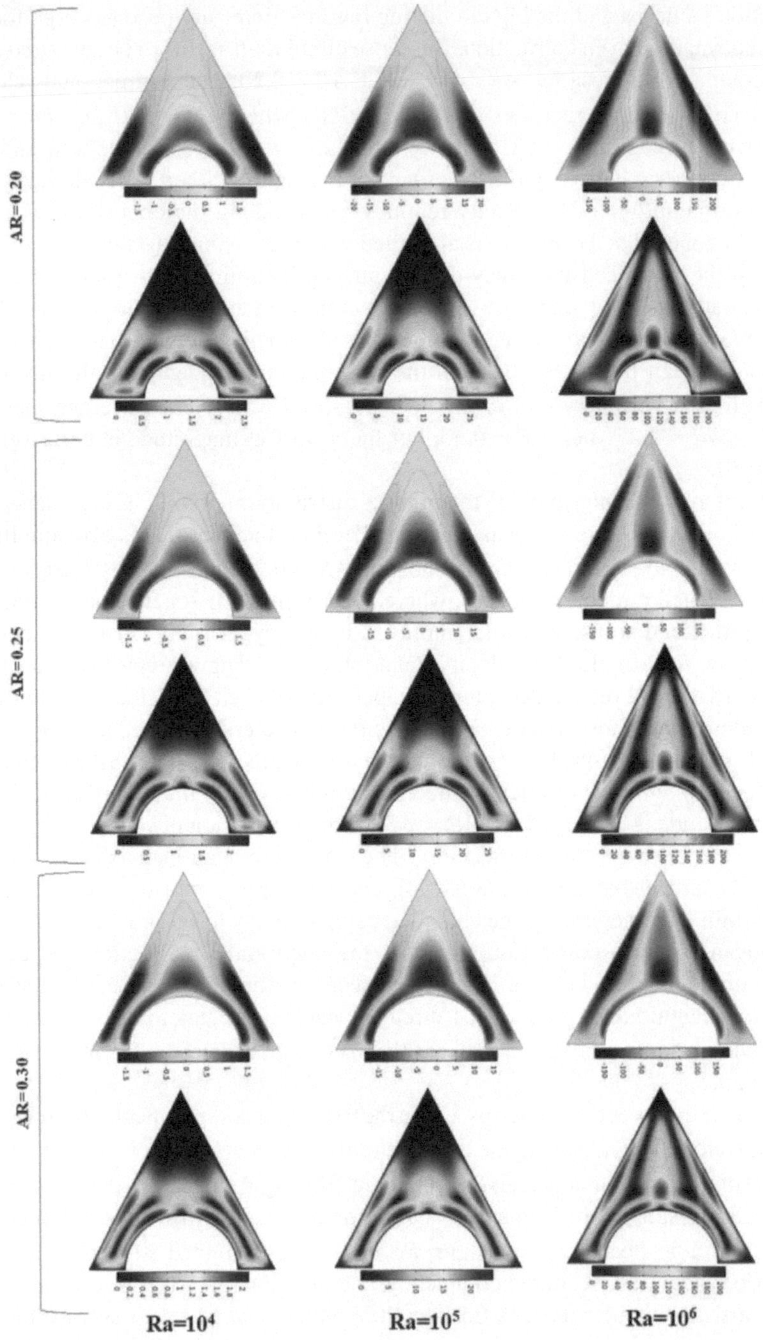

FIGURE 6.7 Velocity V variation (1st, 3rd, and 5th rows) and velocity magnitude distribution (2nd, 4th, and 6th rows) for varied AR when Ra varies and $Ha = 40$, $\phi_1 = 0.01$, and $\phi_2 = 0.01$.

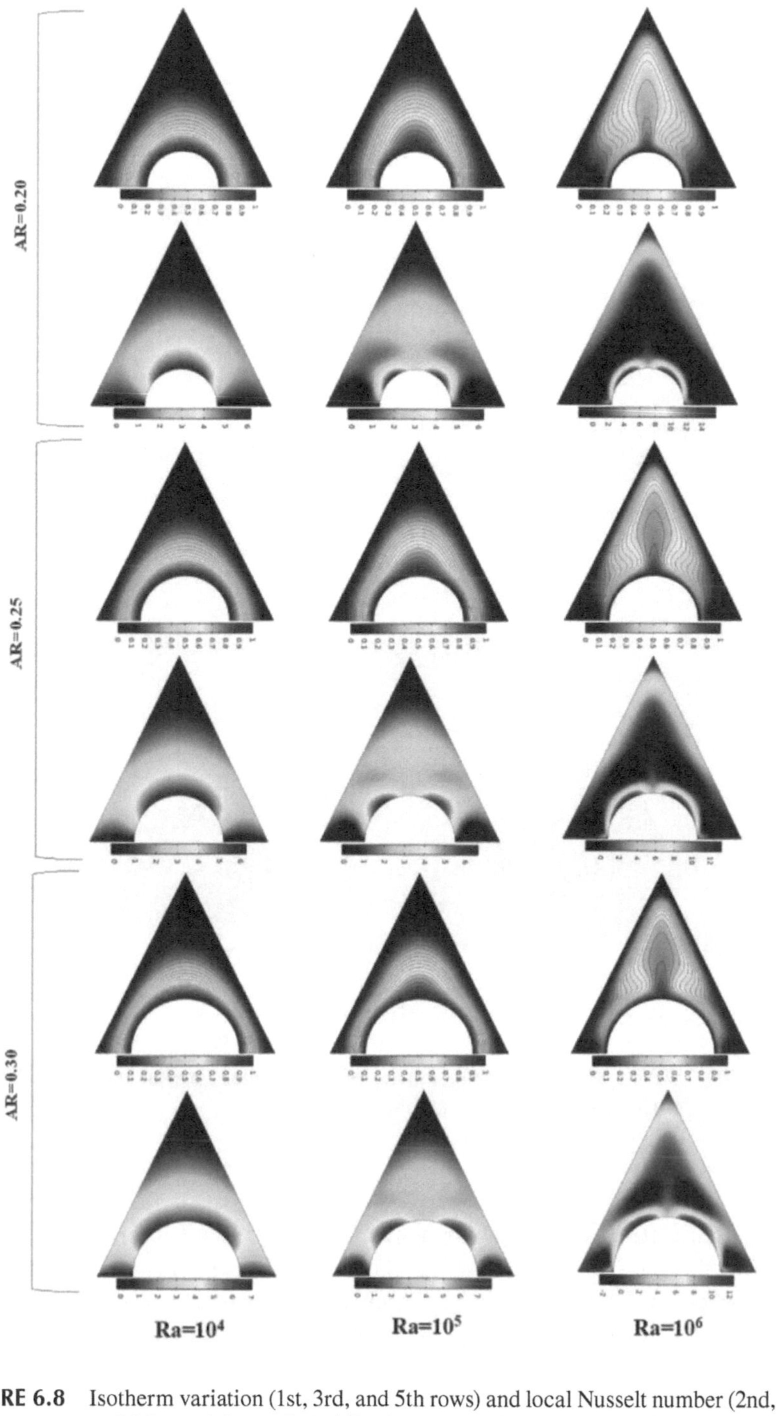

FIGURE 6.8 Isotherm variation (1st, 3rd, and 5th rows) and local Nusselt number (2nd, 4th, and 6th rows) for varied AR when Ra varies and $Ha = 40$, $\phi_1 = 0.01$, and $\phi_2 = 0.01$.

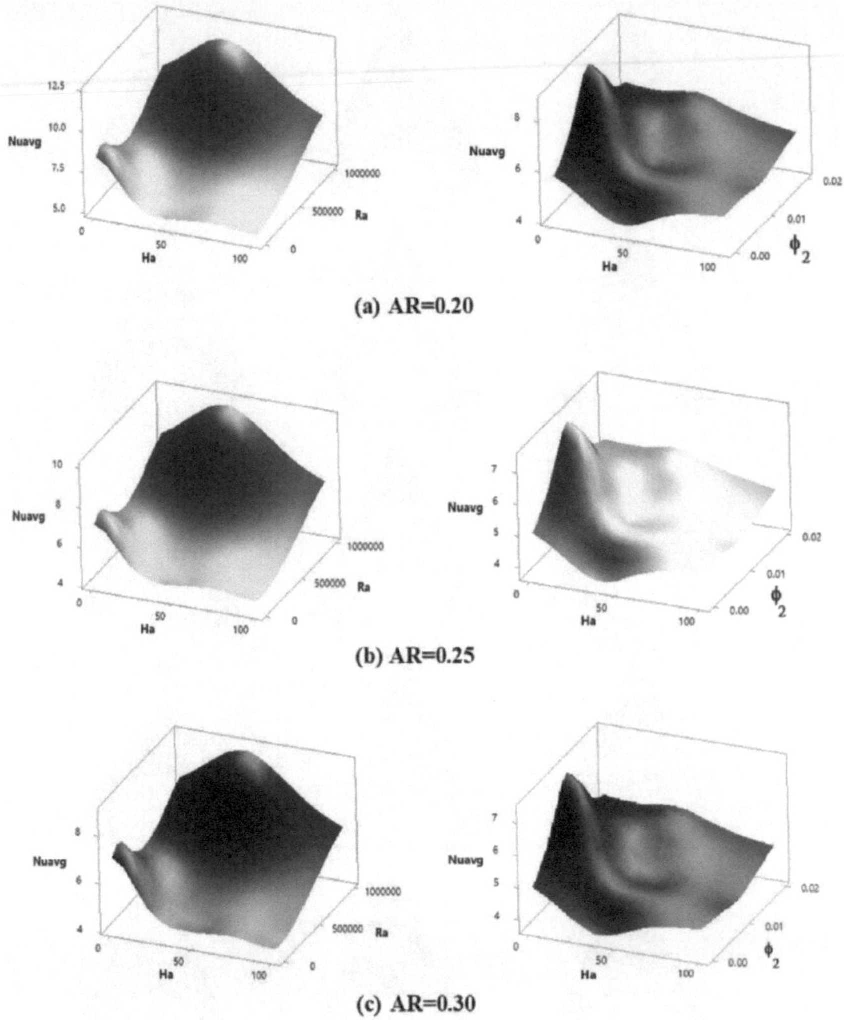

FIGURE 6.9 Effects of Rayleigh number, Hartmann number, and nanoparticle concentrations on average Nusselt number.

for $10^4 \leq Ra \leq 10^5$ a slightly high magnitude in Nu_{loc} is noted for a large circular heater, but for $Ra > 10^5$ maximum magnitude in Nu_{loc} is predicted for $AR = 0.20$. The same outcomes are assured in Figure 6.9 for Nu_{avg}.Approximately 18.22% higher Nu_{loc} is predicted for $Ra = 10^4$ and $AR = 0.30$ compared to $AR = 0.20$, while the same is calculated as 12.46% higher than that for $AR = 0.25$. Again, the highest Nu_{avg} is recorded for a smaller heater; i.e., for $AR = 0.20$, a 31.35% enhancement in Nu_{avg} is predicted at $Ra = 10^6$.

6.4.2 Impact of Hartmann Number

Figure 6.10 delineates the streamline and velocity-U variations for the magnetic field. Two main intense circulations are noted for $Ha = 0$ and $AR = 0.20$, while for $AR = 0.25$ and $AR = 0.30$ those circulating regimes are observed to be less intense during magnetic absence. Moreover, two weak circulations near the lower boundary are observed for $AR = 0.30$ and those connected with the central circulating zones. When $Ha = 0$ is switched to $Ha = 50$, then those circulations seem to squeeze. For $AR = 0.20$ and $AR = 0.25$, those circulations are squeezed in downward directions, and those small circulating zones for $AR = 0.30$ seem to be smaller. The uniform circulation patterns break down completely when $Ha = 50$ is switched to $Ha = 100$, the streamlines are noted to be more squeezed at the lower walls. The small circulating regimes are separated from the core circulating sections for $AR = 0.30$. Such a scenario approves the buoyancy-reduced convection within the chamber. The Lorentz effect originated from applied magnetic fields resists the buoyancy current and thereby declines the streamlines' intensity. The U-profiles exhibit a decreasing trend. Initially, for $Ha = 0$, the left circular half and right upper inclined sections consume the higher velocities, while the immediate opposite faces illustrate the minimum velocity zones. The Lorentz effect and reduced buoyancy convection completely squeeze the U-velocity magnitude. The maximum drop is spotted along the circular periphery, and the previous regularity pattern has almost collapsed for $Ha = 100$. Comparatively, $AR = 0.30$ renders high reduction and $AR = 0.20$ sustains maximum magnitude. The *V*-velocity trajectories in Figure 6.11 initially produce high magnitude on the circular heater's tip and central mid position within the triangular section for $AR = 0.20$ and $Ha = 0$, while for large circular heaters, the effects are only confined within the tip portion. But enhancing magnetic strength to $Ha = 50$ and $Ha = 100$ reduces the convective motion, thus the plume-shaped high-velocity region is squeezed and mostly covers up the circular arc and minimum regions aligned near inclined walls. The resistive Lorentz force causes a huge reduction in buoyant motion, and the maximum reducing scenario is captured for larger circular heaters, i.e., $AR = 0.30$. The overall velocity magnitudes explore decreasing trends for intense magnetic strength. For $AR = 0.20$ and $Ha = 0$, the maximum velocity magnitude is perceived near the upper half of the inclined triangular walls and along the circular heater periphery, and a plume-shaped red-shaded high magnitude is spotted there, while for $AR = 0.25$ and 0.30 the maximum magnitude mostly distributes near the circular tip. But the stronger effect of the applied magnetic strength, i.e., $Ha = 0 \rightarrow 50$ and 100, destroys the regularity magnitude distribution, and a downward motion is observed. One can notice the disappearance of a high-magnitude plume-shaped red zone from the circular tip portion. The higher magnitude regimes are squeezed and only reside along the lower portions of inclined walls, circular arc, and lower bottom walls. Such a pattern ensures reduced buoyancy-driven motion within the chamber. The extreme lessening is witnessed for $AR = 0.30$.

The isotherms in Figure 6.12 seem to be distorted highly for $AR = 0.20$ and $Ha = 0$. The part near the circular heater's tip and the mid-section of the chamber exhibit high thermal distortion. But the enhancement in the acting magnetic

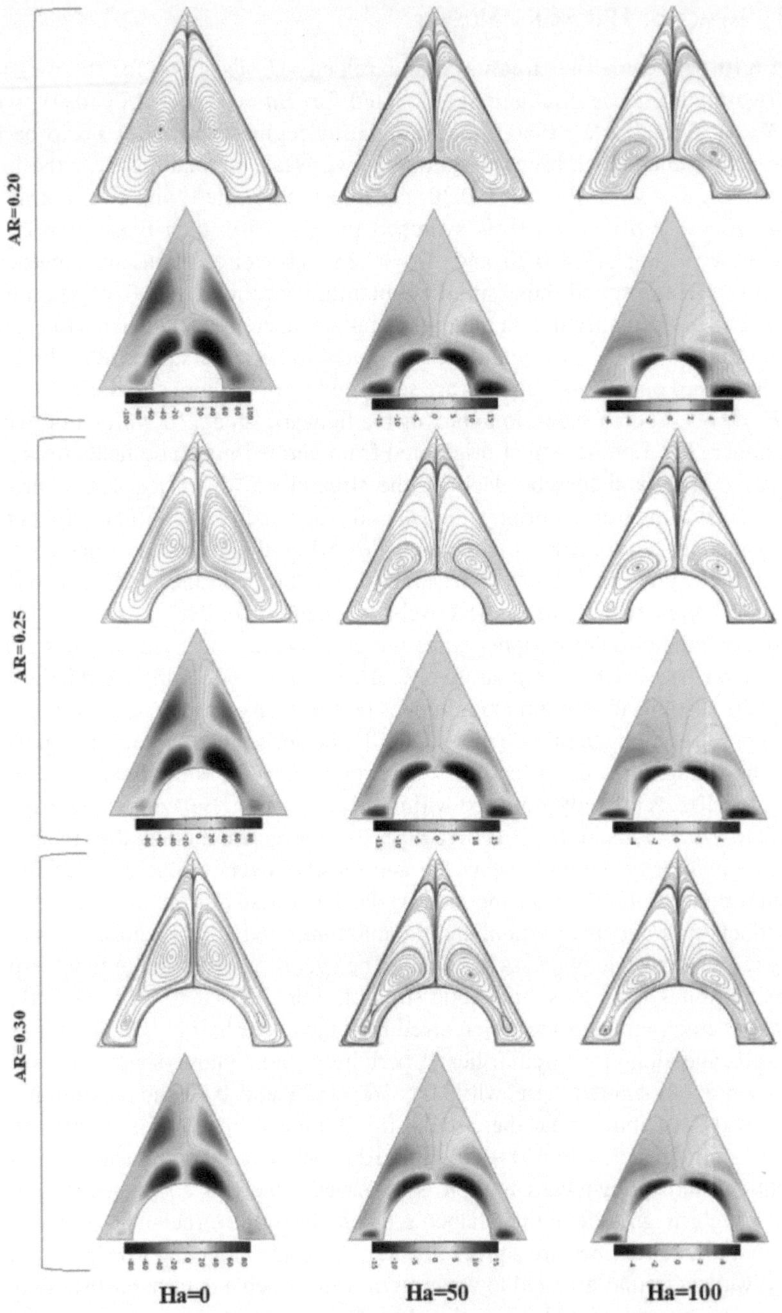

FIGURE 6.10 Streamline variation (1st, 3rd, and 5th rows) and velocity U variation (2nd, 4th, and 6th rows) for varied AR when Ha varies and $Ra = 10^5$, $\phi_1 = 0.01$, and $\phi_2 = 0.01$.

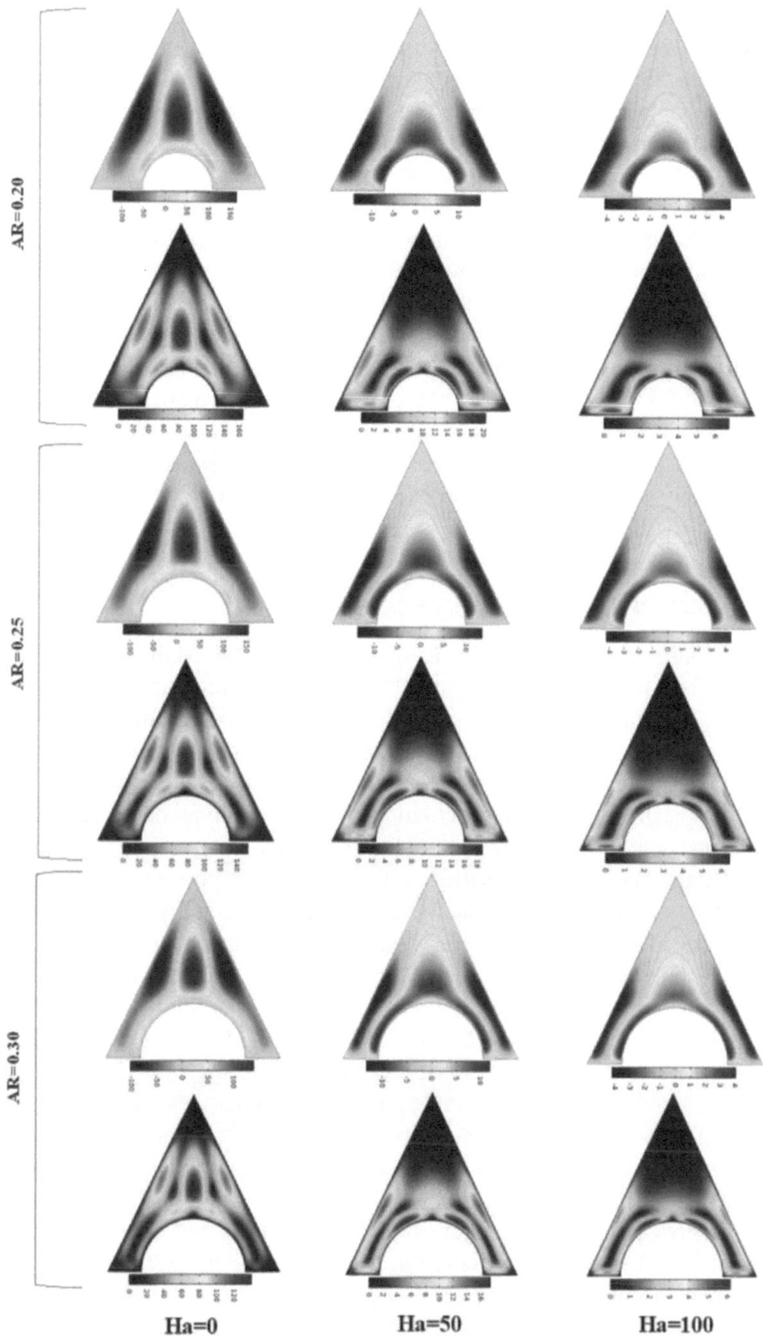

FIGURE 6.11 Velocity *V* variation (1st, 3rd, and 5th rows) and velocity magnitude distribution (2nd, 4th, and 6th rows) for varied AR when Ha varies and $Ra = 10^5$ $\phi_1 = 0.01$, and $\phi_2 = 0.01$.

field declines the isotherm distortions, and the plume-shaped distortion vanishes. Increasing the magnetic effect forces the isotherm lines to be aligned along the heated circular arc. Such a reducing scenario ensures the conductive hear transfer mode supremacy within the chamber. The interacting Lorentz force declines the vortices' strength, and thus the buoyancy-induced convective motion collapses. Hence, the thermal profile deteriorates. Since the thermal gradient due to magnetic strength is also reduced for the decrease in isotherms' distortion, heat transfer is noted to drop off, as shown in Figure 6.12. Numeric outcomes and related plots convey some noteworthy switchover results in Nu_{loc} and Nu_{avg}. For $0 \leq Ha < 50$, the smaller size circular heater, i.e., $AR = 0.20$, exhibits high Nu_{loc} compared to others, but for $50 \leq Ha \leq 100$ the reverse outcomes are predicted, i.e., high Nu_{loc} is seen for $AR = 0.30$. The Nu_{avg} in Figure 6.9 agrees with such a trend, but an additional remark is that for $0 \leq Ha \leq 60$ the Nu_{avg} declines, while a slight enhancement is spotted for $60 < Ha \leq 100$. When $0 \leq Ha < 50$, Nu_{avg} is observed to enhance by 18.49% for $AR = 0.20$ compared to $AR = 0.30$. But the opposite outcome is noted when $50 \leq Ha \leq 100$. Nu_{avg} is observed to enhance 18.49% for $AR = 0.30$ compared to $AR = 0.20$.

6.4.3 Impact of Nanoparticle Concentration

Figure 6.13 enlightens the streamline ànd U-velocity variations for nanoparticle concentration. For $AR = 0.20$ and $\phi_2 = 0.00$, two main circulating zones are noted within the triangular compartment, but enhancing nanoparticle concentrations leads to a decrease in the streamline intensity, and they are noted to squeeze at the lower boundary. Thus, mono-nanofluids explore higher intensity compared to hybrid ones. Also, for $AR = 0.25$, two weak circulating cells are received on either side of the circular heater, but intensity decreases with increased nanoparticle concentrations. The circulating zones become smaller a little bit with high concentrations. Again for $AR = 0.30$, two inner weak circulating regions are connected with two other small circulating zones at the lower boundary, but gradual enhancement in concentrations weakens the buoyancy-induced convection. The inner smooth circulating pattern seems to collapse as the resulting hybrid nanofluids' viscosity amplifies for high nanoparticle concentrations. The *U*-velocity in Figure 6.13, *V*-velocity profiles, and velocity magnitude distributions in Figure 6.14 capture the same locations within the cavity as mentioned previously in Sections 6.4.1 and 6.4.2, so further description is ignored. The mentioned profiles exhibit a deteriorating trend for higher concentrations. The maximum reduction is noted for the large circular heater and hybrid nanofluids.

The isotherm distortions in Figure 6.15 are high for small circular heaters and mono-nanofluids because higher concentrations collapse the smooth orientation of buoyancy-driven convection, and the resulting higher viscosity prevents thermal enhancement within the chamber. Heat transport amplifies with increased ϕ_2. One interesting outcome is that higher Nu_{loc} is perceived for a larger circular heater in Figure 6.15. The same was presented in Figure 6.9 for Nu_{avg}. Nu_{avg} increases 19.03% for $AR = 0.30$ compared to $AR = 0.20$.

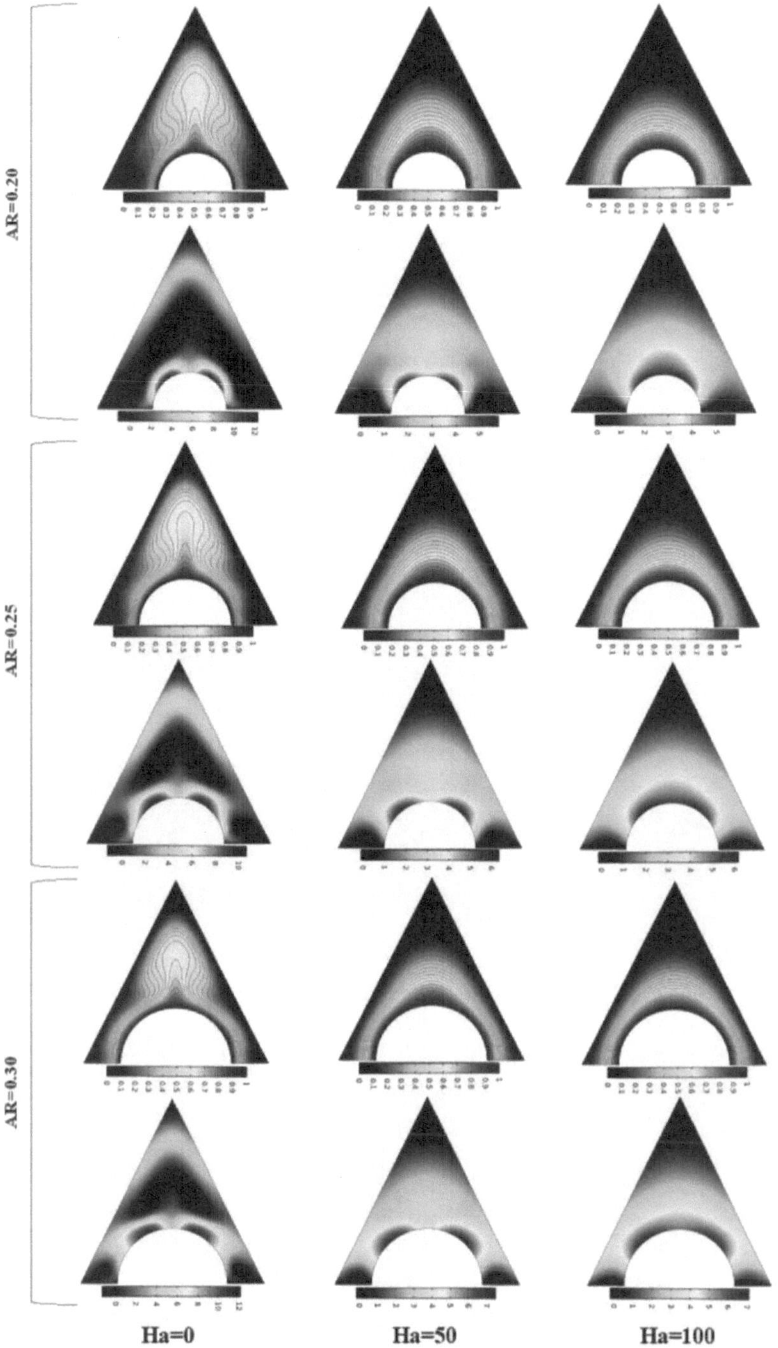

FIGURE 6.12 Isotherm variation (1st, 3rd, and 5th rows) and local Nusselt number (2nd, 4th, and 6th rows) for varied AR when Ha varies and $Ra = 10^5$, $\phi_1 = 0.01$, and $\phi_2 = 0.01$.

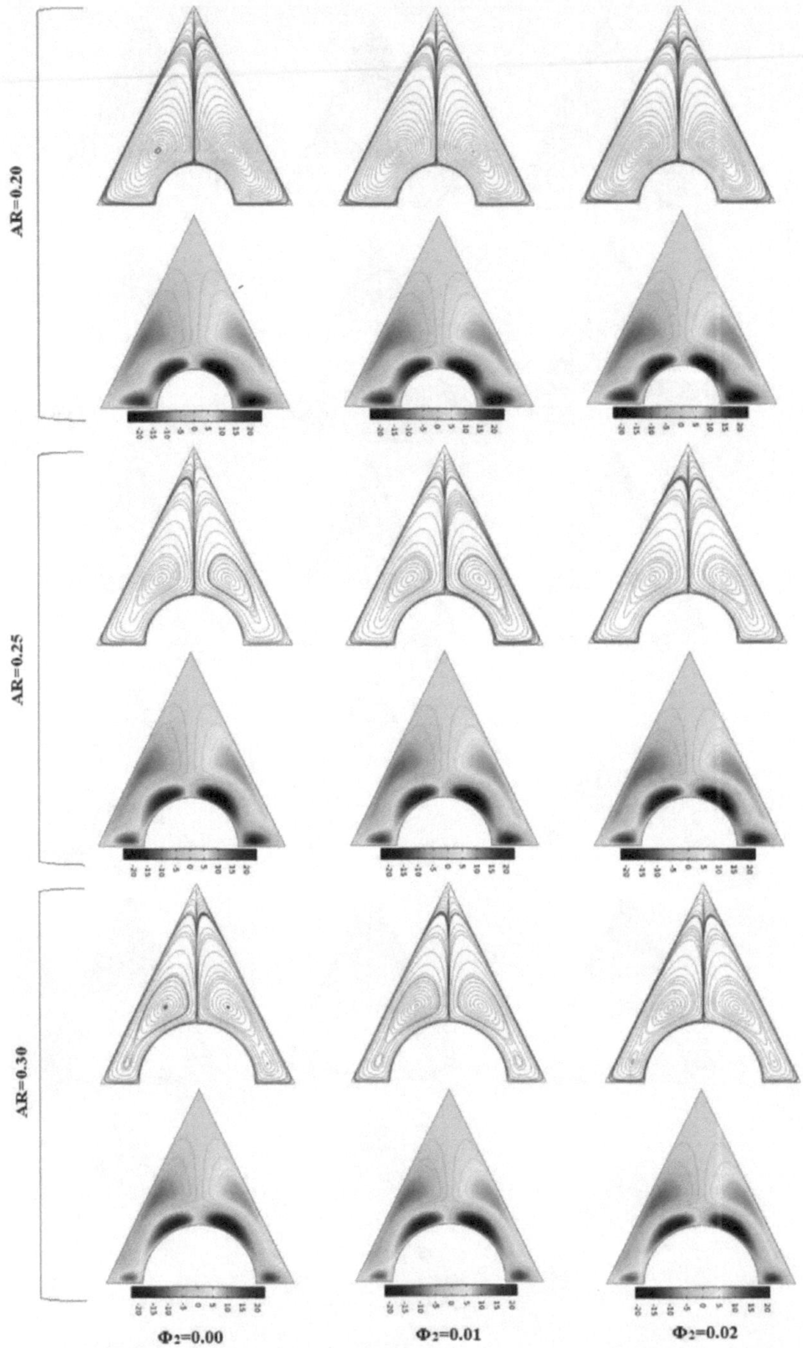

FIGURE 6.13 Streamline variation (1st, 3rd, and 5th rows) and velocity U variation (2nd, 4th, and 6th rows) for varied AR when ϕ_2 varies and $Ra = 10^5$, $Ha = 40$, and $\phi_1 = 0.01$.

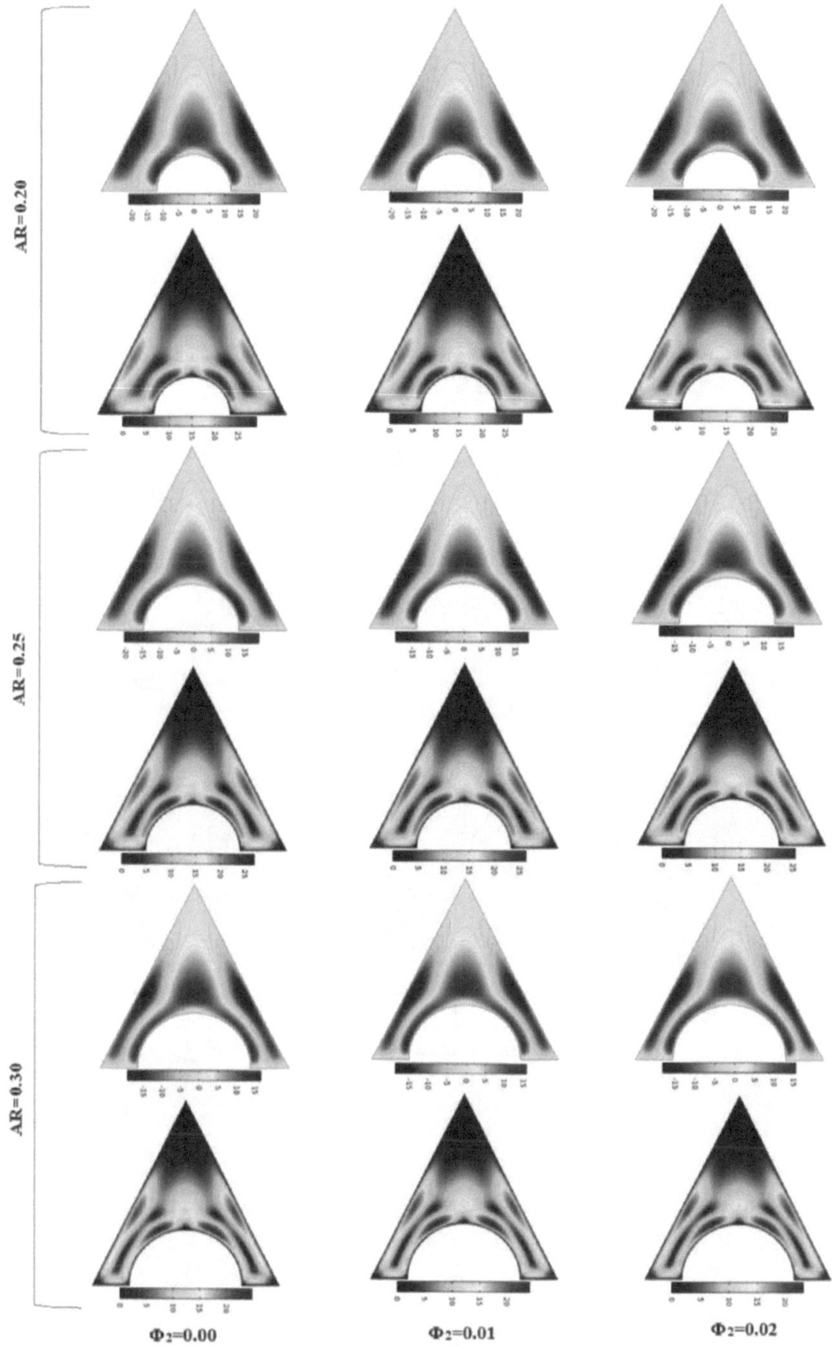

FIGURE 6.14 Velocity V variation (1st, 3rd, and 5th rows) and velocity magnitude distribution (2nd, 4th, and 6th rows) for varied AR when ϕ_2 varies and $Ra = 10^5$, $Ha = 40$, and $\phi_1 = 0.01$.

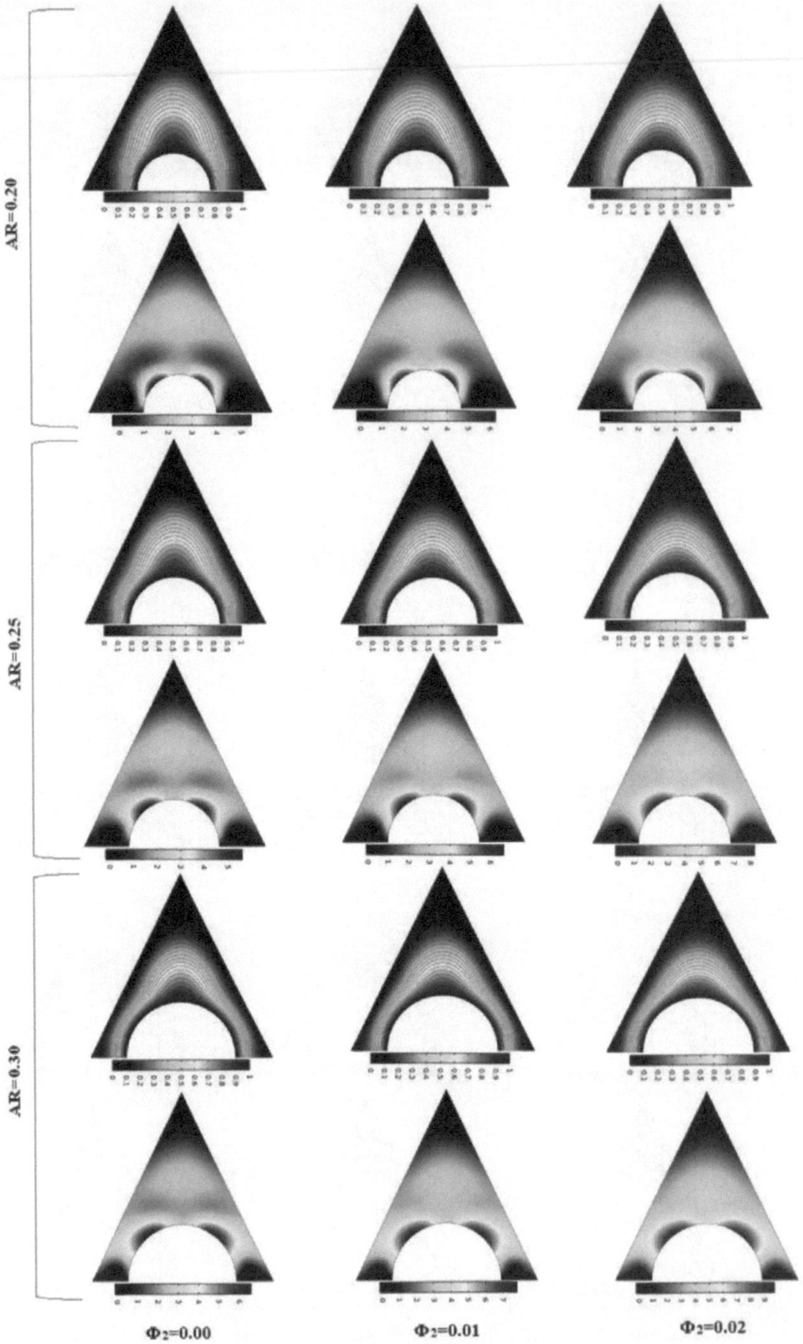

FIGURE 6.15 Isotherm variations (1st, 3rd, and 5th rows) and local Nusselt number (2nd, 4th, and 6th rows) for varied AR when ϕ_2 varies and $Ra = 10^5$, $Ha = 40$, and $\phi_1 = 0.01$.

6.5 MULTIPLE LINEAR REGRESSION ANALYSIS

A regression study examines the association or relation among the predictors (independent variables) and response (dependent variables). The multiple linear regression study is clutched to assess Nu_{avg} as the correlations are significant [66]. Now, the overall correlation of Nu_{avg} for $AR = 0.20,\ 0.25$, and 0.30 is assessed by

$$\left.\begin{aligned} Nu_{avg}\Big|_{AR=0.20} &= a^*_0 + a^*_{Ra}Ra + a^*_{Ha}Ha + a^*_{\phi_2}\phi_2 \\ Nu_{avg}\Big|_{AR=0.25} &= b^*_0 + b^*_{Ra}Ra + b^*_{Ha}Ha + b^*_{\phi_2}\phi_2 \\ Nu_{avg}\Big|_{AR=0.30} &= c^*_0 + c^*_{Ra}Ra + c^*_{Ha}Ha + c^*_{\phi_2}\phi_2 \end{aligned}\right\}, \tag{6.18}$$

where $a^*_0, a^*_{Ra}, a^*_{Ha}, a^*_{\phi_2}, b^*_0, b^*_{Ra}, b^*_{Ha}, b^*_{\phi_2}, c^*_0, c^*_{Ra}, c^*_{Ha}, c^*_{\phi_2}$ describes the regression coefficients. The estimation of Nu_{avg} is executed from the 30 sets of parametric values: $[10^4, 10^6]$ for Ra, $[0,100]$ for Ha, and $[0.00, 0.02]$ for ϕ_2. On receiving $a^*_0, a^*_{Ra}, a^*_{Ha}, a^*_{\phi_2}, b^*_0, b^*_{Ra}, b^*_{Ha}, b^*_{\phi_2}, c^*_0, c^*_{Ra}, c^*_{Ha}, c^*_{\phi_2}$ values, the desired model for $AR = 0.20,\ 0.25$, and 0.30 is acknowledged as

$$\left.\begin{aligned} Nu_{avg}\Big|_{AR=0.20} &= 4.765883 + \left(7.99\times10^{-6}\right)Ra - 0.03408Ha + 83.0739\phi_2 \\ Nu_{avg}\Big|_{AR=0.25} &= 4.255975 + \left(6.09\times10^{-6}\right)Ra - 0.020804Ha + 96.25364\phi_2 \\ Nu_{avg}\Big|_{AR=0.30} &= 4.637 + \left(4.14\times10^{-6}\right)Ra - 0.01655Ha + 113.1343\phi_2 \end{aligned}\right\}. \tag{6.19}$$

The negative sign of $a^*_0, a^*_{Ra}, a^*_{Ha}, a^*_{\phi_2}, b^*_0, b^*_{Ra}, b^*_{Ha}, b^*_{\phi_2}, c^*_0, c^*_{Ra}, c^*_{Ha}, c^*_{\phi_2}$ ensures a negative relationship with Nu_{avg}, while the positive sign assures a positive association with Nu_{avg}. Thus, Hartmann numbers decline Nu_{avg}, while nanoparticle concentrations and Rayleigh numbers boost Nu_{avg}. Figure 6.16 highlights the regression model's accuracy. A good association is noted with actual and estimated outcomes. For the numeric outcomes, Table 6.5 presents the goodness of the fitted polynomial against each circular heater's size, i.e., for $AR = 0.20,\ 0.25$, and 0.30.

TABLE 6.5
Goodness of Fit

	$AR = 0.20$	$AR = 0.25$	$AR = 0.30$
R-square	0.978044	0.978112	0.976402
RMSE	0.214694	0.218342	0.229498

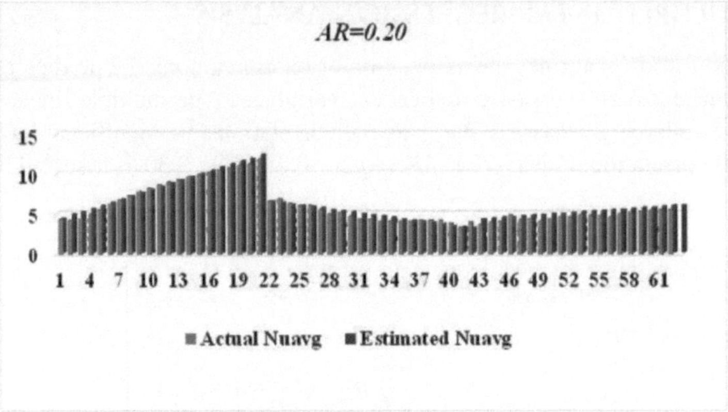

(a) AR=0.20

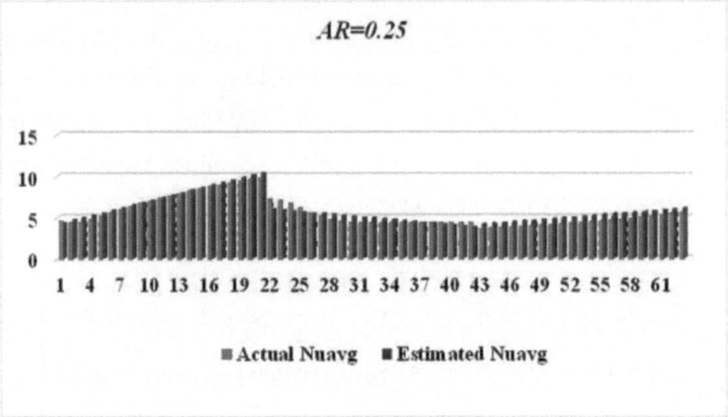

(b) AR=0.25

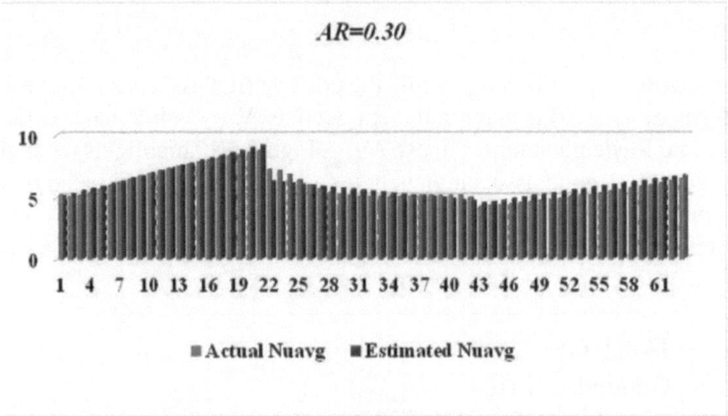

(c) AR=0.30

FIGURE 6.16 Estimated vs. actual Nu_{avg}.

6.6 CONCLUSIONS

A numeric simulation is executed to enlighten the hydrothermal variations of Al_2O_3-TiO_2-water hybrid nanofluids through a triangular chamber in which a heated semi-circular heater is attached at the lowermost wall of the triangular chamber. The inclined faces of the triangular compartment are cold, while the remaining parts of the triangular sector are fixed as adiabatic. The numeric simulation is performed using the Galerkin finite element discretization method. Several heaters' sizes were set to inspect the outcomes. Based on the numeric investigation, the following core outcomes are summarized:

- The streamlines' intensity within the chamber boosts for Rayleigh numbers, while the same is inversely interrelated to nanoparticle concentrations and Hartmann numbers. The extremely intensified trajectory is numerically suitable for the small circular heater, i.e., $AR = 0.20$.
- The *U*- and *V*-velocity profiles and velocity magnitude distributions are negatively associated with magnetic parameters and nanoparticle concentration, whereas Rayleigh numbers promote such profiles. The lower profile is ensured for a large-size circular heater, i.e., $AR = 0.30$.
- Magnetic impacts and nanoparticle concentrations decline the isotherm distortion within the chamber, but Rayleigh numbers support such thermal distortions. Consequently, the thermal scenario exhibits the same trend. On average, a lower-size heater, i.e., $AR = 0.20$, addresses the high distortions.
- The heat transmission seems to escalate for Rayleigh numbers and nanoparticle concentrations, but the magnetic effect declines such phenomenon. Moreover, a parametric range limitation is predicted for heat transmission. Although initially for $10^4 \le Ra \le 10^5$ a slightly high magnitude in Nu_{loc} is foreseen for a large circular heater, i.e., $AR = 0.30$, for $Ra > 10^5$ the maximum magnitude in Nu_{loc} is predicted for $AR = 0.20$. For $0 \le Ha < 50$, the smaller size circular heater, i.e., $AR = 0.20$, exhibits high Nu_{loc} compared to others, but for $50 \le Ha \le 100$ the reverse outcome is predicted, i.e., high Nu_{loc} is seen for $AR = 0.30$. The Nu_{avg} agrees with such a trend, but an additional remark is that for $0 \le Ha \le 60$ the Nu_{avg} declines, while a slight enhancement is spotted for $60 < Ha \le 100$. One exciting consequence is that higher heat transfer is perceived for a larger circular heater.
- The assessed regression model for $AR = 0.20,\ 0.25$, and 0.30 is established as

$$\left.\begin{aligned} Nu_{avg}\Big|_{AR=0.20} &= 4.765883 + \left(7.99\times10^{-6}\right)Ra - 0.03408Ha + 83.0739\phi_2 \\ Nu_{avg}\Big|_{AR=0.25} &= 4.255975 + \left(6.09\times10^{-6}\right)Ra - 0.020804Ha + 96.25364\phi_2 \\ Nu_{avg}\Big|_{AR=0.30} &= 4.637 + \left(4.14\times10^{-6}\right)Ra - 0.01655Ha + 113.1343\phi_2 \end{aligned}\right\}.$$

REFERENCES

1. Sankar, M., Do, Y., Ryu, S., & Jang, B. (2015). Cooling of heat sources by natural convection heat transfer in a vertical annulus. *Numerical Heat Transfer, Part A: Applications*, 68(8), 847–869.

2. Sheremet, M. A., Pop, I., & Mahian, O. (2018). Natural convection in an inclined cavity with time-periodic temperature boundary conditions using nanofluids: Application in solar collectors. *International Journal of Heat and Mass Transfer*, 116, 751–761.
3. Hussien, A. A., Al-Kouz, W., El Hassan, M., Janvekar, A. A., & Chamkha, A. J. (2021). A review of flow and heat transfer in cavities and their applications. *The European Physical Journal Plus*, 136(4), 353.
4. Ghani, A. A., Farid, M. M., Chen, X. D., & Richards, P. (1999). Numerical simulation of natural convection heating of canned food by computational fluid dynamics. *Journal of Food Engineering*, 41(1), 55–64.
5. Mallya, N., & Haussener, S. (2021). Buoyancy-driven melting and solidification heat transfer analysis in encapsulated phase change materials. *International Journal of Heat and Mass Transfer*, 164, 120525.
6. Williams, J. G., Castrejon-Pita, A. A., Turney, B. W., Farrell, P. E., Tavener, S. J., Moulton, D. E., & Waters, S. L. (2020). Cavity flow characteristics and applications to kidney stone removal. *Journal of Fluid Mechanics*, 902, A16.
7. Chen, L., & Zhang, X. R. (2017). Natural convection supercritical fluid systems for geothermal, heat transfer, and energy conversion. *Energy Solutions to Combat Global Warming*, 391–433.
8. Li, M. (1999). Numerical solutions for the incompressible Navier–Stokes equations. Simon Fraser University.
9. McQuain, W. D., Ribbens, C. J., Wang, C. Y., & Watson, L. T. (1994). Steady viscous flow in a trapezoidal cavity. *Computers & Fluids*, 23(4), 613–626.
10. Haese, P. M., & Teubner, M. D. (2002). Heat exchange in an attic space. *International Journal of Heat and Mass Transfer*, 45(25), 4925–4936.
11. Yesiloz, G., & Aydin, O. (2013). Laminar natural convection in right-angled triangular enclosures heated and cooled on adjacent walls. *International Journal of Heat and Mass Transfer*, 60, 365–374.
12. Aich, W. (2018). 3D buoyancy induced heat transfer in triangular solar collector having a corrugated bottom wall. *Engineering, Technology & Applied Science Research*, 8(2).
13. Milyani, A. H., Attar, E. T., Abdulaal, M. J., Ajour, M. N., Abu-Hamdeh, N. H., & Karimipour, A. (2022). Thermal analysis of battery cells placed in triangular enclosures filled with PCM in the presence of forced airflow in an air duct. *Journal of Building Engineering*, 57, 104887.
14. Li, M., & Tang, T. (1996). Steady viscous flow in a triangular cavity by efficient numerical techniques. *Computers & Mathematics with Applications*, 31(10), 55–65.
15. Gaskell, P. H., Thompson, H. M., & Savage, M. D. (1999). A finite element analysis of steady viscous flow in triangular cavities. *Proceedings of the Institution of Mechanical Engineers, Part C: Journal of Mechanical Engineering Science*, 213(3), 263–276.
16. Nazeer, M., Ali, N., & Javed, T. (2019). Numerical simulations of MHD forced convection flow of micropolar fluid inside a right-angled triangular cavity saturated with porous medium: Effects of vertical moving wall. *Canadian Journal of Physics*, 97(1), 1–13.
17. Ali, N., Nazeer, M., Javed, T., & Razzaq, M. (2019). Finite element analysis of bi-viscosity fluid enclosed in a triangular cavity under thermal and magnetic effects. *The European Physical Journal Plus*, 134(1), 2.
18. Shah, I. A., Bilal, S., Akgül, A., Omri, M., Bouslimi, J., & Khan, N. Z. (2022). Significance of cold cylinder in heat control in power law fluid enclosed in isosceles triangular cavity generated by natural convection: A computational approach. *Alexandria Engineering Journal*, 61(9), 7277–7290.
19. Turkyilmazoglu, M., & Duraihem, F. Z. (2023). Fully developed flow in a long triangular channel under an applied magnetic field. *Journal of Magnetism and Magnetic Materials*, 578, 170803.

20. Ali, N., Nazeer, M., Javed, T., & Abbas, F. (2018). A numerical study of micropolar flow inside a lid-driven triangular enclosure. *Meccanica*, 53, 3279–3299.
21. Li, Z., Shahsavar, A., Niazi, K., Al-Rashed, A. A., & Talebizadehsardari, P. (2020). The effects of vertical and horizontal sources on heat transfer and entropy generation in an inclined triangular enclosure filled with non-Newtonian fluid and subjected to magnetic field. *Powder Technology*, 364, 924–942.
22. Bhattacharya, M., & Basak, T. (2021). Analysis of multiple steady states for natural convection of Newtonian fluids in a square enclosure. *Physics of Fluids*, 33(10).
23. Schelling, P. K., Shi, L., & Goodson, K. E. (2005). Managing heat for electronics. *Materials Today*, 8(6), 30–35.
24. Choi, S. U., & Eastman, J. A. (1995). Enhancing thermal conductivity of fluids with nanoparticles. Argonne, IL: Argonne National Lab (ANL).
25. Pordanjani, A. H., Aghakhani, S., Afrand, M., Mahmoudi, B., Mahian, O., & Wongwises, S. (2019). An updated review on application of nanofluids in heat exchangers for saving energy. *Energy Conversion and Management*, 198, 111886.
26. Wole-Osho, I., Okonkwo, E. C., Abbasoglu, S., & Kavaz, D. (2020). Nanofluids in solar thermal collectors: Review and limitations. *International Journal of Thermophysics*, 41(11), 157.
27. Singh, I., Sehgal, S. S., & Khullar, V. (2022). Nanofluid filled enclosures: Potential photo-thermal energy conversion and sensible heat storage devices. *Thermal Science and Engineering Progress*, 33, 101376.
28. Sonawane, S. S., Mohammed, H. A., Mungray, A. K., & Sonawane, S. (Eds.). (2022). Applications of nanofluids in chemical and bio-medical process industry. Elsevier.
29. Moldoveanu, G. M., Minea, A. A., Iacob, M., Ibanescu, C., & Danu, M. (2018). Experimental study on viscosity of stabilized Al_2O_3, TiO_2 nanofluids and their hybrid. *Thermochimica Acta*, 659, 203–212.
30. Moldoveanu, G. M., Minea, A. A., Huminic, G., & Huminic, A. (2019). Al_2O_3/TiO_2 hybrid nanofluids thermal conductivity: An experimental approach. *Journal of Thermal Analysis and Calorimetry*, 137, 583–592.
31. Ali, H. M. (Ed.). (2020). Hybrid nanofluids for convection heat transfer. Academic Press.
32. Adun, H., Wole-Osho, I., Okonkwo, E. C., Ruwa, T., Agwa, T., Onochie, K., Ukwu, H., Bamisile, O., & Dagbasi, M. (2022). Estimation of thermophysical property of hybrid nanofluids for solar thermal applications: Implementation of novel optimizable gaussian process regression (O-GPR) approach for viscosity prediction. *Neural Computing and Applications*, 34(13), 11233–11254.
33. Jamei, M., Karbasi, M., Mosharaf-Dehkordi, M., Olumegbon, I. A., Abualigah, L., Said, Z., & Asadi, A. (2022). Estimating the density of hybrid nanofluids for thermal energy application: Application of non-parametric and evolutionary polynomial regression data-intelligent techniques. *Measurement*, 189, 110524.
34. Zahan, I., Nasrin, R., & Alim, M. A. (2019). Hybrid nanofluid flow in combined convective lid-driven sinusoidal triangular enclosure. *AIP Conference Proceeding*, 2121, 070001.
35. Izadi, M. (2020). Effects of porous material on transient natural convection heat transfer of nano-fluids inside a triangular chamber. *Chinese Journal of Chemical Engineering*, 28(5), 1203–1213.
36. Chabani, I., Mebarek-Oudina, F., & Ismail, A. A. I. (2022). MHD flow of a hybrid nanofluid in a triangular enclosure with zigzags and an elliptic obstacle. *Micromachines*, 13(2), 224.
37. Redouane, F., Jamshed, W., Devi, S. S. U., Prakash, M., Nasir, N. A. A. M., Hammouch, Z., Eid, M. R., Nisar, K. S., Mahammed, A. B., Abdel-Aty, A. H., Yahia, I. S., & Eed, E. M. (2022). Heat flow saturate of Ag/MgO-water hybrid nanofluid in heated trigonal

enclosure with rotate cylindrical cavity by using Galerkin finite element. *Scientific Reports*, 12(1), 2302.
38. Ahmed, S. E., & Raizah, Z. A. (2022). Magnetic mixed convection of a Casson hybrid nanofluid due to split lid driven heat generated porous triangular containers with elliptic obstacles. *Journal of Magnetism and Magnetic Materials*, 559, 169549.
39. Islam, Z., Azad, A. K., Hasan, M. J., Hossain, R., & Rahman, M. M. (2022). Unsteady periodic natural convection in a triangular enclosure heated sinusoidally from the bottom using CNT-water nanofluid. *Results in Engineering*, 14, 100376.
40. Abdelhak, R., Abderrazak, A., Redouane, F., & Khelili, Y. (2023). Mixed convection magnetohydrodynamics of different forms of triangular cavity involving CuO/water nanofluid. *Journal of Nanofluids*, 12(4), 1082–1094.
41. Acharya, N. (2022). Effects of different thermal modes of obstacles on the natural convective Al_2O_3-water nanofluidic transport inside a triangular cavity. *Proceedings of the Institution of Mechanical Engineers, Part C: Journal of Mechanical Engineering Science*, 236(10), 5282–5299.
42. Acharya, N. (2022). On the magnetohydrodynamic natural convective alumina nanofluidic transport inside a triangular enclosure fitted with fins. *Journal of the Indian Chemical Society*, 99(12), 100784.
43. Abderrahmane, A., Qasem, N. A., Younis, O., Marzouki, R., Mourad, A., Shah, N. A., & Chung, J. D. (2022). MHD hybrid nanofluid mixed convection heat transfer and entropy generation in a 3-D triangular porous cavity with zigzag wall and rotating cylinder. *Mathematics*, 10(5), 769.
44. Munawar, S., Saleem, N., & Khan, W. A. (2022). Entropy generation minimization EGM analysis of free convective hybrid nanofluid flow in a corrugated triangular annulus with a central triangular heater. *Chinese Journal of Physics*, 75, 38–54.
45. Keyhani, M., Prasad, V., & Cox, R. (1988). An experimental study of natural convection in a vertical cavity with discrete heat sources. *Journal of Heat Transfer*, 110(3), 616–624.
46. Zhan, N., Ding, L., Chai, Y., Wu, J., & Xu, Y. (2019). Experimental study on natural convective heat transfer in a closed cavity. *Thermal Science and Engineering Progress*, 9, 132–141.
47. Izadi, M., Mohebbi, R., Chamkha, A., & Pop, I. (2018). Effects of cavity and heat source aspect ratios on natural convection of a nanofluid in a C-shaped cavity using Lattice Boltzmann method. *International Journal of Numerical Methods for Heat & Fluid Flow*, 28(8), 1930–1955.
48. Wyttenbach, J., Bougard, J., Descy, G., Skrylnyk, O., Courbon, E., Frère, M., & Bruyat, F. (2018). Performances and modelling of a circular moving bed thermochemical reactor for seasonal storage. *Applied Energy*, 230, 803–815.
49. Ghritlahre, H. K., Sahu, P. K., & Chand, S. (2020). Thermal performance and heat transfer analysis of arc shaped roughened solar air heater–an experimental study. *Solar Energy*, 199, 173–182.
50. Liu, K., Yang, F., Wang, S., Gao, B., & Xu, C. (2018). The research on the heat source characteristics and the equivalent heat source of the arc in gaps. *International Journal of Heat and Mass Transfer*, 124, 177–189.
51. Roy, N. C. (2022). MHD natural convection of a hybrid nanofluid in an enclosure with multiple heat sources. *Alexandria Engineering Journal*, 61(2), 1679–1694.
52. Manna, N. K., Biswas, N., Mandal, D. K., Sarkar, U. K., Öztop, H. F., & Abu-Hamdeh, N. (2023). Impacts of heater-cooler position and Lorentz force on heat transfer and entropy generation of hybrid nanofluid convection in quarter-circular cavity. *International Journal of Numerical Methods for Heat & Fluid Flow*, 33(3), 1249–1286.

53. Gul, T., Nasir, S., Berrouk, A. S., Raizah, Z., Alghamdi, W., Ali, I., & Bariq, A. (2023). Simulation of the water-based hybrid nanofluids flow through a porous cavity for the applications of the heat transfer. *Scientific Reports*, 13(1), 7009.
54. Mahalakshmi, T. (2023). A numerical analysis on MHD mixed convective hybrid nanofluid flow inside enclosure with heat sources. *Journal of Nanofluids*, 12(4), 942–954.
55. Thirumalaisamy, K., & Ramachandran, S. (2023). Comparative heat transfer analysis on Fe_3O_4–H_2O and Fe_3O_4–Cu–H_2O flow inside a tilted square porous cavity with shape effects. *Physics of Fluids*, 35(2).
56. Mansour, M. A., & Bakier, M. A. Y. (2023). Magnetohydrodynamic mixed convection of TiO_2–Cu/water between the double lid-driven cavity and a central heat source surrounding by a wavy tilted domain of porous medium under local thermal non-equilibrium. *SN Applied Sciences*, 5(2), 51.
57. Yan, S. R., Aghakhani, S., & Karimipour, A. (2020). Influence of a membrane on nanofluid heat transfer and irreversibilities inside a cavity with two constant-temperature semicircular sources on the lower wall: Applicable to solar collectors. *Physica Scripta*, 95(8), 085702.
58. Acharya, N. (2021). On the flow patterns and thermal control of radiative natural convective hybrid nanofluid flow inside a square enclosure having various shaped multiple heated obstacles. *The European Physical Journal Plus*, 136(8), 889.
59. Qureshi, M. A., Hussain, S., & Sadiq, M. A. (2021). Numerical simulations of MHD mixed convection of hybrid nanofluid flow in a horizontal channel with cavity: Impact on heat transfer and hydrodynamic forces. *Case Studies in Thermal Engineering*, 27, 101321.
60. Armaghani, T., Sadeghi, M. S., Rashad, A. M., Mansour, M. A., Chamkha, A. J., Dogonchi, A. S., & Nabwey, H. A. (2021). MHD mixed convection of localized heat source/sink in an Al_2O_3-Cu/water hybrid nanofluid in L-shaped cavity. *Alexandria Engineering Journal*, 60(3), 2947–2962.
61. Farhana, K., Kadirgama, K., Rahman, M. M., Noor, M. M., Ramasamy, D., Samykano, M., Najafi, G., Sidik, N. A. C., & Tarlochan, F. (2019). Significance of alumina in nanofluid technology: An overview. *Journal of Thermal Analysis and Calorimetry*, 138, 1107–1126.
62. Rangasamy, S., Raghavan, R. R. V., Elavarasan, R. M., & Kasinathan, P. (2023). Energy analysis of flattened heat pipe with nanofluids for sustainable electronic cooling applications. *Sustainability*, 15(6), 4716.
63. Yang, L., & Hu, Y. (2017). Toward TiO_2 nanofluids—part 2: Applications and challenges. *Nanoscale Research Letters*, 12, 1–21.
64. Murtadha, T. K. (2023). Effect of using Al_2O_3/TiO_2 hybrid nanofluids on improving the photovoltaic performance. *Case Studies in Thermal Engineering*, 47, 103112.
65. Roy, S., & Basak, T. (2005). Finite element analysis of natural convection flows in a square cavity with non-uniformly heated wall (s). *International Journal of Engineering Science*, 43(8–9), 668–680.
66. Neethu, T. S., Sabu, A. S., Mathew, A., Wakif, A., & Areekara, S. (2022). Multiple linear regression on bioconvective MHD hybrid nanofluid flow past an exponential stretching sheet with radiation and dissipation effects. *International Communications in Heat and Mass Transfer*, 135, 106115.

7 A Numerical Study of Micropolar Hybrid Nanofluid Flow over the Wedge with the Impacts of Hall and Ion Slip Using PINN

Sai Ganga, Ziya Uddin, and Himanshu Upreti

Nomenclature

u, v, w	Velocity components	m	Falkner–Skan parameter
β	Hartree pressure gradient parameter	β_i	Ion slip parameter
j	Micro inertia	ν	Kinematic viscosity
β_e	Hall effect parameter	κ	Vortex viscosity
N	Component of microrotation	B_0	Constant magnetic field
T	Temperature	γ	Spin gradient viscosity
T_w	Surface temperature	T_∞	Ambient temperature

7.1 INTRODUCTION

In the last several decades, a lot of attention has been paid to the fluid flows across wedge-shaped surfaces due to its extensive range of applications [1]. Falkner and Skan [2] studied and pioneered the problem on wedges, generalizing Blasius solution to fluid flows over wedge. Natural fluids come in a variety of forms. Some of them fall under the category of Newtonian fluids – those that obey Newton's viscosity

 DOI: 10.1201/9781003595786-7

law. However, those fluids that does not obey Newton's viscosity law are categorized into non-Newtonian fluids. Most of the applications use fluids with complex nature, those that are mostly non-Newtonian fluids. This study focuses on micropolar fluids, a particular class of non-Newtonian fluids. Eringen [3] established the theory of micropolar fluids. Since then, many scientists have begun to pay increasing attention to these fluids. Uddin et al. investigated the flow of micropolar fluids through a wedge under the influence of magnetic field along with the impacts of Hall and ion-slip effects, viscous dissipation, and Joule heating [4]. Extension of this work by considering radiation effects and heat generation or absorption can be found in [5]. Singh et al. [6] studied the effect of micropolar fluid flowing over a wedge using an analytical approach. Recently, a similar study of heat transfer flow across a wedge by considering the effect of heat source or sink and chemical reaction was done by Sharma et al. [7].

The significance of heat transfer lies in its vital role within the industrial realm, facilitating the flow of energy in systems. There is a substantial industrial requirement for a working fluid with a better heat transfer rate. Because of this, researchers are continuously striving to develop new heat transfer fluids that have the potential to significantly improve heat transmission. A widely studied fluid that improves heat transfer is nanofluids [8]. The study of micropolar fluids over a wedge containing gamma nanoparticles was conducted by Zaib et al. [9]. By considering the impact of chemical reaction, Zulkifli et al. [10] investigated micropolar nanofluid flow over a wedge. El-dawy et al. [11] elucidated the impact of nanoparticles on a wedge surface when a magnetic field and heat generation or absorption were present. The entropy generation of a micropolar nanofluid passing through a wedge under a bioconvective flow along with the impact of oxytactic microorganisms was studied by Patil et al. [12]. Other similar works on the investigation of micropolar nanofluid flow over a wedge can be found in [13, 14].

Researchers have also looked at hybrid nanofluids, a new type of fluid that enhances heat transmission more effectively than nanofluids. A hybrid nanofluid is produced by mixing two different nanoparticles with the base fluid. The increased thermal conductivity of hybrid nanofluids is primarily responsible for their superior heat transfer rate, as supported by several studies [15]. A number of elements influencing heat transfer improvements have been found through the examination of hybrid nanofluids in various fields. Few of these aspects include thermal conductivity, preparation technique, optimal preparation techniques, and components impacting thermal transfer performance [16]. A few research on the flow of hybrid nanofluids over a wedge are included below. Dinarvand et al. [17] found in their study that hybrid nanofluids performed better than the other nanofluids they examined with regard to their thermophysical properties. They also provided a comparison of the effects of different nanoparticle shape factor values on the local heat transfer rate. Under the influence of magnetic field and radiation, Waini et al. [18] compared the performance of hybrid nanofluids and nanofluids numerically. A study over an isothermal wedge can be found in [19]. Khashi'ie et al. [20] used a statistical data analysis on the obtained solutions for drawing conclusions. Zainal et al. [21] carried out a stability analysis for the case with a convective boundary condition and observed that the thermal performance was improved by the parameters representing wedge angle and dual-type nanoparticle. Kho et al. [1] explored the behaviour of an $Ag-TiO_2$ hybrid nanofluid flowing across a permeable wedge using magnetohydrodynamics, accounting for factors such as heat radiation and viscous dissipation. Lately, Baby et al. [22] conducted research with the goal of examining the thermal properties of hybrid nanoparticles when Brownian motion and thermophoresis were considered.

Al-Hanaya et al. [23] studied about the single-wall carbon nanotube and multiwall carbon nanotube over a curved surface along with the influence of magnetic field. Over an inclined channel, Raju [24] discussed about the impact of ohmic heating and radiation by considering micropolar and hybrid nanofluid flow in two different cases. Tassaddiq [25] uses the Cattaneo–Christov model for a micropolar fluid over a stretching body to investigate the effect of hybrid nanofluid on the thermal performance of the fluid. Algehyne et al. [26], who also employed the Cattaneo–Christov flux model, examined the effects of the micropolar and magnetic field on the hybrid nanofluid flow's velocity and micro-rotational distributions.

Inspired by the aforementioned research, we investigated the micropolar hybrid nanofluid flow in this work where magnetic field, Hall, and ion-slip effects are present. We considered water-based graphene oxide–molybdenum disulphide ($GO-MoS_2$) as the hybrid nanofluid. Over the past decade, graphene nanofluids gained a lot of attention because of their remarkable features [27]. MoS_2, similar to graphene, possesses a layered structure, exhibits sufficiently wide band gaps, and is known to demonstrate better thermophysical characteristics [28]. Modelling this we obtain a system of partial differential equations. The system is converted into a system of ordinary differential equations with the help of similarity variables, which is then solved using a deep learning technique. In recent decades, physics-informed neural networks (PINN) have been widely employed for the purpose of solving various differential equations. The versatility and effectiveness in the approximation of neural networks have drawn this attention. While there are works that use PINN to solve real-world applications, researchers are also interested in the study of the design of neural networks. Combining the capabilities of neural networks and wavelets is one such study, i.e., using wavelet as an activation function in the neural networks [28]. In this work, we use this methodology for solving the resulting system of non-linear differential equations.

7.2 MATHEMATICAL FORMULATION

An incompressible steady viscous hybrid nanofluid fluid flow over a non-conducting wedge for an electrically conducting fluid is considered. The Cartesian coordinate system (x, y, z) is used to model the governing flow. where x is considered as the distance measured along the wedge and y is that of normal to the surface of the wedge. Along the wedge's leading edge, z is measured. It is assumed that a magnetic field B exists along the y axis. The effects of ion-slip and hall currents are also considered. The velocity components u, v, w are considered in the x, y, z direction, respectively. The free stream velocity is assumed as $U = ax^m$, $m = \frac{\beta}{2-\beta}$, $\beta = \frac{\Omega}{\pi}$, where a and Ω represent a positive constant and angle of the wedge, respectively [4].

Following [4], the governing equations are given as follows:

$$\frac{\partial u}{\partial x}+\frac{\partial v}{\partial y}=0, \tag{7.1}$$

$$u\frac{\partial u}{\partial x}+v\frac{\partial u}{\partial y}=\left(\frac{\mu_{hnf}+\kappa}{\rho_{hnf}}\right)\frac{\partial^2 u}{\partial y^2}+\frac{\kappa}{\rho_{hnf}}\frac{\partial N}{\partial y}+\frac{1}{\rho_{hnf}}\left[\frac{\sigma_{hnf}B_y^2}{\alpha_e^2+\beta_e^2}\left(\alpha_e\left(U-u\right)-\beta_e w\right)\right]+U\frac{dU}{dx}, \tag{7.2}$$

$$u\frac{\partial w}{\partial x}+v\frac{\partial w}{\partial y}=\left(\frac{\mu_{hnf}+\kappa}{\rho_{hnf}}\right)\frac{\partial^2 w}{\partial y^2}-\frac{1}{\rho_{hnf}}\frac{\sigma_{hnf}B_y^2}{\alpha_e^2+\beta_e^2}\left(\alpha_e w-\beta_e u\right), \tag{7.3}$$

$$\rho_{hnf}j\left(u\frac{\partial N}{\partial x}+v\frac{\partial N}{\partial y}\right)=\gamma\frac{\partial^2 N}{\partial y^2}-\kappa\left(2N+\frac{\partial u}{\partial y}\right), \tag{7.4}$$

$$\left(\rho C_p\right)_{hnf}\left(u\frac{\partial T}{\partial x}+v\frac{\partial T}{\partial y}\right)=k_{hnf}\frac{\partial^2 T}{\partial y^2}+\left(\mu_{hnf}+\kappa\right)\left[\left(\frac{\partial u}{\partial y}\right)^2+\left(\frac{\partial w}{\partial y}\right)^2\right]$$
$$+\frac{\sigma_{hnf}B_y^2}{\alpha_e^2+\beta_e^2}\left(u^2+w^2\right). \tag{7.5}$$

Along with the boundary conditions:

$$u=v=w=0, N=-n\frac{\partial u}{\partial y}, q_w=-k_{hnf}\frac{\partial T}{\partial y}, \quad \text{at} \quad y=0, \tag{7.6}$$

$$u\to U, w\to 0, N\to 0, T\to T_\infty, \quad \text{as} \quad y\to\infty \tag{7.7}$$

Here, $\gamma=\left(\mu_{hnf}+\kappa/2\right)$, $\alpha_e=1+\beta_i\beta_e, B_y=B_0 x^{\frac{m-1}{2}}, 0\le m\le 1, n$ is the constant that lies between 0 and 1, and in this study, we choose $n=1/2$. [4].

The following similarity transformation [4] is applied to Equations (7.1)–(7.7):

$$\begin{gathered}\psi=f(\eta)\sqrt{\frac{2\nu_f xU}{m+1}}, \quad N=Uh(\eta)\sqrt{\frac{(m+1)U}{2\nu_f x}}, w=Ug(\eta), \\ \eta=\sqrt{\frac{(m+1)U}{2\nu_f x}}y, \theta(\cdot)=\frac{k_f(T-T_\infty)}{q_w}\sqrt{\frac{(m+1)U}{2\nu_f x}}\end{gathered} \quad . \tag{7.8}$$

And Equations (7.2)—(7.5) reduce to the following system of equations:

$$(1/C_1)(C_2+K)f'''+ff''+\left(\frac{2m}{m+1}\right)(1-f'^2)+(1/C_1)Kh', \\ +\frac{C_3}{C_1}\frac{M}{\alpha_e^2+\beta_e^2}\left[\alpha_e(1-f')-\beta_e g\right]=0 \tag{7.9}$$

$$(1/C_1)(C_2+K)g''+g'f-\left(\frac{2m}{m+1}\right)gf'-\frac{C_3}{C_1}\frac{M}{\alpha_e^2+\beta_e^2}\left[\alpha_e g-\beta_e f'\right]=0, \tag{7.10}$$

$$\left(C_2+\frac{K}{2}\right)h''-C_1\left[\left(\frac{3m-1}{m+1}\right)hf'-fh'\right]-\frac{2KI}{m+1}(2h+f'')=0, \tag{7.11}$$

$$\frac{1}{Pr}\theta''+\frac{C_5}{C_4}(m+1)f\theta'+\frac{C_5}{C_4}\left(\frac{m-1}{m+1}\right)f'\theta+Ec \\ \frac{C_2+K}{C_4}\left(f''^2+g'^2\right)+\frac{C_3}{C_1}\frac{M.Ec}{\alpha_e^2+\beta_e^2}\left(f'^2+g^2\right)=0. \tag{7.12}$$

And Equations (7.6)—(7.7) take the following form:

$$f(0)=0, f'(0)=0, g(0)=0, h(0)=-\frac{1}{2}f''(0), \theta'(0)=-\frac{1}{C_4}, \tag{7.13}$$

$$\text{As } \eta\to\infty, f'(\eta)\to 1, g(\eta)\to 0, h(\eta)\to 0, \theta(\eta)\to 0, \tag{7.14}$$

where ψ is defined as the stream function.

$$u=\frac{\partial\psi}{\partial y}, v=-\frac{\partial\psi}{\partial x}, K(\text{material parameter})=\frac{\kappa}{\mu_f},$$

$$Ec(\text{Eckert number})=\frac{U^2}{(C_p)_f(T_w-T_\infty)},$$

$$Pr(\text{Prandtl number})=\frac{(\mu C_p)_f}{k_f}, M(\text{magnetic parameter})=\frac{2\sigma_f B_0^2}{a\rho_f(m+1)},$$

$$Re(\text{Reynolds number})=\frac{Ux}{\nu_f}, I(\text{inertial parameter})=\frac{\nu_f^2 Re}{jU^2}, C_1=\frac{\rho_{hnf}}{\rho_f}, C_2=\frac{\mu_{hnf}}{\mu_f},$$

$$C_3=\frac{\sigma_{hnf}}{\sigma_f}, C_4=\frac{k_{hnf}}{k_f}, C_5=\frac{(\rho C_P)_{hnf}}{(\rho C_P)_f}.$$

The suffixes *hnf* and *f* correspond to the hybrid nanofluid and base fluid, respectively.

The thermophysical properties of the hybrid nanofluids [29] are given as follows:

$$\rho_{hnf} = (1-\phi_2)\left[(1-\phi_1)\rho_f + \phi_1\rho_{s1}\right] + \phi_2\rho_{s2},$$

$$k_{hnf} = \left[\frac{k_{s2} + 2k_{bf} - 2\phi_2\left(k_{bf} - k_{s2}\right)}{k_{s2} + 2k_{bf} + \phi_2\left(k_{bf} - k_{s2}\right)}\right] k_{bf}, \text{ where}$$

$$k_{bf} = \left[\frac{(k_{s1} + 2k_f - 2\phi_1\left(k_f - k_{s1}\right)}{k_{s1} + 2k_f + \phi_1\left(k_f - k_{s1}\right)}\right] k_f,$$

$$\mu_{hnf} = \frac{\mu_f}{(1-\phi_1)^{2.5}(1-\phi_2)^{2.5}},$$

$$\sigma_{hnf} = \left[\frac{\sigma_{s2} + 2\sigma_{bf} - 2\phi_2\left(\sigma_{bf} - \sigma_{s2}\right)}{\sigma_{s2} + 2\sigma_{bf} + \phi_2\left(\sigma_{bf} - \sigma_{s2}\right)}\right] \sigma_{bf}, \text{ where} \tag{7.15}$$

$$\sigma_{bf} = \left[\frac{\left(\sigma_{s1} + 2\sigma_f - 2\phi_1\left(\sigma_f - \sigma_{s1}\right)\right)}{\sigma_{s1} + 2\sigma_f + \phi_1\left(\sigma_f - \sigma_{s1}\right)}\right] \sigma_f,$$

$$\left(\rho C_p\right)_{hnf} = (1-\phi_2)\left[(1-\phi_1)\left(\rho C_p\right)_f + \phi_1\left(\rho C_p\right)_{s1}\right] + \phi_2\left(\rho C_p\right)_{s2}.$$

Here, ϕ_1 and ϕ_2 represent the nanofluid concentration of MoS_2 and GO in the base fluid, respectively. ρ is the density, k is the thermal conductivity, μ is the dynamic viscosity, C_p is the specific heat, and σ is the electrical conductivity. The suffixes s_1, s_2 represent MoS_2 and GO nanoparticles, respectively. The thermophysical properties of the hybrid nanoparticles and base fluid considered in this study are tabulated in Table 7.1. We have chosen the concentration of both nanoparticles (ϕ_1, ϕ_2) as 0.02 for the numerical calculation.

TABLE 7.1
Thermophysical Properties of MoS_2, GO, and H_2O [30]

Properties	MoS_2	GO	H_2O
ρ	5060	1800	997.1
C_p	397.21	717	4179
k	904.4	5000	0.613
σ	2.09E4	6.3E7	0.005

We are interested in determining the following physical quantities:

$$C_{fx} = \frac{\tau_w}{(1/2)\rho_{hnf}U^2} \quad \text{(skin friction coefficient in } x \text{ direction)}$$
$$C_{fz} = \frac{\tau_z}{(1/2)\rho_{hnf}U^2} \quad \text{(skin friction coefficient in } z \text{ direction)} \tag{7.16}$$
$$Nu = \frac{xq_w}{k_{hnf}(T - T_\infty)} \quad \text{(Nusselt number)}$$

where

$$\tau_w = \left.\left((\mu_{hnf} + \kappa)\frac{\partial u}{\partial y} + \kappa N\right)\right|_{y=0} \quad \text{(wall shear stress)}$$
$$\tau_z = \left.\left((\mu_{hnf} + \kappa)\frac{\partial w}{\partial y}\right)\right|_{y=0} \quad \text{(wall shear stress)} \tag{7.17}$$
$$q_w = \left.\left(-k_f\frac{\partial T}{\partial y}\right)\right|_{y=0} \quad \text{(wall heat flux).}$$

Hence,

$$C_{fx} = \frac{1}{C_1}\sqrt{\frac{2(m+1)}{Re}}\left(C_2 + \frac{K}{2}\right)f''(0)$$
$$C_{fz} = \frac{1}{C_1}\sqrt{\frac{2(m+1)}{Re}}(C_2 + K)g'(0) \tag{7.18}$$
$$Nu = \frac{1}{C_4}\sqrt{\frac{(m+1)Re}{2}}\frac{1}{\theta(0)}.$$

7.3 SOLUTION METHODOLOGY

An effective deep learning method for handling data-driven problems is the artificial neural network. Neural networks take an input, then process it through hidden layers, and give an output. It is a mathematical model where the solution is the neural network approximation. Weights and biases are the unknown parameters in this approximation, and these parameters are randomly initialized. Neural network predicts the output, and a back propagation algorithm is performed to minimize the error occurred while predicting the output. The parameters are iteratively adjusted to minimize the error. This process is called as the training of the neural network. The function that quantifies the error occurred is called as the loss function. Physics-informed neural networks (PINN) make use of neural networks and the physical laws that can be described by differential equations. To approximate the differential equation's solution, a neural network is utilized. The input fed into the neural network is random data points from the domain defined. In PINN, the error that we minimize is the residual of differential equation and the residual of the defined constraints of differential equations. This is called as loss function in PINN. The neural network is trained to minimize this loss function, which is expected to produce the solution of differential equation as the output. Design of the neural network is often very important for a successful training. We use wavelet as an activation because of its advantages [31]. In the considered problem of study, Gaussian wavelet is observed to perform better in comparison to Mexican wavelet and Morlet wavelet. Gaussian wavelet is given by

$$\Psi(x) = -xe^{\frac{-x^2}{2}}. \tag{7.19}$$

Input is initialized by taking random points from the uniform distribution, and Xavier's initialization is used for the unknown parameters. For input initialization, we used 10,000 collocation points from the considered domain for the residual of differential equation and 50 collocation points for the boundary conditions. The problem is defined over a semi-infinite interval, and in the present study, we assumed the right boundary to be 3. Loss function is defined by taking the residual of the system of differential equations and the constraints from Equations (7.9)—(7.14). During the back propagation, optimization is the crucial step, and here we have used Adam optimizer, and for the learning rate, we used a decaying learning rate. The details and relevance of using this choice can be found in [31]. In our study, we observed that a number of hidden layers and number of neurons have a greater impact on obtaining the correct prediction. The following were the chosen values for which we obtained lower loss values and accurate predictions:

For approximating f, we chose 7 hidden layers with 2, 5, 5, 3, 8, 5, and 6 neurons in each of those hidden layers.

For approximating g, we chose 5 hidden layers with 3, 5, 8, 1, and 4 neurons in each of those hidden layers.

For approximating h, we chose 5 hidden layers with 6, 2, 7, 8, and 6 neurons in each of those hidden layers.

And for approximating θ, we chose 1 hidden layer with 6 neurons in that hidden layer.

7.4 RESULTS AND DISCUSSION

We analysed the influence of the following parameters considered in this study, including $m, \beta_i, \beta_e, Pr, I, K, \phi_1, \phi_2$, and Ec. For the same, the influence of some of these parameters on temperature distribution (θ), horizontal velocity (f'), transverse velocity (g), and angular velocity (h) is illustrated in Figures 7.1–7.7. Also, the influence of these parameters on the values of $\frac{1}{2}C_{fx}\sqrt{Re}$, $\frac{1}{2}C_{fz}\sqrt{Re}$, $\frac{Nu}{\sqrt{Re}}$ are displayed in Tables 7.3–7.6. Table 7.2 shows the results of the validation of the PINN output using the wavelet activation function against the available data.

Figure 7.1(a–c) presents the influence of m on f', θ, and g. The horizontal surface is represented by $m=0$ and the wedge geometries by $0<m<1$. $m=1$ represents the forward stagnation flow. The effects of changing the value of m on horizontal velocity are shown in Figure 7.1(a). At $m=0$, the horizontal velocity is lowest, while at $m=1$, it is highest. It is noted that the momentum boundary layer thickness reduces as the value of m increases. The effect of m on temperature distribution is presented in Figure 7.1(b). For $m=1$, the temperature is found to be higher close to the wall. But as we can see, the temperature is higher for $m=0$ as we go away from the wall. As the value of m increases, we see a decrease in the thickness of the thermal boundary layer, indicating that the heat transfer rate is optimal for forward stagnation flow and minimal for flat surfaces. As m increases, we can observe

TABLE 7.2

Comparison of the Values of $\frac{1}{2}C_{fx\sqrt{Re}}$ and $\frac{Nu}{\sqrt{Re}}$ with the Available Results [32]

Values of $\frac{1}{2}C_{fx}\sqrt{Re}$ for varying m when $C_1=C_2=C_3=C_4=C_5=1, K=0,$ $M=0, I=0.5, Ec=0, \beta_e=0, \beta_i=0.4, Pr=1$			Values of $\frac{Nu}{\sqrt{Re}}$ for varying Pr when $C_1=C_2=C_3=C_4=C_5=1, K=0,$ $M=0,\ I=0.5, Ec=0, \beta_e=0, \beta_i=0.4, m=0$		
m	[32]	Present results	Pr	[32]	Present results
0	0.3321	0.3464	0.1	0.2007	0.2705
1/3	0.7575	0.7586	1	0.459	0.4674
1	1.2326	1.2321	10	0.998	1.013

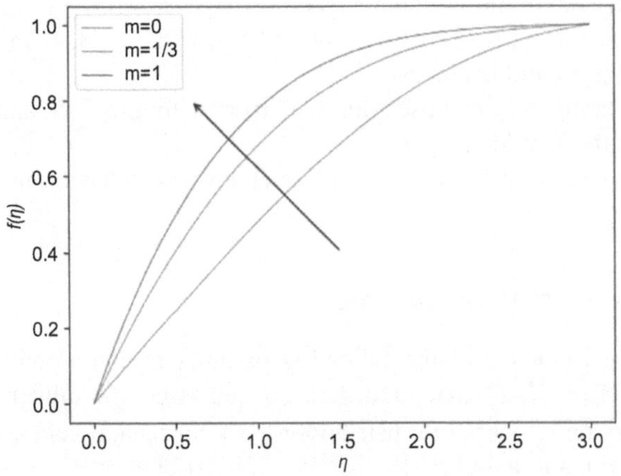

FIGURE 7.1(a) Profiles of f' for multiple values of m.

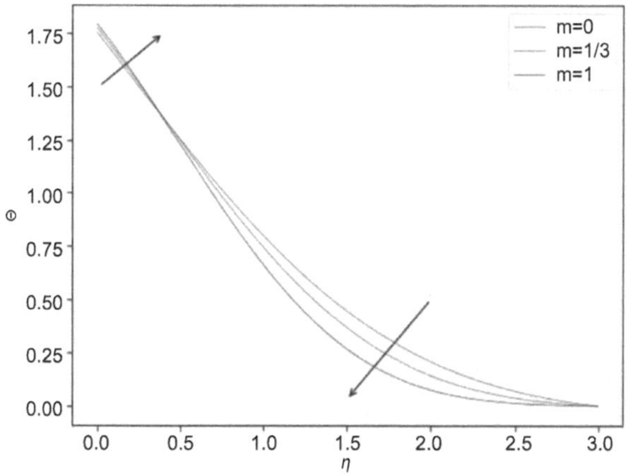

FIGURE 7.1(b) Profiles of θ for multiple values of m .

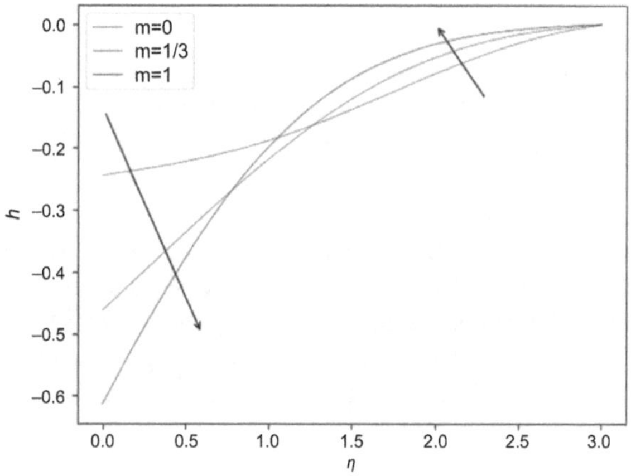

FIGURE 7.1(c) Profiles of h for multiple values of m .

a reduction in the magnitude of angular velocity away from the wall and an increase closer to the wall. For the flat surface, the highest magnitude of angular velocity is observed farther from the wall; for a forward stagnation flow, highest magnitude is observed closer to the wall. The effect that m has on angular velocity is shown in Figure 7.1(c). It is found that the angular velocity is significantly influenced by m. From Table 7.3, we can see an increase in the values of skin friction coefficients and Nusselt number with increasing values of m. That is largest for the forward stagnation flow and least for the flat surface.

Now we discuss the impact of Hall effect and ion-slip on horizontal velocity. From Figure 7.2(a), we can see that the horizontal velocity declines as β_e increases. However, in Figure 7.3(a) we can see an increase in horizontal velocity with the increase in β_i. The angular velocity of a conducting fluid can be influenced by the Hall effect and ion-slip, two phenomena related to the behaviour of charged particles in the presence of a magnetic field. Figures 7.2(c) and 7.3(c) illustrate how the absolute value of angular velocity increases with a decrease in β_e and increases with an increase in β_i. The transverse velocity decreases as β_e and β_i increase, as seen in Figures 7.2(d) and 7.3(d), which can be attributed to the decrease in force in the z direction because of the increase in Hall and ion-slip currents. The temperature distribution of the fluid is determined by its velocity gradients. From Figures 7.2(a) and 7.3(a), we can see that the horizontal velocity is only slightly affected by changes in β_e and β_i . Consequently, the temperature change is mostly caused by the transverse velocity, which decreases as β_e and β_i increase. There is a slight decrease in temperature with an increase in β_e and β_i, as can be seen from both Figures 7.2(b) and Figure 7.3(b). A similar conclusion was drawn by [4], wherein it is suggested that the slight temperature increase might be because of the small magnitude of transverse velocity. Table 7.3 displays a decrease in skin friction coefficient with an increase in β_e. However, we can see an increase in skin friction coefficient in the x direction with an increase in in β_i, but a decrease in skin friction coefficient in the z direction. Nusselt number increases with increasing β_e, and a slight increase is seen for β_i.

We now go into detail about how K affects f',g,h, and θ. The fluid exhibits Newtonian behaviour when $K=0$. Figure 7.4(a) shows that the greatest horizontal velocity is attained at $K=0$ and diminishes as K increases. The fluid's resistance to rotating movements inside the fluid is measured by the vortex viscosity, which increases the viscous effects on the fluid flow and causes a drop in horizontal velocity. A similar finding for transverse velocity presented in Figure 7.4(d) suggests a predominance of micropolar effects in the fluid flow. We can see from Figure 7.4(c) that there is a reverse behaviour of angular velocity in varying material parameters near the surface of the wedge and away from it. The magnitude of angular velocity decreases with increasing K; however near the free stream, the magnitude of angular

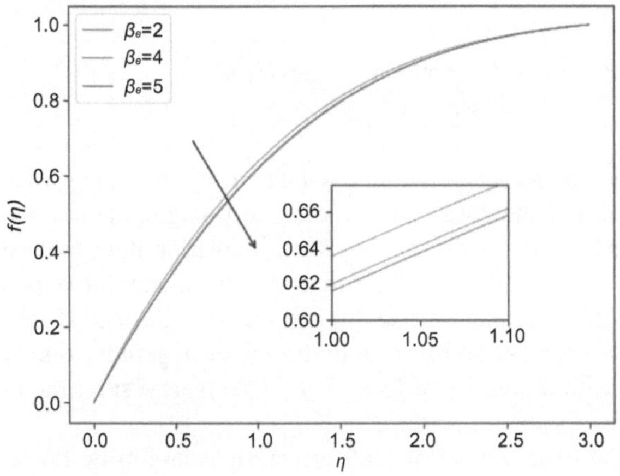

FIGURE 7.2(a) Profiles of f' for multiple values of β_e

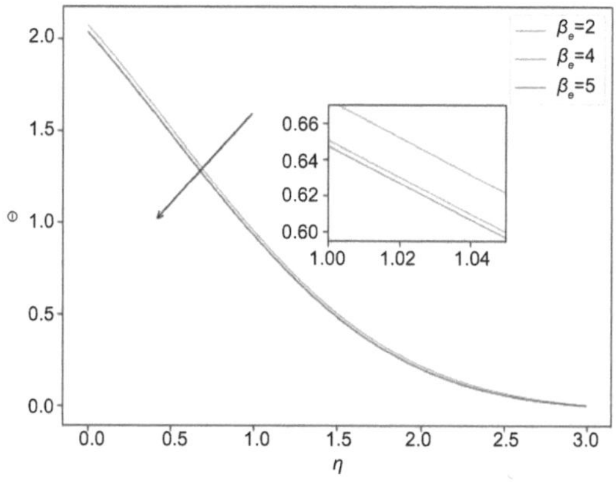

FIGURE 7.2(b) Profiles of θ for multiple values of β_e.

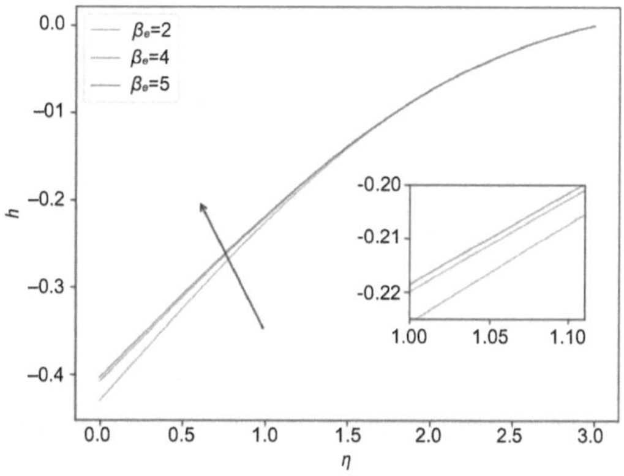

FIGURE 7.2(c) Profiles of h for multiple values of β_e.

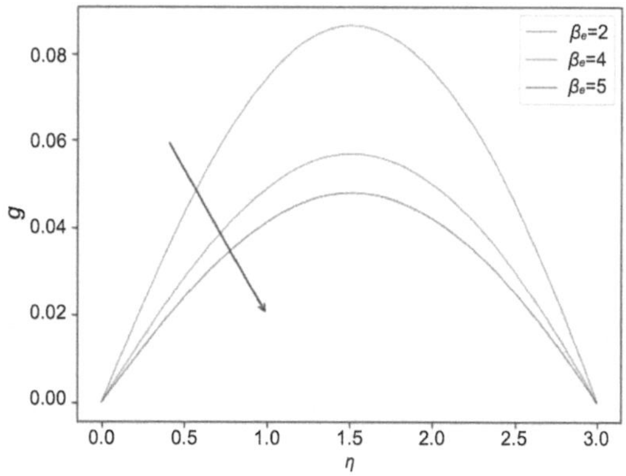

FIGURE 7.2(d) Profiles of g for multiple values of β_e.

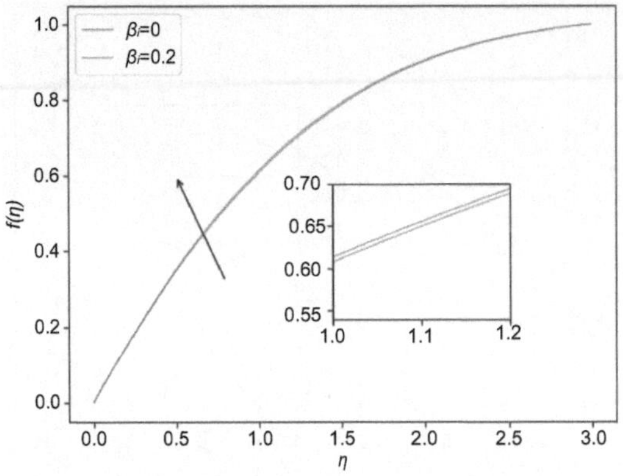

FIGURE 7.3(a) Profiles of f' for multiple values of β_i.

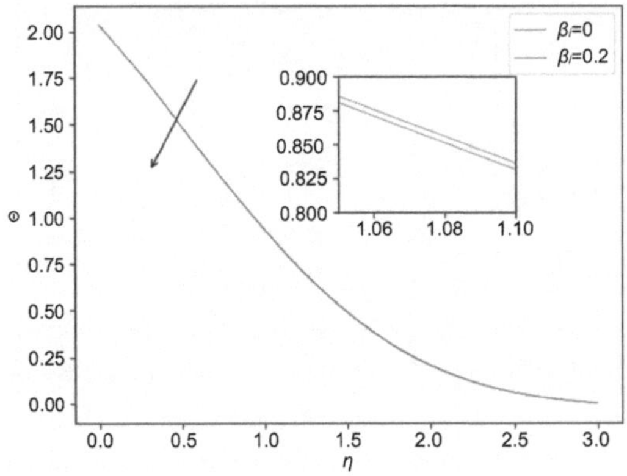

FIGURE 7.3(b) Profiles of θ for multiple values of β_i.

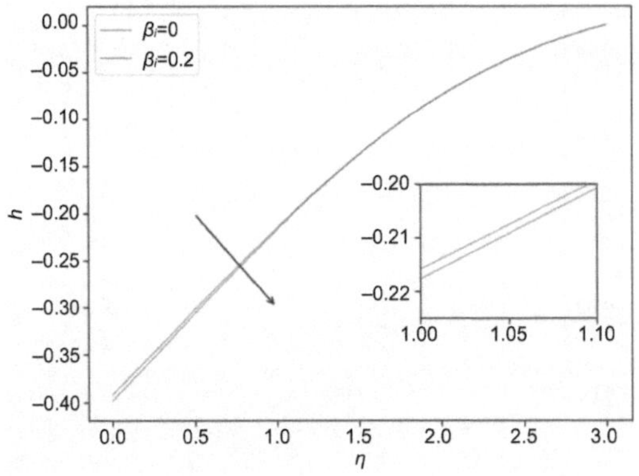

FIGURE 7.3(c) Profiles of h for multiple values of β_i.

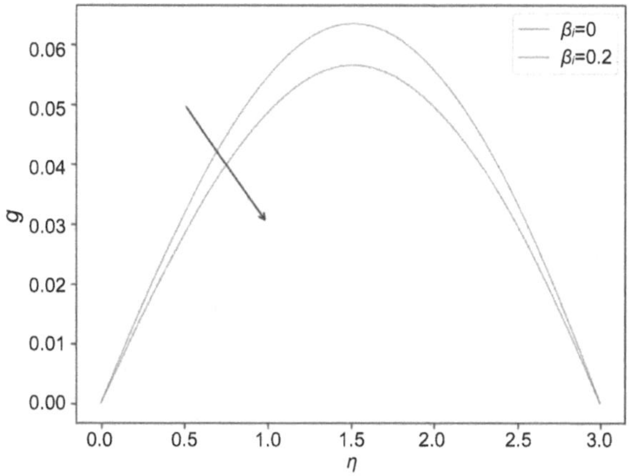

FIGURE 7.3(d) Profiles of g for multiple values of β_i.

TABLE 7.3

Values of $\frac{1}{2}C_{fx\sqrt{Re}}$, $\frac{1}{2}C_{fz\sqrt{Re}}$, $\frac{Nu}{\sqrt{Re}}$ for Varying β_e and β_i

$\phi_1 = \phi_2 = 0.02, K = 0, M = 0, l = 0.5,$ $Ec = 0.5, \beta_i = 0.4, Pr = 1, m = 1/3$				$\phi_1 = \phi_2 = 0.02, K = 1, M = 1, l = 0.5,$ $Ec = 0.5, \beta_e = 2, Pr = 1, m = 1/3$			
β_e	$\frac{1}{2}C_{fx}\sqrt{Re}$	$\frac{1}{2}C_{fz}\sqrt{Re}$	$\frac{Nu}{\sqrt{Re}}$	β_i	$\frac{1}{2}C_{fx}\sqrt{Re}$	$\frac{1}{2}C_{fz}\sqrt{Re}$	$\frac{Nu}{\sqrt{Re}}$
2	1.0302	0.1426	0.4035	0	0.9402	0.104	0.411
4	0.977	0.0936	0.4105	0.2	0.9561	0.0927	0.4113
5	0.9659	0.0788	0.4117	0.4	0.9661	0.079	0.4119

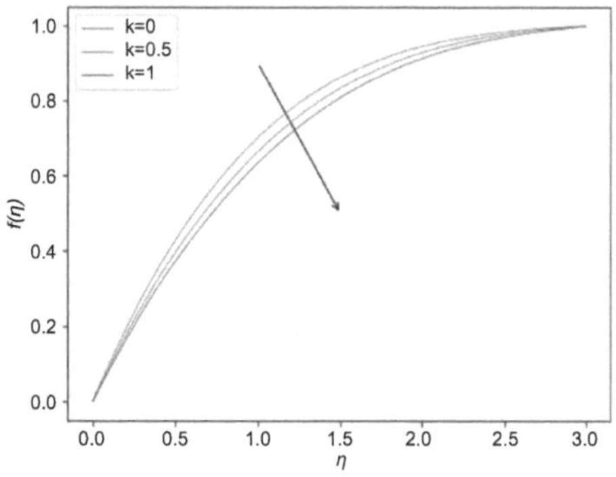

FIGURE 7.4(a) Profiles of f' for multiple values of K.

velocity increases with increasing K. As the fluid's viscosity increases, so do the viscous forces, producing greater heat. And, Figure 7.4(b) shows that as the value of K increases, so does the temperature. From Table 7.3, we can conclude an increase in the value of skin friction coefficient and decrease in Nusselt number with increasing value of the material parameter.

Figure 7.5 illustrates how temperature increases as Ec increases. This phenomenon can be attributed to the increase in the viscous dissipation as we increase Eckert number. It is also evident that when the value of Ec increases, the thickness of the thermal boundary layer increases, leading to a decrease in the rate of heat transmission.

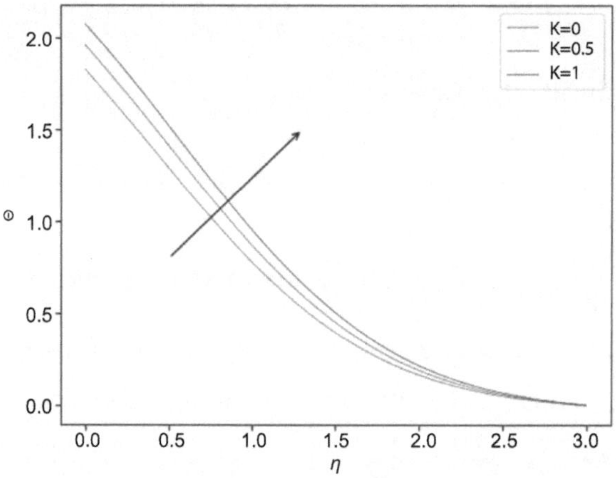

FIGURE 7.4(b) Profiles of θ for multiple values of K.

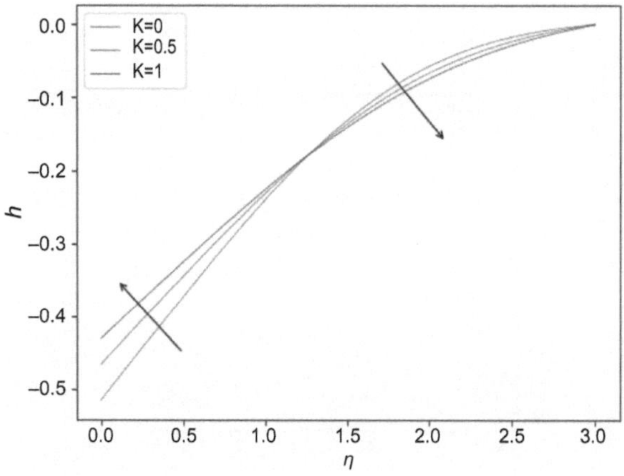

FIGURE 7.4(c) Profiles of h for multiple values of K.

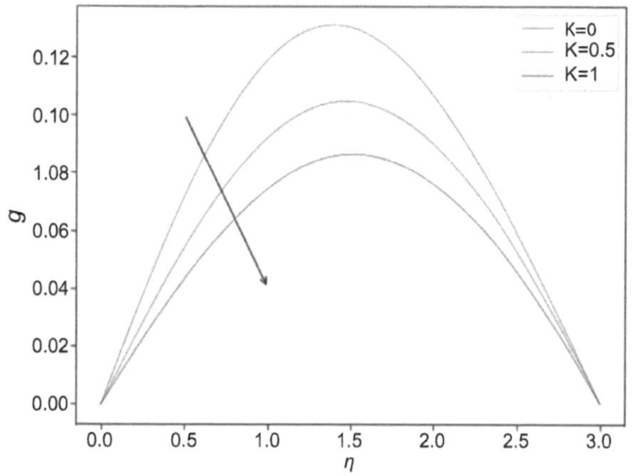

FIGURE 7.4(d) Profiles of g for multiple values of K.

TABLE 7.4
Values of $\frac{1}{2}C_{fz\sqrt{Re}}$, $\frac{1}{2}C_{fz\sqrt{Re}}$, and $\frac{Nu}{\sqrt{Re}}$, for Varying m and K

$\phi_1 = \phi_2 = 0.02, K = 0, M = 0, I = 0.5, Ec = 0.5, \beta_e = 0, \beta_i = 0.4, Pr = 1$				$\phi_1 = \phi_2 = 0.02, M = 1, I = 0.5, Ec = 0, \beta_e = 2, \beta_i = 0.4, Pr = 1, m = 1/3$			
m	$\frac{1}{2}C_{fx}\sqrt{Re}$	$\frac{1}{2}C_{fz}\sqrt{Re}$	$\frac{Nu}{\sqrt{Re}}$	K	$\frac{1}{2}C_{fx}\sqrt{Re}$	$\frac{1}{2}C_{fz}\sqrt{Re}$	$\frac{Nu}{\sqrt{Re}}$
0	0.5063	0.0003	0.4134	0	0.848	0.1262	0.4586
1/3	1.1069	0.0003	0.4711	0.5	0.9407	0.1372	0.4276
1	1.8001	0.0004	0.5715	1	1.0302	0.1426	0.4035

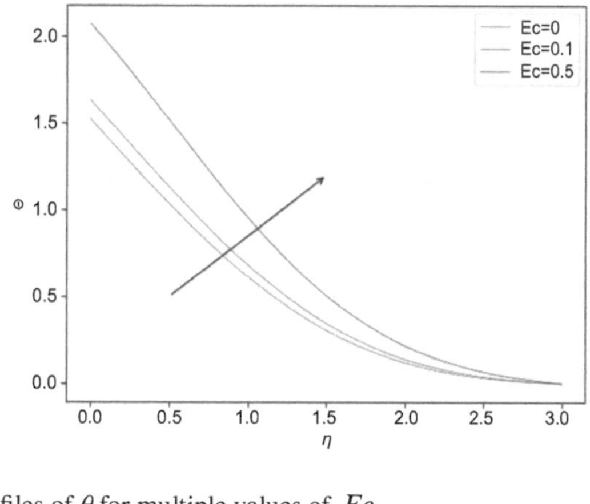

FIGURE 7.5 Profiles of θ for multiple values of Ec.

Figures 7.6 and 7.7 reveal the impact of the nanofluid concentration of MoS_2 and GO on the velocity and temperature. We can see that by keeping the concentration of GO fixed and increasing the concentration of MoS_2, temperature, horizontal velocity, transverse velocity, and magnitude of angular velocity increase. Increasing the concentration of GO, keeping the concentration of MoS_2 fixed, horizontal velocity is noted to decrease and transverse velocity is noted to increase; however, angular velocity is not significantly affected. Also, there is an increment in temperature with an increase in GO concentration. Table 7.5 reveals that the value of skin friction coefficient and Nusselt number reduces with the increasing concentration of MoS_2. For the case of GO, the value of skin friction coefficient increases with increasing GO concentration, while the value of Nusselt number decreases.

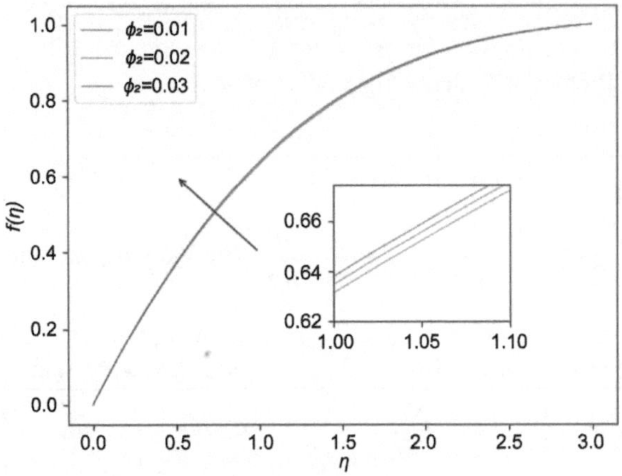

FIGURE 7.6(a) Profiles of f' for multiple values of ϕ_1 .

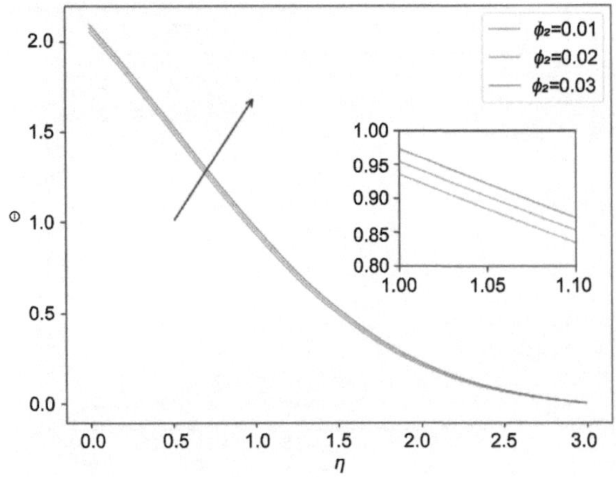

FIGURE 7.6(b) Profiles of θ for multiple values of ϕ_1 .

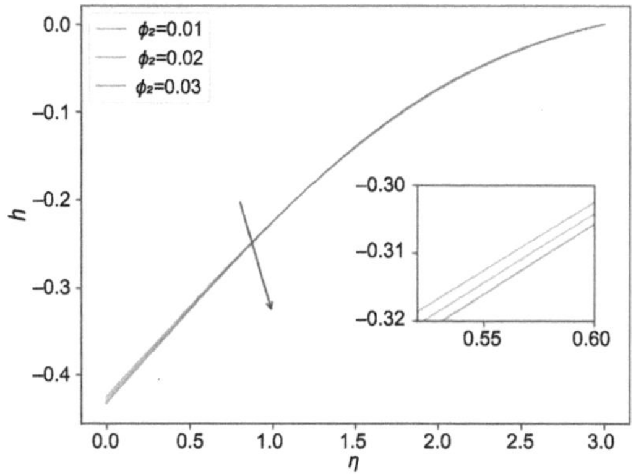

FIGURE 7.6(c) Profiles of h for multiple values of ϕ_1.

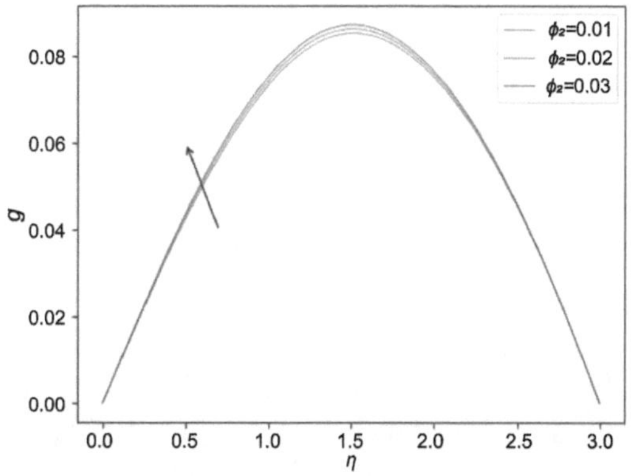

FIGURE 7.6(d) Profiles of g for multiple values of ϕ_1.

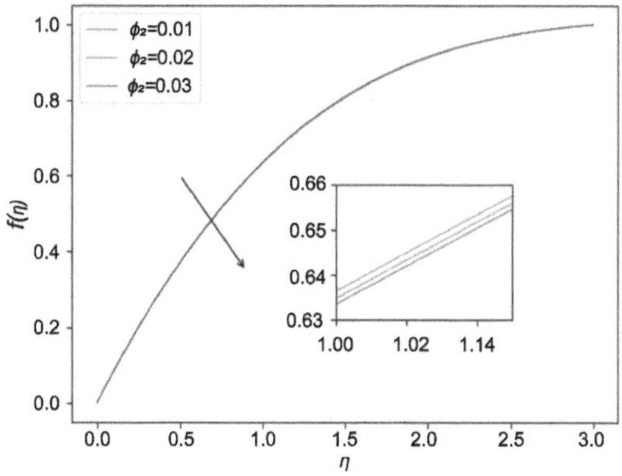

FIGURE 7.7(a) Profiles of f' for multiple values of ϕ_2.

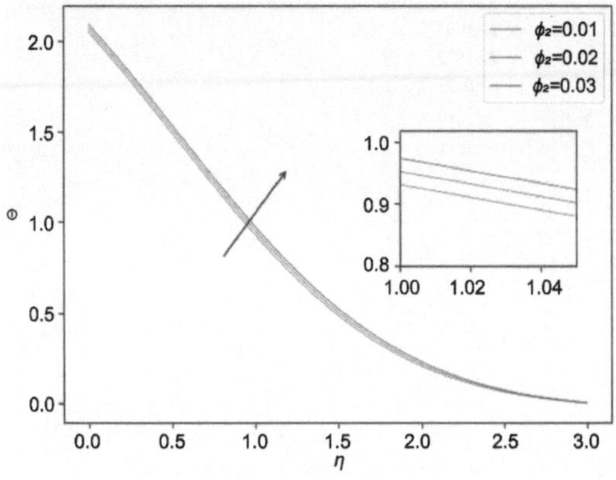

FIGURE 7.7(b) Profiles of θ for multiple values of ϕ_2.

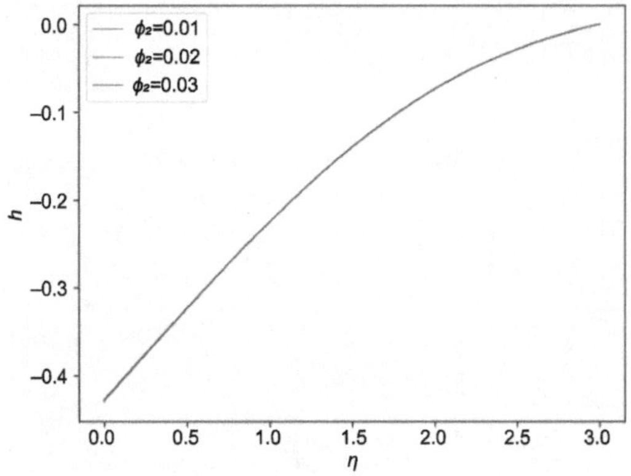

FIGURE 7.7(c) Profiles of h for multiple values of ϕ_2.

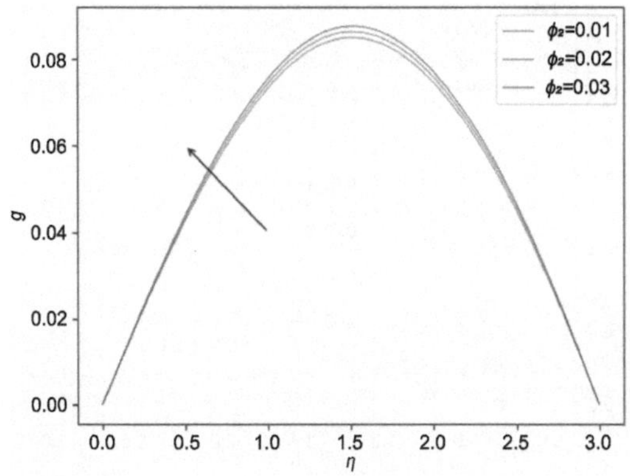

FIGURE 7.7(d) Profiles of g for multiple values of ϕ_2.

TABLE 7.5

Values of $\frac{1}{2}C_{fx\sqrt{Re}}$, $\frac{1}{2}C_{fz\sqrt{Re}}$, and $\frac{Nu}{\sqrt{Re}}$ for Varying ϕ_1 and ϕ_2

$\phi_2 = 0.02, K = 1, I = 0.5, Ec = 0.5,$ $\beta_e = 2, \beta_i = 0.4, Pr = 1, m = 1/3$				$\phi_1 = 0.02, K = 1, M = 1, Ec = 0.5$ $\beta_e = 2, \beta_i = 0.4, Pr = 1, m = 1/3$			
ϕ_1	$\frac{1}{2}C_{fx}\sqrt{Re}$	$\frac{1}{2}C_{fz}\sqrt{Re}$	$\frac{Nu}{\sqrt{Re}}$	ϕ_2	$\frac{1}{2}C_{fx}\sqrt{Re}$	$\frac{1}{2}C_{fz}\sqrt{Re}$	$\frac{Nu}{\sqrt{Re}}$
0.01	1.0413	0.1438	0.4054	0.01	1.0216	0.1396	0.4047
0.02	1.0296	0.1425	0.4035	0.02	1.0296	0.1425	0.4035
0.03	1.0191	0.1414	0.4016	0.03	1.0381	0.1455	0.4024

TABLE 7.6

Values of $\frac{1}{2}C_{fx\sqrt{Re}}$, $\frac{1}{2}C_{fz\sqrt{Re}}$, and $\frac{Nu}{\sqrt{Re}}$ for Varying M and I

$\phi_1 = \phi_2 = 0.02, K = 1, I = 0.5, Ec = 0.5,$ $\beta_e = 5, \beta_i = 0.4, Pr = 1, m = 1/3$				$\phi_1 = \phi_2 = 0.02, K = 1, M = 1, Ec = 0.5$ $\beta_e = 2, \beta_i = 0.4, Pr = 1, m = 1/3$			
M	$\frac{1}{2}C_{fx}\sqrt{Re}$	$\frac{1}{2}C_{fz}\sqrt{Re}$	$\frac{Nu}{\sqrt{Re}}$	I	$\frac{1}{2}C_{fx}\sqrt{Re}$	$\frac{1}{2}C_{fz}\sqrt{Re}$	$\frac{Nu}{\sqrt{Re}}$
0.5	0.9489	0.0406	0.413	0.5	1.0302	0.1426	0.4035
2	0.9959	0.154	0.4085	2	1.028	0.1422	0.4041
5	1.0459	0.3486	0.3969	4	1.0217	0.1416	0.4044

Table 7.6 displays the influence of M and I on the skin friction coefficient and Nusselt number. As M increases, the skin friction coefficient increases and the Nusselt number decreases, whereas the opposite behaviour is seen for I, i.e., the skin friction coefficient reduces with the increasing value of I and the Nusselt number decreases.

7.5 CONCLUSION

A machine learning approach of PINN using wavelet activation function was used to numerically study the micropolar hybrid nanofluid flow over a wedge. The influence of Hall and ion-slip effect was studied for investigating the electrically conducting fluid flow through a magnetic field. We considered a water-based GO-MoS_2 hybrid nanofluid. In the considered methodology, Gaussian wavelet was chosen, and we obtained reliable results. We analysed the impact of influencing parameters on velocity and temperature, and correspondingly on the skin friction coefficient and Nusselt number. It was observed that m and K have a significant impact on angular velocity, whereas the concentration of GO does not have any impact. In addition, the heat transfer rate was found to be only slightly influenced by β_i.

REFERENCES

1. Bing Kho, Y., Jusoh, R., Zuki Salleh, M., Hisyam Ariff, M., & Zainuddin, N. (2023). Magnetohydrodynamics flow of Ag-TiO_2 hybrid nanofluid over a permeable wedge with thermal radiation and viscous dissipation. Journal of Magnetism and Magnetic Materials, 565, 170284.
2. Falkneb, V. M., & Skan, S. W. (1931). LXXXV. Solutions of the boundary-layer equations. The London, Edinburgh, and Dublin Philosophical Magazine and Journal of Science, 12(80), 865–896.
3. Eringen, A. C. (1966). Theory of micropolar fluids. Journal of Mathematics and Mechanics, 1–18.
4. Uddin, Z., & Kumar, M. (2013). Hall and ion-slip effect on MHD boundary layer flow of a micro polar fluid past a wedge. Scientia Iranica, 20(3), 467–476.
5. Uddin, Z., Kumar, M., & Harmand, S. (2014). Influence of thermal radiation and heat generation/absorption on MHD heat transfer flow of a micropolar fluid past a wedge with hall and ion slip currents. Thermal Science, 18(2), 489–502.
6. Singh, K., Pandey, A. K., & Kumar, M. (2020). Slip flow of micropolar fluid through a permeable wedge due to the effects of chemical reaction and heat source/sink with hall and ion-slip currents: An analytic approach. Propulsion and Power Research, 9(3), 289–303.
7. Sharma, R. P., Mishra, S. R., Pattnaik, P. K., Tinker, S., & Allipudi, S. R. (2023). Numerical study of slip flow of a micropolar fluid through a porous wedge surface with the impact of a chemical reaction and heat source/sink. International Journal of Modern Physics B, 2450314.
8. Choi, S. U., & Eastman, J. A. (1995). Enhancing thermal conductivity of fluids with nanoparticles (No. ANL/MSD/CP84938; CONF-951135-29). Argonne, IL: Argonne National Lab (ANL).
9. Zaib, A., Haq, R. U., Sheikholeslami, M., & Khan, U. (2020). Numerical analysis of effective Prandtl model on mixed convection flow of γAl_2O_3–H_2O nanoliquids with micropolar liquid driven through wedge. Physica Scripta, 95(3), 035005.
10. Zulkifli, S. N., Sarif, N. M., Salleh, M. Z., & Samsudin, A. (2020). Magnetohydrodynamic micropolar nanofluid flow over a wedge with chemical reaction. In IOP Conference Series: Materials Science and Engineering (Vol. 991, No. 1, p. 012145). IOP Publishing.
11. El-Dawy, H. A., & Gorla, R. S. R. (2021). The flow of a micropolar nanofluid past a stretched and shrinking wedge surface with absorption. Case Studies in Thermal Engineering, 26, 101005.
12. Patil, P. M., Goudar, B., & Momoniat, E. (2023). Magnetized bioconvective micropolar nanofluid flow over a wedge in the presence of oxytactic microorganisms. Case Studies in Thermal Engineering, 49, 103284.
13. Umavathi, J. C. (2022). Electrically conducting micropolar nanofluid with heat source/sink over a wedge: Ion and hall currents. Journal of Magnetism and Magnetic Materials, 559, 169548.
14. Umavathi, J. C., Kumar, M. A., & Bég, O. A. (2023). Computation of micropolar nanofluid from a wedge with heterogeneous carbon/metallic nanoparticles, viscous dissipation and heat sink/source: Rheological nanocoating flow simulation. International Journal of Modelling and Simulation, 1–17.
15. Kakar, N., Khalid, A., Al-Johani, A. S., Alshammari, N., & Khan, I. (2022). Melting heat transfer of a magnetized water-based hybrid nanofluid flow past over a stretching/shrinking wedge. Case Studies in Thermal Engineering, 30, 101674.

16. Salman, S., Talib, A. R. A., Saadon, S., & Sultan, M. T. H. (2020). Hybrid nanofluid flow and heat transfer over backward and forward steps: A review. Powder Technology, 363, 448–472.
17. Dinarvand, S., Rostami, M. N., & Pop, I. (2019). A novel hybridity model for TiO_2-CuO/water hybrid nanofluid flow over a static/moving wedge or corner. Scientific Reports, 9(1), 16290.
18. Waini, I., Ishak, A., & Pop, I. (2020). MHD flow and heat transfer of a hybrid nanofluid past a permeable stretching/shrinking wedge. Applied Mathematics and Mechanics (English Edition), 41(3), 507–520.
19. Mahanthesh, B., Shehzad, S. A., Ambreen, T., & Khan, S. U. (2021). Significance of Joule heating and viscous heating on heat transport of MoS_2–Ag hybrid nanofluid past an isothermal wedge. Journal of Thermal Analysis and Calorimetry, 143(2), 1221–1229.
20. Khashi'ie, N. S., Waini, I., Mukhtar, M. F., Zainal, N. A., Hamzah, K. B., Arifin, N. M., & Pop, I. (2022). Response surface methodology (RSM) on the hybrid nanofluid flow subject to a vertical and permeable wedge. Nanomaterials, 12(22), 4016.
21. Zainal, N. A., Nazar, R., Naganthran, K., & Pop, I. (2022). Stability analysis of unsteady hybrid nanofluid flow over the Falkner-Skan wedge. Nanomaterials, 12(10), 1771.
22. Baby, R., Puneeth, V., Narayan, S. S., Khan, M. I., Anwar, M. S., Bafakeeh, O. T., Oreijah, M., & Geudri, K. (2023). The impact of slip mechanisms on the flow of hybrid nanofluid past a wedge subjected to thermal and solutal stratification. International Journal of Modern Physics B, 37(15), 2350145.
23. Al-Hanaya, A. M., Sajid, F., Abbas, N., & Nadeem, S. (2020). Effect of SWCNT and MWCNT on the flow of micropolar hybrid nanofluid over a curved stretching surface with induced magnetic field. Scientific Reports, 10(1), 8488.
24. Raju, S. S. K. (2023). Dynamical dissipative and radiative flow of comparative an irreversibility analysis of micropolar and hybrid nanofluid over a Joule heating inclined channel. Scientific Reports, 13(1), 5356.
25. Tassaddiq, A. (2021). Impact of Cattaneo-Christov heat flux model on MHD hybrid nano-micropolar fluid flow and heat transfer with viscous and Joule dissipation effects. Scientific Reports, 11(1), 67.
26. Algehyne, E. A., Haq, I., Raizah, Z., Alduais, F. S., Saeed, A., & Galal, A. M. (2023). A passive control strategy of a micropolar hybrid nanofluid flow over a convectively heated flat surface. Journal of Magnetism and Magnetic Materials, 567, 170355.
27. Uddin, Z., Hassan, H., Harmand, S., & Ibrahim, W. (2022). Soft computing and statistical approach for sensitivity analysis of heat transfer through the hybrid nanoliquid film in rotating heat pipe. Scientific Reports, 12(1), 14983.
28. Asifa, Anwar, T., Kumam, P., Shah, Z., & Sitthithakerngkiet, K. (2021). Significance of shape factor in heat transfer performance of molybdenum-disulfide nanofluid in multiple flow situations; A comparative fractional study. Molecules, 26(12), 3711.
29. Khashi'ie, N. S., Arifin, N. M., Nazar, R., Hafidzuddin, E. H., Wahi, N., & Pop, I. (2020). Magnetohydrodynamics (MHD) axisymmetric flow and heat transfer of a hybrid nanofluid past a radially permeable stretching/shrinking sheet with Joule heating. Chinese Journal of Physics, 64, 251–263.
30. Chu, Y. M., Nisar, K. S., Khan, U., Daei Kasmaei, H., Malaver, M., Zaib, A., & Khan, I. (2020). Mixed convection in MHD water-based molybdenum disulfide-graphene oxide hybrid nanofluid through an upright cylinder with shape factor. Water, 12(6), 1723.
31. Uddin, Z., Ganga, S., Asthana, R., & Ibrahim, W. (2023). Wavelets based physics informed neural networks to solve non-linear differential equations. Scientific Reports, 13(1), 2882.
32. Ishak, A., Nazar, R., & Pop, I. (2009). MHD boundary-layer flow of a micropolar fluid past a wedge with constant wall heat flux. Communications in Nonlinear Science and Numerical Simulation, 14(1), 109–118.

8 Heat Diffusion in Non-Newtonian Magnetized Hybrid Blood Flow over a Wedge-Shaped Region

Niraj Rathore and N. Sandeep

8.1 INTRODUCTION

The human body's intricate network of arteries and blood vessels is vital in supplying oxygen and nutrients to all the tissues and organs. However, when these pathways encounter obstructions or irregularities, it can lead to various health issues. One particular concern is the development of wedge-shaped defects or infarcts in arteries, often associated with blood flow problems that can have severe consequences. A wedge-shaped artery defect may indicate an area of the brain that has suffered from inadequate blood flow, leading to ischaemic damage. Ischaemic strokes are the most common type of stroke and can result from a blood clot or an embolism that blocks blood flow to a specific brain region.

Wedge-shaped artery defects are a radiological or pathological sign of restricted blood flow in the brain, often caused by underlying health conditions and diseases. Recognizing these patterns is crucial for early diagnosis and appropriate treatment to prevent further damage and improve patient outcomes.

Mathematical modelling is crucial in understanding the haemodynamics (blood flow) of the arteries. Computational fluid dynamics (CFD) simulations, which involve solving complex mathematical equations, can help researchers and clinicians visualize how blood flows through arteries. Studying these models can help identify regions of disturbed flow, turbulence, or areas with low shear stress that may be more prone to plaque formation and thrombosis, contributing to wedge artery diseases. This knowledge can guide interventions and treatments to improve blood flow. Advanced medical imaging systems like magnetic resonance imaging (MRI) and CT scans rely deeply on mathematical processes for image reconstruction and investigation [1–3]. These tools enable early detection of arterial defects and contribute to the diagnosis of wedge artery diseases. Additionally, mathematical algorithms can help improve image quality and reduce noise in medical imaging, making it easier to identify and characterize arterial abnormalities.

Nanoparticles are important in the biomedical field due to their unique properties and versatile applications. Nuzhat et al. [4] proved that small-size nanoparticles, typically in the nanometre range (1–100 nm), allow them to interact with biological

DOI: 10.1201/9781003595786-8

systems in ways that larger materials cannot. Biomedical applications are utilized in the cure of arterial diseases by many researchers. Shafiq et al. [5] used gold (Au) nanoparticles in thermofluids and proved that the energy transfer speed in the blood flow can be enhanced by using gold nanoparticles. Similarly, ferromagnetic nanoparticles are important in the biomedical field, particularly in MRI and targeted cancer treatment; the authors of [6, 7] used ferro-nanoparticles and analysed the energy and blood flow for different external physical parameters. The researchers studied various nanoparticles, such as gold (Au), alumina (Al_2O_3), and combined nanoparticles TiO_2-Ag, in the blood flow for improved drug delivery results. Nanoparticles offer precision, versatility, and control in biomedical applications, allowing researchers and clinicians to develop innovative therapies, diagnostics, and treatments that significantly improve patient outcomes while minimizing side effects.

The geometry of an artery has a significant impact on the blood flow dynamics within it. Blood flow is subjective to numerous elements, including the size, shape, and structure of the artery. The major challenge in developing a mathematical model for blood flow is that it varies according to the geometrical shape and nature of stenosis of the artery. The mathematical models for cylindrical, curved shape, and wedge-shaped arteries are considered by different researchers [8–11]. Blood flow through cylindrical arteries over cosine-shaped stenosis and the blood flow in combination with different nanoparticles were analysed by Anber et al. [12]. They noticed that hybrid nanofluids produced more significant and favourable results in drug delivery.

Arteries often branch into smaller vessels, and these branching points can affect blood flow. At the bifurcations, blood flow may experience changes in velocity and pressure, which can impact flow patterns and contribute to turbulence. The flow behaviour and thermal nature of blood over a flat region, wedge, and stagnation point was analysed by Basha and Sivaraj [13]. They perceived that the energy transmission speed of nano-blood flow was more on the stagnation point than on the wedge and flat region. Exploration of this study provides a view of the energy and mass transmission behaviour of the bloodstream in the vascular system, which is helpful in drug delivery and hyperthermia treatment. Blood movement in curve-shaped arteries is kept vertically, and Zaman et al. studied the flow and thermal profiles for different physical parameters [14]. The symmetry of the bloodstream is majorly dependent on the shape of the artery. On growing the magnetic effect, the magnitude of motion and wall shear stress decrease. A study on six types of different stenosed regions in circular cylinders was done by Ashfaq and Sohil [15]. The speed, flow rate, stress on the wall, and resistance impedance were observed in this study. The flow speed was found to be high in trapezoid-shaped stenosis in comparison to that in irregular arteries. At the same time, resistance impedance was more elevated in the bell-shaped stenosis region and lower in the trapezoidal region. Hence, the nanoparticles and geometry of the object play a significant role in regulating the flow and thermal nature of the blood flow. Different nanoparticles express different natures together with the fluids. According to Poiseuille's law, stream speed (Q) is relative to the fourth power of the radius (r) of the artery. This means that even small changes in arterial diameter can have a substantial impact on blood flow. A narrower artery has a higher resistance to blood flow, while a wider artery allows for increased flow.

Al_2O_3, also known as alumina, is a versatile material with various applications in the biomedical field, particularly in the form of nanoparticles. Al_2O_3 nanoparticles are attractive for biomedical applications, due to their biocompatibility, high surface area, durability, and chemical stability. The biomedical applications given in [16] for Al_2O_3 nanoparticles include drug delivery [17], imaging, tissue engineering, cancer therapy [18], and antimicrobial therapy.

Graphene oxide (GO) nanoparticles are a type of nanomaterial derived from GO, which is a single layer of carbon atoms arranged in a hexagonal lattice with oxygen-containing functional groups (such as epoxides, hydroxyls, and carboxyls) attached to its surface. Graphene itself is a two-dimensional carbon allotrope with exceptional electrical, mechanical, and thermal properties. GO nanoparticles have found applications in a wide range of fields, including materials science, electronics, nanotechnology, biomedicine, environmental remediation, and more. GO nanoparticles have gained significant attention in various biomedical applications due to their unique properties. The biomedical properties of GO nanoparticles are biocompatibility, wide surface area, and two-dimensional planar structure, making them useful in biomedical fields [20]. The applications of GO nanoparticles in biomedicine include drug delivery, imaging, cancer therapy, biosensing, antibacterial agents, and gene delivery [21].

Inspired by the extraordinary properties and biomedical applications of Al_2O_3 and GO nanoparticles, the authors of this chapter analysed the thermal, momentum, drag, and heat transmission rate in nanofluids and hybrid nanofluids. The main focus of this study was to analyse the behaviour of nanofluids and hybrid nanoparticles in blood flow and its thermal changes. Blood is considered in wedge-shaped arteries and assumed to be a non-Newtonian fluid; in this study, a viscoelastic flow model is used to describe the flow phenomenon. The novel results for nanofluids and hybrid nanofluids were obtained, and the comparison is shown graphically. Numerical examination was performed to investigate the outcomes, and a comparison was made between nanofluids and hybrid nanofluids.

8.2 WEDGE-SHAPED ARTERY

Arteries often branch into smaller vessels, and these branching points can affect blood flow. At the bifurcations, blood may experience changes in velocity and pressure, which can impact flow patterns and contribute to turbulence. Medical professionals and researchers study bifurcated arteries and their haemodynamics to better understand the factors that contribute to vascular diseases and to develop treatment strategies that can address these challenges. Techniques such as CFD and medical imaging are used to assess blood flow patterns and identify risk of disease development in bifurcated arteries in wedge-shaped regions. The schematic diagram (see Figure 8.1) of a bifurcated artery and wedge area is given below.

8.3 MATHEMATICAL MODELLING OF FLOW IN A WEDGE-SHAPED REGION

Consider 2D blood flow over a wedge-shaped region in the artery. It is assumed that the flow takes place along the x-direction, and the y-axis is perpendicular to

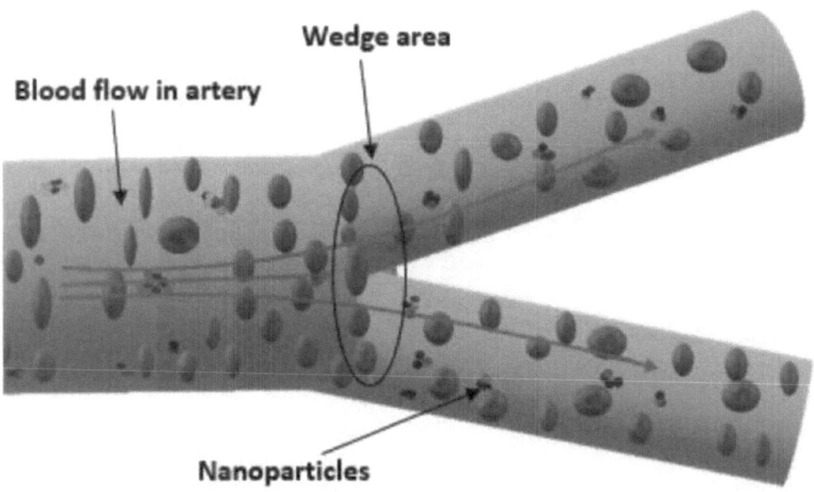

FIGURE 8.1 Schematic of a bifurcated artery in which a wedge-shaped region is highlighted. The movement of blood together with nanoparticles is shown in this figure.

it. The flow velocity $U_w = ax^n$ is assumed to be adjacent to the artery wall, and $U_e = bx^n$ is free stream velocity, where b is a positive constant (see Figure 8.2(a, b)). It is assumed that the boundary of the arterial wall is stretchable, and it allows the flow to slip and the velocity term is given as $U = U_w + \lambda \frac{\partial U}{\partial y}$. Here, λ is the slip factor and $U_w = ax^n$ is the stretch/shrunk velocity of the wedge, where $U_w > 0$ represents the stretching boundary, $U_w < 0$ signifies the shrinking boundary, and $U_w = 0$ corresponds for static boundary layers. The blood flow field is influenced by variable magnetic $B = B_0 x^{\frac{n-1}{2}}$ [10], where B_0 denotes the magnetic strength. The buoyancy effect at an angle α, the Darcy–Forchheimer effect, the Joule effect, thermal radiation effect, and heat sink/source parameters are applied in the energy equation together with convective boundary conditions for external thermal variation.

With the assumptions made above, the modelling terms are given as [10]

$$\frac{\partial U}{\partial x} + \frac{\partial V}{\partial y} = 0, \tag{8.1}$$

$$\rho_{nf}\left(V\frac{\partial U}{\partial y} + U\frac{\partial U}{\partial x}\right) = \mu_{nf}\left(\frac{\partial^2 U}{\partial y^2}\right) + k_0\left(\begin{array}{c} U\dfrac{\partial^3 U}{\partial x \partial y^2} + V\dfrac{\partial^3 U}{\partial y^3} \\ -\dfrac{\partial U}{\partial y}\dfrac{\partial^2 U}{\partial x \partial y} + \dfrac{\partial U}{\partial y}\dfrac{\partial^2 U}{\partial y^2} \end{array}\right) - \frac{c^* U^2}{\sqrt{k}} - \sigma_{nf} B^2 U \left.\vphantom{\int}\right\} + g(\rho\beta)_{nf}(T - T_\infty)\sin(\alpha), \tag{8.2}$$

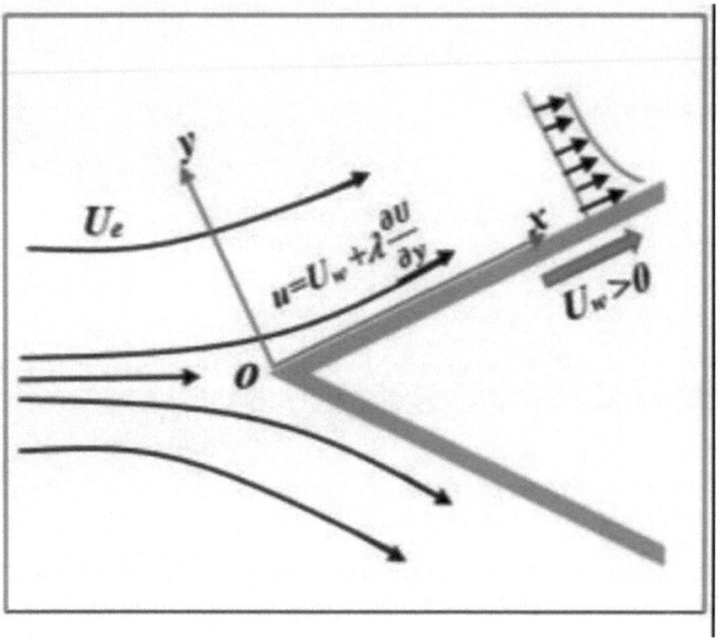

FIGURE 8.2(a) Stretching wedge-shaped geometry.

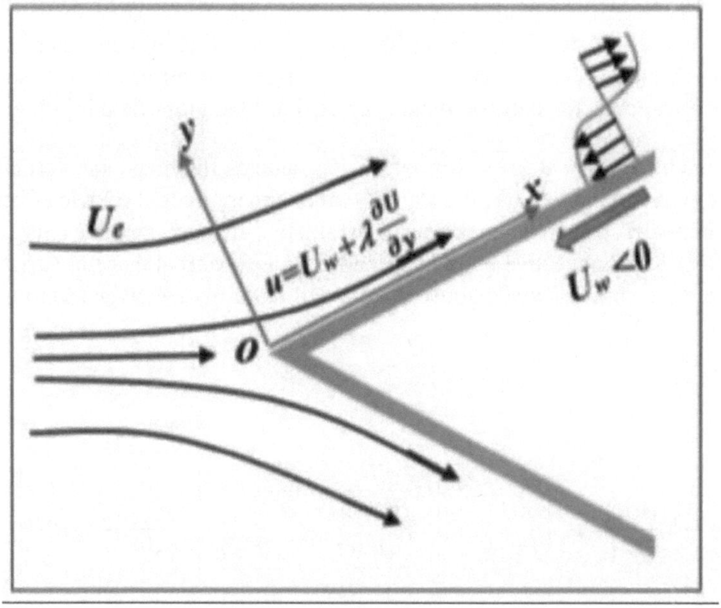

FIGURE 8.2(b) Shrinking wedge-shaped geometry.

$$(\rho C_p)_{nf}\left(U\frac{\partial T}{\partial x}+V\frac{\partial T}{\partial y}\right)=k_{nf}\frac{\partial^2 T}{\partial y^2}+\mu_{nf}\left(\frac{\partial U}{\partial y}\right)^2+\frac{16\sigma^* T_\infty^3}{3k^*}\frac{\partial^2 T}{\partial y^2}+\sigma_{nf}B_0^2U^2+q''' \quad (8.3)$$

With the border restrictions given as:

$$\left.\begin{aligned}&U=\lambda\frac{\partial U}{\partial y}+U_w,\ V=V_w, at\ y=0, U=U_e(x),\text{ as } y\to\infty,\\&-\frac{k_f}{h_f}\frac{\partial T}{\partial y}=-(T_w-T), at\ y=0, T\to T_\infty,\text{ as } y\to\infty\end{aligned}\right\} \quad (8.4)$$

Similarity and hybrid nanoparticle parameters are given by

$$\left.\begin{aligned}&U=ax^n\,F'(\eta),\ V=-\sqrt{\frac{\upsilon_f(n+1)ax^{n-1}}{2}}\left(\frac{(n-1)}{(n+1)}\eta F'(\eta)+F\right),\\&\eta=y\sqrt{\frac{(n+1)ax^{n-1}}{2\upsilon_f}},\ T-T_\infty=\Theta(T_w-T_\infty),\end{aligned}\right\} \quad (8.5)$$

where $U_w=ax^n$ and $U_e=bx^n$ are the velocity at the wall and free stream, respectively. $V_w=\sqrt{\frac{n+1}{2}}\sqrt{\upsilon_f ax^{n-1}}$ denotes the suction/injection velocity of the flow. T is the temperature of the liquid, and T_∞ is the ambient temperature, T_w is the temperature at the wall, and q''' is the heat generation and absorption parameter, which given as:

$$q'''=\frac{k_fU_w}{x\nu_f}(T-T_\infty)\left(\frac{(T_w-T_\infty)}{(T-T_\infty)}A^*f'+B^*\right).$$

Here, ϕ_1, ϕ_2 are the nanoparticle's volume fractions. Suffixes f, and s represent fluid and solid elements respectively.

Now, using Equation (8.5) and Table 8.1, Equations (8.1)–(8.4) become

$$\left.\begin{aligned}&A\left(\left(\frac{n+1}{2}\right)FF''-nF'^2\right)+B\left(\frac{n+1}{2}\right)F'''+K_1\left(\frac{n+1}{2}\right)\\&\begin{pmatrix}(3n-1)F'F'''-\left(\frac{n+1}{2}\right)\\FF^{iv}-\left(\frac{3n-1}{2}\right)F''^2\end{pmatrix}+C\frac{Gr}{\text{Re}^2}\sin(\alpha)\theta-FrF'^2-DMF'=0\end{aligned}\right\} \quad (8.6)$$

TABLE 8.1
Thermophysical Traits of Nanofluids [24, 25]

Feature	Mono and hybrid nanofluid parameters
Density	$A=\frac{\rho_{hnf}}{\rho_f}=(1-\phi_2)\left((1-\phi_1)+\frac{\phi_1\rho_{1s}}{\rho_f}\right)+\phi_2\frac{\rho_{2s}}{\rho_f}$
Viscosity	$B=\frac{\mu_{hnf}}{\mu_f}=\frac{1}{((1-\phi_1)(1-\phi_2))^{2.5}}$
Thermal expansion	$C=\frac{(\rho\beta)_{nf}}{(\rho\beta)_f}=(1-\phi_2)\left((1-\phi_1)+\phi_1\frac{(\rho\beta)_{1s}}{(\rho\beta)_f}\right)+\phi_2\frac{(\rho\beta)_{2s}}{(\rho\beta)_f}$
Electrical conductivity	$D=\frac{\sigma_{hnf}}{\sigma_f}=1\frac{3(\sigma_{1s}\phi_1-(\phi_1+\phi_2)\sigma_f)+\phi_{2s}\sigma_{2s}}{\sigma_{1s}(1-\phi_1)+\sigma_{2s}(1-\phi_2)+(2+\phi_1+\phi_2)\sigma_f}$
Heat capacity	$E=\frac{(\rho C_p)_{hnf}}{(\rho C_p)_f}=(1-\phi_2)\left((1-\phi_1)+\frac{\phi_1(\rho C_p)_{1s}}{(\rho C_p)_f}\right)+\phi_2\frac{(\rho C_p)_{2s}}{(\rho C_p)_f}$
Thermal conductivity	$F_1=\frac{k_{hnf}}{k_f}=\frac{k_{2s}+2k_f-2\phi_2(k_f-k_{2s})}{k_{2s}+2k_f+\phi_2(k_f-k_{2s})}\times k_{nf}$, where $k_{nf}=\frac{k_{1s}+2k_f-2\phi_1(k_f-k_{1s})}{k_{1s}+2k_f+\phi_1(k_f-k_{1s})}$

$$E\,\mathrm{Pr}\,F\Theta'+F_1\Theta''+Ra\,\Theta''+B\,Ec\,\mathrm{Pr}\,F''^2+\left(\frac{2}{n+1}\right)MEc\,\mathrm{Pr}\,F'^2$$
$$+\left(\frac{2}{n+1}\right)\mathrm{Pr}\left(A^*F'+B^*\Theta\right)=0 \tag{8.7}$$

The non-dimensional limit settings are

$$\left.\begin{aligned}&F(0)=1,\ F'(0)-\delta_1 F''(0)=1,\ \text{and}\ F'(\eta)\to\frac{b}{a},\ as\ \eta\to\infty\\&-\Theta'(0)=Bi(1-\Theta(0)),\text{and}\ \Theta(\eta)\to 0,\ as\ \eta\to\infty\end{aligned}\right\} \tag{8.8}$$

The dimensionless constraints in Equations (8.6)–(8.8) are mentioned in the following details:

Here, the velocity slip parameter $\delta_1 = \lambda\sqrt{\frac{(n+1)U_w}{2x\nu_f}}$ is [22], Biot number $Bi = \frac{h_f}{k_f}\sqrt{\frac{2\upsilon_f x}{(n+1)U_w}}$, viscoelastic parameter $K_1 = \frac{k_0 x^{n-1}}{\mu_f}$ [23], Grashoff number $Gr = \frac{g\beta_f (T_w - T_\infty)x^3}{\upsilon_f^2}$, Reynolds number $\mathrm{Re} = \frac{U_w x}{\upsilon_f}$, Forchheimer number $Fr = \frac{C^* x}{\rho_f \sqrt{k}}$ (where k is the porosity of the blood vessel wall and C^* is the resistance factor), magnetic quantity $M = \frac{\sigma_f B^2}{a\rho_f}$, Rayleigh number $Ra = \frac{16\sigma^* T_\infty^3}{3k^* k_f}$, Prandtl number $\Pr = \frac{(\mu C_p)_f}{k_f}$, Eckert number $Ec = \frac{U_w^2 (T_w - T_\infty)^{-1}}{(C_p)_f}$, and heat source parameter $Q_1 = \frac{Q_0}{a(\rho C_p)_f}$.

The physical quantities, i.e., resistance measurement C_{fx} and heat transmission quantity Nu_x of blood flow, are given as

$$C_{fx} = \frac{2\tau_w}{\rho_f U_w^2}, \tag{8.9}$$

$$Nu_x = \frac{xq_w (T_w - T_\infty)^{-1}}{k_f}. \tag{8.10}$$

The non-dimensional form of Equations (8.9) and (8.10) becomes [22]

$$\sqrt{\mathrm{Re}}C_{fx} = \frac{\mu_{hnf}}{\mu_f}\sqrt{\frac{n+1}{2}}F''(0), \tag{8.11}$$

$$\frac{1}{\sqrt{\mathrm{Re}}}Nu_x = -\frac{k_{hnf}}{k_f}(1+Ra)\sqrt{\frac{n+1}{2}}\Theta'(0). \tag{8.12}$$

8.4 OUTCOMES AND DISCOURSE OF RESULTS

Results for non-dimensional Equations (8.6) and (8.7) together with limit settings (8.8) are solved by means of the bvp5c method and computed using MATLAB software, and outcomes are drawn graphically and in a tabular form. For favourable results and better convergence of the velocity and thermal profiles, the physical constraints are in the range $\text{Re} = 100, 0.5 \le M \le 1.5,\ 0.4 \le \delta \le 0.8,\ 0.3 \le \text{Ec} \le 0.7,\ 0.1 \le K_1 \le 0.3,\ 1 \le \text{Fr} \le 5,$ $0.2 \le \text{Ra} \le 1, 0.1 \le A^* \le 0.8, 0.1 \le B^* \le 0.8, 0.1 \le \text{Bi} \le 0.9$. The effect of varying physical factors on the fluid stream profiles is displayed in Figures 8.3–8.17. The change in heat transfer rate is presented in Figures 8.18–8.23. Table 8.2 presents the physical properties of nanofluids.

TABLE 8.2
Thermophysical Belongings of Base Liquids and Nanoparticles [26, 27]

Type of particles	$\rho\left(Kgm^{-3}\right)$	$\beta\ (K^{-1})$	$C_p\left(JK^{-1}/kg\right)$	$K\left(WK^{-1}m^{-1}\right)$	$\sigma\left(Sm^{-1}\right)$
Blood	1050	0.8×10^{5}	3594	0.492	0.8
Al_2O_3	3970	0.85×10^{-5}	765	40	1×10^{-10}
GO	3600	1.25×10^{-5}	765	3000	6.30×10^{7}

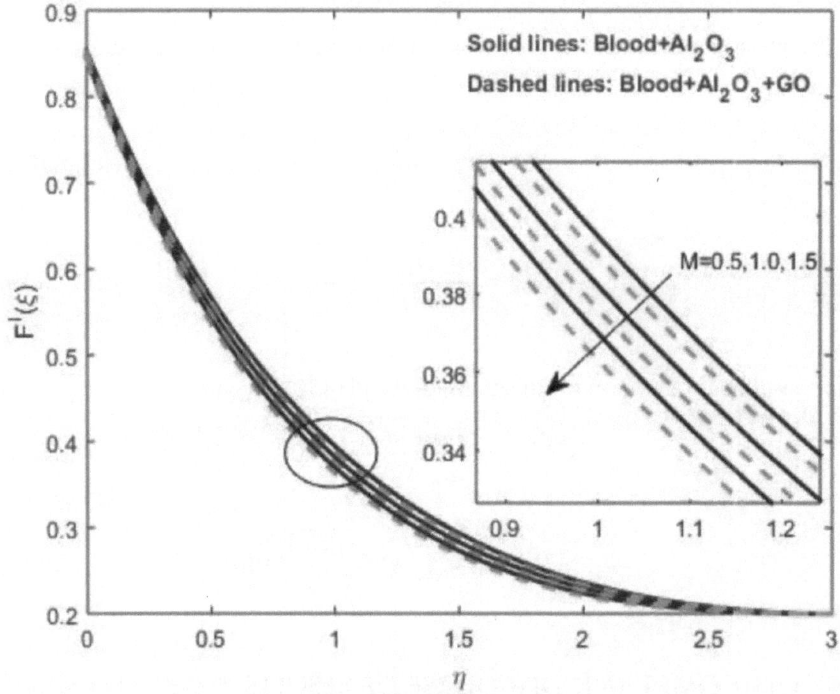

FIGURE 8.3 Velocity of nanofluids and hybrid nanofluids for increasing values of magnetic numbers *M*.

Figures 8.3 and 8.4 reveal the variation in thermal and velocity profiles of nanofluids and hybrid nanoliquids for increasing values of magnetic parameters. Here, an increment in M increases the heat and diminishes the velocity of both fluids. Also, it can be seen that hybrid nanofluids have less velocity than the nanofluids, while the temperature is the same in both fluids for increasing values of magnetic parameters. Enforcing a magnetic field in an electrically guiding fluid can persuade electric currents within the fluid. These currents, known as induced or Lorentz currents, generate their own magnetic fields. According to the Lorentz force law, these currents experience a force with the existence of an exterior magnetic field. This force opposes the motion of the fluid, resulting in a decrease in fluid velocity. In other words, the magnetic field exerts a drag force on the fluid, which decelerates it. In the case of temperature, the energy lost by the fluid as it slows down due to the magnetic field-induced drag must go somewhere. In many cases, this energy is converted into heat through various mechanisms, including viscous dissipation (internal friction) and resistive heating (due to the electrical currents induced in the fluid). As the fluid loses kinetic energy, its temperature can increase due to this conversion of mechanical energy into thermal energy.

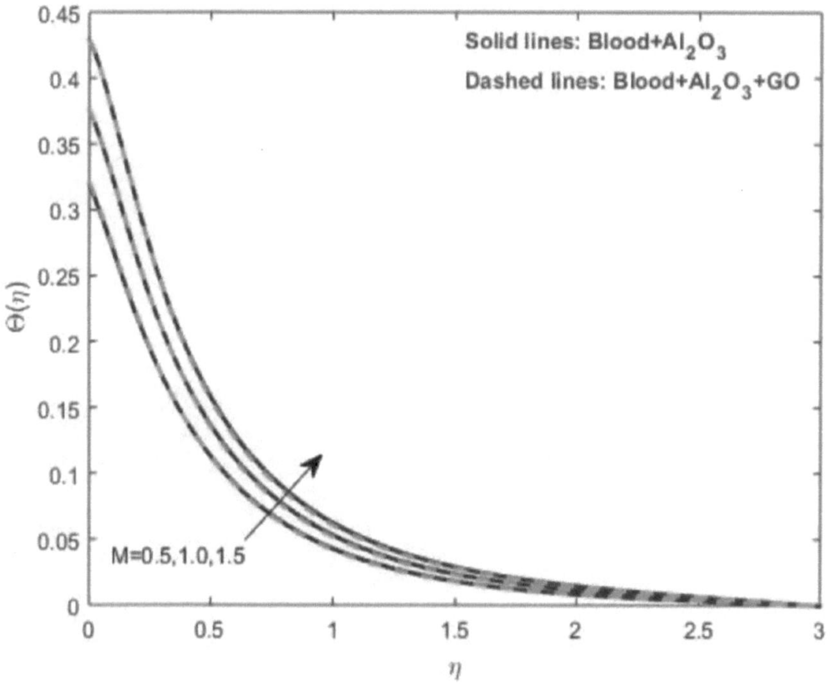

FIGURE 8.4 Thermal profiles for nanofluids and hybrid nanofluids for increasing values of *M*.

Figures 8.5–8.6 display the alteration in velocity and thermal outlines for nanofluids and hybrid nanofluids for increasing values of slip parameters. It is observed that the stream and thermic outlines are decreased to improve the standards of slip parameters. Here, velocity profiles in the case of hybrid nanofluid parameters are less than those of the nanofluid profiles.

Non-Newtonian fluid parameters and their impact on velocity and thermal profiles are displayed in Figures 8.7–8.8, which shows the velocity upsurges for growing standards of non-Newtonian fluid parameters. For increasing values of non-Newtonian fluid parameters, the temperature profiles near the boundary region decrease, while the temperature declines at a small distance of the boundary layer. Physically, blood is a non-Newtonian fluid that reveals shear-thinning behaviour, meaning its viscosity reductions with growing nature depend on the shear rate. As we apply more shear force to the liquid, its viscosity decreases, which allows for higher velocity.

Temperature profiles grow in both nanofluids and hybrid nanofluids for increasing standards of Eckert number Ec, radiation parameters Ra, and Prandtl number Pr shown in Figures 8.9–8.11, respectively. It is observed that the temperatures of nanofluids and hybrid nanofluids are almost the same in all three cases. Generally, when the Eckert number is increased, it signifies that the fluid's kinetic energy is more significant than the heat input, and the flow becomes more turbulent. As a result, the kinetic energy is converted into heat within the fluid, which results in an

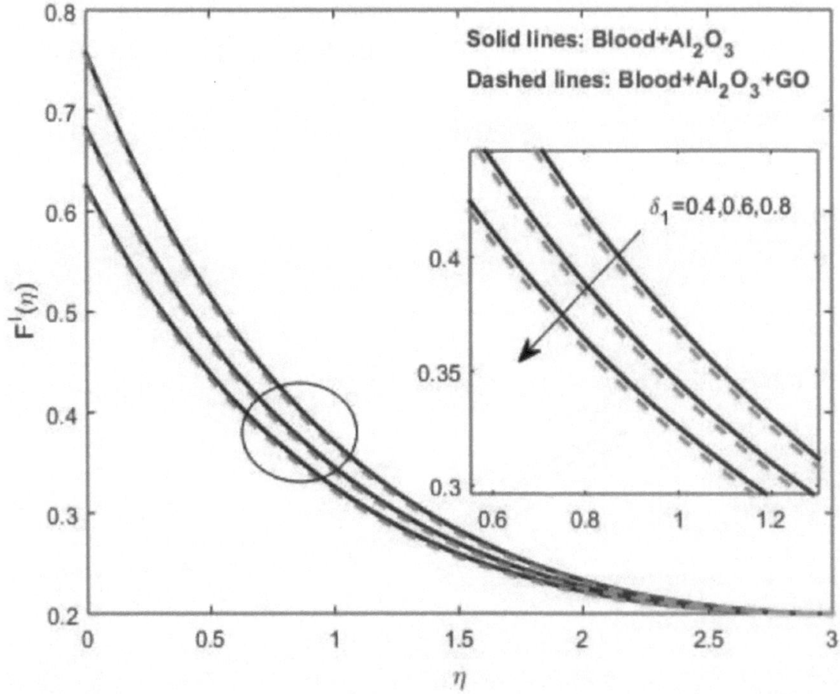

FIGURE 8.5 Velocity of nanofluids and hybrid nanofluids for increasing values for slip parameters.

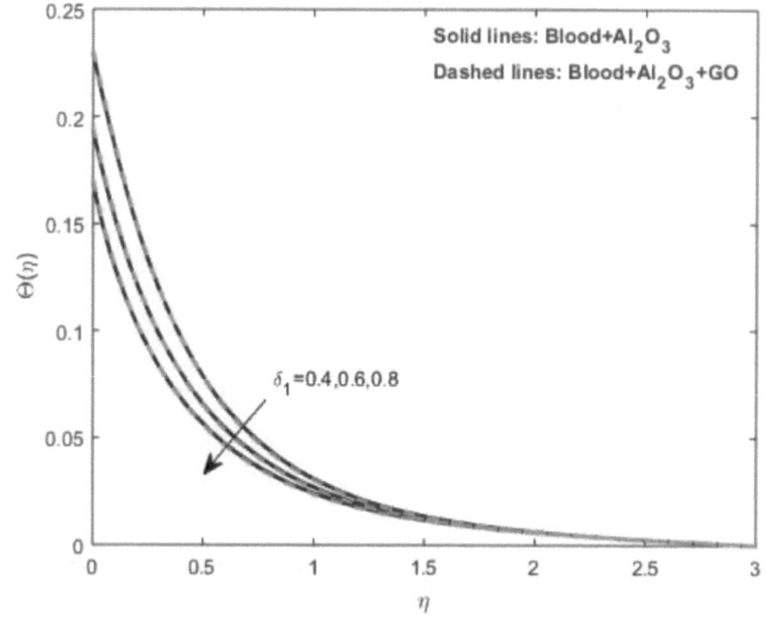

FIGURE 8.6 Thermal profile for nanofluids and hybrid nanofluids for increasing values for slip parameters.

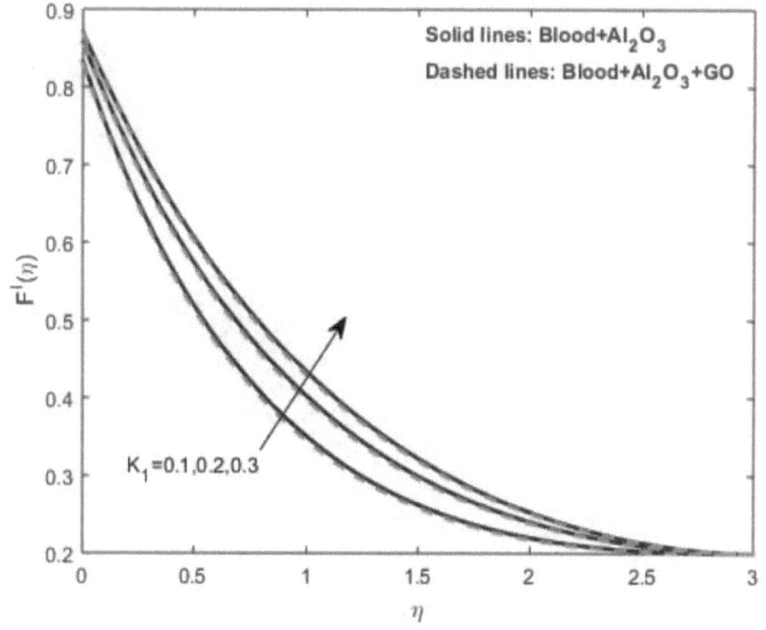

FIGURE 8.7 Velocity of nanofluids and hybrid nanofluids for increasing values for visco-elastic parameters.

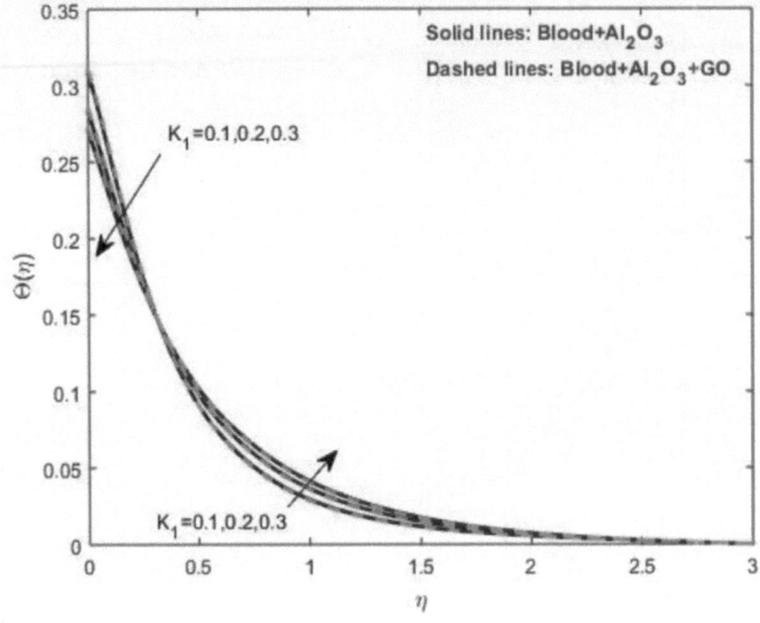

FIGURE 8.8 Thermal profile of nanofluids and hybrid nanofluids for viscoelastic parameters.

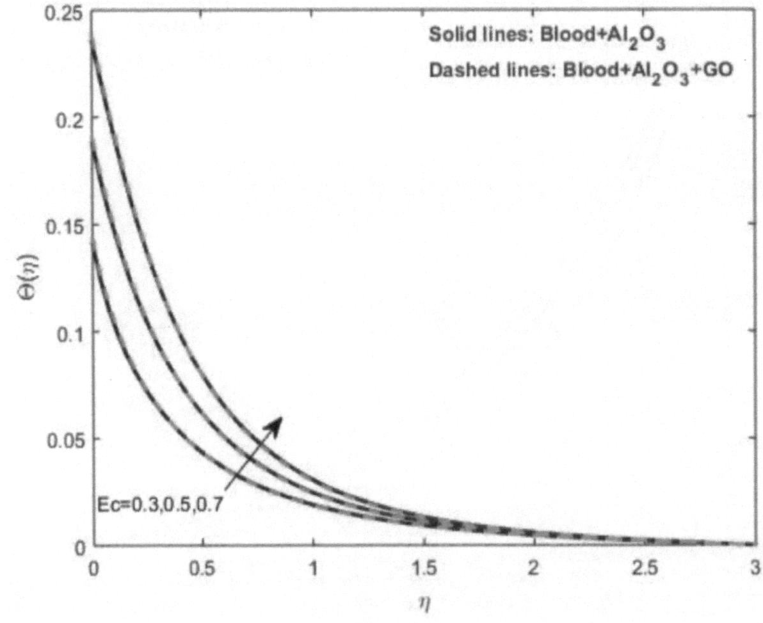

FIGURE 8.9 Thermal profile of nanofluids and hybrid nanofluids for an increase in Eckert number.

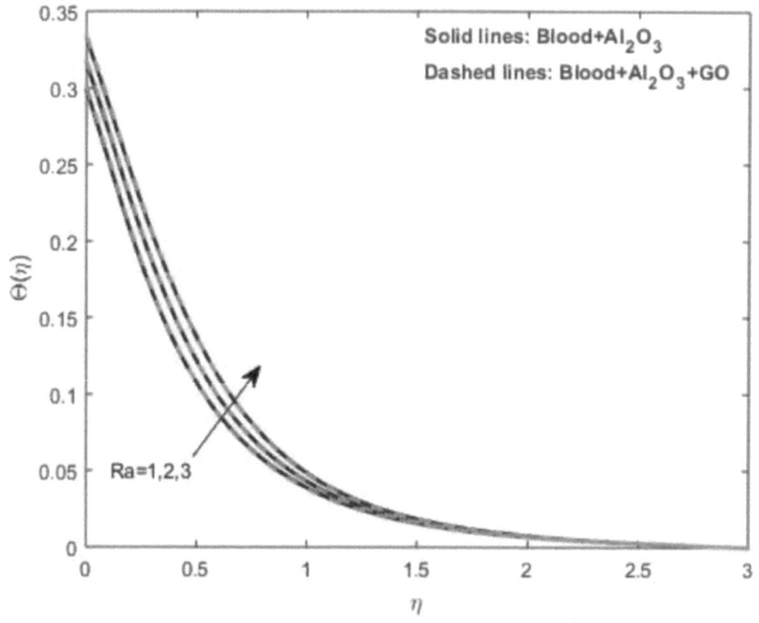

FIGURE 8.10 Thermal outline of nanofluids and hybrid nanofluids for an increase in standards of *Ra*.

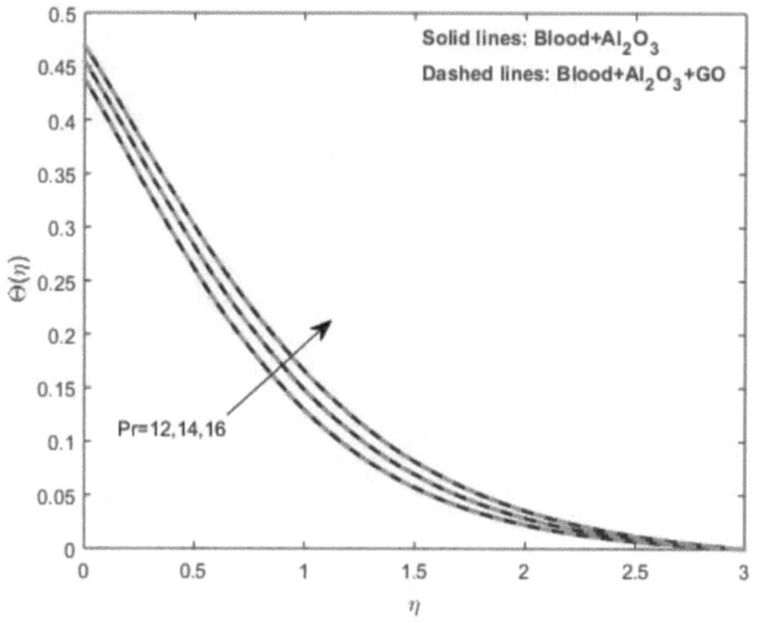

FIGURE 8.11 Thermal outline of nanofluids and hybrid nanofluids for increasing values of Pr.

increase in its temperature. The reason for increasing the temperature with respect to thermal radiation parameter value can be that growing the thermic radiation factor enhances the efficiency of radiative heat transmission within a system. This leads to a developed rate of heat absorption by the material or surface, causing an increase in its temperature until thermal equilibrium is achieved.

Figure 8.12 presents the relation between Forchheimer number Fr and the flow velocity of nanofluids and hybrid nanofluids. Here, the author noticed the improvement in velocity for increasing values of Forchheimer number. Since, with reduced viscous damping, the fluid can maintain a higher velocity as it flows through the porous medium, the fluid particles experience less resistance, allowing it to move faster. This results in an improvement in the fluid's velocity for higher Forchheimer numbers.

For increasing heat sink parameter values, the temperature decreases, while for source parameters, the temperature increases, as shown by Figures 8.13 and 8.14. Physically, the heat sink parameters typically lead to lower temperatures in a system because it enhances the heat dissipation and transfer capabilities of the heat sink. Increasing the heat source parameters results in an increased rate of energy production within the system. This, in turn, causes the temperature of the system to increase until a new thermal equilibrium is reached.

Figure 8.15 displays the relation between Biot number and thermal profiles. For increasing values of Biot number Bi, the thermal profiles are enhanced very near the boundary and are almost the same in other regions. Also, it is noticed that the temperature is greater in the case of a hybrid nanofluid near the borderline layer.

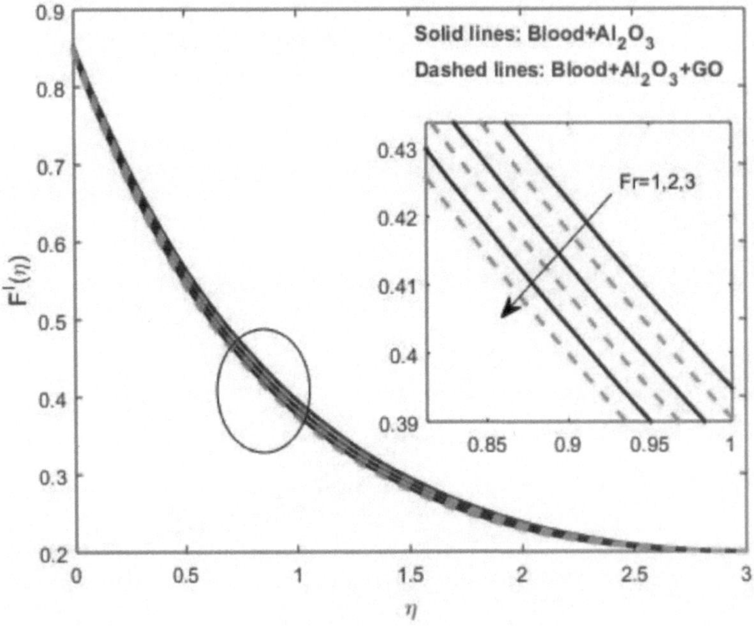

FIGURE 8.12 Velocity outline of nanofluids and hybrid nanofluids for increasing values of Fr.

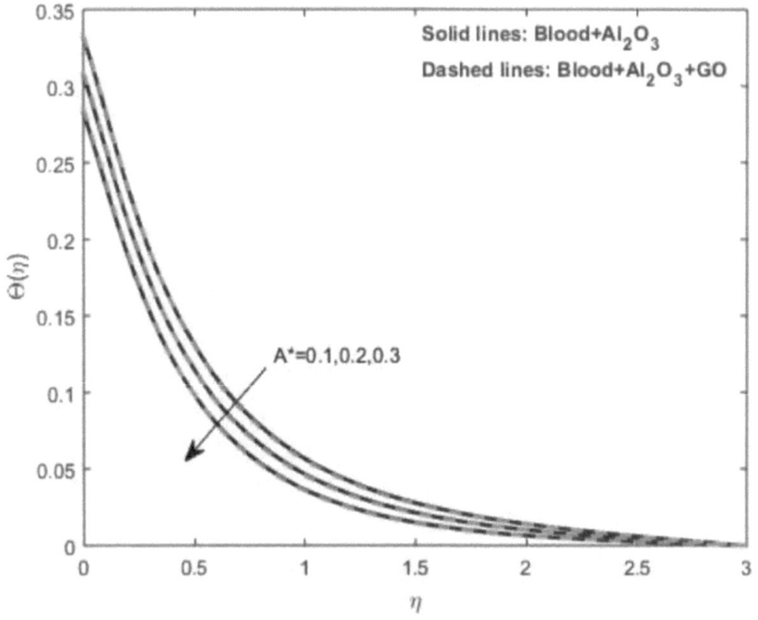

FIGURE 8.13 Temperature profile for increasing standards of energy source/sink parameter A^*.

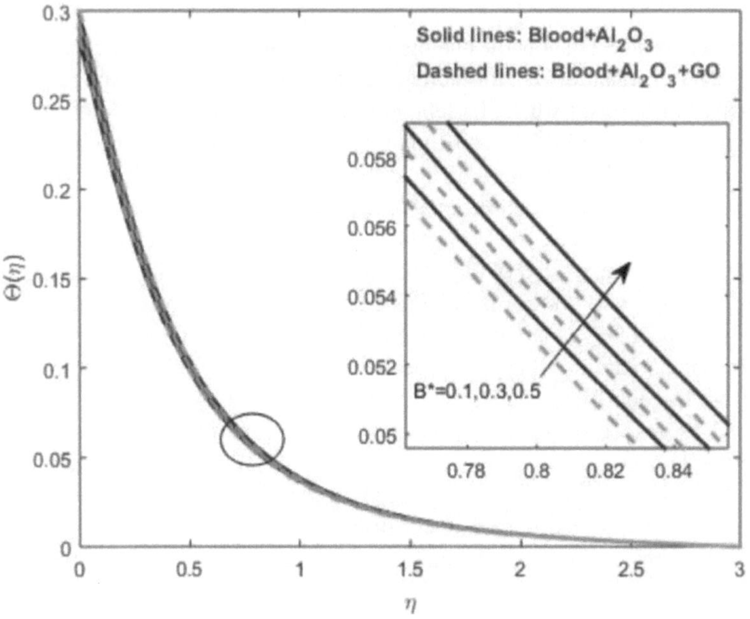

FIGURE 8.14 Thermal outline for growing standards of heat source/sink parameter B^*.

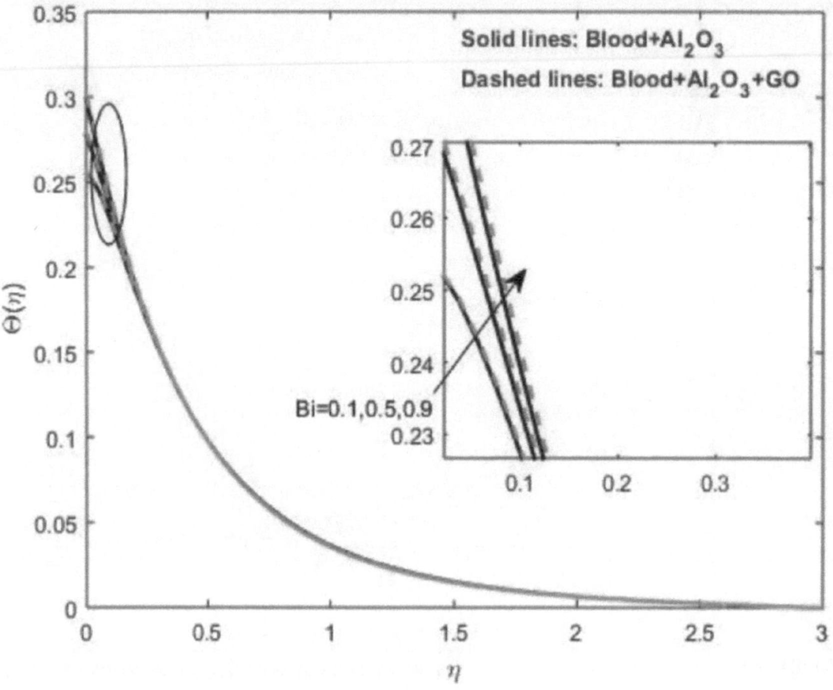

FIGURE 8.15 Thermal profile for increasing standards of Biot number Bi.

Figures 8.16 and 8.17 show the change in velocity and energy for increasing values of the velocity exponent. Here, for increasing values of the velocity exponent ($n \geq 1$), the velocity is improved while the temperature is diminished. It is noticed that the velocity of hybrid nanofluids is less than that of the nanofluids, while in the case of the thermal profile, both nanofluids have the same temperature. This phenomenon can help understand the suitable choice of the path of nanofluids in arteries.

Figures 8.18–8.25 show the variation in heat transfer rate for different physical parameters and compare nanofluids and hybrid nanofluids. The heat transfer rate declines for improving values of M, Fr, A^*, and B^* as shown in Figures 8.18 and 8.22–8.24, respectively. Also, it is noticed that the heat transfer rate is higher in the case of hybrid nanofluids when compared with nanofluids. While for improving values of δ, K_l, Ra, and Bi, the heat transfer rate is increased as shown in Figures 8.19–8.21 and 8.25, respectively. Also, it is important to note that the heat transfer rate is higher for hybrid nanofluids (i.e., blood + Al_2O_3 + GO) when compared with bio nanofluid (i.e., blood + Al_2O_3), as shown in Figures 8.18–8.25.

The detailed numerical outcomes for drag and energy transmission rate for distinct values of physical constraints and comparison of nanofluids and hybrid nanofluids are shown in Table 8.3.

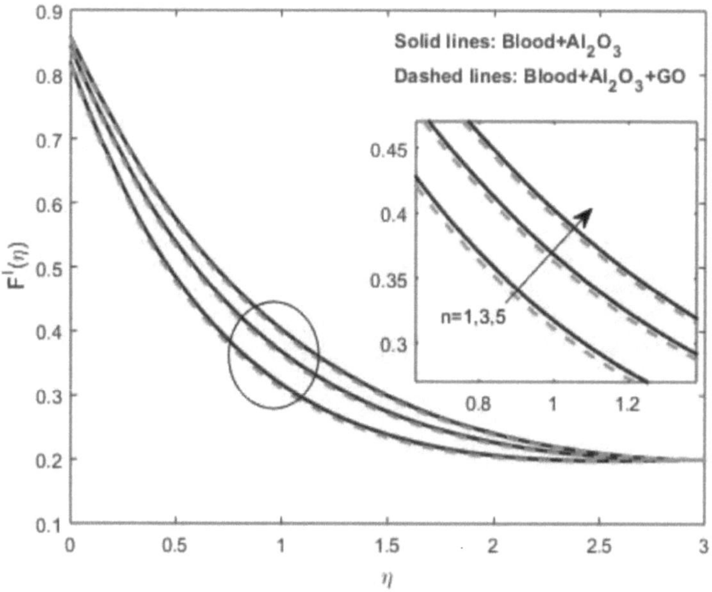

FIGURE 8.16 Fluid flow outlines for distinct values of *n*.

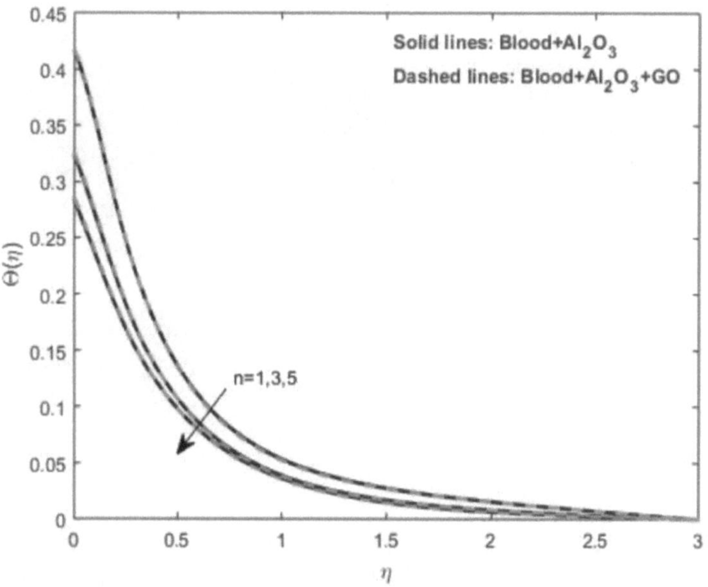

FIGURE 8.17 Thermal outlines for distinct values of *n*.

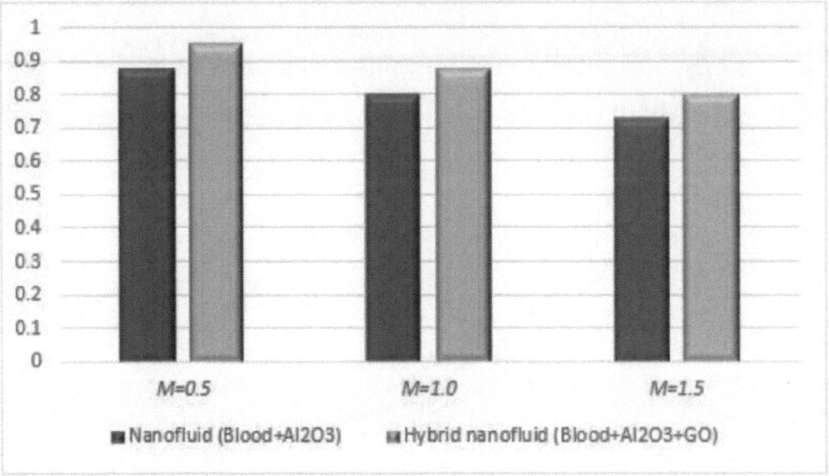

FIGURE 8.18 Heat transfer rate corresponding to *M*.

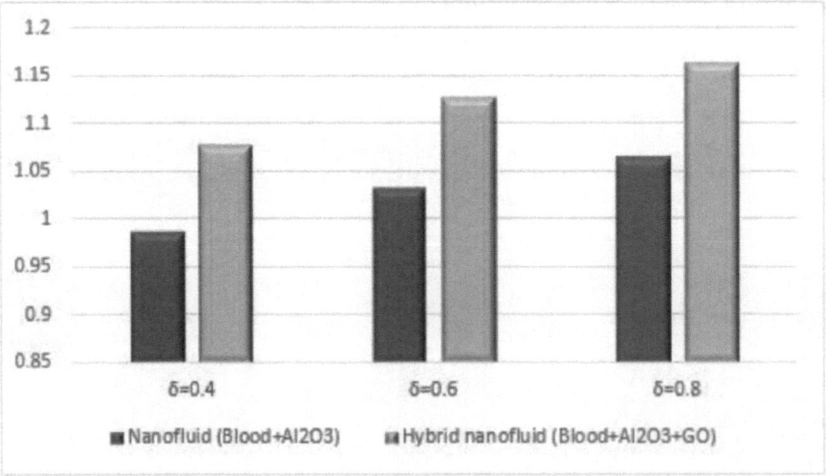

FIGURE 8.19 Heat transfer rate corresponding to δ.

8.5 CONCLUSIONS

The authors considered 2D, steady, incompressible nano- and hybrid nano-blood flow in wedge-shaped arteries. The flow equations are considered according to the geometry and boundary condition; therefore, the motion equation is equipped with the magnetic influence, Darcy–Forchheimer situation, and Buoyance effect. The energy equation is fulfilled with viscous dissipation, thermic radiation, Joule heating, heat

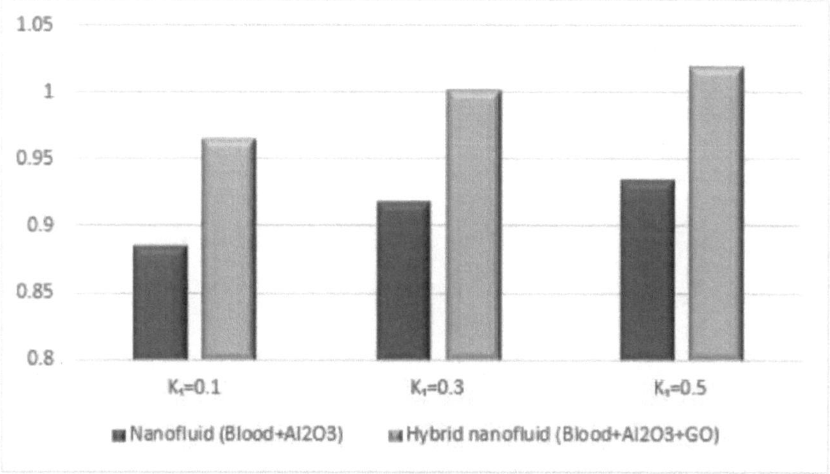

FIGURE 8.20 Heat transfer rate corresponding to K_1.

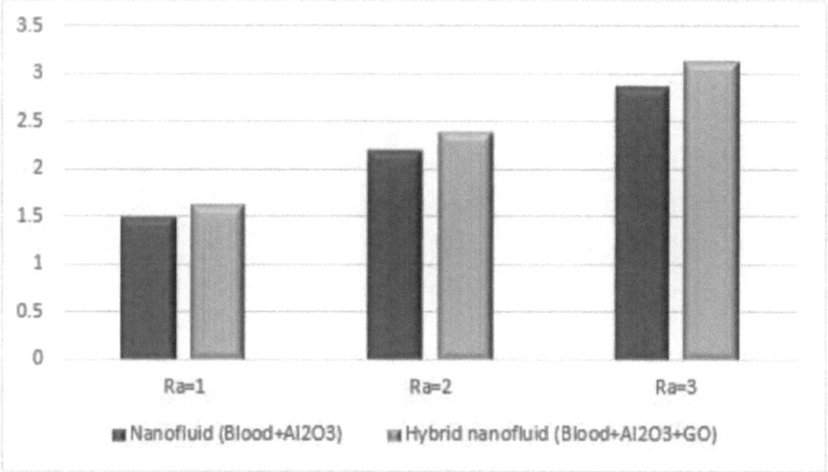

FIGURE 8.21 Heat transfer rate corresponding to Ra.

source/sink effects, and convective borderline conditions. The main equations are converted into non-dimensional forms and solved using the bvp5c method, and the MATLAB package is used for computational and graphical purposes. In this chapter, we compared the flow of nano- and hybrid nano-blood flow in wedge-shaped arteries and studied the momentum, thermal, drag, and heat transmission rate for different ranges of parameters.

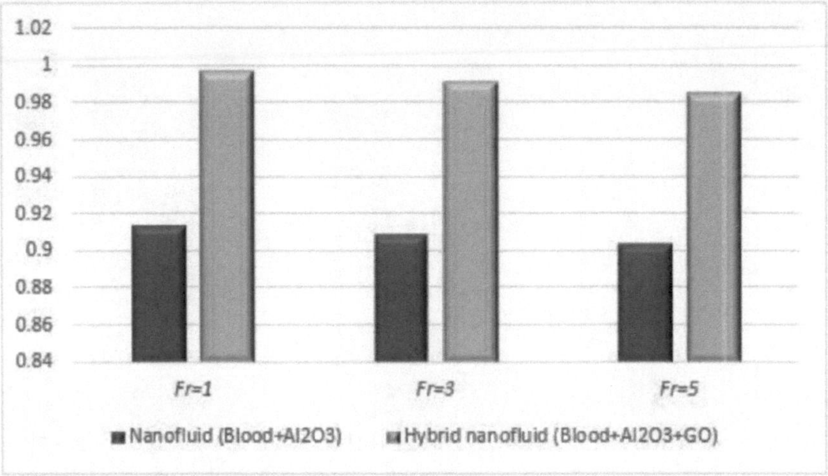

FIGURE 8.22 Heat transfer rate corresponding to Fr.

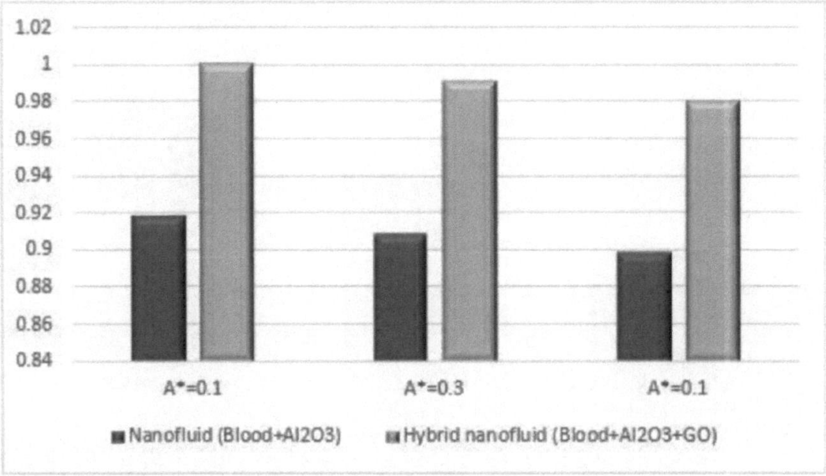

FIGURE 8.23 Heat transfer rate corresponding to A^*.

The following noteworthy specifics are perceived during the study:

- The heat transmission rate is higher in hybrid nanofluids when compared with mono-nanofluids. The increased heat transfer rate can enhance the efficiency of hyperthermia treatment, making it more effective at killing cancer cells while sparing healthy tissue.

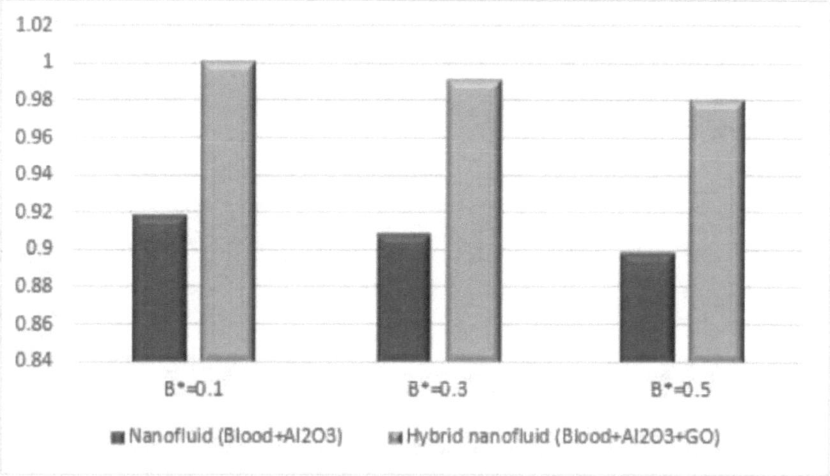

FIGURE 8.24 Heat transfer rate corresponding to B^*.

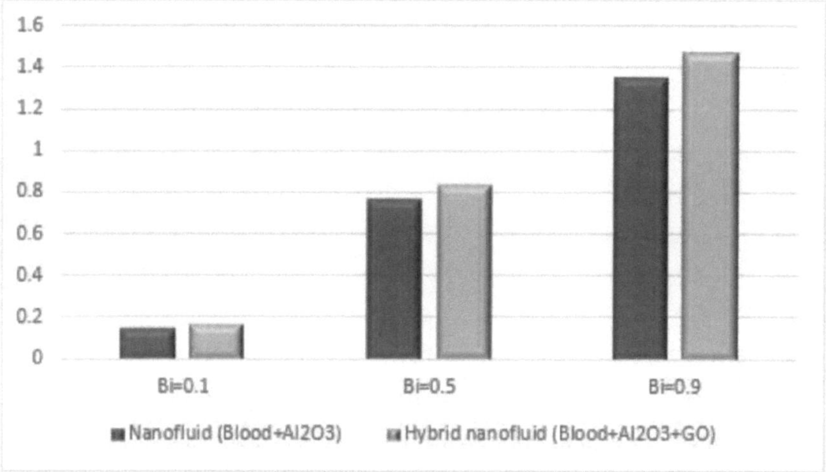

FIGURE 8.25 Heat transfer rate corresponding to Bi.

- Increasing values of thermic emission parameter (Ra) and Biot number help improve the heat allocation rate in both fluids. This result can be helpful in imaging techniques like MRI or photoacoustic imaging, and nanoparticles can be used as contrast agents. When these nanoparticles are heated, they can produce distinct signals that enhance the quality of imaging, providing better visualization of tumours or specific tissues.

TABLE 8.3
Difference in the Nusselt Number and Drag Coefficient for Different Parameters

Physical constraint										Drag numeral		Heat transfer number	
M	δ_1	K_1	Ec	Ra	Pr	Fr	A^*	B^*	Bi	Nano-blood flow	Hybrid nano-blood flow	Nano-blood flow	Hybrid nano-blood flow
0.5										−1.396156	−1.417377	0.871857	0.951428
1.0										−1.441937	−1.466829	0.799478	0.873666
1.5										−1.486728	−1.515107	0.731120	0.800515
	0.4									−1.163958	−1.178028	0.986299	1.076462
	0.6									−1.014004	−1.024958	1.031758	1.126336
	0.8									−0.898999	−0.907821	1.063463	1.160995
		0.1								−1.599993	−1.624333	0.885804	0.965228
		0.2								−1.368219	−1.387151	0.917312	1.000421
		0.3								−1.239067	−1.254527	0.933099	1.018176
			0.3							−1.368219	−1.387151	1.099436	1.199163
			0.5							−1.368219	−1.387151	1.038728	1.132916
			0.7							−1.368219	−1.387151	0.978020	1.066669
				1						−1.368219	−1.387151	1.496238	1.632252
				2						−1.368219	−1.387151	2.188058	2.387651
				3						−1.368219	−1.387151	2.848350	3.108927
					12					−1.368218	−1.387151	7.793663	8.516932
					13					−1.368218	−1.387150	8.728021	9.539452
					14					−1.368218	−1.387150	9.626043	10.52230
						0.4				−1.404124	−1.422429	0.913146	0.995943
						0.6				−1.447019	−1.464601	0.908000	0.990415
						0.8				−1.487878	−1.504798	0.902934	0.984976

(Continued)

TABLE 8.3
(Continued)

Physical constraint										Drag numeral		Heat transfer number	
M	δ_1	K_l	Ec	Ra	Pr	Fr	A^*	B^*	Bi	Nano-blood flow	Hybrid nano-blood flow	Nano-blood flow	Hybrid nano-blood flow
							0.1			−1.368219	−1.387151	0.917312	1.000421
							0.3			−1.368219	−1.387151	0.908112	0.990470
							0.5			−1.368219	−1.387151	0.898408	0.979976
								0.1		−1.368219	−1.387151	0.917312	1.000421
								0.3		−1.368219	−1.387151	0.908112	0.990470
								0.5		−1.368219	−1.387151	0.898408	0.979976
									0.4	−1.368219	−1.387151	0.158886	0.173720
									0.6	−1.368219	−1.387151	0.770245	0.840441
									0.8	−1.368219	−1.387151	1.345478	1.465292

- Hybrid nanoparticles diminish the velocity of the blood stream as compared to mono-nanofluids, while the temperature remains almost the same in both fluids.
- Slowing down blood velocity can be advantageous in drug delivery. When blood moves more slowly, it gives nanoparticles more time to interact with the surrounding tissue and cells. This can enhance the effectiveness of targeted drug delivery systems. Nanoparticles loaded with drugs can remain in the bloodstream longer, improving the chances of reaching their target location. Hence, manipulation of blood velocity using nanoparticles in biomedical applications can be advantageous for targeted drug delivery, imaging, and therapeutic purposes.
- The direction of the blood flow can be regulated by increasing the exponent of the velocity term.
- This work can benefit in understanding the nature of the blood flow in wedge-shaped arteries and heat transfer variations due to Al_2O_3 and GO nanoparticles.
- The outcomes and results of this study are useful in various biomedical applications.

REFERENCES

1. Hormuth, D. A., Phillips, C. M., Wu, C., Lima, E. A. B. F., Lorenzo, G., Jha, P. K., Jarrett, A. M., Oden, J. T., & Yankeelov, T. E. (2021). Biologically based mathematical modeling of tumor vasculature and angiogenesis via time-resolved imaging data. *Cancers*, 13(12), 3008.
2. Zhang, Q., Ouyang, H., Ye, F., Chen, S., Xie, L., Zhao, X., & Yu, X. (2020). Multiple mathematical models of diffusion-weighted imaging for endometrial cancer characterization: Correlation with prognosis-related risk factors. *European Journal of Radiology*, 130, 109102.
3. Hussain, S., Mubeen, I., Ullah, N., Shah, S. S. U. D., Khan, B. A., Zahoor, M., Ullah, R., Khan, F. A., & Sultan, M. A. (2022). Modern diagnostic imaging technique applications and risk factors in the medical field: A review. *BioMed Research International*, 2022.
4. Zahin, N., Anwar, R., Tewari, D., Kabir, M. T., Sajid, A., Mathew, B., Uddin, M. S., Aleya, L., & Abdel-Daim, M. M. (2020). Nanoparticles and its biomedical applications in health and diseases: Special focus on drug delivery. *Environmental Science and Pollution Research*, 27(16), 19151–19168.
5. Ahmad, S., Ali, F., Khan, I., & Haq, S. U. (2023). Biomedical applications of gold nanoparticles in thermofluids flow through a porous medium. *International Journal of Thermofluids*, 20, 100425.
6. Njingang Ketchate, C. G., Tiam Kapen, P., Madiebie-Lambou, I., Fokwa, D., Chegnimonhan, V., Tchinda, R., & Tchuen, G. (2023). Chemical reaction, Dufour and Soret effects on the stability of magnetohydrodynamic blood flow conveying magnetic nanoparticle in presence of thermal radiation: A biomedical application. *Heliyon*, 9(1), e12962.
7. Dharmaiah, G., Prasad, J. L. R., Balamurugan, K. S., Nurhidayat, I., Fernandez-Gamiz, U., & Noeiaghdam, S. (2023). Performance of magnetic dipole contribution on ferromagnetic non-Newtonian radiative MHD blood flow: An application of biotechnology and medical sciences. *Heliyon*, 9(2), e13369.

8. Al Nuwairan, M., & Souayeh, B. (2022). Simulation of gold nanoparticle transport during MHD electroosmotic flow in a peristaltic micro-channel for biomedical treatment. *Micromachines*, 13(3), 374.
9. Shojaie Chahregh, H., & Dinarvand, S. (2020). TiO_2-Ag/blood hybrid nanofluid flow through an artery with applications of drug delivery and blood circulation in the respiratory system. *International Journal of Numerical Methods for Heat and Fluid Flow*, 30(11), 4775–4796.
10. Kakar, N., Khalid, A., Al-Johani, A. S., Alshammari, N., & Khan, I. (2022). Melting heat transfer of a magnetized water-based hybrid nanofluid flow past over a stretching/shrinking wedge. *Case Studies in Thermal Engineering*, 30, 101674.
11. Karvelas, E., Sofiadis, G., Papathanasiou, T., & Sarris, I. (2020). Effect of micropolar fluid properties on the blood flow in a human carotid model. *Fluids*, 5(3), 1–16.
12. Saleem, A., Akhtar, S., Nadeem, S., Issakhov, A., & Ghalambaz, M. (2020). Blood flow through a catheterized artery having a mild stenosis at the wall with a blood clot at the centre. *Computer Modeling in Engineering and Sciences*, 125(2), 565–577.
13. Basha, H. T., & Sivaraj, R. (2021). Numerical simulation of blood nanofluid flow over three different geometries by means of gyrotactic microorganisms: Applications to the flow in a circulatory system. *Proceedings of the Institution of Mechanical Engineers, Part C: Journal of Mechanical Engineering Science*, 235(2), 441–460.
14. Zaman, A., Mabood, F., Khan, A. A., Abbasi, A., Nadeem, M. F., & Badruddin, I. A. (2021). Simulations of unsteady blood flow through curved stenosed channel with effects of entropy generations and magneto-hydrodynamics. *International Communications in Heat and Mass Transfer*, 127, 105569.
15. Ahmed, A., & Nadeem, S. (2017). Effects of magnetohydrodynamics and hybrid nanoparticles on a micropolar fluid with 6-types of stenosis. *Results in Physics*, 7, 4130–4139.
16. Gudkov, S. V., Burmistrov, D. E., Smirnova, V. V., Semenova, A. A., & Lisitsyn, A. B. (2022). A mini review of antibacterial properties of Al_2O_3 nanoparticles. *Nanomaterials*, 12(15), 1–17.
17. Borbane, S., & Pande, V. (2015). Design and fabrication of ordered mesoporous alumina scaffold for drug delivery of poorly water soluble drug. *Austin Therapeutics*, 2(1), 1015.
18. Wang, Y., Kaur, G., Chen, Y., Santos, A., Losic, D., & Evdokiou, A. (2015). Bioinert anodic alumina nanotubes for targeting of endoplasmic reticulum stress and autophagic signaling: A combinatorial nanotube-based drug delivery system for enhancing cancer therapy. *ACS Applied Materials and Interfaces*, 7(49), 27140–27151.
19. Hassanpour, P., Panahi, Y., Ebrahimi-Kalan, A., Akbarzadeh, A., Davaran, S., Nasibova, A. N., Khalilov, R., & Kavetskyy, T. (2018). Biomedical applications of aluminium oxide nanoparticles. *Micro and Nano Letters*, 13(9), 1227–1231.
20. Singh, D. P., Herrera, C. E., Singh, B., Singh, S., Singh, R. K., & Kumar, R. (2018). Graphene oxide: An efficient material and recent approach for biotechnological and biomedical applications. *Materials Science and Engineering C*, 86, 173–197.
21. Ghosal, K., & Sarkar, K. (2018). Biomedical applications of graphene nanomaterials and beyond. *ACS Biomaterials Science and Engineering*, 4(8), 2653–2703.
22. Mabood, F., Shafiq, A., Khan, W. A., & Badruddin, I. A. (2022). MHD and nonlinear thermal radiation effects on hybrid nanofluid past a wedge with heat source and entropy generation. *International Journal of Numerical Methods for Heat and Fluid Flow*, 32(1), 120–137.
23. Rashidi, M. M., Ali, M., Freidoonimehr, N., Rostami, B., & Hossain, M. A. (2014). Mixed convective heat transfer for MHD viscoelastic fluid flow over a porous wedge with thermal radiation. *Advances in Mechanical Engineering*, 2014.

24. Hou, E., Wang, F., Nazir, U., Sohail, M., Jabbar, N., & Thounthong, P. (2022). Dynamics of tri-hybrid nanoparticles in the rheology of pseudo-plastic liquid with Dufour and Soret effects. *Micromachines*, 13(2), 1–17.
25. Alwawi, F. A., Swalmeh, M. Z., Qazaq, A. S., & Idris, R. (2021). Heat transmission reinforcers induced by MHD hybrid nanoparticles for water/water-EG flowing over a cylinder. *Coatings*, 11(6), 223.
26. Cui, J., Munir, S., Raies, S. F., Farooq, U., & Razzaq, R. (2022). Non-similar aspects of heat generation in bioconvection from flat surface subjected to chemically reactive stagnation point flow of Oldroyd-B fluid. *Alexandria Engineering Journal*, 61(7), 5397–5411.
27. Safaei, M. R., Goshayeshi, H. R., & Chaer, I. (2019). Solar still efficiency enhancement by using graphene oxide/paraffin nano-PCM. *Energies*, 12(10), 2002.

9 Sensitivity Analysis in Hybridized Casson Nanofluid Near a Perforated Riga Plate

Soumitra Sarkar, Asgar Ali, and Sanatan Das

Nomenclature

$(u, v)(m.sec^{-1})$	Non-dimensional velocity components in (x, y)	k^*	Mean absorption coefficient
τ_0	Casson yield stress	κ	Boltzmann's constant
π	Permutation of deformation rate component with itself	b	Chemotaxis constant
$\dot{\sigma}$	Yield stress of fluid	δ	Temperature difference factor
e_{ij}	$(i,j)^{th}$ component of deformation rate	$k(m.kg.sec^{-3}.K)$	Thermal conductivity
π_c	Casson fluid critical value of π	f	Dimensionless velocity
$\mu(Pa.s)$	Dynamic viscosity	θ	Dimensionless temperature
$\mu_B(Pa.s)$	Plastic dynamic viscosity	ϕ	Dimensionless nanoparticle (NP) concentration
$\rho(kg.m^{-3})$	Density	$\nu(m^2.sec^{-1})$	Viscosity
$\sigma(sec^3.m^2.kg^{-1})$	Electrical conductivity	$\overline{\tau}$	Fraction of heat capacities of NPs and nanofluids
β_T	Thermal expansion coefficient	σ^*	Stefan–Boltzmann constant
β_C	Solutal volumetric coefficient	q_r	Radiative heat flux

DOI: 10.1201/9781003595786-9

$T(K)$	Temperature of Casson nanofluid (CNF)	R_d	Thermal radiation parameter
$T_\infty(K)$	Reference temperature	N_r	Concentration flux of nanomaterial
$T_W(K)$	Surface temperature of CNF	Sc	Schmidt number
$U_w(m.sec^{-1})$	Surface escalating velocity	N_t	Thermophoresis parameter
$U_0(m.sec^{-1})$	Reference velocity	N_b	Brownian motion parameter
C	NP concentration	E_a	Activating energy
C_w	NP concentration near the plate	D_T	Thermophoresis diffusive coefficient
C_∞	Concentration of NPs away from the plate	D_B	Random (Brownian) diffusive coefficient
$B_0(kg.sec^{-2}.m^{-2})$	Strength of magnetic field	k_r	Chemical reacting rate
g	Resultant gravity	γ	CNF parameter
a	Widths of magnets and electrodes	E	Activating energy parameter
M_0	Magnetization parameter	K	Chemical reacting parameter
$j_0(A.m^2)$	Density of current in electrodes	H_a	Modified Hartmann parameter
$q_w(kg.sec^{-3})$	Heat flux at the surface	λ	Richardson number
ϕ_1	Cu nanoparticle solid volume fraction	Pr	Prandtl number
ϕ_2	Al_2O_3 nanoparticle solid volume fraction	m	Fitting constant rate
s_1	Suffice for copper	c_p	Specific heat capacity
s_2	Suffice for alumina	hnf	Bi-hybrid nanofluid
f	Casson fluid		

9.1 INTRODUCTION

In recent decades, scientists have extensively explored non-Newtonian fluids for their distinct properties, finding applications in various fields. These fluids range from lubricants and shampoos to drilling muds and organic substances like blood. Understanding the behaviour of viscoelastic fluids, especially when the fluid behaves like a solid substance, requires the use of the Casson model. It effectively accounts for yield stress, making it a widely recognized approach for studying rheological properties in previous research. This model, initially introduced by N. Casson [1] for ink dispersion in printing, exhibits shear-thinning behaviour with solid-like elasticity due to applied yield stress. Beyond a certain threshold, it transitions from non-Newtonian to Newtonian behaviour [2]. This versatile model finds applications in medicines, food and chemical production, biodiesel manufacturing, and bioreactors. Ali et al. [3] investigated Casson fluid flow across an elongated surface under magnetohydrodynamics (MHD), discovering that increasing the Casson parameter improved thermal property of the fluid. Recent articles [4–9] further contribute to the understanding of Casson nanofluid (CNF) flows.

The crucial function that nanofluids play in both medical and commercial investigations has caused their popularity to soar. The addition of nanomaterials, usually non-metallic or metallic particles smaller than 100 nm, to ordinary fluids produces nanofluids. This infusion of nanomaterials greatly enhances the base fluids' thermal characteristics, a concept initially coined by Choi [10]. However, the quest for superior fluids continues, leading to the development of hybrid nanofluids. These novel fluids exhibit superior thermal properties compared to traditional nanofluids and hold great promise for a wide range of applications, including nuclear chilling systems, vehicle and engine thermal management, electronics cooling, generator cooling, lubrication, machining, solar heating, pharmaceutical processes, transformer chilling, and medical uses. Krishna et al. [11] showed increased heat transfer in an electronic washbasin using a hybrid nanofluid, which outperformed ionized water. Jamshed and Aziz [12] applied the Cattaneo–Christov model to study TiO_2CuO/ethylene glycol (EG) Casson hybrid nanofluid flow near a stretching plate, emphasizing the augmented heat transfer capabilities of hybrid nanofluids. Recent studies by various researchers [13–18] have extensively explored the miscellaneous assets of nanofluids.

Activation energy, a key determinant in chemical processes, represents the least amount of energy needed for a reaction. Svante Arrhenius introduced this idea, which finds widespread use in industries ranging from food production to synthetic manufacturing. Bestman's work [19] highlighted the effect of activation and chemical kinetics on mass transfer. Further research on the effects of activation energy on flow exposed to heat radiation and reactions of chemicals was conducted by Makinde et al. [20] and Ahmad et al. [21]. Numerous ongoing research works related to Arrhenius chemical kinetics can be found in references [22–25].

The boundary and the lateral electromagnetic field align, increasing the pressure gradient in the flow. The boundary layer flow and the Lorentz force formed by the perpendicular electric and magnetic fields magnify the pressure differential. Gailitis [26] initiated an effective magnetoelectric actuator, shown in Figure 9.1, named the Riga plate, designed especially for low conductivity liquids. By minimizing the separation of the boundary layer, the Riga plate has the potential to reduce pressure and drag friction in submarines. Research on MHD nanoliquid flow with swimming creatures close to a Riga plate was performed by Abbas et al. [27]. Ganesh et al. [28] investigated the mass and heat transfer of nanofluids across a Riga surface. Their research revealed a significant increase in nanofluid velocity due to magnetic attraction. Recent studies on flow features over the Riga plate can be found in [29–32].

This study addresses a research gap by investigating Casson bi-hybrid nanofluids (CHNFs) near a Riga surface with Arrhenius effects. The aim was to analyse their flow behaviour by employing an appropriate similarity approach to convert governing equations into a set of ordinary differential equations (ODEs). To tackle this, an amalgamation of the fourth-order Runge–Kutta–Fehlberg procedure and a shooting technique was utilized. Additionally, a sensitivity analysis using response surface methodology (RSM; based on [22, 33, 34]) revealed the relationship between input and response parameters. Adjustments were made to optimize Sherwood and Nusselt numbers. This study's novelty lies in its comprehensive examination of CHNFs over a Riga plate, with chemical reactions and activation energy. The positive findings are anticipated to encourage further investigational study in this area.

9.2 MATHEMATICAL FORMULATION

In Figure 9.1, we present the motion of a CHNF over a Riga surface. With the help of persistent magnets and alternating electrodes, this plate that is oriented along the x-axis produces flow through a Lorentz force f. It is assumed that together the nanomaterials and Casson liquid start with identical velocities and are in thermodynamic stability. Table 9.1 provides the thermophysical properties of the working fluid, which is considered a mixture of Cu and Al_2O_3 (nanoparticles in EG). Notably, a heightened nanomaterial concentration can elevate the fluid's thickness [35].

The Casson fluid's thermophysical properties are defined by Equation (9.1) [22, 36]:

$$\tau_{ij} = \tau_0 + \mu\dot{\sigma} = \begin{cases} 2\left(\mu_B + \dfrac{p_y}{\sqrt{2\pi}}\right)e_{ij}, & \pi > \pi_c \\ 2\left(\mu_B + \dfrac{p_y}{\sqrt{2\pi_c}}\right)e_{ij}, & \pi < \pi_c. \end{cases} \tag{9.1}$$

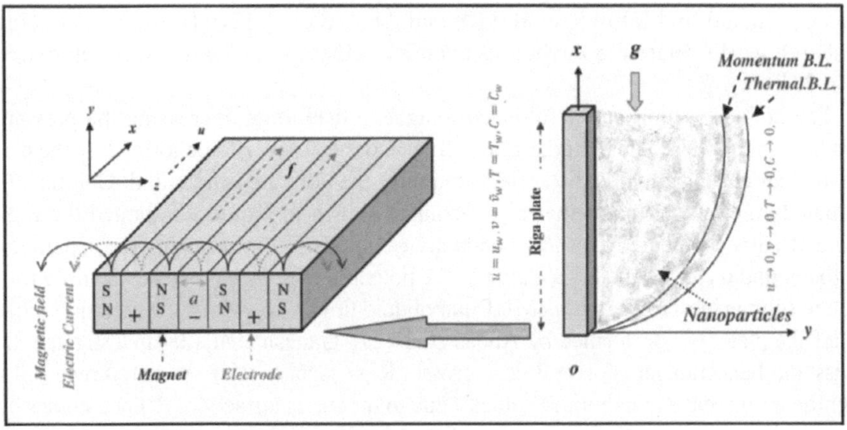

FIGURE 9.1 Flow geometry.

We obtain for $\pi < \pi_c$

$$\tau_{ij} = \mu_B \left(1+\frac{1}{\gamma}\right) 2e_{ij}. \tag{9.2}$$

The primary equation governing CHNF flow over a Riga surface can be obtained through the Boussinesq approximation, building upon the previously stated assumptions [22, 27, 31]:

$$\frac{\partial u}{\partial x}+\frac{\partial v}{\partial y}=0, \tag{9.3}$$

$$\begin{aligned}\rho_{hnf}\left(u\frac{\partial u}{\partial x}+v\frac{\partial u}{\partial y}\right) &= \mu_{hnf}\left(1+\frac{1}{\gamma}\right)\frac{\partial^2 u}{\partial y^2}+\frac{j_0\pi M_0}{8}\exp\left(-\frac{\pi}{a}y\right)\\ &+(\rho\beta_T)_{hnf}\,g\left(T-T_\infty\right)-(\rho\beta_C)_{hnf}\,g\left(C-C_\infty\right),\end{aligned} \tag{9.4}$$

$$\begin{aligned}u\frac{\partial T}{\partial x}+v\frac{\partial T}{\partial y} &= \frac{k_{hnf}}{(\rho c_p)_{hnf}}\frac{\partial^2 T}{\partial y^2}+\overline{\tau}\left[\frac{D_T}{T_\infty}\left(\frac{\partial T}{\partial y}\right)^2+D_B\frac{\partial T}{\partial y}\frac{\partial C}{\partial y}\right]\\ &-\frac{1}{(\rho c_p)_{hnf}}\frac{\partial q_r}{\partial y},\end{aligned} \tag{9.5}$$

$$u\frac{\partial C}{\partial x}+v\frac{\partial C}{\partial y}=D_B\frac{\partial^2 C}{\partial y^2}+\frac{D_T}{T_\infty}\frac{\partial^2 T}{\partial y^2}-k_r^2\left(C-C_\infty\right)\left(\frac{T}{T_\infty}\right)^m\exp\left(\frac{-E_a}{\kappa T}\right). \tag{9.6}$$

With boundary conditions (BCs) [27, 36],

$$u=u_w,\quad v=v_w,\quad T=T_w,\quad C=C_w \quad \text{at} \quad y=0 \tag{9.7}$$

$$u\to 0,\quad v\to 0,\quad T\to T_\infty,\quad C\to C_\infty \quad \text{as} \quad y\to\infty, \tag{9.8}$$

where

$$\mu_{hnf}=\frac{\mu_f}{(1-\phi_1)^{2.5}(1-\phi_2)^{2.5}},$$

$$\rho_{hnf}=\left(1-\phi_2\right)\left[\left(1-\phi_1\right)\rho_f+\phi_1\,\rho_{s_1}\right]+\phi_2\,\rho_{s_2},$$

$$(\rho\beta_T)_{hnf}=\left(1-\phi_2\right)[\left(1-\phi_1\right)(\rho\beta_T)_f+\phi_1\left(\rho\beta_T\right)_{s_1}]+\phi_2(\rho\beta_T)_{s_2},$$

$$(\rho\beta_C)_{hnf}=\left(1-\phi_2\right)[\left(1-\phi_1\right)(\rho\beta_C)_f+\phi_1\left(\rho\beta_C\right)_{s_1}]+\phi_2(\rho\beta_C)_{s_2},$$

$$\sigma_{hnf}=\sigma_{bf}\left[\frac{\sigma_{s_2}(1+2\phi_2)+2\sigma_{s_{bf}}(1-\phi_2)}{\sigma_{s_2}(1-\phi_2)+\sigma_{s_{bf}}(2+\phi_2)}\right],$$

$$\sigma_{bf}=\sigma_f\left[\frac{\sigma_{s_1}(1+2\phi_1)+2\sigma_{s_f}(1-\phi_1)}{\sigma_{s_1}(1-\phi_1)+\sigma_{s_f}(2+\phi_1)}\right],$$

$$(\rho c_p)_{hnf}=(1-\phi_2)\left[\left((1-\phi_1)(\rho c_p)_f+\phi_1\right)(\rho c_p)_{s_1}\right]+\phi_2(\rho c_p)_{s_2}.$$

$$k_{hnf}=k_{bf}\left[\frac{k_{s_2}+2k_{bf}-2\phi_2\left(k_{bf}-k_{s_2}\right)}{k_{s_2}+2k_{bf}+\phi_2\left(k_{bf}-k_{s_2}\right)}\right],$$

$$k_{bf}=k_f\left[\frac{k_{s_1}+2k_f-2\phi_1\left(k_f-k_{s_1}\right)}{k_{s_1}+2k_f+\phi_1\left(k_f-k_{s_1}\right)}\right]. \tag{9.9}$$

The combination of thick and Rosseland approximations accurately represents temperature and the total heat flux for highly optically thick cases. Equation (9.5) in [37, 38] estimates the radiative heat flux (q_r):

$$q_r=-\frac{4\sigma^*}{3k^*}\frac{\partial T^4}{\partial y}=-\frac{16\sigma^* T_w^3}{3k^*}\frac{\partial T}{\partial y}. \tag{9.10}$$

Using non-dimensionalized parameters [27, 37],

$$\left.\begin{array}{l}
\chi=\frac{x}{l},\ \eta=\frac{y}{l},\ u_1=\frac{u}{u_w},\ v_1=\frac{v}{v_0},\ \theta(\eta)=\frac{T-T_\infty}{T_w-T_\infty},\ \phi(\eta)=\frac{C-C_\infty}{C_w-C_\infty},\\
\lambda=\frac{a^2 g(\rho\beta_T)_f(T_w-T_\infty)}{\pi^2 \upsilon u_w},\ \gamma=\frac{\mu_B\sqrt{2\pi_c}}{p_y},\ N_b=\frac{\overline{\tau}D_B(C_w-C_\infty)}{\nu},\ H_a=\frac{a^2 j_0 M_0}{8\pi\nu u_w\rho_f},\\
N_t=\frac{\overline{\tau}D_T(T_w-T_\infty)}{\nu T_\infty},\ \overline{\tau}=\frac{(\rho c)_p}{(\rho c)_f},\ Pr=\frac{\nu(\rho c)_f}{k_f},\ N_r=\frac{a^2 g(\rho\beta_C)_f(C_w-C_\infty)}{\pi^2\nu u_w},\\
Sc=\frac{\nu}{D_B},\ R_d=\frac{4\sigma^* T_w^3}{\rho c_p K^*},\ E=\frac{E_a}{\kappa T_\infty},\ \delta=\frac{T_w-T_0}{T_\infty},\ Pe=\frac{bW_c}{D_N},\ K=\frac{k_r^2 l^2}{\nu}.
\end{array}\right\} \quad 9.11)$$

By substituting Equation (9.11) in Equations (9.4)–(9.6) and considering the BCs (9.7) and (9.8), the resulting expression are obtained as

$$\epsilon_1\left(u_1\frac{\partial u_1}{\partial\chi}+v_1\frac{\partial u_1}{\partial\eta}\right)=\epsilon_2\left(1+\frac{1}{\gamma}\right)\frac{\partial^2 u_1}{\partial\eta^2}+H_a e^{-\eta}+\epsilon_3\lambda\theta-\epsilon_4 N_r\phi, \tag{9.12}$$

TABLE 9.1
Characteristics of EG, Copper (Cu), and Al_2O_3 (Alumina) [13]

Physical properties	Ethylene glycol (EG)	Cu	Al_2O_3
$\rho(\text{kg.m}^{-3})$	1115	8933	3970
$k(W.m^{-1}.K)$	0.253	401	40
$C_p\ k(J.kg^{-1}.K)$	2430	385	765
$\beta_T \times 10^5 (K^{-1})$	5.7	1.67	0.85
$\sigma(s.m^{-1})$	1.07 x 10^{-4}	59.6 x 10^6	35 x 10^6

$$\epsilon_6\left(u_1\frac{\partial\theta}{\partial\chi}+v_1\frac{\partial\theta}{\partial\eta}\right)=\left(\frac{\epsilon_7}{Pr}+\frac{4}{3}R_d\right)\frac{\partial^2\theta}{\partial\eta^2}+N_t\left(\frac{\partial\theta}{\partial\eta}\right)^2+N_b\frac{\partial\theta}{\partial\eta}\frac{\partial\phi}{\partial\eta}, \quad (9.13)$$

$$u_1\frac{\partial\phi}{\partial\chi}+v_1\frac{\partial\phi}{\partial\eta}=\frac{1}{Sc}\frac{\partial^2\phi}{\partial\eta^2}+\frac{N_t}{ScN_b}\frac{\partial^2\theta}{\partial\eta^2}-K(1+\delta\theta)^m\phi e^{-\frac{E}{1+\delta\theta}}, \quad (9.14)$$

with

$$u_1=1,\quad v_1=v_w,\quad \theta=1,\quad \phi=1\quad \text{when}\quad \eta=0 \quad (9.15)$$

$$u_1\to 0,\quad v_1\to 0,\quad \theta\to 0,\quad \phi\to 0,\quad \text{as}\quad \eta\to\infty. \quad (9.16)$$

Using BCs $v_1=v_w$ at $\eta=0$, it can be concluded that $v_1=v_w$. Furthermore, Equations. (9.12)–(9.14) can be rewritten as

$$\epsilon_1 v_w\frac{\partial u_1}{\partial\eta}=\epsilon_2\left(1+\frac{1}{\gamma}\right)\frac{\partial^2 u_1}{\partial\eta^2}+H_a e^{-\eta}+\epsilon_3\lambda\theta-\epsilon_4 N_r\phi, \quad (9.17)$$

$$\epsilon_6 v_w\frac{\partial\theta}{\partial\eta}=\left(\frac{\epsilon_7}{Pr}+\frac{4}{3}R_d\right)\frac{\partial^2\theta}{\partial\eta^2}+N_t\left(\frac{\partial\theta}{\partial\eta}\right)^2+N_b\frac{\partial\theta}{\partial\eta}\frac{\partial\phi}{\partial\eta}, \quad (9.18)$$

$$v_w\frac{\partial\phi}{\partial\eta}=\frac{1}{Sc}\frac{\partial^2\phi}{\partial\eta^2}+\frac{N_t}{ScN_b}\frac{\partial^2\theta}{\partial\eta^2}-K(1+\delta\theta)^m\phi e^{-\frac{E}{1+\delta\theta}}, \quad (9.19)$$

9.3 NUMERICAL SOLUTION

This study introduces a model to investigate the behaviour of CHNFs adjacent to a Riga plate surface. Solving the resulting set of highly non-linear ODEs, along with BCs, calls for a numerical approach. To tackle this, an amalgamation of the fourth-order Runge–Kutta–Fehlberg procedure and a shooting technique is utilized. To address this complexity, a new variable is proposed:

$$(u_1, u_1', \theta, \theta', \phi, \phi')^T = (h_1, h_2, h_3, h_4, h_5, h_6)^T. \tag{9.20}$$

Using Equations (9.17)–(9.19) and BCs (9.15):

$$\begin{pmatrix} h_1' \\ h_2' \\ h_3' \\ h_4' \\ h_5' \\ h_6' \end{pmatrix} = \begin{pmatrix} h_2 \\ \dfrac{1}{\epsilon_2\left(1+\dfrac{1}{\gamma}\right)}[\epsilon_1\, \upsilon_w h_2 - H_a e^{-\eta} - \epsilon_3\, \lambda h_3 + \epsilon_4\, N_r h_5] \\ h_4 \\ \dfrac{1}{\left(\dfrac{\epsilon_7}{Pr}+\dfrac{4}{3}Rd\right)}[\epsilon_6\, \upsilon_w h_4 - N_t h_4^2 - N_b h_4 h_6] \\ h_6 \\ Sc\left[\upsilon_w h_6 + K(1+\delta h_3)^m\, h_5 e - \dfrac{E}{1+\delta h_3}\right] - \dfrac{N_t}{N_b} h_4' \end{pmatrix} \tag{9.21}$$

with the BCs

$$h_1 = 1,\quad h_3 = 1,\quad h_5 = 1 \quad \text{at} \quad \eta = 0$$

$$h_1 \to 0,\quad \theta \to h_3,\quad h_5 \to 0 \quad \text{as} \quad \eta \to \infty. \tag{9.22}$$

9.4 THEORETICAL FINDINGS AND ANALYSIS

The research employed numerical simulations to inspect the control of diverse physical parameters on physical characteristics. Results are illustrated using tables and graphics created with Mathematics 12.

9.4.1 Velocity Profile

In Figure 9.2(a), the fluid velocity varies with nanoparticle concentration flux (N_r). Higher N_r shows a decrease in velocity profile due to increased buoyancy forces, promoting the growing move of hotter fluid and subsequently reducing fluid velocity. Figure 9.2(b) shows the influence of moderated Hartmann number (Ha) on the flow field. Increased Ha results in higher velocity and momentum boundary layer (MBL) thickness, attributed to enhanced external and internal forces, including electric and adhesive forces. Figure 9.2(c) demonstrates that higher Casson parameter (γ) values lead to increased velocity profile and MBL thickness, as they indicate lower yield stress and higher fluid velocity. Higher Richardson numbers (λ) correspond to increased velocity profiles, as seen in Figure 9.2(d). Flow regime is dictated by λ, signifying the balance between flow gradient and buoyant force. When $\lambda \gg 1$, buoyancy dominates, while $\lambda \ll 1$ indicates its lesser significance. This implies insufficient kinetic energy for uniform nanofluid mixing. Figures 9.2(a–d) show that bi-hybrid CNFs have an inferior fluid velocity when compared with CNFs. Given the higher viscosity of the hybrid nanofluid, it is anticipated that this outcome would occur. Additionally, the use of hybrid nanomaterials in EG more effectively decreases fluid velocity compared to the use of single nanoparticles.

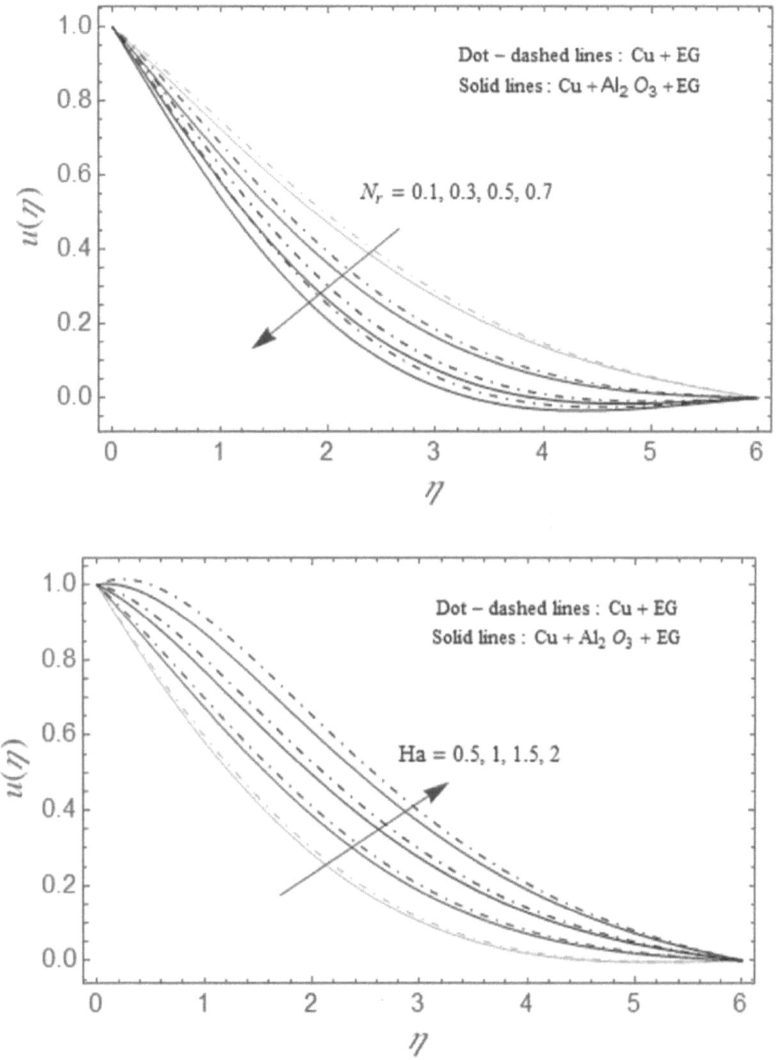

FIGURE 9.2 Velocity profile for changing values of (a) N_r, (b) Ha, (c) γ, and (d) λ.

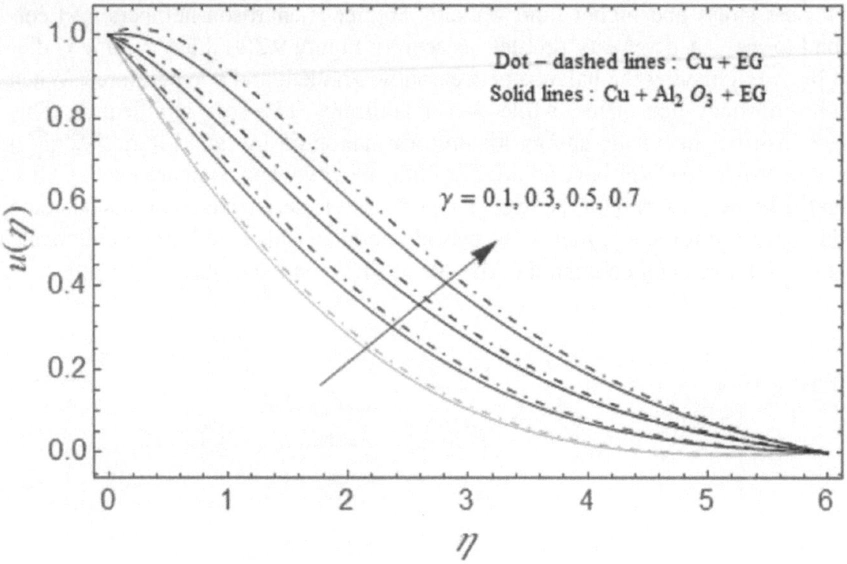

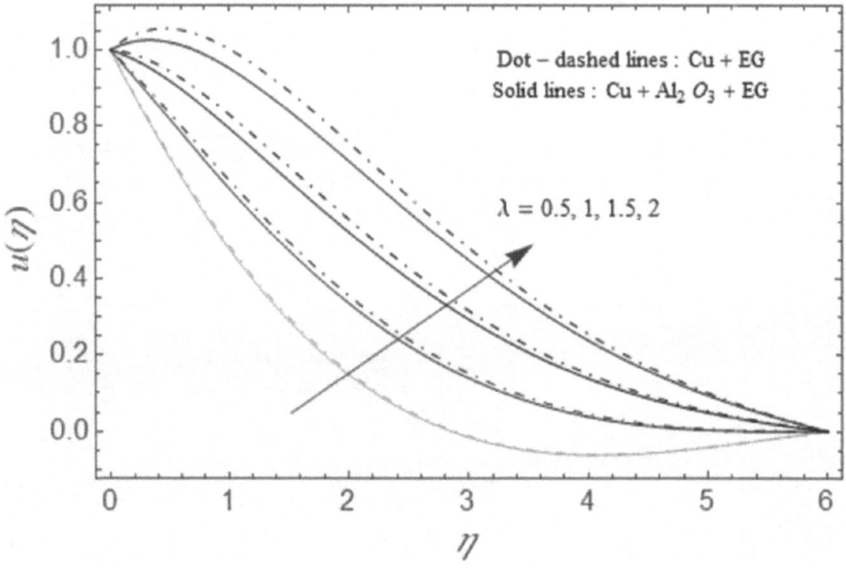

FIGURE 9.2 (Continued)

9.4.2 Temperature Profile

Figure 9.3(a) shows that variations in the thermophoresis factor (N_t) notably amplify fluid temperature. These phenomena engage the movement of nanomaterials from a heated area to a cooler one due to heat disparities, resulting in heightened temperatures. Figure 9.3(b) shows that alterations in the Brownian motion parameter (N_b)

significantly impact the temperature field. Elevated values of N_b lead to marked improvements in nanofluid temperature. This is attributed to the increased haphazard movement of nanomaterials inside the flow area, resulting in collisions and interactions that ultimately enhance the temperature field and increase the thickness of the temperature boundary layer (TBL). Higher Prandtl number (Pr) leads to a compressed temperature field and thinner TBL (Figure 9.3(c)). This is due to lower thermal diffusivity, resulting in reduced heat transfer over a narrower range. Elevated Pr also indicates greater nanofluid heat capacity, offering the potential for enhanced cooling efficiency in applications like bioreactors, bioconvection, and fuel cells. Figure 9.3(d) demonstrates the effect of radiating factor R_d on nanofluid temperature. As R_d increases, so does the temperature. This is because increasing R_d reduces the thermal absorption coefficient, which leads to a boost in heat in nanofluid. CHNFs display higher temperatures than CNFs due to the enhanced thermal properties of EG with hybridized nanoparticles (Figures 9.3(a–d)). This underscores the potential of integrating hybrid nanoparticles in cooling/heating procedures.

9.4.3 Nanoparticle Concentration

In Figure 9.4(a), we observe how activation energy (E) influences nanoparticle concentration. The data indicates that elevated values of E lead to an increased concentration of nanoparticles. Activation energy is the least amount of energy necessary to begin a chemical process, stimulating molecules in the system to engage in the reaction. This promotes the generation of nanoparticles. Figure 9.4(b) shows the influence of the reaction parameter (K) on the nanomaterial's concentration. The consequences demonstrate that an elevated K corresponds to a higher concentration

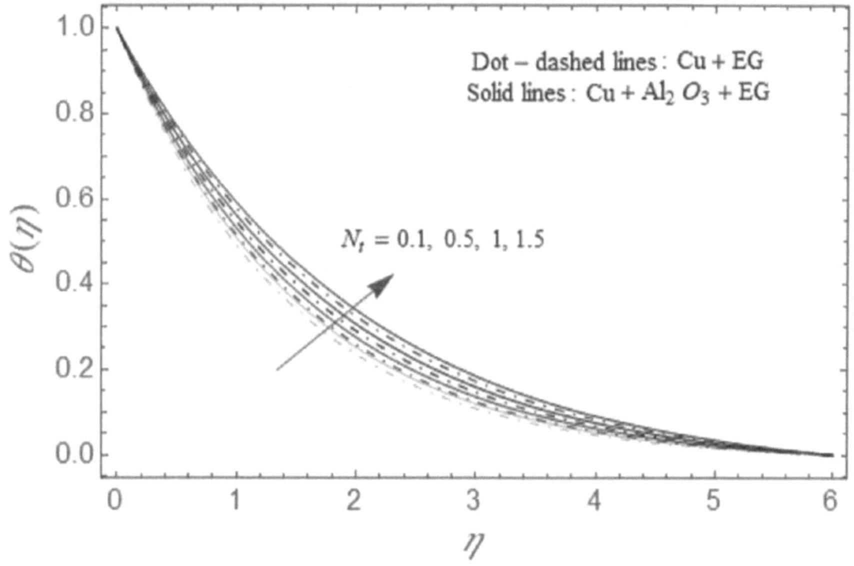

FIGURE 9.3 Temperature profile for changing values of (a) N_t , (b) N_b , (c) Pr, and (d) R_d .

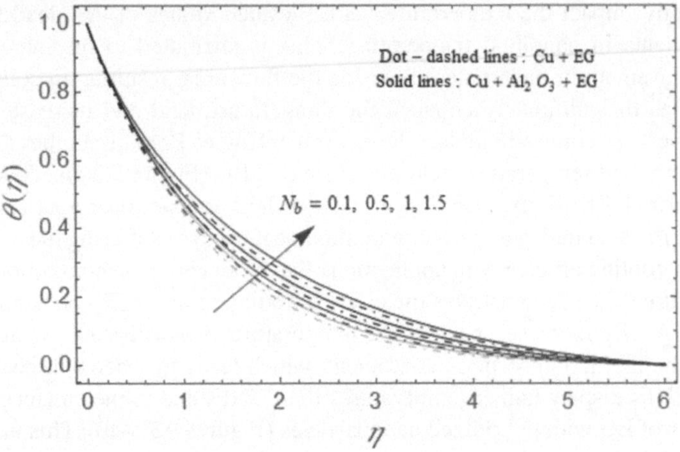

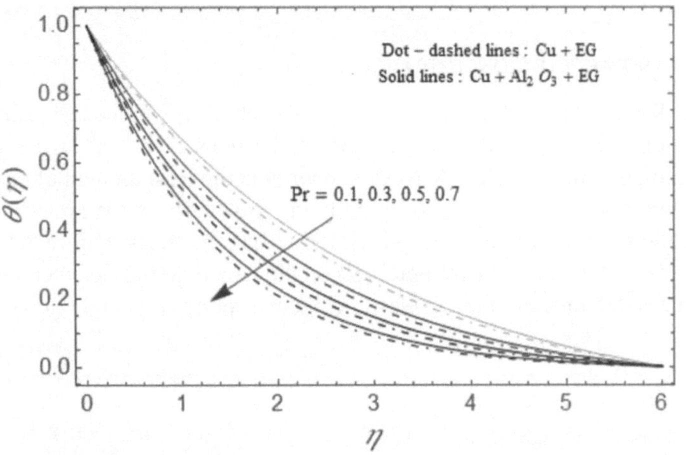

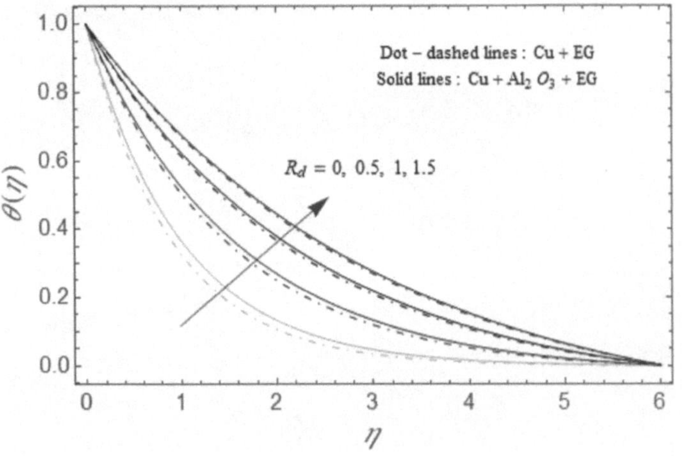

FIGURE 9.3 (Continued)

of nanomaterials. Consequently, an increase in the reaction factor leads to an augmented production rate of the reactive species. Examining Figures 9.4(a–b), it becomes evident that the CHNF (Cu-Al_2O_3-EG) exhibits a notably higher nanoparticle concentration than CNF (Cu-EG).

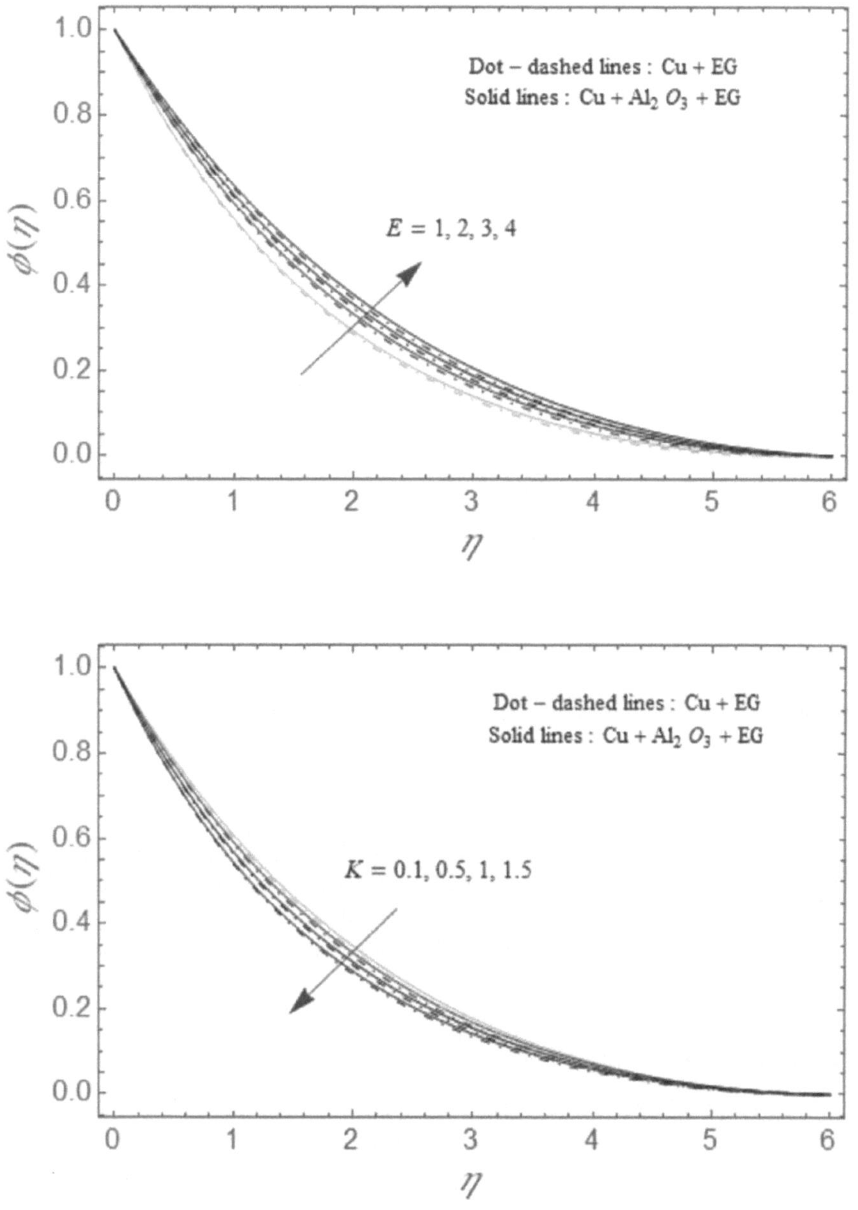

FIGURE 9.4 Concentration profile for changing values of (a) E and (b) K .

9.4.4 Engineering Factors

In this section, we delve into the mathematical representation of engineering elements.

9.4.4.1 Nusselt Number

$$Nu_x = \frac{x q_w}{k\left(T_w - T_\infty\right)}, \tag{9.23}$$

$$q_w = -\left(k + \frac{16\sigma^* T_0^3}{3k^*\left(\rho c_p\right)}\right)\frac{\partial T}{\partial y}\Big|_{y=0} . \tag{9.24}$$

The dimensionless Nusselt number is given as

$$Nu_x = -\left(1 + \frac{4}{3}R_d\right)\theta'(0). \tag{9.25}$$

9.4.4.2 Sherwood Number

$$Sh_x = \frac{x q_m}{D_B\left(C_w - C_\infty\right)}, \tag{9.26}$$

$$q_m = -D_B \frac{\partial C}{\partial y}\Big|_{y=0} . \tag{9.27}$$

The Sherwood number in its non-dimensional form is given as

$$Sh_x = -\phi'(0). \tag{9.28}$$

Table 9.2 shows the impact of physical attributes (N_t, K , N_b, and E) on Sherwood (Sh_x) and Nusselt (Nu_x) numbers. Higher values of N_b, N_t, and K lead to lower Nu_x but higher Nu_x for higher E . The Sherwood number increases with an increase in N_b, N_t, or K , but this effect diminishes with higher E. CNFs show higher Sh_x and Nu_x values compared to CHNFs.

9.5 RESPONSE SURFACE METHODOLOGY

The study utilizes RSM to establish relationships between input variables (encoded as P , Q , and R) representing thermophoresis (N_t), activation energy (E), and Brownian movement (N_b)) (see Table 9.3), and response factors, i.e., Sherwood number (Sh_x) and Nusselt number (Nu_x) for all possible combinations of the input variables (see Table 9.4).

TABLE 9.2
Numeric Data for Nusselt and Sherwood Numbers

		Nu_x		Sh_x	
		Cu + EG	Cu + Al_2O_3 + EG	Cu + EG	Cu + Al_2O_3 + EG
N_t	0.1	0.35288	0.33068	0.06312	0.02435
	0.3	0.32416	0.30649	0.31306	0.24353
	0.5	0.29824	0.28441	0.48838	0.40207
N_b	0.1	0.34289	0.32201	0.07596	0.06411
	0.3	0.25747	0.28345	0.12291	0.10357
	0.5	0.29824	0.24281	0.16627	0.14071
K	1	0.35021	0.32841	0.18277	0.17587
	2	0.34801	0.32657	0.25535	0.24213
	3	0.32341	0.32499	0.32116	0.30187
E	0.5	0.34993	0.32819	0.22306	0.21064
	1	0.35145	0.32945	0.18742	0.16979
	1.5	0.35218	0.33007	0.16170	0.14412

The second-order model represents for linear, quadratic, and interactive elements in the connections between Nu_x and Sh_x, providing a comprehensive analysis of the relationships [22, 39]:

$$Nu_x = \alpha_0 + \alpha_1 P + \alpha_2 Q + \alpha_3 R + \alpha_4 PQ + \alpha_5 PR + \alpha_6 QR + \alpha_7 P^2 + \alpha_8 Q^2 + \alpha_9 R^2, \tag{9.29}$$

$$Sh_x = \beta_0 + \beta_1 P + \beta_2 Q + \beta_3 R + \beta_4 PQ + \beta_5 PR + \beta_6 QR + \beta_7 P^2 + \beta_8 Q^2 + \beta_9 R^2, \tag{9.30}$$

where α_i and β_i $(0 \le i \le 9)$ are often called unidentified regression coefficients.

For the analysis, a central composite design (CCD) technique was employed for the statistical experiment. This approach involves 20 runs and entails selecting three variable stages, resulting in 19 degrees of freedom.

9.5.1 The Model's Accuracy

Tables 9.5 and 9.6 present the significance of RSM models through *P*-values compared to a 0.05 threshold. In Table 9.5, the square-term *E* lacks significance, as its *P*-value exceeds 0.05. Additionally, Table 9.6 shows the insignificance of certain product terms. As a result, these terms are excluded from the RSM model. Consequently, Equations (9.29) and (9.30) are adjusted accordingly based on these findings:

TABLE 9.3
Efficient Factor Levels on Central Composite Design [33]

		Level		
Factor	Symbol	Low (−1)	Mid (0)	High (+1)
N_t	P	0.2	0.4	0.6
E	Q	1	2	3
N_b	R	0.1	0.3	0.5

TABLE 9.4
Process of Designing Experiments and Analysing the Outcomes Obtained [39]

		Coded values			Real values				
Run	Point type	P	Q	R	N_t	E	N_b	Nu_x	Sh_x
1.	Factorial	−1	−1	−1	0.2	1	0.1	0.35539	0.07837
2.		1	−1	−1	0.6	1	0.1	0.33937	0.73142
3.		−1	1	−1	0.2	3	0.1	0.35597	0.16032
4.		1	1	−1	0.6	3	0.1	0.34032	0.86115
5.		−1	−1	1	0.2	1	0.5	0.33620	0.22991
6.		1	−1	1	0.6	1	0.5	0.32071	0.11451
7.		−1	1	1	0.2	3	0.5	0.33807	0.16703
8.		1	1	1	0.6	3	0.5	0.31896	0.17333
9.	Axial	−1	0	0	0.2	2	0.3	0.34655	0.13424
10.		1	0	0	0.6	2	0.3	0.33102	0.08170
11.		0	−1	0	0.4	1	0.3	0.33779	0.07072
12.		0	1	0	0.4	3	0.3	0.33919	0.00358
13.		0	0	−1	0.4	2	0.1	0.34785	0.49196
14.		0	0	1	0.4	2	0.5	0.32976	0.12365
15.	Centre	0	0	0	0.4	2	0.3	0.33874	0.02099
16.		0	0	0	0.4	2	0.3	0.33874	0.02099
17.		0	0	0	0.4	2	0.3	0.33874	0.02099
18.		0	0	0	0.4	2	0.3	0.33874	0.02099
19.		0	0	0	0.4	2	0.3	0.33874	0.02099
20.		0	0	0	0.4	2	0.3	0.33874	0.02099

$$Nu_x = 0.3377 + 0.0045P + 0.0014Q - 0.0042R + 0.0033PQ + 0.0076PR - 0.0013QR - 0.0019P^2 - 0.0021R^2, \tag{9.31}$$

$$Sh_x = -0.0176 + 0.1678P + 0.0289Q - 0.3132R - 0.0258QR - 0.0209P^2 - 0.1968R^2. \tag{9.32}$$

Tables 9.5 and 9.6 show that the adjustment R^2 is 99.86% for Nu_x and 94.91% for Sh_x. These high values affirm the accuracy and suitability of the presented model.

9.6 SENSITIVITY INVESTIGATION

To compute sensitivity function, we calculate the partial derivative of the response parameter with respect to input factors:

TABLE 9.5
Nusselt Number's Regression Coefficients

Source	Degree of freedom	Adj. sum of squares	Adj. mean squares	Regression coefficient	F-value	P-value
Model	9	0.1212	0.0134		1226.11	0.0000
Linear terms	3	0.1164	0.0430		352.22	0.0000
N_t	1	0.0002	0.0002	0.0045	0.800	0.0016
E	1	0.0431	0.0431	0.0014	0.936	0.0000
N_b	1	0.0761	0.0761	−0.0042	0.366	0.0000
Square terms	3	0.0005	0.0005		0.028	0.0000
N_t^2	1	0.0006	0.0006	−0.0019	0.0562	0.0407
E^2	1	0.0003	0.0003	0.0072	0.0426	0.2886
N_b^2	1	0.0005	0.0005	−0.0021	0.0612	0.0056
Interaction terms	3	0.0015	0.0004		0.0473	0.0000
$N_t \times E$	1	0.0001	0.0001	0.0033	0.0134	0.0125
$N_t \times N_b$	1	0.0007	0.0007	0.0076	0.0087	0.0385
$E \times N_b$	1	0.0006	0.0006	−0.0013	0.1412	0.0002
Constant				0.3377		0.0000
Errors	10	0.0002	0.0002	0.0189		
Total	19	0.1095				
Adj. $R^2 = 99.97\%$				$R^2 = 99.81\%$		

TABLE 9.6
Sherwood Number's Regression Coefficients

Source	Degree of freedom	Adj. sum of squares	Adj. mean squares	Regression coefficient	F-value	P-value
Model	9	0.8142	0.0155		9.2599	0.0000
Linear terms	3	0.6649	0.3544		17.0036	0.0000

(Continued)

TABLE 9.6
(Continued)

Source	Degree of freedom	Adj. sum of squares	Adj. mean squares	Regression coefficient	F-value	P-value
N_t	1	0.0252	0.0252	0.1678	0.3901	0.0149
E	1	0.0682	0.0682	0.0289	1.0081	0.0181
N_b	1	0.7012	0.7012	−0.3132	36.1155	0.0002
Square terms	3	0.1200	0.0305		0.6505	0.0000
N_t^2	1	0.0001	0.0001	−0.0136	0.0014	0.0267
E^2	1	0.0005	0.0005	−0.0209	0.0113	0.5170
N_b^2	1	0.0033	0.0033	0.1968	1.0142	0.0077
Interaction terms	3	0.0135	0.0135		0.2097	0.0000
$N_t \times E$	1	0.0079	0.0079	−0.0091	0.1436	0.3894
$N_t \times N_b$	1	0.0068	0.0068	−0.1555	0.1554	0.3716
$E \times N_b$	1	0.0220	0.0220	−0.0258	0.1703	0.0003
Constant				−0.0176		0.0000
Errors	10	0.0998	0.0099			
Total	19	0.9145				
Adj. $R^2 = 94.48\%$				$R^2 = 98.98\%$		

$$\frac{\partial Nu_x}{\partial P} = 0.0045 + 0.0033Q + 0.0076R - 0.0038P, \tag{9.33}$$

$$\frac{\partial Nu_x}{\partial Q} = 0.0014 + 0.0033P - 0.0013R, \tag{9.34}$$

$$\frac{\partial Nu_x}{\partial R} = -0.0042 + 0.0076P - 0.0013Q - 0.0042R, \tag{9.35}$$

$$\frac{\partial Sh_x}{\partial P} = 0.1678 - 0.0418P, \tag{9.36}$$

$$\frac{\partial Sh_x}{\partial Q} = 0.0289 - 0.0258R, \tag{9.37}$$

$$\frac{\partial Sh_x}{\partial R} = -0.3132 - 0.0258Q - 0.3936R. \tag{9.38}$$

The sensitivity analysis reveals how changes in input coefficients affect response variables, either positively or negatively. Positive sensitivity implies a boost in input factors to achieve a positive effect on the response, while negative sensitivity indicates the opposite effect. Figures 9.5(a–c) and 9.6 (a–c) illustrate the sensitivity results for Nu_x and Sh_x using vertical bar charts.

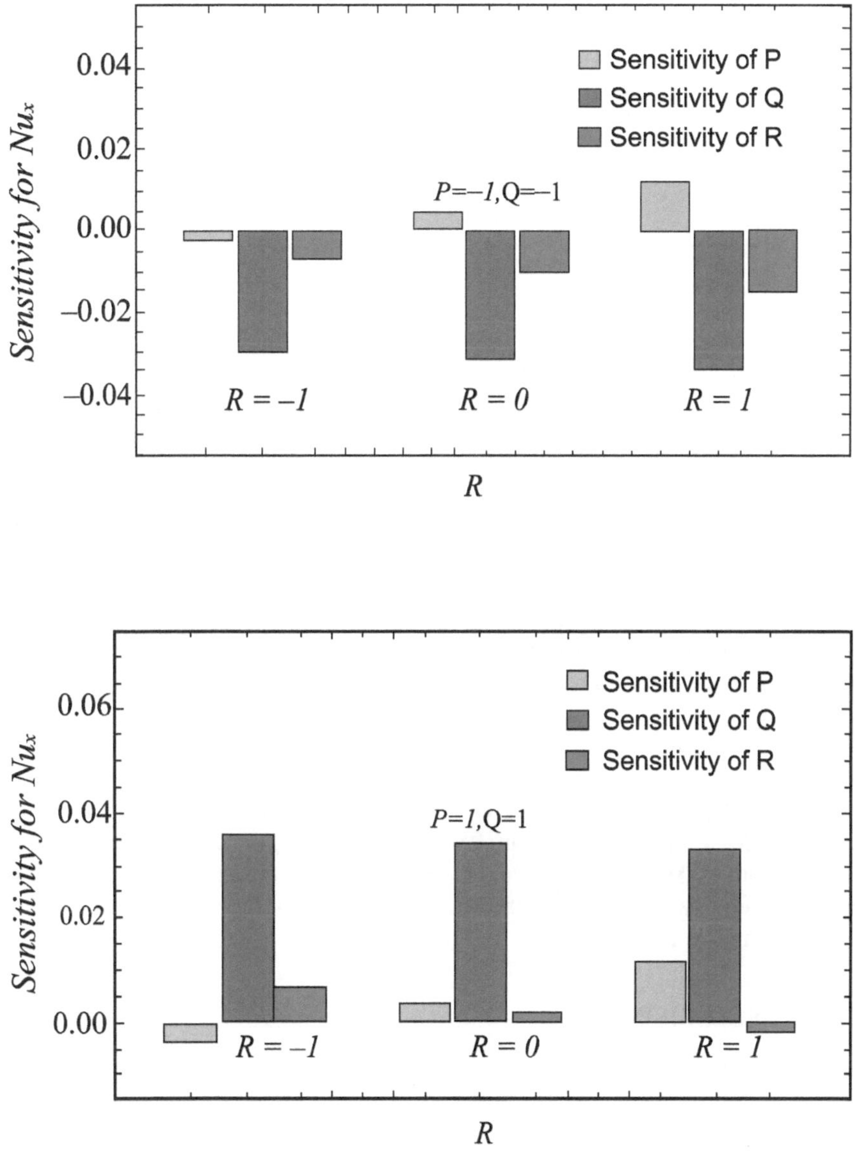

FIGURE 9.5 Sensitivity of Nu_x for (a) $P=-1, Q=-1$, (b) $P=1, Q=1$, and (c) $P=1, Q=-1$.

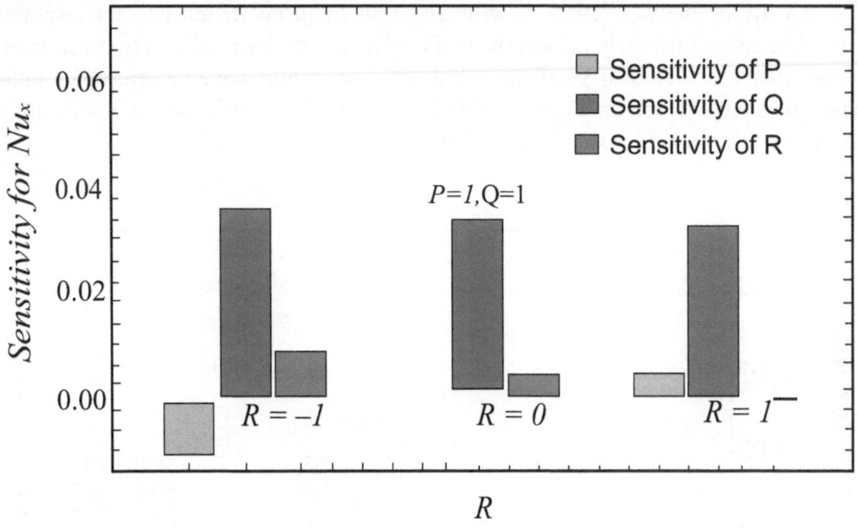

FIGURE 9.5 (Continued)

In Figure 9.5(a), Nu_x responds positively to changes in N_t (P) and the opposite response is achieved with E and N_b (Q and R) across all three points of N_b. In Figure 9.5(b), with E and N_t at their highest points ($E = 3$ and $N_t = 0.5$), Nu_x is absolutely reactive to changes in E and N_b, while it is negatively receptive to changes in N_t for every N_b point. The highest responsiveness of Nu_x is observed at the highest points of N_t and E. Figure 9.5(c) shows a consistent trend where Nu_x displays positive responsiveness to all governing variables: E, N_t, and N_b. Overall, it is evident from Figures 9.5(a–c) that Nu_x is more sensitive to N_b and N_t compared to E.

Figure 9.6 presents the sensitivity analysis of Sh_x with respect to N_b levels. Figure 9.6(a) shows that Sh_x exhibits positive sensitivity to N_t and E but negative sensitivity to N_b. As E and N_t increase, Sh_x shows positive sensitivity, apart from heightened N_b as observed in Figure 9.6(b). Similarly, Figure 9.6(c) shows the positive sensitivity of Sh_x towards N_t and E, apart from that for all levels of N_b. These figures collectively highlight that the sensitivity of Sh_x to E is higher compared to that to N_b and N_t.

9.7 CONCLUSIONS

The major results of this simulation are as follows:

- The concentration of nanoparticles, represented by N_r, causes the fluid velocity to compress, although the Casson parameter (γ) and modified Hartmann parameter (Ha) cause the fluid velocity to decompress.

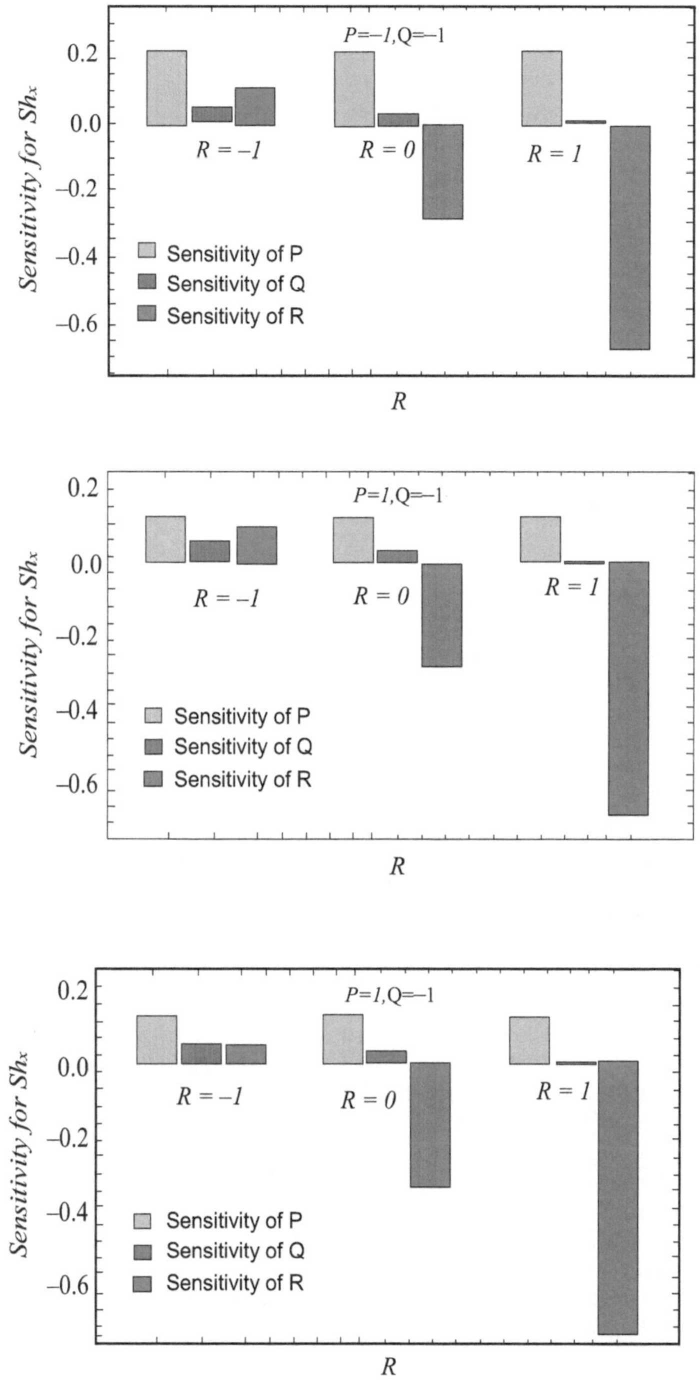

FIGURE 9.6 Sensitivity of Sh_x for (a) $P=-1, Q=-1$, (b) $P=1, Q=1$, and (c) $P=1, Q=-1$.

- The distribution of temperature increases as the Brownian movement factor (N_b), radiating factor (R_d), and thermophoresis factor (N_t) increase, while the Prandtl number (*Pr*) decreases.
- The existence of activating energy factor (E) is beneficial in increasing the nanoparticle concentration, whereas an increase in reacting factor (K) has a positive impact on the nanoparticle concentration.
- The velocity profile of CHNFs (Cu-Al_2O_3-EG) is lower than that of the CNF (Cu-EG).
- The concentration of nanomaterials and temperature in CHNFs are greater than those of CNFs.
- The thermophoresis factor (N_t) and Brownian movement factor (N_b) have the greatest impact on Nusselt number sensitivity than the activation parameter (E).
- The consequence of the activating energy factor (E) on the Sherwood number is greater than the effect of the thermophoresis factor (N_t) and Brownian movement factor (N_b).

REFERENCES

1. Casson, N. (1959). Flow equation for pigment-oil suspensions of the printing ink-type. Rheology of Disperse Systems, 84–104.
2. Subba Rao, A., Ramachandra Prasad, V., Bhaskar Reddy, N., & Anwar Bég, O. (2015). Heat transfer in a Casson rheological fluid from a semi-infinite vertical plate with partial slip. Heat Transfer—Asian Research, 44(3), 272–291.
3. Ali, L., Ali, B., & Ghori, M. B. (2022). Melting effect on Cattaneo–Christov and thermal radiation features for aligned MHD nanofluid flow comprising microorganisms to leading edge: FEM approach. Computers & Mathematics with Applications, 109, 260–269.
4. Abdal, S., Hussain, S., Siddique, I., Ahmadian, A., & Ferrara, M. (2021). On solution existence of MHD Casson nanofluid transportation across an extending cylinder through porous media and evaluation of priori bounds. Scientific Reports, 11(1), 7799.
5. Abo-Dahab, S. M., Abdelhafez, M. A., Mebarek-Oudina, F., & Bilal, S. M. (2021). MHD Casson nanofluid flow over nonlinearly heated porous medium in presence of extending surface effect with suction/injection. Indian Journal of Physics, 95(12), 2703–2717.
6. Ali, B., Naqvi, R. A., Haider, A., Hussain, D., & Hussain, S. (2020). Finite element study of MHD impacts on the rotating flow of Casson nanofluid with the double diffusion Cattaneo–Christov heat flux model. Mathematics, 8(9), 1555.
7. Al-Mamun, A., Arifuzzaman, S. M., Rabbi, S. R., Alam, U. S., Islam, S., & Khan, M. (2021). Numerical simulation of periodic MHD Casson nanofluid flow through porous stretching sheet. SN Applied Sciences, 3, 271.
8. Sarkar, S., Jana, R. N., & Das, S. (2020). Time-dependent entropy analysis in magnetized Cl-Al_2O_3/ethylene glycol hybrid nanofluid flow due to a vibrating vertical plate. International Journal of Fluid Mechanics Research, 47, 419–443.
9. Shah, Z., Kumam, P., & Deebani, W. (2020). Radiative MHD Casson nanofluid flow with activation energy and chemical reaction over past nonlinearly stretching surface through entropy generation. Scientific Reports, 10(1), 4402.
10. Choi, S. U. S. (1995). Enhancing thermal conductivity of fluid with nanoparticles. Developments and Application of non-Newtonian Flows, 66, 99–105.

11. Krishna, V. M., Kumar, M. S., Muthalagu, R., Kumar, P. S., & Mounika, R. (2022). Numerical study of fluid flow and heat transfer for flow of Cu-Al_2O_3-water hybrid nanofluid in a microchannel heat sink. Materials Today: Proceedings, 49, 1298–1302.
12. Jamshed, W., & Aziz, A. (2018). Cattaneo–Christov based study of TiO_2–CuO/EG Casson hybrid nanofluid flow over a stretching surface with entropy generation. Applied Nanoscience, 8(4), 685–698.
13. Das, S., Sarkar, S., & Jana, R. N. (2020). Feature of entropy generation in Cu-Al_2O_3/ethylene glycol hybrid nanofluid flow through a rotating channel. Bionanoscience, 10(4), 950–967.
14. Mahabaleshwar, U. S., Aly, E. H., & Anusha, T. (2022). MHD slip flow of a Casson hybrid nanofluid over a stretching/shrinking sheet with thermal radiation. Chinese Journal of Physics, 80, 74–106.
15. Manzoor, U., Imran, M., Muhammad, T., Waqas, H., & Alghamdi, M. (2021). Heat transfer improvement in hybrid nanofluid flow over a moving sheet with magnetic dipole. Waves in Random and Complex Media, 1–15.
16. Manzoor, U., Muhammad, T., Farooq, U., & Waqas, H. (2022). Investigation of thermal stratification and nonlinear thermal radiation in Darcy-Forchheimer transport of hybrid nanofluid by rotating disk with Marangoni convection. International Journal of Ambient Energy, 43(1), 6724–6731.
17. Sneha, K. N., Mahabaleshwar, U. S., & Bhattacharyya, S. (2023). An effect of thermal radiation on inclined MHD flow in hybrid nanofluids over a stretching/shrinking sheet. Journal of Thermal Analysis and Calorimetry, 148(7), 2961–2975.
18. Waqas, H., Farooq, U., Muhammad, T., & Manzoor, U. (2022). Importance of shape factor in Sisko nanofluid flow considering gold nanoparticles. Alexandria Engineering Journal, 61(5), 3665–3672.
19. Bestman, A. (1990). Natural convection boundary layer with suction and mass transfer in a porous medium. International Journal of Energy Research, 14, 389–396.
20. Makinde, O. D., Olanrewaju, P. O., & Charle, W. M. (2011). Unsteady convection with chemical reaction and radiative heat transfer past a flat porous plate moving through a binary mixture. Afrika Matematika, 22, 65–78.
21. Ahmad, I., Qureshi, N., Al-Khaled, K., Aziz, S., Chammam, W., & Khan. S. (2021). Magnetohydrodynamic time dependent 3-D simulations for Casson nano-material configured by unsteady stretched surface with thermal radiation and chemical reaction aspects. Journal of Nanofluids, 10, 232–245.
22. Abdelmalek, Z., Mahanthesh, B., Basir, M. F. M., Imtiaz, M., Mackolil, J., Khan, N. S., & Tlili, I. (2020). Mixed radiated magneto Casson fluid flow with Arrhenius activation energy and Newtonian heating effects: Flow and sensitivity analysis. Alexandria Engineering Journal, 59(5), 3991–4011.
23. Sarkar, S., & Das, S. (2023). Gyrotactic microbes' movement in a magneto-nano-polymer induced by a stretchable cylindrical surface set in a DF porous medium subject to non-linear radiation and Arrhenius kinetics. International Journal of Modelling and Simulation, 1–18.
24. Sarkar, S., Jana, R. N., & Das, S. (2020). Activation energy impact on radiated magneto-Sisko nanofluid flow over a stretching and slipping cylinder: Entropy analysis. Multidiscipline Modeling in Materials and Structures, 16(5), 1085–1115.
25. Waqas, H., Manzoor, U., Hussain, S., & Bhatti, M. M. (2022). Maxwell time-dependent nanofluid flow over a wedge covered with gyrotactic microorganism: An activation energy process. International Journal of Ambient Energy, 43(1), 5560–5570.
26. Gailitis, A. (1961). On a possibility to reduce the hydrodynamical resistance of a plate in an electrolyte. Applied Magnetohydrodynamics, 12, 143–146.

27. Abbas, T., Hayat, T., Ayub, M., Bhatti, M. M., & Alsaedi, A. (2019). Electromagnetohydrodynamic nanofluid flow past a porous Riga plate containing gyrotactic microorganism. Neural Computing and Applications, 31, 1905–1913.
28. Ganesh, N. V., Al-Mdallal, Q., Fahel, S., & Dadoa, S. (2019). Riga-plate flow of gamma Al_2O_3-water/ethylene glycol with effective Prandtl number impacts. Heliyon, 5, 01651.
29. Mburu, Z., Mondal, S., Sibanda, P., & Sharma, R. (2021). A numerical study of entropy generation on Oldroyd-B nanofluid flow past a Riga plate. Journal of Thermal Engineering, 845–866.
30. Sarkar, S., Ali, A., & Das, S. (2022). Bioconvection in non-Newtonian nanofluid near a perforated Riga plate induced by haphazard motion of nanoparticles and gyrotactic microorganisms in the attendance of thermal radiation, and Arrhenius chemical reaction: Sensitivity analysis. International Journal of Ambient Energy, 43, 1–34.
31. Sarkar, S., Pal, T., Ali, A., & Das, S. (2022). Themo-bioconvection of gyrotactic microorganisms in a polymer solution near a perforated Riga plate immersed in a DF medium involving heat radiation, and Arrhenius kinetics. Chemical Physics Letters, 797, 139557.
32. Vaidya, H., Prasad, K. V., Tlili, I., Makinde, O. D., Rajashekhar, C., Khan, S. U., Kumar, R., & Mahendra, D. L. (2021). Mixed convective nanofluid flow over a non-linearly stretched Riga plate. Case Studies in Thermal Engineering, 24, 100828.
33. Chan, S. Q., Aman, F., & Mansur, S. (2018). Sensitivity analysis on thermal conductivity characteristics of a water-based bionanofluid flow past a wedge surface. Mathematical Problems in Engineering, 2018, 1–12.
34. Sarkar, S., & Das, S. (2023). Magneto-thermo-bioconvection of a chemically sensitive Cross nanofluid with an infusion of gyrotactic microorganisms over a lubricious cylindrical surface: Statistical analysis. International Journal of Modelling and Simulation, 43(6), 980–1001.
35. Mutuku, W., & Makinde, O. D. (2014). Hydromagnetic bioconvection of nanofluid over a permeable vertical plate due to gyrotactic microorganisms. Computers & Fluids, 95, 88–97.
36. Das, S., Sarkar, S., & Jana, R. N. (2018). Entropy generation analysis of MHD slip flow of non-Newtonian Cu-Casson nanofluid in a porous microchannel filled with saturated porous medium considering thermal radiation. Journal of Nanofluids, 7, 1217–1232.
37. Bhatti, M. M., Abbas, T., & Rashidi, M. M. (2016). Effects of thermal radiation and electromagnetohydrodynamics on viscous nanofluid through a Riga plate. Multidiscipline Modeling in Materials and Structures, 12(4), 605–618.
38. Ali, B., Shafiq, A., Siddique, I., Al-Mdallal, Q., & Jarad, F. (2021). Significance of suction/injection, gravity modulation, thermal radiation, and magnetohydrodynamic on dynamics of micropolar fluid subject to an inclined sheet via finite element approach. Case Studies in Thermal Engineering, 28, 101537.
39. Shafiq, A., Sindhu, T. N., & Khalique, C. M. (2020). Numerical investigation and sensitivity analysis on bioconvective tangent hyperbolic nanofluid flow towards stretching surface by response surface methodology. Alexandria Engineering Journal, 59(6), 4533–4548.

10 Characteristic of Heat-Induced Ferrofluid Flow on a Riga Sensor Plate with the Effect of Viscous Dissipation

S.R. Mishra, P.K. Pattnaik, and Subhajit Panda

10.1 INTRODUCTION

Incorporating ferrofluids into heat exchangers allows for increased thermal conductivity and improved heat transfer efficiency. Their ability to respond to external magnetic fields allows for the precise regulation of fluid flow and distribution, resulting in a significant improvement in heat dissipation capabilities. Hassan et al. [1] deliberated the impact of thermal slip flow behaviour and analysed a hybrid ferrofluid comprising ethylene glycol/water. This analysis took into account the imposition of nanoparticle shapes and slippage constraints. Yilbas et al. [2] explored the manner in which ferrofluid droplets interact with a hydrophobic surface in the presence of magnetism. Sivakumar et al. [3] explored the association of slip and dissipation with radiation-induced ferrofluid dynamics in a permeable convectively heated widening sheet. Zeeshan et al. [4] explored the influence of a magnetic dipole on the behaviour of a viscous ferrofluid as it flows across a stretched surface, taking heat radiation into account. Numerous studies [5–8] have included ferrofluids in their research due to their ability to improve heat transfer performance.

Riga plates have the ability to improve both condensation and evaporation processes in refrigeration and air conditioning systems. The increased surface area and controlled fluid flow caused by the plate's properties speed up phase change heat transfer. As a result, thermal exchange is faster and more efficient. Ali et al. [9] showed an extensive work that focused on thermal transmission analysis. A hybrid nanofluid was flowed across a Riga plate in this study. The study takes a complete approach, including the impacts of Lorentz forces and entropy formation, with the goal of providing a thorough understanding of the underlying process. Das et al. [10] explored the dynamic tendencies exhibited by a radioactive graphene oxide (rGO)–magnetite–water mixture while passing over a vibrating Riga plate sensor. The study encompassed deliberate manipulation of temperature and concentration gradients to systematically scrutinize how the system reacted. Wahid et al. [11] conducted a study to investigate how a hybrid nanofluid performs towards a stagnation point when it

DOI: 10.1201/9781003595786-10

encounters a diminishing Riga plate with slip circumstances. Many researchers [12–15] directed their attention towards the Riga plate owing to its ability to enhance heat transfer properties.

The use of viscous dissipation in heat transmission is a critical aspect of thermal processes, finding application in various engineering and scientific contexts. Viscous dissipation is the conversion of mechanical energy into heat energy caused by internal friction within a fluid during its flow. This occurrence is especially significant in situations involving high-velocity flows, complex geometries, and fluids that depart from Newtonian behaviour. Awati et al. [16] focused on exploring the occurrence of viscous dissipation in a polar nanofluids under the constraints of magnetohydrodynamic (MHD) flow. Akram et al. [17] looked into how convection theory may be used to analyse the radiative peristaltic movement of a magnetic nanofluid within an asymmetric channel. The research looked into a variety of parameters, including partial slide, viscous dissipation, and the effect of magnetic fields. Mahesh et al. [18] discussed how radiation influences the movement of an MHD couple stress hybrid nanofluid over a porous sheet, as well as the impacts of viscous dissipation. Numerous researchers [19, 20] have considered the effects of viscous dissipation in their investigations.

The innovative aspect of the existing study involves delving into the behaviour of hybrid nanofluids within a stagnation point flow scenario. The versatility of magnetic nanofluids, owing to their external controllability and the capacity to manipulate physical properties through adjustments in nanoparticle volume percentage and magnetic field strength, opens the way for a wide range of potential applications. This study advances our understanding of the behaviour of hybrid nanofluids in intricate flow scenarios by considering a combination of magnetic nanoparticles, a shrinking surface, slip conditions, viscous dissipation, and the introduction of radiative heat flux. The utilization of advanced mathematical models and sophisticated numerical methods bolsters the accuracy and reliability of the study's findings.

10.2 MATHEMATICAL DESCRIPTION

Consider a hybridized ferrofluid comprising Fe_3O_4 (magnetite) and $CoFe_2O_4$ (cobalt ferrite) dispersed in water (H_2O) with stagnation point movement under slip velocity across a heated shrinking Riga surface. The velocity of the diminishing surface in the viscous fluid is $u_e(x)$. The mass transmission velocity over the wall's permeable surface is represented by $v_w(x)$.

The standard governing equations are as follows:

$$u_x + v_y = 0, \tag{10.1}$$

$$uu_x + vu_y = u_e(u_e)_x + \frac{\mu_{hnf}}{\rho_{hnf}} u_{yy} + \frac{1}{8}\frac{\pi j_0 \bar{M}_0}{\rho_{hnf}} \exp(-y\pi / \alpha_1), \tag{10.2}$$

$$uT_x + vT_y = \left(\frac{k}{\rho c_p}\right)_{hnf} T_{yy} + \frac{Q_0}{(\rho c_p)_{hnf}}(T - T_\infty) - \frac{(q_r)_y}{(\rho c_p)_{hnf}} + \left(\frac{\mu}{\rho c_p}\right)_{hnf} (u_y)^2, \tag{10.3}$$

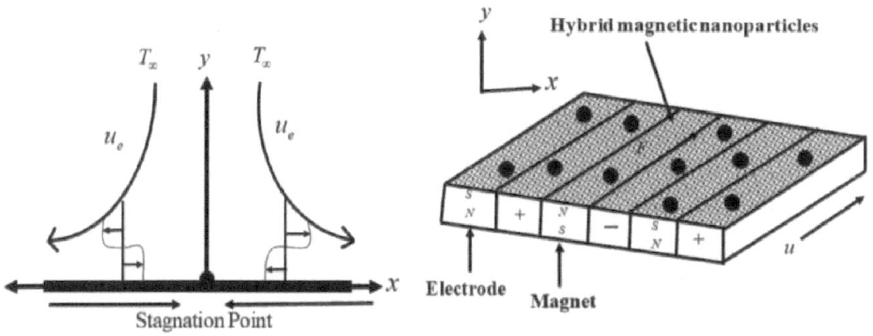

FIGURE 10.1 Schematic view of the present flow model.

$$\left.\begin{array}{ll} \text{at } y=0; \; u=u_w(x)=A_1\dfrac{\mu_{hnf}}{\rho_{hnf}}u_y; v=v_w(x); T=T_w; \\ \text{as } y\to\infty; \quad u\to u_e(x)=ax; T\to T_\infty \end{array}\right\}. \tag{10.4}$$

Here, $\bar{M}_0=(xM_0/l)$ is the magnetization and $c=0$ designates the static plate, and $c<0$ designates the diminishing surface.

The similarity variables are as follows:

$$u=axf'(\eta), T=\theta(\eta)(T_w-T_\infty)+T_\infty, v=-\sqrt{av_f}f(\eta),\ \eta=y\left(\frac{a}{v_f}\right)^{0.5}. \tag{10.5}$$

Utilizing equation (10.5) into equations (10.2) – (10.4) the transformed equations are:

$$\left(\frac{\chi_1}{\chi_2}\right)f'''+1+ff''-f'^2+Q\left(\frac{1}{\chi_2}\right)\exp(-d\eta)=0, \quad 10.(6)$$

$$\frac{1}{Pr}\frac{(\chi_3+Rd)}{\chi_4}\theta''+f\theta'+\left(\frac{\chi_1}{\chi_4}\right)Ec\,f''^2+\frac{H\theta}{\chi_4}=0, \tag{10.7}$$

$$\begin{array}{l} f'(0)=\lambda+A\left(\dfrac{\chi_1}{\chi_2}\right)f''(0), f(0)=S, \theta(0)=1 \\ f'(\eta)\to 1,\ \theta(\eta)\to 0 \text{ as } \eta\to\infty \end{array}, \tag{10.8}$$

$$\frac{\mu_{hnf}}{\mu_f}=\chi_1, \frac{\rho_{hnf}}{\rho_f}=\chi_2, \frac{k_{hnf}}{k_f}=\chi_3, \frac{(\rho c_p)_{hnf}}{(\rho c_p)_f}=\chi_4.$$

The dimensionless parameters applied in the aforementioned equations are as follows:

$A = A_1\sqrt{a/v_f}$ (slip velocity parameter), $Q = (\pi j_0 M_0)/(8\rho_f a^2 l)$ (modified Hartmann number), $Rd = \dfrac{16\sigma^* T_\infty^3}{3k^* k_f}$ (radiation parameter), $\Pr = (v\rho c_p / k)_f$ (Prandtl number), $H = \dfrac{Q_0}{a(\rho C_P)_f}$ (heat source parameter), $d = \left(\pi\sqrt{v_f / a}\right)/\alpha_1$, $S = -v_w(x)/\sqrt{av_f}$, and $Ec = \dfrac{u_w^2}{(T_w - T_\infty)(C_P)_f}$ (Eckert number).

In the scenario of suction $S > 0$ and injection $S < 0$, the constant $\lambda = c/a$ with $\lambda < 0$ and $\lambda = 0$ according to contracting and static surfaces.

The engineering aspects Nu_x and C_f are

$$C_f = \frac{\mu_{hnf}}{u_e^2(x)}(u_y)_{y=0}\frac{1}{\rho_f},\ Nu_x = \frac{k_{hnf}x}{(T_w - T_\infty)k_f}(T_y)_{y=0} + q_r|_{y=0}. \tag{10.9}$$

Equations (10.5) and (10.9) are given as

$$C_f \mathrm{Re}_x^{1/2} = \chi_1 f''(0), Nu_x \mathrm{Re}_x^{-1/2} = -[\chi_3 + Rd]\theta'(0), \tag{10.10}$$

where $\mathrm{Re}_x = u_e(x)x/v_f$ indicates the local Reynolds number.

TABLE 10.1
Thermophysical Features of the Hybrid Nanofluid

Attribute	Hybrid nanofluid
Effective density	$\rho_{hnf} = \rho_{nf}\left[(1-\phi_{CoFe_2O_4})\left[1-\phi_{Fe_3O_4}+\phi_{Fe_3O_4}\left(\frac{\rho_{Fe_3O_4}}{\rho_{H_2O}}\right)\right]+\phi_{CoFe_2O_4}\left(\frac{\rho_{CoFe_2O_4}}{\rho_{H_2O}}\right)\right]$
Dynamic viscosity	$\mu_{hnf} = \mu_{nf}(0.904)^2 e^{14.8(\phi_{Fe_3O_4}+\phi_{CoFe_2O_4})}$
Heat capacity	$(\rho c_p)_{hnf} = (1-\phi_{CoFe_2O_4})\left[(1-\phi_{Fe_3O_4})(\rho c_p)_{H_2O}+\phi_{Fe_3O_4}(\rho c_p)_{Fe_3O_4}\right]+\phi_{CoFe_2O_4}(\rho c_p)_{CoFe_2O_4}$
Thermal conductivity	$\frac{k_{hnf}}{k_{nf}} = \left[\frac{k_{CoFe_2O_4}+2k_{nf}-2\phi_{CoFe_2O_4}(k_{nf}-k_{CoFe_2O_4})}{k_{CoFe_2O_4}+2k_{nf}-\phi_{CoFe_2O_4}(k_{nf}-k_{CoFe_2O_4})}\right]$, where $\frac{k_{nf}}{k_f} = \left[\frac{k_{Fe_3O_4}+2k_{H_2O}-2\phi_{Fe_3O_4}(k_{H_2O}-k_{Fe_3O_4})}{k_{Fe_3O_4}+2k_{H_2O}-\phi_{Fe_3O_4}(k_{H_2O}-k_{Fe_3O_4})}\right]$

TABLE 10.2
Thermophysical Characteristics of the Base Fluid and Nanoparticles

Base fluid and nanoparticle	$\rho\left[\text{kg m}^{-3}\right]$	$C_p\left[\text{Jkg}^{-1}\text{K}^{-1}\right]$	$k\left[\text{Wm}^{-1}\text{K}^{-1}\right]$
Water	997.1	4179	0.613
Magnetite (Fe_3O_4)	5180	670	9.7
Cobalt ferrite ($CoFe_2O_4$)	4907	700	3.7

TABLE 10.3
Comparison Value of $C_f\,\text{Re}_x^{1/2}$ at $\phi_2 = A = S = Q = H = Rd = \lambda = 0$

	$C_f\,\text{Re}_x^{1/2}$			
	Previous study [11]		Present study	
ϕ_1	Magnetite and water	Cobalt ferrite and water	Magnetite and water	Cobalt ferrite and water
0.05	1.44547875	1.43727688	1.445472	1.43727
0.1	1.675264477	1.65902959	1.675264	1.65902
0.15	1.927699497	1.90324846	1.927699	1.90324
0.2	1.927699497	2.17612436	1.927693	2.17612

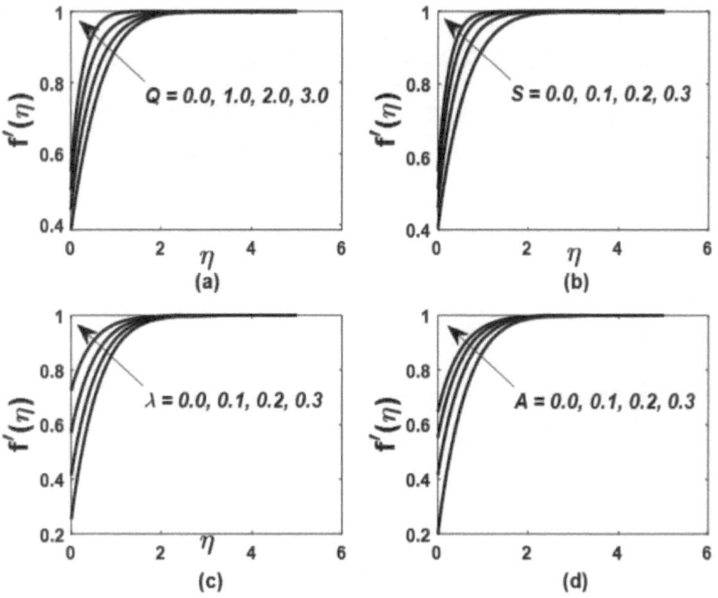

FIGURE 10.2 Momentum profile for (a) Hartmann number (Q), (b) suction (S), (c) velocity ratio (λ), and (d) slip velocity (A).

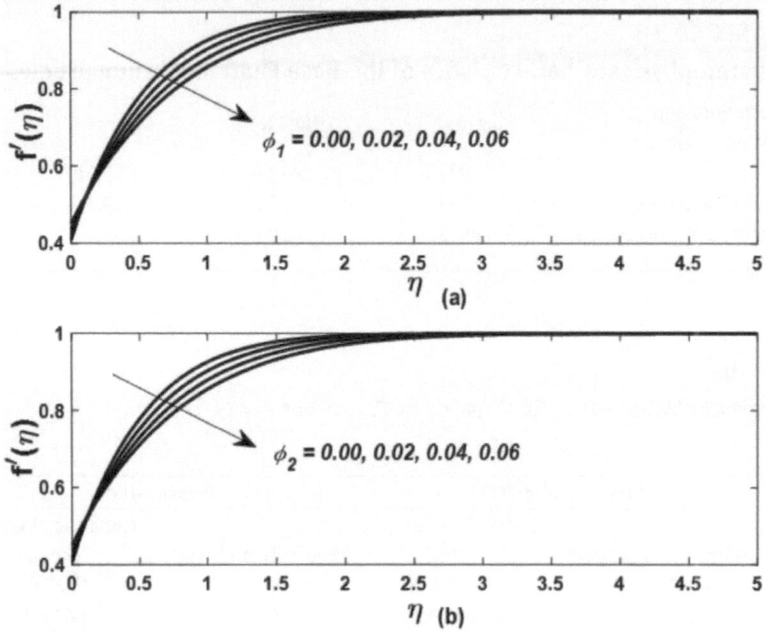

FIGURE 10.3 Momentum profile for (a) ϕ_1 and (b) ϕ_2.

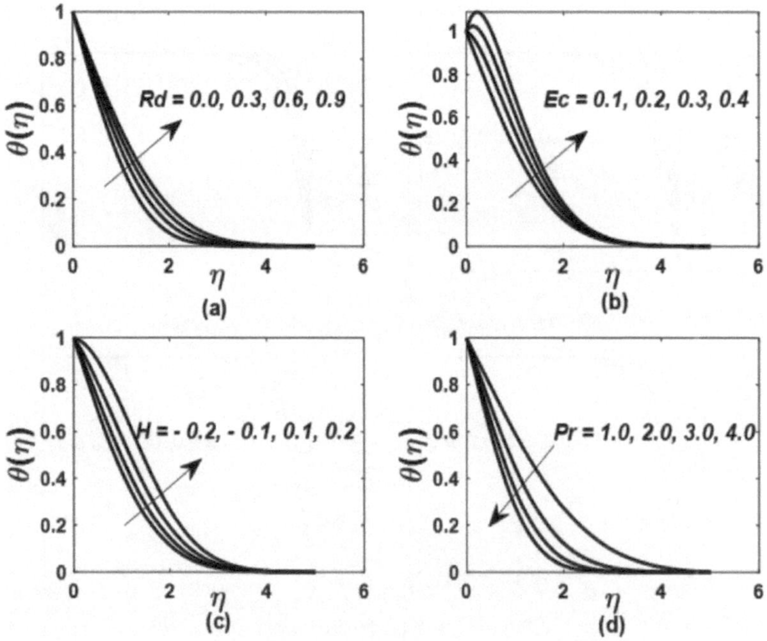

FIGURE 10.4 Temperature profile for (a) radiation parameter (Rd), (b) Eckert number (Ec), and (c) heat source/sink parameter (H).

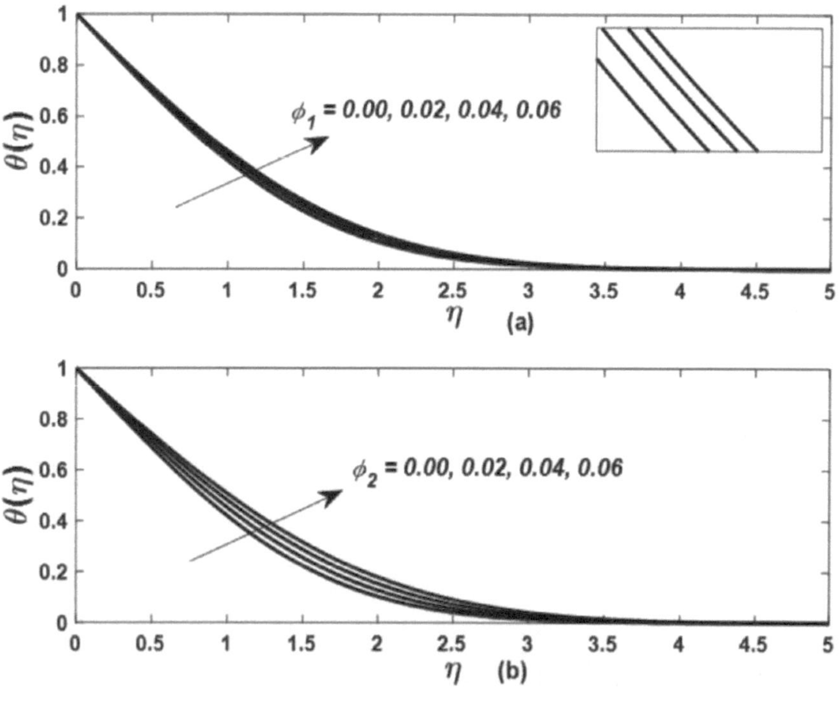

FIGURE 10.5 Temperature profile for (a) ϕ_1 and (b) ϕ_2.

TABLE 10.4
Computational Values of Skin Friction Coefficient and Nusselt Number

Q	S	A	λ	Rd	Ec	H	ϕ_1	ϕ_2	C_f	Nu_x
0.5									1.1855	0.9221
1									1.3621	0.9505
1.5									1.5339	0.977
0.5	0.2								1.3621	0.9505
	0.4								1.4276	1.0942
	0.6								1.4929	1.244
	0.2	0.1							1.5247	0.9216
		0.2							1.3621	0.9505
		0.3							1.2259	0.9727
		0.1	0.1						1.4782	0.9301
			0.2						1.3621	0.9505
			0.3						1.2422	0.9702
			0.1	0.3					1.3621	0.8893
				0.6					1.3621	0.9796

(Continued)

TABLE 10.4
(Continued)

Q	S	A	λ	Rd	Ec	H	ϕ_1	ϕ_2	C_f	Nu_x
				0.9					1.3621	1.0626
				0.3	0.1				1.3621	0.9135
					0.2				1.3621	0.8724
					0.3				1.3621	0.8314
					0.1	0.1			1.3621	0.9505
						0.2			1.3621	0.8762
						0.3			1.3621	0.7976
						0.1	0.01		1.3621	0.9505
							0.02		1.4495	0.943
							0.03		1.5418	0.9358
							0.01	0.01	1.3621	0.9505
								0.02	1.4479	0.9484
								0.03	1.5384	0.9466

10.3 RESULTS AND DISCUSSION

In-depth analysis of the complex dynamics of heat-induced ferrofluid flow on a Riga sensor plate was done in this study. Understanding the influence of viscous dissipation was the main focus of the investigation. The thermophysical features for the hybrid nanofluid is presented in Table 10.1. Table 10.2 presents the thermophysical characteristics of the base fluid and nanoparticles at 298° K. Likewise, the simulation is obtainable by taking fixed values of the influencing factors as $Rd = 0.5$, $Q = 1$, $\phi_1 = 0.01$, $d = 3$, $S = 0.3$, $Pr = 1$, $H = 0.1$, $\lambda = 0.1$, $A = 0.2$, $n = 3$, and $\phi_2 = 0.01$, whereas the variation of each constraints deployed in corresponding figure. Table 10.3 presents the validation and conformity of the solution, demonstrating a reasonable relationship between the previous and current studies. The main emphasis of this work is to examine the influence of numerous crucial factors on the motion and heat transmission features. Figure 10.2(a) shows the association between the modified Hartmann number (Q) and $f'(\eta)$. The hydrodynamic boundary layer thickness decreases as the Hartmann number increases due to the intensified impact of the Lorentz force on the fluid flow. As the Hartmann number increases, the strength of magnetism augments, leading to the exertion of more potent Lorentz forces on the liquid. These Lorentz forces strongly retard the fluid's motion, leading to a pronounced reduction in momentum boundary layer thickness. The effect of variations in suction (S) on $f'(\eta)$ is shown in Figure 10.2(b). Suction causes the fluid momentum to increase close to the surface. As the fluid passes towards the surface due to suction, its velocity in that particular area increases. This intensified velocity gradient nearby the surface leads to a decrease in the thickness of the hydrodynamic boundary layer. Figure 10.2(c) shows the influence of the velocity ratio factor (λ) on

$f'(\eta)$. Analysing the graph reveals a notable trend: an upsurge in the velocity ratio factor value corresponds to a reduction in the thickness of the momentum boundary layer. The increase in the velocity ratio parameter signifies that the free stream velocity exceeds the extending velocity. As a result, there is an intensification in pressure and straining motion adjacent the stagnation point, leading to a reduction in the boundary layer thickness. Figure 10.2(d) showcases the influence of the slip parameter (A) on $f'(\eta)$. The momentum profile experiences a reduction due to the presence of the slip parameter, which exerts its influence on fluid flow behaviour in close proximity to surfaces. When the slip parameter is elevated, it signifies an enhancement in slip effects. Slip effects denote a decreased interaction between the fluid and the solid surface, enabling the fluid molecules to encounter less resistance at the boundary. This diminished resistance facilitates smoother movement of the fluid across the surface, resulting in increased fluid mobility. The influence of the volume fraction on $\theta(\eta)$ is shown in Figures 10.3(a, b). Here $\phi_1 \approx \phi_{Fe_3O_4}$ and $\phi_2 \approx \phi_{COFe_2O_4}$ are taken into consideration. The current figures designate the case of pure fluid ($(\phi_1 = \phi_2 = 0)$, the case of nanofluid ($((\phi_1 = 0, \phi_2 \neq 0 \; or \, \phi_1 \neq 0, \phi_2 = 0)$, and the hybridized nanofluid $(\phi_1 \neq 0, \phi_2 \neq 0)$. Ferrofluids enhance velocity profiles due to the interplay between the applied magnetic field and suspended nanoparticles. When a magnetized field is applied to a ferromagnetic nanofluid, nanoparticles can magnetize and align with the field's orientation. This alignment process can lead to the formation of nanoparticle clusters, thus contributing to enhancements in fluid flow characteristics. Figure 10.4(a) shows the association between the radiation parameter (Rd) and $\theta(\eta)$. Thermal radiation denotes the conveyance of heat through electromagnetic waves, taking place without reliance on any intervening medium. Within the realm of ferrofluids, the phenomenon of thermal radiation holds significant importance in facilitating energy exchange between the liquid and the solid surface. As an outcome, an upsurge in the radiative heat flux value leads to an elevation in $\theta(\eta)$. The influence of the Eckert number (Ec) on $\theta(\eta)$ is depicted in Figure 10.4(b). The Eckert number computes the ratio between fluid kinetic energy and fluid thermal energy at the interface between the fluid and the solid. A higher Eckert number indicates a greater prevalence of the fluid's kinetic energy relative to its thermal energy. This shift can lead to an enhancement in the $\theta(\eta)$ near the boundary. The influence of the heat source/sink parameter H on $\theta(\eta)$ is presented in Figure10.4(c). The heat source factor contributes to an enhancement in $\theta(\eta)$ by increasing the thermal energy delivered into a system. An increase in the heat source parameter means the system is producing or absorbing more heat. The association between the Prandtl number (Pr) and $\theta(\eta)$ is shown in Figure 10.4(d). The Prandtl number characterizes the ratio of momentum diffusivity to thermal diffusivity. An increase in the Prandtl number leads to a reduction in the thickness of the thermal boundary layer. Figures 10.5(a, b) illustrate the variation in temperature distribution resulting from alterations in ferromagnetic volume fraction. Ferromagnetic nanofluids enhance temperature profiles through magnetic heating. Upon applying a magnetic field to the ferromagnetic nanofluid, the magnetic nanoparticles within the fluid convert the magnetic energy into heat, inducing heating within the nanofluid. Consequently, this results in a more uniform temperature distribution throughout the fluid. As observed in Table 10.4, it becomes evident that elevated values of the modified Hartmann number correspond to a significant increase in the shear rate. The outcomes further indicate that a higher stretching parameter along with slip velocity values leads to lower shear rates. Additionally, the findings reveal that heightened thermal radiation correlates with an enhanced heat transfer rate. The observations also indicate that elevated heat source and Eckert number values contribute to a reduction in the heat transmission rate.

10.4 CONCLUSION

This study provides an extensive investigation into the heat-induced ferrofluid flow on a Riga sensor plate. The central focus of this inquiry is understanding the consequences of viscous dissipation. Additionally, a special emphasis is placed on considering the impact of radiative heat flux on the analysed system. The study resulted in several significant findings, including the following:

- Increasing the quantity of ferrofluid nanoparticles leads to improved thermal features of the fluid, including improved viscosity and conductivity.
- Highlighting the importance, it is worth noting that the modified Hartmann number, slip velocity, and velocity ratio together contribute to the decrease in the thickness of the momentum boundary layer.
- It is crucial to emphasize that the radiation parameter, heat source parameter, and Eckert number contribute to an increase in the thickness of the thermal boundary layer, while the Prandtl number exhibits the opposite trend.
- The heat transfer rate is amplified by increasing the radiation parameter, modified Hartmann number, suction, and velocity slip; however, the Eckert number and heat source parameter have the reverse effect and reduce the heat transfer rate.

REFERENCES

1. Hassan, A., Alsubaie, N., Alharbi, F. M., Alhushaybari, A., & Galal, A. M. (2023). Scrutinization of Stefan suction/blowing on thermal slip flow of ethylene glycol/water based hybrid ferro-fluid with nano-particles shape effect and partial slip. Journal of Magnetism and Magnetic Materials, 565, 170276.
2. Yilbas, B. S., Abubakar, A. A., Hassan, G., Al-Qahtani, H., Al-Sharafi, A., Alzahrani, A. A., & Mohammed, A. S. (2022). Ferro-fluid droplet impact on hydrophobic surface under magnetic influence. Surfaces and Interfaces, 29, 101731.
3. Sivakumar, N., Prasad, P. D., Raju, C. S. K., Varma, S. V. K., & Shehzad, S. A. (2017). Partial slip and dissipation on MHD radiative ferro-fluid over a non-linear permeable convectively heated stretching sheet. Results in Physics, 7, 1940–1949.
4. Zeeshan, A., Majeed, A., & Ellahi, R. (2016). Effect of magnetic dipole on viscous ferro-fluid past a stretching surface with thermal radiation. Journal of Molecular Liquids, 215, 549–554.
5. Elsaid, E. M., & Abdel-Wahed, M. S. (2022). MHD mixed convection ferro Fe_3O_4/Cu-hybrid-nanofluid runs in a vertical channel. Chinese Journal of Physics, 76, 269–282.
6. Khan, S. A., Yasmin, S., Waqas, H., Az-Zo'bi, E. A., Alhushaybari, A., Akgül, A., Hassan, A. M., & Imran, M. (2023). Entropy optimized ferro-copper/blood based nanofluid flow

between double stretchable disks: Application to brain dynamic. Alexandria Engineering Journal, 79, 296–307.

7. Abdel-Wahed, M. S. (2019). Magnetohydrodynamic ferro-nano fluid flow in a semi-porous curved tube under the effect of hall current and nonlinear thermal radiative. Journal of Magnetism and Magnetic Materials, 474, 347–354.
8. Majeed, A., Zeeshan, A., & Ellahi, R. (2017). Chemical reaction and heat transfer on boundary layer Maxwell ferro-fluid flow under magnetic dipole with Soret and suction effects. Engineering Science and Technology, an International Journal, 20(3), 1122–1128.
9. Ali, A., Ahmed, M., Ahmad, A., & Nawaz, R. (2023). Enhanced heat transfer analysis of hybrid nanofluid over a Riga plate: Incorporating Lorentz forces and entropy generation. Tribology International, 188, 108844.
10. Das, S., Mahato, N., Ali, A., & Jana, R. N. (2023). Dynamics pattern of a radioactive rGO-magnetite-water flowed by a vibrated Riga plate sensor with ramped temperature and concentration. Chemical Engineering Journal Advances, 15, 100517.
11. Wahid, N. S., Arifin, N. M., Khashi'ie, N. S., Pop, I., Bachok, N., & Hafidzuddin, M. E. H. (2022). Hybrid nanofluid stagnation point flow past a slip shrinking Riga plate. Chinese Journal of Physics, 78, 180–193.
12. Otman, H. A., Mahmood, Z., Khan, U., Eldin, S. M., Fadhl, B. M., & Makhdoum, B. M. (2023). Mathematical analysis of mixed convective stagnation point flow over extendable porous Riga plate with aggregation and joule heating effects. Heliyon, 9(6).
13. Khan, U., Mahmood, Z., Eldin, S. M., Makhdoum, B. M., Fadhl, B. M., & Alshehri, A. (2023). Mathematical analysis of heat and mass transfer on unsteady stagnation point flow of Riga plate with binary chemical reaction and thermal radiation effects. Heliyon, 9(3).
14. Darvesh, A., Wahab, H. A., Sarakorn, W., Sánchez-Chero, M., Apaza, O. A., Villarreyes, S. S. C., & Palacios, A. Z. (2023). Infinite shear rate viscosity of cross model over Riga plate with entropy generation and melting process: A numerical Keller box approach. Results in Engineering, 17, 100942.
15. Rana, S., Tabassum, R., Mehmood, R., Tag-Eldin, E. M., & Shah, R. (2024). Influence of Hall current & Lorentz force with nonlinear thermal radiation in an inclined slip flow of couple stress fluid over a Riga plate. Ain Shams Engineering Journal, 15(1), 102319.
16. Awati, V. B., Goravar, A., & Kumar, M. (2024). Spectral and Haar wavelet collocation method for the solution of heat generation and viscous dissipation in micro-polar nanofluid for MHD stagnation point flow. Mathematics and Computers in Simulation, 215, 158–183.
17. Akram, S., Saeed, K., Athar, M., Razia, A., Hussain, A., & Naz, I. (2023). Convection theory on thermally radiative peristaltic flow of Prandtl tilted magneto nanofluid in an asymmetric channel with effects of partial slip and viscous dissipation. Materials Today Communications, 35, 106171.
18. Mahesh, R., Mahabaleshwar, U. S., Kumar, P. V., Öztop, H. F., & Abu-Hamdeh, N. (2023). Impact of radiation on the MHD couple stress hybrid nanofluid flow over a porous sheet with viscous dissipation. Results in Engineering, 17, 100905.
19. Kho, Y. B., Jusoh, R., Salleh, M. Z., Ariff, M. H., & Zainuddin, N. (2023). Magnetohydrodynamics flow of Ag-TiO_2 hybrid nanofluid over a permeable wedge with thermal radiation and viscous dissipation. Journal of Magnetism and Magnetic Materials, 565, 170284.
20. Riaz, S., Naheed, N., Farooq, U., Lu, D., & Hussain, M. (2023). Non-similar investigation of magnetized boundary layer flow of nanofluid with the effects of Joule heating, viscous dissipation and heat source/sink. Journal of Magnetism and Magnetic Materials, 574, 170707.

11 Impact of Thermal Radiation on Electrically Conducting Hybrid Nanofluid Flow through an Expanding/Contracting Wedge Surface with Heat Generation

J.R. Pattnaik, P.K. Pattnaik, and R.S. Tripathy

Nomenclature

λ	Stretching/shrinking factor
Rd	Thermal radiation
Q	Heat source/sink
$(\rho C_p)_{hnf}$	Specific heat capacitance
Ec	Eckert number
Pr	Prandtl number
k_{hnf}	Thermal conductivity
μ_{hnf}	Dynamic viscosity
S	Suction/injection parameter
ρ_{hnf}	Density
q_r	Radiative heat flux
σ_{hnf}	Electrical conductivity

DOI: 10.1201/9781003595786-11

11.1 INTRODUCTION

Hybrid nanofluids belong to a class of nanofluids that involve a combination of nanoparticles made from different materials to enhance their heat transfer properties. Nanofluids are composed of colloidal suspensions of nanoparticles within a base fluid. When these nanoparticles are blended in a hybrid configuration, they can yield even more significant improvements in heat transfer efficiency, outperforming the advantages seen in nanofluids containing only a single type of nanoparticle. Alfellag et al. [1] looked at how creating green hybrid nanofluids based on carbon can spur innovation in nanomaterial applications and renewable energy technology. In this context, the ongoing experimental study aims to improve the thermal efficiency of flat-plate solar collectors (FPSCs) by employing eco-friendly (clove) working fluids composed of nanocomposite materials. The use of dangerous chemicals in conventional ways of manufacturing nanofluids raises questions about their effect on the environment and human health. Hussein et al. [2] looked at how basic methods like floating nanoparticles in water may improve their thermophysical properties and improve the efficiency of solar distillation systems. The improved features that amplify the temperature differential between the nanofluid and condensing cover of the solar still allow for the absorption of more solar energy. Guo et al. [3] investigated how stability affected the flow boiling heat transfer performance of hybrid nanofluids, or Al_2O_3-TiO_2/W, which are composed of Al_2O_3-TiO_2 nanoparticles dissolved in water. Response surface methodology (RSM) was used to optimize the stability of the Al_2O_3-TiO_2/W hybrid nanofluid. Dai et al. [4] discovered that when compared to mono-nanofluids, hybrid nanofluids perform better in terms of heat conductivity without appreciably increasing viscosity. The process of improved heat transmission in nanofluidic systems must be understood by examining the microstructures created by the nanoparticles in the base fluid. Ouyang et al. [5] investigated the use of hybrid nanofluids to accelerate chemical breakdown and heat transfer to enhance water quality. Water systems can benefit from the effective removal of pollutants, optimal heat transmission, control over pollution sources, and regulation of fluid dynamics that nanofluids provide. Sreekumar et al. [6] researched how notable improvements in thermophysical and optical properties of nanofluids over base fluids have made them a potential heat transfer fluid (HTF). Comparing hybrid nanofluids containing several nanomaterials to nanocomponent nanofluids, the former has synergistic qualities. Sun et al. [7] researched the effective technique of water-alternating CO_2 injection (CO_2-WAG) for improving geologic CO_2 storage and heavy oil recovery. Nevertheless, there are still several issues with CO_2-WAG, including gravity overriding, water blocking, and viscous fingering of injected CO_2. In order to enhance the impact of CO_2-WAG, a novel technique known as hybrid nanofluid-alternating-CO_2 microbubble injection is put forth in this work. Fahliyany et al. [8] investigated the properties of nanofluids, which are liquids containing particles smaller than 100 nm. Nanofluids are used in many different disciplines, including the pharmaceutical industry, electrical engineering, nuclear engineering, chemical engineering, and petroleum engineering. These qualities include electrical properties, heat capacity, lubricating effect, and viscous behaviour. Tawade et al. [9] worked on boundary layer flow, and heat transfer across a linearly stretched plate has been enormously successful in recent years due to its wide variety of applications in research and industry. Metal spinning, metal extrusion, glass fibre, wire drawing, hot rolling, artificial fibres, continuous stretching of plastic films, copper wire drawing, and polymer extrusion are only a few examples. The current study is concerned with boundary layer flows, heat transport, and the chemical reaction of Casson nanofluids on a stretched surface. Fluid refers to the liquid that flows. A nanofluid is made up of a base fluid and many nanoparticles. Nanoparticles range in size from 1 to 100 nm. Hamid [10] investigated scientific studies on the thermal conductivity of various fluids. The main hindrance to the heat transmission of these operating fluids has been revealed to be a lack of thermal conductivity in technology and diverse technological divisions. Thermal conductivity may increase if we examine solids rather than fluids. Amalgamation of solid particles in a field is a novel way of enhancing thermal conductivity. This fluid used is referred to as nanofluid. The solids or particles utilized for this purpose are known as nanoparticles due to their minuscule sizes. Nanofluids are used in a variety of applications, including chemical reactors, biomedicine, paper manufacturing, and industrial cooling. Li et al. [11] investigated a range of heat exchangers utilized by various industries. The development of more efficient heat exchangers will lead to cost reductions in manpower, resources, and energy. Suspending small solid metallic, oxide, or polymeric particles in ordinary fluids is the proper technique to improve the process of heat transfer. Khan et al. [12] showed how the porous laminar stream has become much more relevant over the past two decades as a result of the widespread usage of technologically based instruments and biodiversity fluxes. Fluid dynamics research has continued to improve, thanks to many applications of nanofluid laminar flow through porous conduits using expanding or contracting permeable channels. The movement of basic biological fluids through porous channels, such as kidney filtration, breathing system, air circulation, and artificial lung function, has been extensively studied in medical engineering. Pala and Mandal [13] researched the industrial applications of lubrication oils and greases in a variety of contexts, including heat exchangers, petroleum reservoirs, material processing systems, molten plastic products, wall paints, food items, and lubrication oils and greases. To explain the characteristics of non-Newtonian fluids, several mathematical models have been developed in the past; yet, they may not completely capture all of nature's non-Newtonian fluids' capabilities. Roy and Pop [14] focused on the dual solutions for the Oldroyd-B fluid, including alumina nanoparticles and mixed convection flow across a contracting sheet while accounting for the magnetic field and heat source/sink effects. By utilizing similarity transformations, the governing equations are transformed into ordinary differential equations (ODEs) that have been resolved using the well-known gunshot approach. Khashi'ie et al.'s [15] primary area of study was magnetohydrodynamic (MHD) hybrid nanofluid flow with heat transfer on a moving plate with Joule heating. The primary fluid for the study is composed of nanoparticles of water (H_2O), copper (Cu), and metal oxide (Al_2O_3). The system of ODEs is numerically solved using the MATLAB function bvp4c for a variety of values of the governing parameters after similarity transformation, which reduces the complexity of the partial differential equations (PDEs). Due to their significance in numerous physiological and commercial applications, including the transport of cilia, valve-less movement of the cardiovascular system, passage of bile in the bile duct, and small blood vessel vasomotor activity, researchers paid close attention to Arafa et al.'s [16] studies on peristaltic mechanisms. An item or thing moves through tubular structures as a result of peristaltic motion, which is a series of wave motions. The importance of peristaltic flow is also well illustrated by the health sectors, which also comprise thermal locomotion, spinning pumps, heart–lung devices, and hazardous fluid transfer. Sajid et al. [17] investigated how boundary layers behave on porous surfaces. Rotative discs are utilized in several engineering processes, such as those that manufacture paper, glass fibres, and extrusion-produced materials; therefore, understanding how they work is essential. A polymer is continually dispensed from a die to an uppercut swing roller in an industrial environment, where it is used to create various sheets and filaments. The operation's cooling rate and lengthening process in these circumstances establish the superior attributes of the end product. The transport of nanofluids across rotative discs is the subject of the current scientific study. Heat transfer is a topic of research due to its importance in engineering and business. Javid et al. [18] looked at whether proper micro-scale operation of any lab-chip device could control fluid flow through channels or tubes. In addition to other methods, micropumps, electro-osmotic devices, and magnetic devices are frequently employed to regulate the mobility of liquids in microchannels. Noranuar et al. [19] examined the fluid flow and heat transport of MHD Casson nanofluid, which is impacted by a moving disc rotating non-coaxially as it travels through a porous medium. The nanoparticles in Casson human blood are combinations of single-wall and multi-wall carbon nanotubes (CNTs). Utilizing the Laplace transform approach, analytical answers to the temperature and velocity profiles are obtained. The results show that the temperature and velocity profiles increase when CNTs are added. The nanofluid velocity decreases with a greater magnetic field but increases with porosity. The addition of CNTs leads to an increase in the Nusselt number and a decrease in primary and secondary skin friction. Researchers [20–24] investigated the flow of magnetized nanofluids across porous, curved, and stretchy surfaces with various characteristics using heat radiation and activation energy. A more complete analysis of thermal performance is carried out using the exothermic/endothermic reactive effect and the Buongiorno nanofluid model.

Objective and novelty of the proposed investigation

The current work investigates the following:

- The heat transport properties of the electrically conducting water-based hybrid nanofluid flow over a wedged surface.
- The interaction of Fe_3O_4 and $CoFe_2O_4$ magnetite nanoparticles with thermal radiation and dissipative heat enhances the fluid properties.
- The influence of various shapes of the nanoparticles on the flow phenomena.

The novelty of the study is due to the following aspects:

- The influential characteristic of the magnetized hybrid nanofluid flow over a wedged surface and the effect of Lorentz force on the flow phenomena show their vital role.

- The contribution of the combined effect of Fe_3O_4 and $CoFe_2O_4$ over the stretching/shrinking surface overshoots the thermal properties significantly.
- The heat transport properties in association with the thermophysical properties affect the choice of various nanoparticle shapes.
- Dissipative heat due to the inclusion of magnetic dissipation and thermal radiation has several industrial applications.

11.2 MATHEMATICAL DESCRIPTION OF EXISTING PROBLEM

A wedge-shaped surface plate that is either stretching or shrinking is experiencing the movement of a steady two-dimensional flow. This movement involves an incompressible hybrid nanofluid composed of Fe_3O_4-$CoFe_2O_4$. The uniform velocity $u_e(x) = ax^m$ characterizes the fluid movement outside the wedge interface's boundary layer, where a is fixed and $m = \beta/(2-\beta)$ corresponds to the power-law factor with $0 \le m \le 1$ and $\beta = \Omega/\Pi$ designated as the Hartree pressure gradient. Here, Ω signifies the total wedge angle presuming two improbable amounts of Hartree β. If the fluid motion occurs across the flat horizontal surface ($\Omega = 0^0$), then $\beta = 0$, and if the boundary layer flow reaches the upward flat plate, then $\beta = 1$ since $\Omega = 180^0$. Additionally, it is taken into consideration that the velocity of the widening/contracting wedge is $u_w(x) = bx^m$, where $b > 0$ correlates to the stretching wedge and $b < 0$ correlates to the diminishing wedge. The magnetic field $B(x) = B_0 x^{(m-1)/2}$ is applied normal to the wedge so that B_0 is the strength of the applied magnetic field. The uniformity of the applied magnetic field is taken into account, and the magnetic Reynolds number is presumed to be low, enabling the omission of the induced magnetic field. Additionally, the external electric field is assumed to be absent, and the electric field stemming from charge polarization is considered insignificant. It is worth noting that Hall and ion slip currents cannot be disregarded in situations of high electron-atom collision frequency. The illustration of the current model's geometry is presented in Figure 11.1.

Following [21], the governing equations from two-dimensional Navier–Stokes equations for the hybrid nanofluid are as follows:

$$\frac{\partial u}{\partial x} + \frac{\partial v}{\partial y} = 0, \tag{11.1}$$

$$\rho_{hnf}\left(u\frac{\partial u}{\partial x} + v\frac{\partial v}{\partial y}\right) - \mu_{hnf}\frac{\partial^2 u}{\partial y^2} - u_e\rho_{hnf}\frac{du_e}{dx} - \sigma_{hnf}B^2\left(u_e - u\right) = 0, \tag{11.2}$$

$$(\rho c_p)_{hnf}\left(u\frac{\partial T}{\partial x} + v\frac{\partial T}{\partial y}\right) - k_{hnf}\frac{\partial^2 T}{\partial y^2} - \mu_{hnf}\left(\frac{\partial u}{\partial y}\right)^2$$
$$+\frac{\partial q_r}{\partial y} - \sigma_{hnf}B^2u^2 - Q_0(T - T_\infty) = 0. \tag{11.3}$$

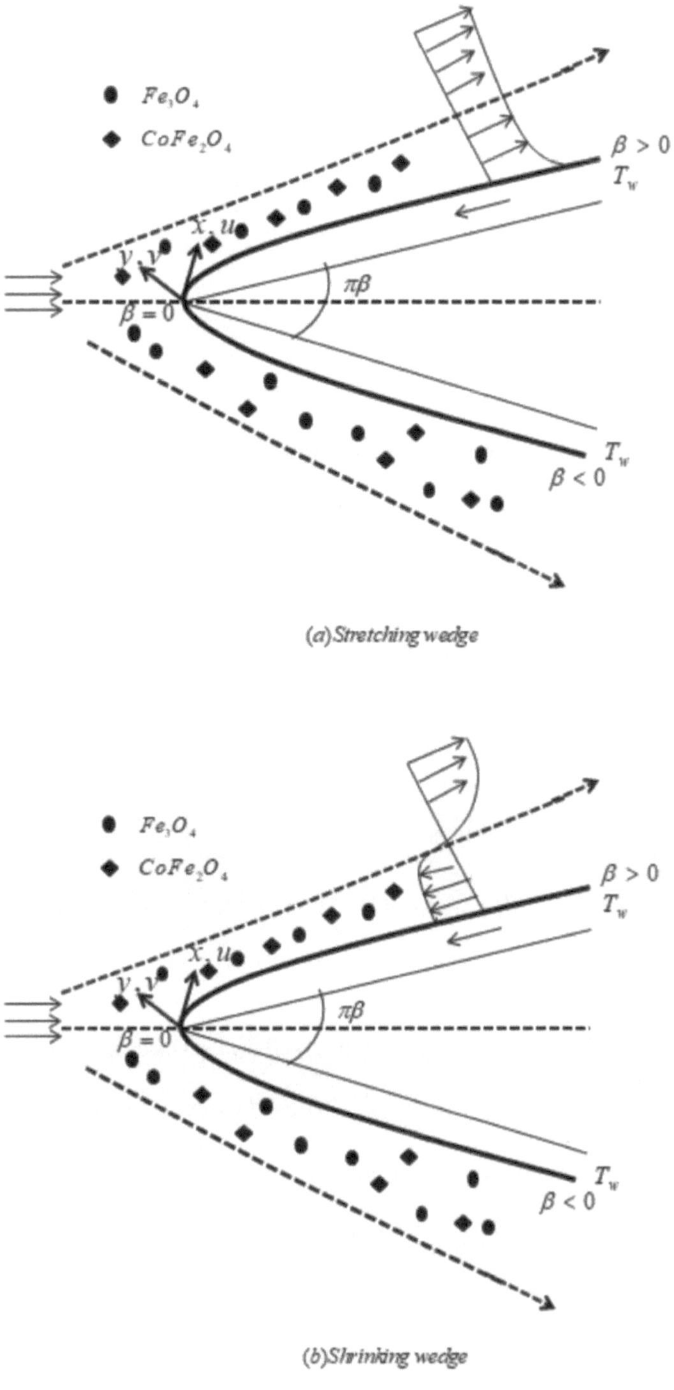

FIGURE 11.1 Physical geometry of hybrid nanofluid across a (a) stretching and (b) a shrinking wedge.

The specific surface constraints are

$$\left.\begin{array}{ll} v = v_w, u = bx^m, T = T_w & \text{at } y = 0, \\ u \to ax^m, T \to T_\infty & \text{as } y \to \infty \end{array}\right\}. \tag{11.4}$$

Here, u and v indicate the components of velocity along the coordinate axes. Also, T_w and T_∞ represent the surface and surrounding temperature.

Mass flux velocity $v_w = -0.5(m+1)S\sqrt{a\upsilon_f x^{m-1}}$, where positive values of S indicate suction.

Equation (11.1) is automatically satisfied by the stream function $\psi = (\upsilon_f a x^{m+1})^{0.5} f(\eta), \eta = \frac{y}{x}\left(\frac{ax^{m+1}}{\upsilon_f}\right)^{0.5}$, with $u = \psi_y, v = -\psi_x$. The other two governing Equations (11.2) and (11.3) are transformed using

$$\begin{gathered} u = ax^m f'(\eta), v = -0.5\big((m+1)f(\eta) + (m-1)\eta f'(\eta)\big) \\ \left(a\,\upsilon_f x^{m-1}\right)^{0.5}, T = T_\infty + (T_w - T_\infty)\theta(\eta). \end{gathered} \tag{11.5}$$

Following the Rosseland approximation, radiative heat flux is defined and denoted as

$$q_r = -\frac{16\sigma^* T_\infty^{\ 3}}{3k^*}\frac{\partial T}{\partial y}. \tag{11.6}$$

So, the subsequent set of highly non-linear ODEs,

$$\begin{gathered} \mu_R f'''(\eta) + m\rho_R\left(1 - f'^2(\eta)\right) + 0.5(m+1) \\ \rho_R f(\eta) f''(\eta) + M\sigma_R\left(1 - f'(\eta)\right) = 0, \end{gathered} \tag{11.7}$$

$$\begin{gathered} \frac{1}{\Pr}(k_R + Rd)\theta''(\eta) + (\rho c_p)_R f'(\eta)\theta(\eta) + 0.5(m+1)(\rho c_p)_R f(\eta)\theta'(\eta) \\ + Ec\mu_R f''^2(\eta) + MEc\sigma_R f'^2(\eta) + Q\theta(\eta) = 0, \end{gathered} \tag{11.8}$$

and boundary circumstances are obtained:

$$\left.\begin{array}{l} f'(0) = \lambda,\ f(0) = S, \theta(0) = 1, \\ f'(\infty) \to 1,\ \ \theta(\infty) \to 0 \end{array}\right\}, \tag{11.9}$$

where $\frac{\mu_{hnf}}{\mu_f} = \mu_R, \frac{\rho_{hnf}}{\rho_f} = \rho_R, \frac{\sigma_{hnf}}{\sigma_f} = \sigma_R, \frac{(\rho c_p)_{hnf}}{(\rho c_p)_f} = (\rho c_p)_R, \frac{k_{hnf}}{k_f} = k_R.$

Here, Fe_3O_4 volume fraction presented as ϕ_1 and $CoFe_2O_4$ volume fraction is ϕ_2.

$$M = \frac{\sigma_f}{a\rho_f} B_0^2, \Pr = \frac{\upsilon_f}{k_f}(\rho c_p)_f, Rd = \frac{16}{3}\frac{\sigma^*}{k^* k_f} T_\infty^3, \lambda = b/a,$$
$$Ec = \frac{u_e^2}{(T_w - T_\infty)(c_p)_f}\ Q = \frac{Q_0}{ax^{m-1}(\rho c_p)_f}. \quad (11.10)$$

The thermal flux emitted from the wedge-shaped surface plate q_w, the Nusselt number $\mathrm{Nu_x}$, and skin friction coefficients C_f are clarified as follows:

$$C_f = \frac{\mu_{hnf}}{\rho_f u_e^2}\left(\frac{\partial u}{\partial y}\right)_{y=0}, \mathrm{Nu}_x = \frac{x}{(T_w - T_\infty)k_f} q_w, q_w = -k_{hnf}\left(\frac{\partial T}{\partial y}\right)_{y=0} + (q_r)_{y=0}. \quad (11.11)$$

Therefore, by incorporating the similarity transformations (11.5) into Equations (11.7) and (11.12), we arrive at the ensuing Equation (11.13):

$$\mathrm{Re}_x^{0.5} C_f = \mu_R f''(0), \mathrm{Re}_x^{-0.5} \mathrm{Nu}_x = -\left(k_R + Rd\right)\theta'(0), \quad (11.12)$$

where $\mathrm{Re_x} = \frac{u_e x}{\upsilon f}$ represents the local Reynolds number.

11.3 THERMAL MODELS [27]

- ***Effective viscosity model***

$$\frac{\mu_{hnf}}{\mu_{nf}} = \left(1-\phi_1\right)^{-2.5}\left(1-\phi_2\right)^{-2.5} \quad (11.13)$$

- ***Density model***

$$\frac{\rho_{hnf}}{\rho_f} = (1-\phi_2)\left(1-\phi_1+\phi_1\frac{\rho_1}{\rho_f}\right)+\phi_2\frac{\rho_2}{\rho_f}, \rho_1$$
$$= \rho_{\mathrm{Fe_3O_4}}, \rho_2 = \rho_{\mathrm{CoFe_2O_4}}, \rho_f = \rho_{\mathrm{H_2O}} \quad (11.14)$$

- ***Heat capacity model***

$$\frac{(\rho c_p)_{hnf}}{(\rho c_p)_f} = (1-\phi_2)\left(1-\phi_1+\phi_1\frac{(\rho c_p)_1}{(\rho c_p)_f}\right)+\phi_2\frac{(\rho c_p)_2}{(\rho c_p)_f},$$
$$(\rho c_p)_1 = (\rho c_p)_{\mathrm{Fe_3O_4}}, (\rho c_p)_2 = (\rho c_p)_{\mathrm{CoFe_2O_4}}, (\rho c_p)_f = (\rho c_p)_{\mathrm{H_2O}} \quad (11.15)$$

- ***Effective thermal conductivity model***

$$\frac{k_{hnf}}{k_{nf}} = \frac{k_2 + (n-1)k_{nf} - (n-1)\phi_2(k_{nf} - k_2)}{k_2 + (n-1)k_{nf} - \phi_2(k_{nf} - k_2)},$$

$$\frac{k_{nf}}{k_f}=\frac{k_1+(n-1)k_f-(n-1)\phi_1(k_f-k_1)}{k_1+(n-1)k_f-\phi_1(k_f-k_1)},$$

where

$$k_1=k_{Fe_3O_4},k_2=k_{CoFe_2O_4},k_f=k_{H_2O} \tag{11.16}$$

- ***Effective electrical conductivity model***

$$\frac{\sigma_{hnf}}{\sigma_{nf}}=\frac{\sigma_2+2\sigma_{nf}-2\phi_2(\sigma_{nf}-\sigma_2)}{\sigma_2+2\sigma_{nf}+\phi_2(\sigma_{nf}-\sigma_2)},$$

where $\frac{\sigma_{nf}}{\sigma_f}=\frac{\sigma_1+2\sigma_f-2\phi_1(\sigma_f-\sigma_1)}{\sigma_1+2\sigma_f+\phi_1(\sigma_f-\sigma_1)},\sigma_1=\sigma_{Fe_3O_4},\sigma_2=\sigma_{CoFe_2O_4},\sigma_f=\sigma_{H_2O}$ (11.17)

11.4 BRIEF DISCUSSION ON THE PARAMETRIC BEHAVIOURS

The flow of hybridized nanofluid combined with the association of ferrite nanoparticles such as Fe_3O_4 and $CoFe_2O_4$ in the water-based liquid is proposed in this article. The effect of thermal radiation along with the Eckert number is due to the interaction of dissipative heat energies and the heat transport phenomenon. In Table 11.1, the thermal characteristics of the nanoparticles and the host water are presented for a standard temperature of 300 K. The thermal enhancement is also characterized by the adaptation of various shapes of the nanoparticles. In particular, the comparative study is deployed for the utilization of spherical, cylindrical, platelet-, brick-, and blade-shaped nanoparticles. Table 11.2 presents the different shapes of the nanoparticles described earlier. The converted standard equations are then solved numerically via Runge–Kutta fourth-order technique combined with a shooting method. The validation of the current result is presented numerically in Table 11.3, showing the results of the heat transmission rate in a particular scenario. The results suggest the convergence criteria of the present methodology. The analysis of the contributing factors is presented briefly, and the results are depicted through graphs.

The significant properties of the magnetic parameter on the momentum profile are presented in Figure 11.2. The introduction of applied magnetic field along the transverse direction of the stream of the hybrid nanofluid causes the impact of magnetization. Here, the standard values of the factor are considered within the range as $M\in[0,2]$.. The numerical value of $M=0$ signifies the non-magnetized effect

TABLE 11.1
Thermophysical Structure of the Base Fluid (H_2O) and Nanoparticles [25, 26]

Base fluid and nanoparticle	ρ	c_p	k	σ
H_2O	997.1	4179	0.613	5.5 x 10^{-6}
Fe_3O_4	5180	670	9.7	2.5 x 10^{-3}
$CoFe_2O_4$	4907	700	3.7	1.1 x 10^{-7}

TABLE 11.2
Nanoparticle Shape

n	Shape of the nanoparticle
3	 Spherical
3.7	 Brick
5	 Cylindrical
5.7	 Platelet
8.9	 Blade

TABLE 11.3
Comparison with Previously Published Work at $m = 1$, $pr = 1$, $Rd = 0.5$, $Ec = 0.1$, $Q = 0$

	$f''(0)$		$-\theta'(0)$	
M	**Bing Kho et al. [28]**	**Present**	**Bing Kho et al. [28]**	**Present**
0.2	1.499709601	1.499707865	0.600961937	0.600959076
0.6	1.598207529	1.598211634	0.592037416	0.592029976
1	1.691084197	1.691093254	0.582022428	0.582019875
2	1.903841128	1.903837401	0.553317768	0.553309987
3	2.095348604	2.095339986	0.520734412	0.520729944

on the flow phenomena. The non-zero numeric deployed in this figure shows the influence of the magnetic field which affects the fluid velocity. The application of the magnetization suddenly produces a resistive force, called "Lorentz force," and this force has the capability to reduce the velocity distribution. This behaviour of the resistive force reduces the thickness of the bounding surface, showing a significant increase in velocity distribution magnitude. The behaviour of the profiles is presented

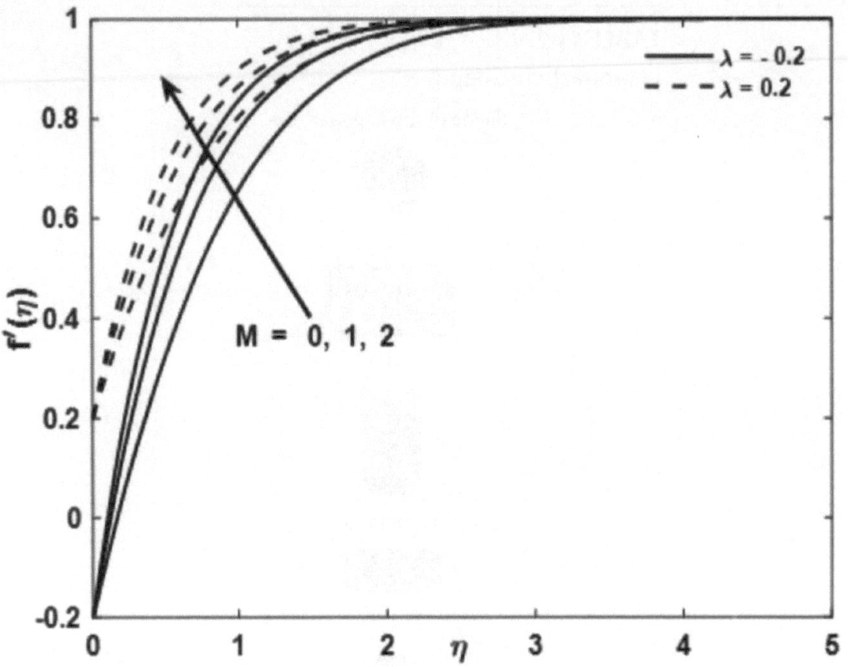

FIGURE 11.2 Impact of M on $f'(\eta)$.

for the variation in the stretching/shrinking ratio, and the graphical view shows that the strength became stronger for the case of the stretching ratio than for the shrinking ratio. Figure 11.3 shows the effects of magnetization on the fluid temperature distribution for the contribution of the ratio parameter. As described earlier, the role of magnetization for the fluid velocity, it is seen that the resistivity attenuates the thickness. Further, near the surface region the fluid element stored energy and gradually strengthen the fluid temperature. Consequently, the profile temperature boosts up significantly in either of the situation of widening/diminishing ratio. However, the case of shrinking shows greater enhancement in the fluid temperature, whereas a reverse impact is rendered in the case of shrinking. The flow properties of the hybrid nanofluid are enhanced due to the inclusion of the solid volume fraction of both the nanoparticles. The thermophysical properties involve in the governing system, all are equipped with volume fraction. Figures 11.4 and 11.5 present the significant effect of solid volume fraction on velocity distribution. Volume fraction is defined as the number of nanoparticles present with respect to the total volume and its percentage. Here, the range of the volume fraction is considered as [0, 0.06]. The numeric values of the concentration $\left(\phi_1 = \phi_2 = 0\right)$ show the behaviour of pure fluid, and the non-zero values signify the behaviour of the hybrid nanofluid. The increasing concentration of both nanoparticles attenuates the velocity distribution irrespective of the variations in the velocity ratio parameter. The higher density of the nanoparticles

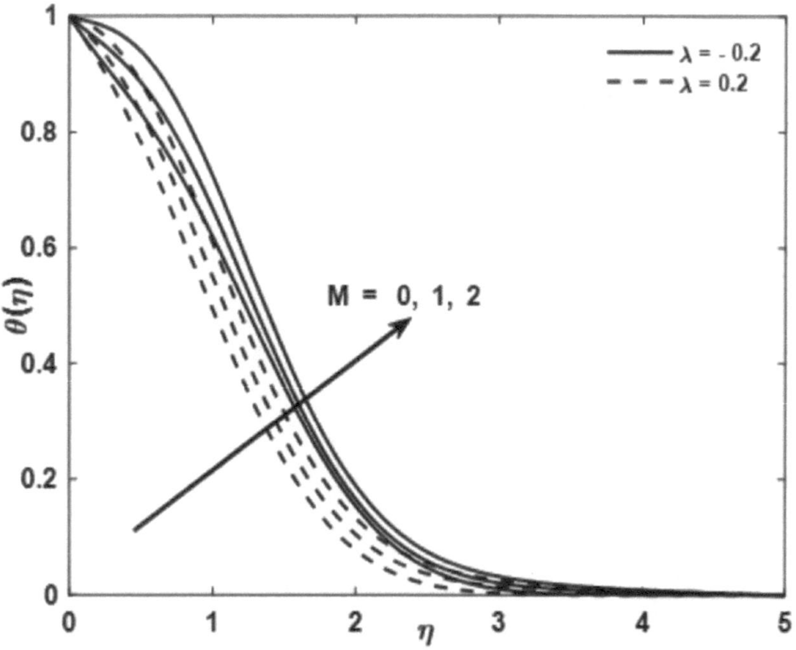

FIGURE 11.3 Impact of M on $\theta(\eta)$.

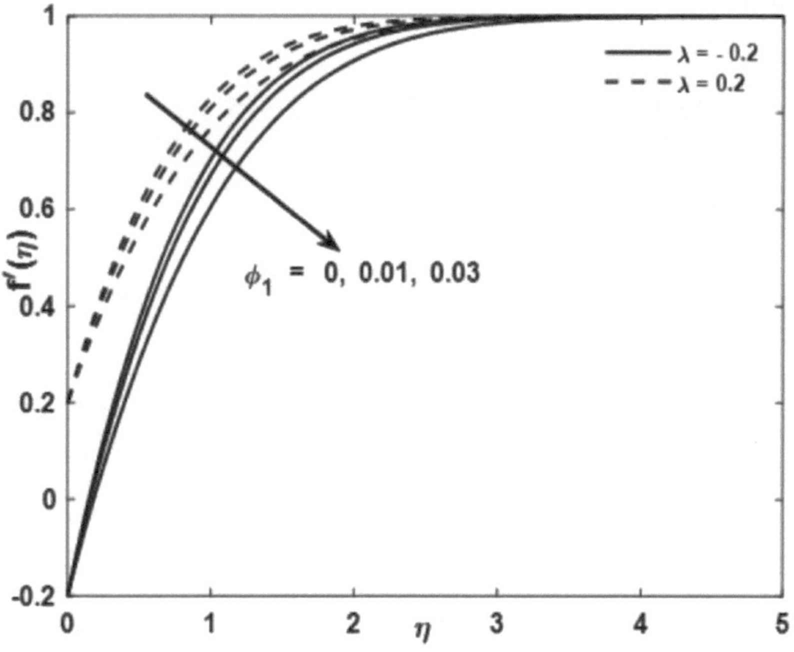

FIGURE 11.4 Impact of ϕ_1 on $f'(\eta)$.

was found to decelerate the profile significantly. The rate of retardation is greater due to the shrinking of the surface. Furthermore, the impact of particle concentration of both nanoparticles on fluid temperature is presented briefly in Figures 11.6 and 11.7. An increase in particle concentration in the host liquid significantly enhances the

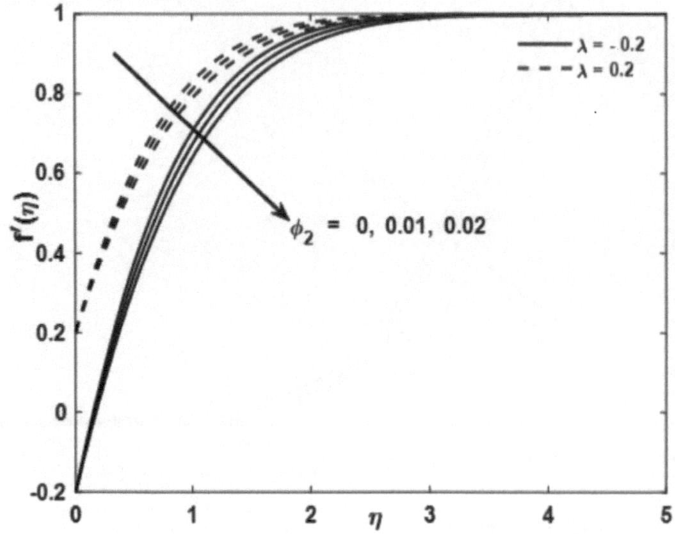

FIGURE 11.5 Impact of ϕ_2 on $f'(\eta)$.

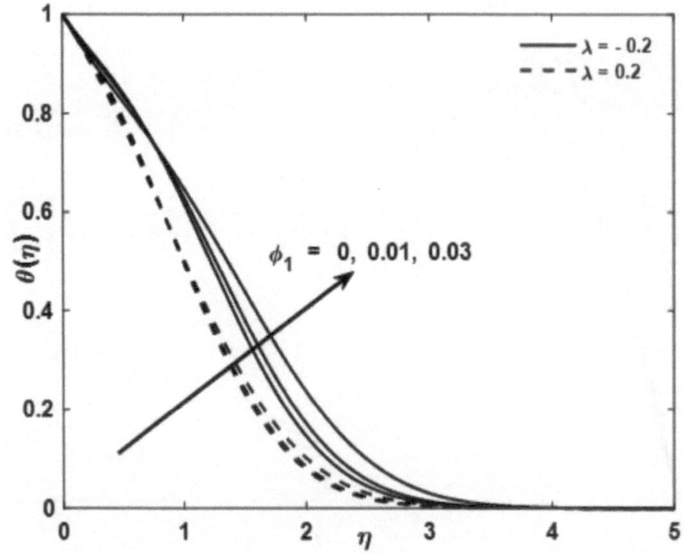

FIGURE 11.6 Impact of ϕ_1 on $\theta(\eta)$.

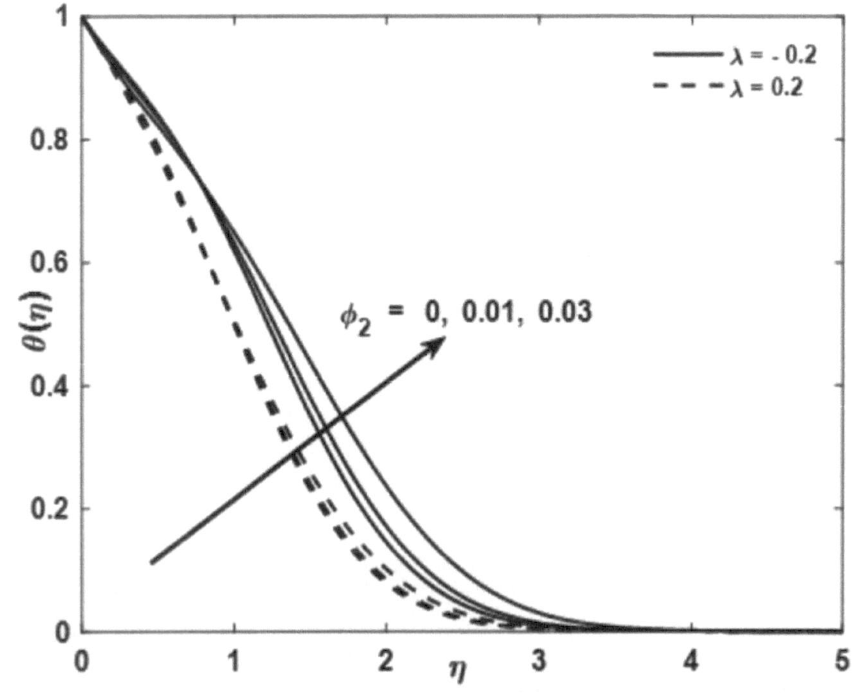

FIGURE 11.7 Impact of ϕ_2 on $\theta(\eta)$.

thermal conductivity of the hybrid nanofluid. Therefore, increasing conductivity is useful in enhancing the fluid temperature throughout the flow domain. The particle clogging closed to surface region enforces to boost the fluid temperature and this causes an increase in the fluid temperature. The result reveals cooling of the surface, which is beneficial in several industrial applications. In particular, during the production of different products, a coolant is useful in achieving the desired shape and size of the product. Similar to that mentioned earlier, it is found that for the case of shrinking, the increase in fluid temperature is higher than that for stretching. The comparative analysis reveals that the fluid temperature diminishes for the case of pure fluid, for which the thermal bounding surface thickness decreases significantly. Figure 11.8 shows the effect of the Eckert number on the temperature profile of the hybrid nanofluid. The coupling constraint Ec appears due to the effect of the dissipative heat energy in the thermal transport phenomena. Physically, it is described as the reciprocal of the enthalpy which occurs due to the difference in temperature. An increase in Ec is due to the loss in enthalpy, i.e., the fluid temperature dominates over the surrounding temperature, and this causes an enhanced heat transport phenomenon. Figure 11.9 displays the characteristic of the thermal radiation affecting the fluid temperature in the presence of the nanoparticles and the parameters involved in the flow phenomena. Thermal radiation is the measure of radiation of the electromagnetic wave from the fluid element. The electromagnetic wave transforms

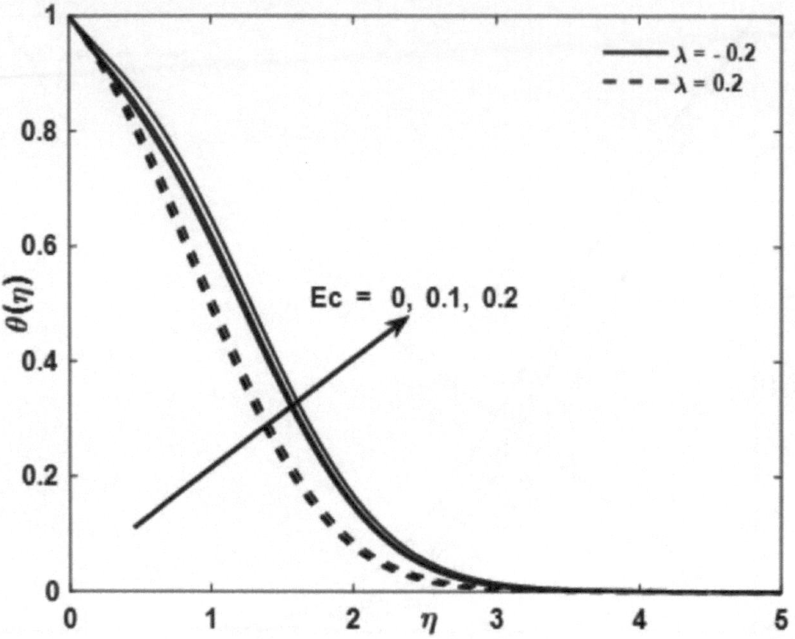

FIGURE 11.8 Impact of Ec on $\theta(\eta)$.

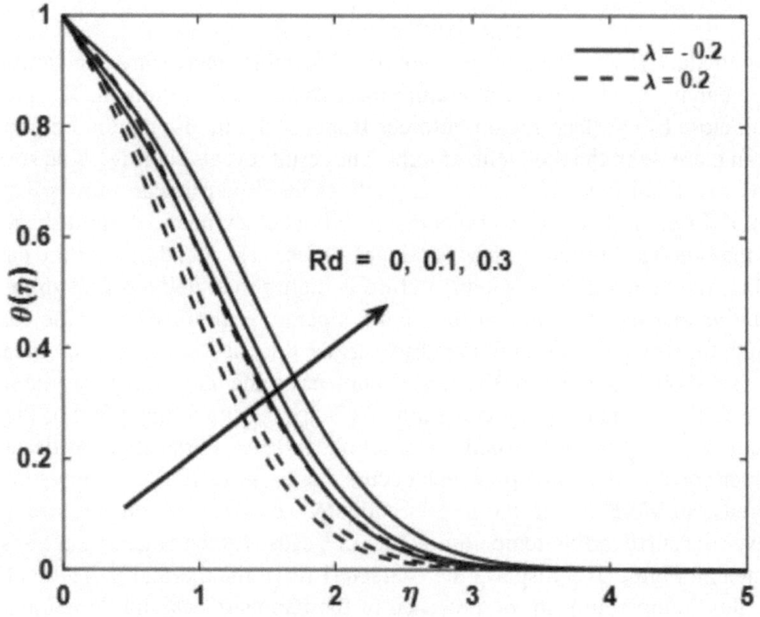

FIGURE 11.9 Impact of Rd on $\theta(\eta)$.

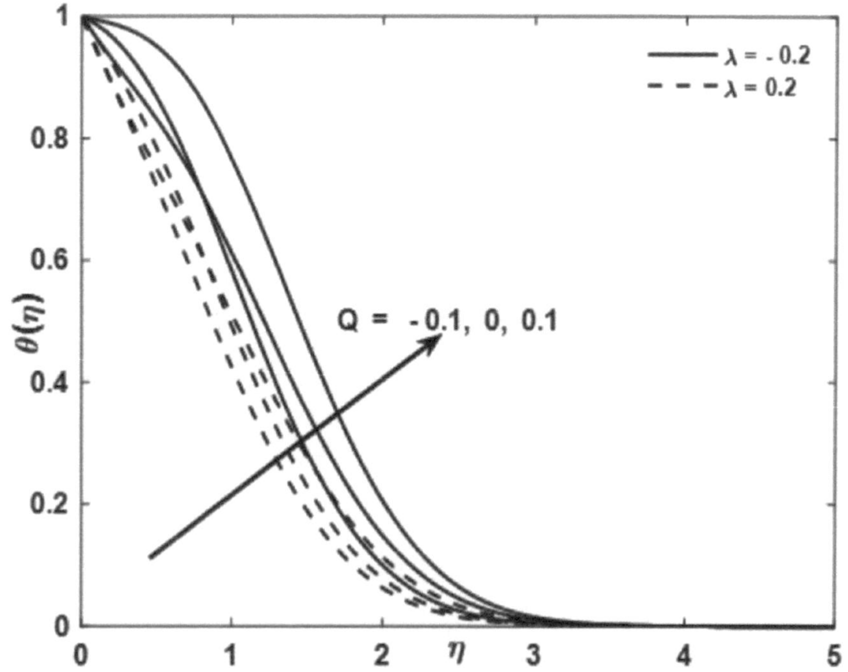

FIGURE 11.10 Impact of Q on $\theta(\eta)$.

to radiating heat in the form of thermal radiation. The heat radiating from the surface makes the surface become cool, and the heat moves away from the surface, causing a substantial intensification in fluid temperature within the entire domain. Figure 11.10 illustrates the significant contribution of the heat source/sink to the fluid temperature profile. The additional heat encountered in the system augments the fluid temperature, which increases with the increasing heat. However, the case of heat sink has an opposite trend; therefore, a reverse impact is rendered. Figure 11.11 presents the properties of particle shape affecting the conductivity of the hybrid nanofluid. The considered Hamilton–Crosser thermal conductivity model, which is based on particle shape, is used for the analysis of spherical, cylindrical, brick-, platelet-, and blade-shaped nanoparticles. The pictorial representation shows that the blade-shaped nanoparticles have greater conductivity than the other shapes.

The numerical results for the drag coefficient and local Nusselt number for the considered stretching/shrinking cases with the variation in other characterizing parameters are presented in Table 11.4. The assigned parametric contribution shows that increasing magnetization, power index, and suction parameter increases the shear stress for both the stretching and shrinking scenarios. However, the enhanced magnetization and suction decelerate the heat transfer rate significantly. The augmentation in the particle concentrations of both the nanoparticles enriches the shear rate, but the heat transmission rate decreases. The increase in the Eckert number, thermal

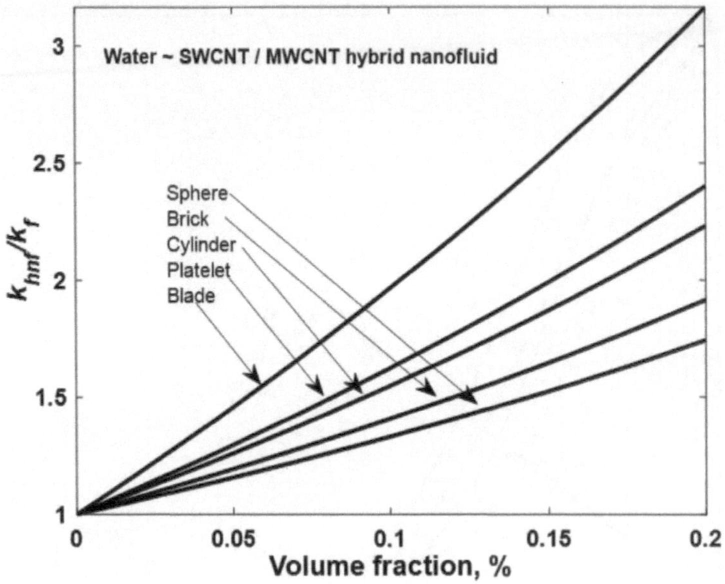

FIGURE 11.11 Variations in the shape of the particles.

TABLE 11.4
Computation of Rate Coefficients When $\lambda > 0$ and $\lambda < 0$

									C_f		Nu_x	
M	m	S	ϕ_1	ϕ_2	Pr	Ec	Q	Rd	$\lambda > 0$	$\lambda < 0$	$\lambda > 0$	$\lambda < 0$
0.1									1.2042	1.6902	0.3197	0.3392
0.5									1.3981	1.8994	0.2397	0.2484
1.5									1.5019	2.1688	0.1491	0.1584
0.1	1								1.4369	1.8987	0.9849	0.9678
	1.5								2.1036	2.6998	1.6985	1.6902
	2								2.3975	3.4009	2.3972	2.3902
	1	0.5							1.0398	1.1977	0.2034	0.2079
		0.75							1.0949	1.4988	0.6052	0.6206
		1							1.1998	1.6776	0.9498	0.9896
		0.5	0.01						1.2134	1.6043	1.6205	0.4198
			0.015						1.4011	1.8664	0.4139	0.4391
			0.02						1.6217	2.2197	0.3981	0.3993
			0.01	0.01					1.2739	1.5889	0.4193	0.4306
				0.015					1.3883	1.8499	0.4198	0.4496
				0.025					1.6215	2.1699	0.4556	0.4695
				0.01	1				1.0794	1.5015	0.0501	0.0507
					2				1.0794	1.5015	0.4009	0.4165

(Continued)

TABLE 11.4
(Continued)

									C_f		Nu_x	
M	m	S	ϕ_1	ϕ_2	Pr	Ec	Q	Rd	$\lambda > 0$	$\lambda < 0$	$\lambda > 0$	$\lambda < 0$
					7				1.0794	1.5015	0.8001	0.8121
					1	0.1			1.0794	1.5015	0.4011	0.4136
						0.2			1.0794	1.5015	0.3865	0.3787
						0.3			1.0794	1.5015	0.3432	0.3404
						0.1	0.01		1.0794	1.5015	0.3911	0.4124
							0.015		1.0794	1.5015	0.3461	0.3877
							0.025		1.0794	1.5015	0.3636	0.3777
							0.01	0.1	1.0794	1.5015	0.3679	0.3826
								0.15	1.0794	1.5015	0.3345	0.3971
								0.2	1.0794	1.5015	0.2871	0.2897

radiation, and heat source increases the heat transfer rate drastically. A comparative analysis revealed that with regard to both shear and heat transfer rate profiles, the results of the shrinking surface case dominate the that of the stretching case.

11.5 CONCLUSIONS

The thermal characteristics of hybrid nanofluid flow over stretching/shrinking wedges were examined in this article. The effects of magnetic field and heat generation rates were investigated for the presence of the particle concentrations. The heat transport phenomena were analysed and formulated as a set of PDEs with the addition of several factors. These equations were then solved numerically using the bvp5c package, and the concluding remarks are summarized as follows:

- The resistivity of the magnetized nanofluid for the increasing agent factor decelerates the fluid velocity.
- The concentration of the hybrid nanoparticles significantly affects fluid velocity in both the stretching and shrinking scenarios.
- The coupling constant, i.e., the Eckert number, due to the insertion of dissipative heat, augments the heat transport phenomena along with thermal radiation.
- The heat transfer rate is significantly controlled by the increase in the thermal radiation and the Eckert number.

REFERENCES

1. Alfellag, M.A., Kamar, H.M., Abidin, U., Kazi, S.N., Alawi, O.A., Muhsan, A.S., Sidik, N.A.C., Shaikh, K. & Khan, W.A. (2024). Green synthesized clove-treated carbon nanotubes/titanium dioxide hybrid nanofluids for enhancing flat-plate solar collector performance. Appl. Therm. Eng., 246, 122982.

2. Hussein, A.K., Rashid, F.L., Rasul, M.K., Basem, A., Younis, O., Homod, R.Z., Attia, M.E.H., Al-Obaidi, M.A., Hamida, M.B.B., Ali, B. & Abdulameer, S.F. (2024). A review of the application of hybrid nanofluids in solar still energy systems and guidelines for future prospects. Sol. Energy, 272, 112485.
3. Guo, W., Zhai, Y., Huang, X. & Li, Z. (2024). Optimizing stability and enhancing flow boiling heat transfer performance of Al_2O_3-TiO_2/water hybrid nanofluids. Exp. Therm. Fluid Sci., 155, 111177.
4. Dai, J., Zhai, Y., Li, Z. & Wang, H. (2024). Mechanism of enhanced thermal conductivity of hybrid nanofluids by adjusting mixing ratio of nanoparticles. J. Mol. Liq., 400, 124518.
5. Ouyang, Y., Md Basir, M.F., Naganthran, K. & Pop, I. (2024). Effects of discharge concentration and convective boundary conditions on unsteady hybrid nanofluid flow in a porous medium. Case Stud. Therm. Eng., 104374.
6. Sreekumar, S., Ganguly, A., Khalil, S., Chakrabarti, S., Hewitt, N., Mondol, J.D. & Shah, N. (2024). Thermo-optical characterization of novel MXene/carbon-dot hybrid nanofluid for heat transfer applications. J. Clean. Prod., 434, 140395.
7. Sun, X., Ning, H., Chen, G., Yu, G., Jia, Z. & Zhang, Y. (2024). Experimental study of hybrid nanofluid-alternating-CO_2 microbubble injection as a novel method for enhancing heavy oil recovery. J. Mol. Liq., 395, 123835.
8. Ghadery-Fahliyany, H., Ansari, S., Mohammadi, M.R., Jafari, S., Schaffie, M., Ghaedi, M. & Hemmati-Sarapardeh, A. (2024). Toward predicting thermal conductivity of hybrid nanofluids: Application of a committee of robust neural networks, theoretical, and empirical models. Powder Technol., 437, 119506.
9. Tawade, J.V., Guled, C.N., Noeiaghdam, S., Gamiz, U.F., Govindan, V. & Balamuralitharan, S. (2022). Effects of thermophoresis and Brownian motion for thermal and chemically reacting Casson nanofluid flow over a linearly stretching sheet. Results Eng., 15.
10. Hamid, A. (2020). Existence of dual solutions for wedge flow of magneto-Williamson nanofluid: A revised model. Alexandria Eng. J., 59(3), 1525–1537.
11. Li, Y.X., Mishra, S.R., Pattnaik, P.K., Baag, S., Li, Y.M., Khan, M.I., Khan, N.B., Alaoui, M.K. & Khan, S.U. (2022). Numerical treatment of time dependent magnetohydrodynamic nanofluid flow of mass and heat transport subject to chemical reaction and heat source. Alexandria Eng. J., 61(3), 2484–2491.
12. Khan, K.A., Raza, N. & Inc, M. (2021). Insights of numerical simulations of magnetohydrodynamic squeezing nanofluid flow through a channel with permeable walls. Propuls. Power Res., 10(4), 412–420.
13. Pal, D. & Mandal, G. (2020). Magnetohydrodynamic stagnation-point flow of Sisko nanofluid over a stretching sheet with suction. Propuls. Power Res., 9(4), 408–422.
14. Roy, N.C. & Pop, I. (2022). Dual solutions of magnetohydrodynamic mixed convection flow of an Oldroyd-B nanofluid over a shrinking sheet with heat source/sink. Alexandria Eng. J., 61(8), 5939–5948.
15. Khashi'ie, N.S., Arifin, N.M. & Pop, I. (2022). Magnetohydrodynamics (MHD) boundary layer flow of hybrid nanofluid over a moving plate with Joule heating. Alexandria Eng. J., 61(3), 1938–1945.
16. Arafa, A.A.M., Ahmed, S.E. & Allan, M.M. (2022). Peristaltic flow of non-homogeneous nanofluids through variable porosity and heat generating porous media with viscous dissipation: Entropy analyses. Case Stud. Therm. Eng., 32.
17. Sajid, T., Pasha, A.A., Jamshed, W., Shahzad, F., Eid, M.R., Ibrahim, R.W. & El Din, S.M. (2022). Radiative and porosity effects of trihybrid Casson nanofluids with Bödewadt flow and inconstant heat source by Yamada-Ota and Xue models. Alexandria Eng. J., 66, 457–473.

18. Javid, K., Ellahi, M., Al-Khaled, K., Raza, M., Khan, S.U., Khan, M.I., El-Zahar, E.R., Gouadria, S., Afzaal, M. & Khan, M.I. (2022). EMHD creeping rheology of nanofluid through a micro-channel via ciliated propulsion under porosity and thermal effects. Case Stud. Therm. Eng., 30.
19. Javid, K., Ellahi, M., Al-Khaled, K., Raza, M., Khan, S.U., Khan, M.I., El-Zahar, E.R., Gouadria, S., Afzaal, M. & Khan, M.I. (2022). EMHD creeping rheology of nanofluid through a micro-channel via ciliated propulsion under porosity and thermal effects. Case Stud. Therm. Eng., 30, 101746.
20. Noranuar, M.N., Mohamad, A.Q., Shafie, S., Khan, I., Jiann, L.Y. & Ilias, M.R. (2021). Non-coaxial rotation flow of MHD Casson nanofluid carbon nanotubes past a moving disk with porosity effect. Ain Shams Eng. J., 12(4), 4099–4110.
21. Ullah, I., Alam, M.M., Rahman, M.M., Pasha, A.A., Jamshed, W. & Galal, A.M. (2022). Theoretical analysis of entropy production in exothermic/endothermic reactive magnetized nanofluid flow through curved porous space with variable permeability and porosity. Int. Commun. Heat Mass Transfer, 139, 106390.
22. Mohanty, B., Jena, S. & Pattnaik, P.K. (2019). MHD nanofluid flow over stretching/shrinking surface in presence of heat radiation using numerical method. Int. J. Emerging Technol., 10(2), 119–125.
23. Pattnaik, P.K., Jena, S., Dei, A. & Sahu, G. (2019). Impact of chemical reaction on micropolar fluid past a stretching sheet. JP J. Heat Mass Transfer, 18(1), 207–223.
24. Pattnaik, P.K., Mohapatra, D.K. & Mishra, S.R. (2021). Influence of velocity slip on the MHD flow of a micropolar fluid over a stretching surface. Recent Trend in Applied Mathematics: Lecture Notes in Mechanical Engineering, Springer. doi:10.1007/978-981-15-9817-3_21
25. Parida, S.K., Mishra, S.R., Dash, R.K., Pattnaik, P.K., Khan, M.I., Chu, Y.M. & Shah, F. (2020). Dynamics of dust particles in a conducting water-based kerosene nanomaterial: A computational approach. Int. J. Chem. React. Eng., 19(8), 787–797.
26. Salawu, S.O., Shamshuddin, M.D. & Anwar Beg, O. (2022). Influence of magnetization, variable viscosity and thermal conductivity on Von Karman swirling flow of H_2O-Fe_3O_4 and H_2O-Mn-$ZnFe_2O_4$ ferromagnetic nanofluids from a spinning disk: Smart spin coating simulations. Mat. Sci. Eng. B., 279, 115659.
27. Mansourian, M., Dinarvand, S. & Pop, I. (2022). Aqua Cobalt ferrite/Mn-Zn ferrite hybrid nanofluid flow over a nonlinearly stretching permeable sheet in a porous medium. J. Nanofluids, 11, 383–391.
28. Kho, Y.B., Jusoh, R., Salleh, M.Z., Ariff, H.M. & Zainuddin, N. (2023). Magnetohydrodynamics flow of Ag-TiO_2 hybrid nanofluid over a permeable wedge with thermal radiation and viscous dissipation. J. Magn. Magn. Mater., 565, 170284.

12 Mixed Convection of Variable Viscosity Hybrid Nanofluid within Two Inclined Concentric Pipes

Oluwole Daniel Makinde, Alok Kumar Pandey, and Himanshu Upreti

12.1 INTRODUCTION

Mixed convection heat transfer refers to the phenomenon in which both forced and natural convections are present. In recent years, the broad and useful applications of mixed convection, including solar collectors, double-layered windows, building insulation, cooling electronic components, and food sterilization processes, have drawn the attention of numerous researchers to investigate it. RamReddy et al. [1] examined the impact of Soret on mixed convection flow of nanofluid (NF) over a flat plate. Malvandi and Ganji [2] studied the magnetized mixed convection flow of a nanofluid inside a vertical microtube in the presence of Brownian motion and thermophoresis. The effect of buoyancy force on the mixed convection Couette flow of a viscous incompressible NF between two cylindrical pipes was discussed by Das et al. [3]. A numerical study on mixed convection flow of a carbon-water NF in an impermeable stretched cylinder was carried out by Hayat et al. [4]. The mixed convection flow of a single-walled carbon nanotube (SWCNT)-Ag/H_2O hybrid NF over a porous stretching surface in the presence of magnetohydrodynamics (MHD) was investigated numerically by Joshi et al. [5]. Pandey and Upreti [6] analysed the heat transfer characteristics of the mixed convection of a magnetized NF on a curved surface with volumetric heat generation. Rasool and Wakif [7] investigated the Cattaneo–Christov model and mixed convection flow towards a Riga plate where the governing dimensionless parameters were the Hartmann, Prandtl, and Reynolds numbers. Hussain et al. [8] addressed the comparative study of SWCNT and multiwalled carbon nanotube (MWCNT) NFs under mixed convection flow effect in a stretchable cylinder.

A fluid's capacity to vary its viscosity in response to many factors, including temperature, pressure, and shear rate, is referred to as variable viscosity (VV). The viscosity quantifies the resistance of a fluid to flow. There are numerous real-world uses for

DOI: 10.1201/9781003595786-12

VV in different industries, such as pharmaceuticals and manufacturing. Tshehla and Makinde [9] studied the concept of the second law of thermodynamics for VV between two concentric pipes. The problem of heat transfer of dusty Jeffrey fluid through a plannar channel in the presence of VV and thermal conductivity was explored by Bhatti and Zeeshan [10]. Oblique stagnation point flow with VV on a surface using the Runge–Kutta–Fehlberg (RKF) method was scrutinized by Mehmood et al. [11]. Vaidya et al. [12] carried out a study on VV of Rabinowitsch fluid flow and heat transfer through a non-uniform tube in the presence of convective surface conditions. The magnetized Couette flow between two pipes with the impact of VV was evaluated by Makinde and Eegunjobi [13]. Idowu and Falodun [14] observed the effect of VV on Casson and Walters' B viscoelastic fluids along a vertical porous plate due to MHD and Soret–Dufour effects. Ryltseva et al. [15] investigated the study of laminar viscoplastic flow through an axisymmetric pipe with temperature-dependent viscosity.

12.2 MODEL PROBLEM

Consider a mixed convection of a variable viscosity water-based hybrid NF containing a mixture of copper (Cu) and alumina (Al_2O_3) nanoparticles in the gap between two inclined concentric pipes as shown in Figure 12.1.

The governing equations, based on the Boussinesq approximation for the buoyancy force, that articulate the physical scenario take the following form [16, 17]:

$$\frac{\partial u}{\partial z}=0, \tag{12.1}$$

$$-\frac{\partial P}{\partial z}+\frac{1}{r}\frac{\partial}{\partial r}\left(r\mu_{hnf}(T)\frac{\partial u}{\partial r}\right)+g\beta_{hnf}(T-T_a)\cos(\varphi)=0, \tag{12.2}$$

$$\frac{\kappa_{hnf}}{r}\frac{\partial}{\partial r}\left(r\frac{\partial T}{\partial r}\right)+\mu_{hnf}(T)\left(\frac{\partial u}{\partial r}\right)^2=0, \tag{12.3}$$

with

$$\left.\begin{array}{l} u=0, T=0 \text{ at } r=r_1 \\ u=0, -\kappa_{hnf}\dfrac{\partial T}{\partial r}=h(T-T_a) \text{ at } r=r_2 \end{array}\right\}. \tag{12.4}$$

The thermophysical correlation of a temperature-dependent dynamical viscosity hybrid NF with respect to the base fluid is given as follows:

$$\left.\begin{array}{l} \mu_{hnf}(T)=\dfrac{\mu_f e^{-m(T-T_a)}}{(1-\phi_1)^{2.5}(1-\phi_2)^{2.5}}, \beta_{hnf}=(1-\phi_2)\left[(1-\phi_1)\beta_f+\phi_1\beta_1\right]+\phi_2\beta_2 \\ \dfrac{\kappa_{hnf}}{\kappa_{bf}}=\dfrac{\kappa_2+2\kappa_{bf}-2\phi_2(\kappa_{bf}-\kappa_2)}{\kappa_2+2\kappa_{bf}+\phi_2(\kappa_{bf}-\kappa_2)}; \dfrac{\kappa_{bf}}{\kappa_f}=\dfrac{\kappa_1+2\kappa_f-2\phi_1(\kappa_f-\kappa_1)}{\kappa_1+2\kappa_f+\phi_1(\kappa_f-\kappa_1)} \end{array}\right\}. \tag{12.5}$$

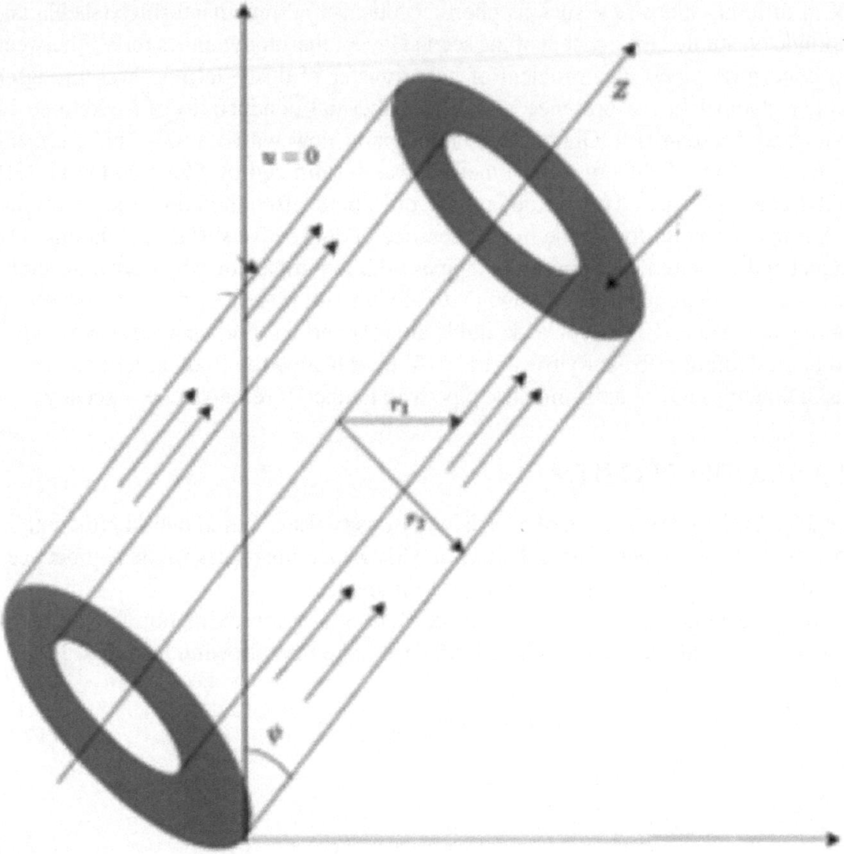

FIGURE 12.1 Geometry of the problem.

The dimensionless variables are

$$\left.\begin{array}{l}
\bar{P}=\dfrac{\rho_f r_1^2 LP}{\left(\mu_f\right)^2},\, w=\dfrac{\rho_f u r_1}{\mu_f},\, \eta=\dfrac{r-r_1}{r_1 L},\, Z=\dfrac{z}{r_1 L},\\
\theta=\dfrac{T-T_a}{T_0-T_a},\, L=\dfrac{r_2-r_1}{r_1},\, \chi=-\dfrac{d\bar{P}}{dZ},\, \gamma=m\left(T_0-T_a\right),\\
Gr=\dfrac{g\rho_f\beta_f r_1^3 L^2\left(T_0-T_a\right)}{\left(\mu_f\right)^2},\, Pr=\dfrac{\mu_f\left(C_p\right)_f}{\kappa_f},\\
Ec=\dfrac{\left(\mu_f\right)^2}{\left(r_1\rho_f\right)^2\left(C_p\right)_f\left(T_0-T_a\right)},\, Bi=\dfrac{h r_1 L}{\kappa_f}.
\end{array}\right\} \quad (12.6)$$

The dimensionless equations are

$$\frac{d^2w}{d\eta^2}-\gamma\frac{dw}{d\eta}\frac{d\theta}{d\eta}+\frac{L}{L\eta+1}\frac{dw}{d\eta}+\left(\frac{\beta_{nf}}{\beta_f}Gr\cos(\varphi)\theta+\chi\right) \\ (1-\phi_1)^{2.5}(1-\phi_2)^{2.5}e^{\gamma\theta}=0, \tag{12.7}$$

$$\frac{\kappa_{nf}}{\kappa_f}\left[\frac{d^2\theta}{d\eta^2}+\frac{L}{L\eta+1}\frac{d\theta}{d\eta}\right]+\frac{PrEc}{(1-\phi_1)^{2.5}(1-\phi_2)^{2.5}}e^{-\gamma\theta}\left(\frac{dw}{d\eta}\right)^2=0\,, \tag{12.8}$$

with

$$\left.\begin{aligned} &w(\eta)=0,\theta(\eta)=1 \text{ at } \eta=0\\ &w(\eta)=0,\frac{d\theta(\eta)}{d\eta}=-\frac{Bi}{(\kappa_{nf}/\kappa_f)}\theta(\eta)\text{ at }\eta=1\end{aligned}\right\}, \tag{12.9}$$

where γ is the rate of decrease in hybrid NF viscosity due to temperature difference and χ is the pressure gradient. The shear stress rate and heat transfer rate are defined as

$$\tau_w=\mu_{hnf}(T)\left.\frac{du}{dr}\right|_{r_1,r_2} \text{ and } q_w=-\kappa_{hnf}\left.\frac{dT}{dr}\right|_{r_1,r_2}. \tag{12.10}$$

By utilizing Equation (12.5) in conjunction with some dimensionless parameters defined in Equation (12.6), we were able to find some parameters of interest, i.e., (i) skin friction and (ii) Nusselt number, which are defined as

$$C_f=\frac{e^{-\gamma\theta(\eta)}}{(1-\phi_1)^{2.5}(1-\phi_2)^{2.5}}\left.\frac{dw(\eta)}{d\eta}\right|_{\eta=1} \text{ and } Nu=-\frac{\kappa_{nf}}{\kappa_f}\left.\frac{d\theta(\eta)}{d\eta}\right|_{\eta=1}. \tag{12.11}$$

12.3 NUMERICAL METHOD

The differential Equations (12.7) and (12.8) along with Equation (12.9) are solved using the RKF method with a shooting technique. The procedures are as follows:

Step 1. Write the non-dimensional Equations (12.7) and (12.8) in the following forms:

$$\frac{d^2w}{d\eta^2}=\gamma\frac{dw}{d\eta}\frac{d\theta}{d\eta}-\frac{L}{L\eta+1}\frac{dw}{d\eta}-\left(\frac{\beta_{nf}}{\beta_f}Gr\cos(\varphi)\theta+\chi\right)(1-\phi_1)^{2.5}(1-\phi_2)^{2.5}e^{\gamma\theta}, \tag{12.12}$$

$$\frac{d^2\theta}{d\eta^2}=\frac{-PrEc}{(1-\phi_1)^{2.5}(1-\phi_2)^{2.5}}\frac{1}{(\kappa_{nf}/\kappa_f)}e^{-\gamma\theta}\left(\frac{dw}{d\eta}\right)^2-\frac{L}{L\eta+1}\frac{d\theta}{d\eta}. \quad (12.13)$$

Step 2. Convert Equations (12.12) and (12.13) to a system of first-order ordinary differential equations (ODEs), denoting the terms as $w=m_1, \frac{dw}{d\eta}=m_2, \theta=m_3$, and $\frac{d\theta}{d\eta}=m_4$.

Now, using them in Equations (12.12) and (12.13), we get the following system of ODEs:

$$\left.\begin{aligned} m_1' &= m_2, \\ m_2' &= \gamma m_2 m_4-\frac{L}{L\eta+1}m_2-\left(\frac{\beta_{nf}}{\beta_f}Gr\cos(\varphi)m_3+\chi\right)(1-\phi_1)^{2.5}(1-\phi_2)^{2.5}e^{\gamma m_3} \\ m_3' &= m_4, \\ m_4' &= \frac{-PrEc}{(1-\phi_1)^{2.5}(1-\phi_2)^{2.5}}\frac{1}{(\kappa_{nf}/\kappa_f)}e^{-\gamma m_3}(m_2)^2-\frac{L}{L\eta+1}m_4 \end{aligned}\right\}, \quad (12.14)$$

and the boundary conditions (12.9) become

$$m_1(0)=0, m_2(0)=h_a, m_3(0)=1, m_4(0)=h_b. \quad (12.15)$$

Step 3: The ODEs (12.14) with Equation (12.15) is an initial value problem (IVP), which is resolved by providing appropriate values of h_a and h_b unknowns. This is attained by applying a shooting technique, continuing this process until error is permissible, i.e., $\max\left\{|m_1(\eta_{\max})-0|, \left|m_4(\eta_{\max})+\frac{Bi}{(\kappa_{nf}/\kappa_f)}m_3(\eta_{\max})\right|\right\}\le 10^{-5}$. Here, for the computation, the authors considered step size $\Delta\eta=0.001$ with $\eta_{\max}=1$.

12.4 DISCUSSION

In the last section, the authors provide the description of the applied methodology followed in a flowchart. This section is devoted to the explanation of graphs obtained for the pertinent parameters involved in the study. The physical properties of the base fluid (H_2O) and nanoparticles (Cu and Al_2O_3) are mentioned in Table 12.1.

The research focused primarily on investigating the flow of a hybrid NF through an inclined concentric pipe. The distance between the pipes is taken as unity, and the surface of the pipe at $\eta=1$ (outer pipe) is exposed to convective heating. Throughout the computation, the Prandtl number is kept fixed at $Pr=6.72$

and the dissipation parameter (Ec)=0.02, and computation is done for parameters $\chi, \gamma, Bi, L, Gr, \varphi$, and ϕ.

Figures 12.2–12.8 show the effect of increasing the parameters $\chi, \gamma, Bi, Gr, \varphi, \phi$, and L on velocity field plotted against the dimensionless distance between the pipes, respectively. Figure 12.2 illustrates the effect of increasing the pressure gradient parameter (χ) on non-dimensional velocity $(w(\eta))$, and it is noted from the figure that velocity increases with an increase in χ. This is because higher pressure gradient assists the flow in the axial direction, and hence the fluid velocity increases. A similar pattern is recorded for γ (see Figure 12.3).

Figure 12.4 shows the consequences of the Biot number on velocity field. It is noted from the plot that fluid velocity decreases as the Biot number increases. Figure 12.5 reveals that fluid velocity increases with and increase in the Grashof

TABLE 12.1
Physical Properties of H_2O, Cu, and Al_2O_3

		Nanoparticles	
Physical properties	**H_2O**	**Cu**	**Al_2O_3**
κ (WmK^{-1})	0.6071	400	40
$\beta \times 10^{-5}$ (K^{-1})	21	1.68	0.85

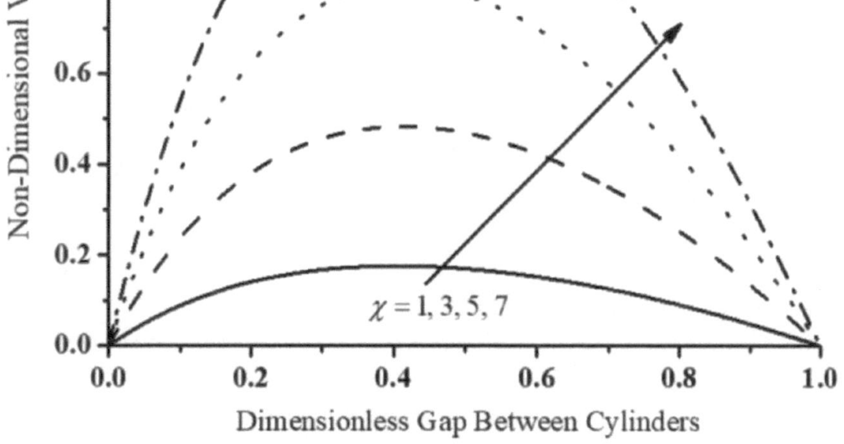

FIGURE 12.2 Change in velocity due to χ.

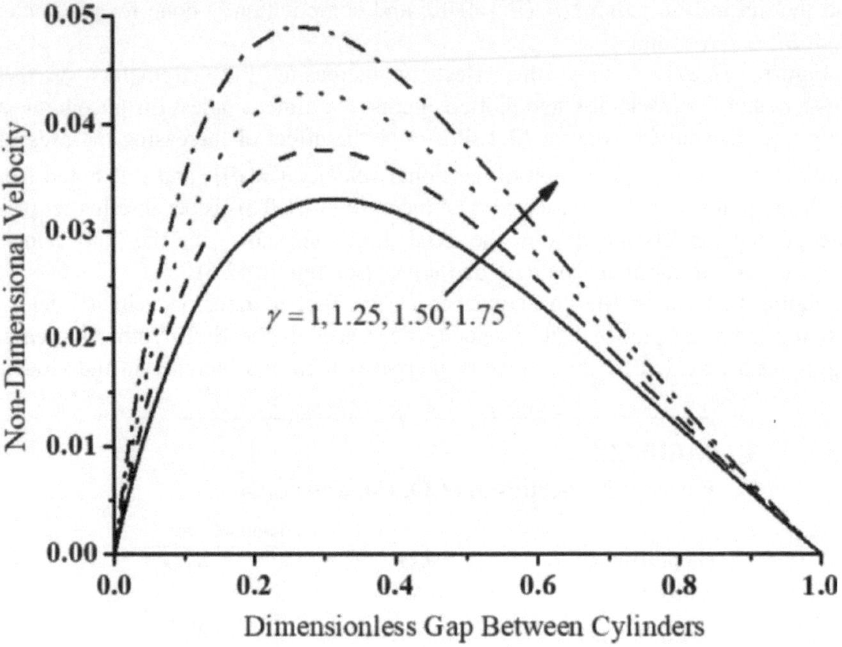

FIGURE 12.3 Change in velocity due to γ.

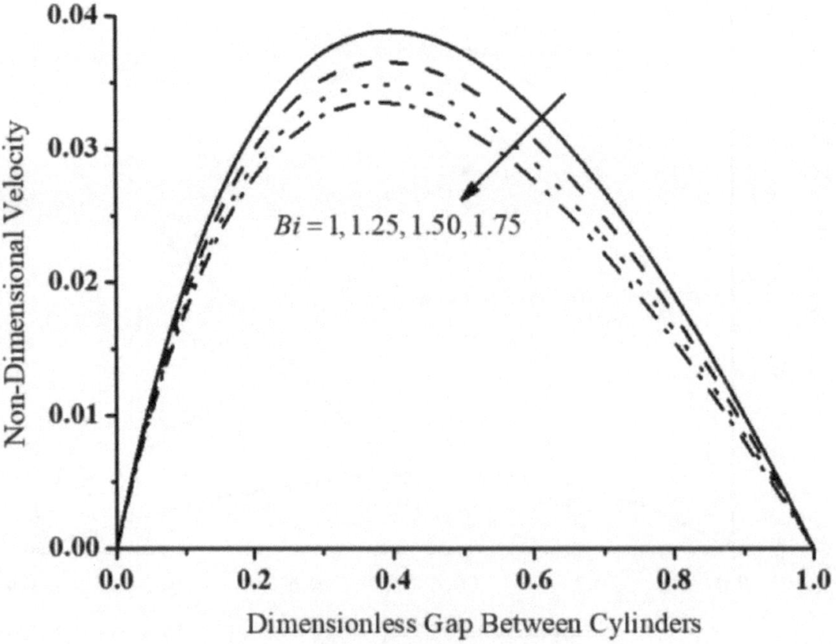

FIGURE 12.4 Change in velocity due to Bi.

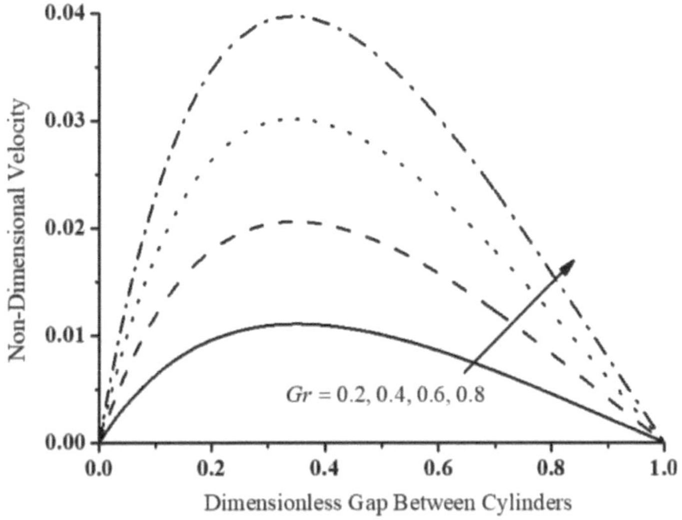

FIGURE 12.5 Change in velocity due to Gr.

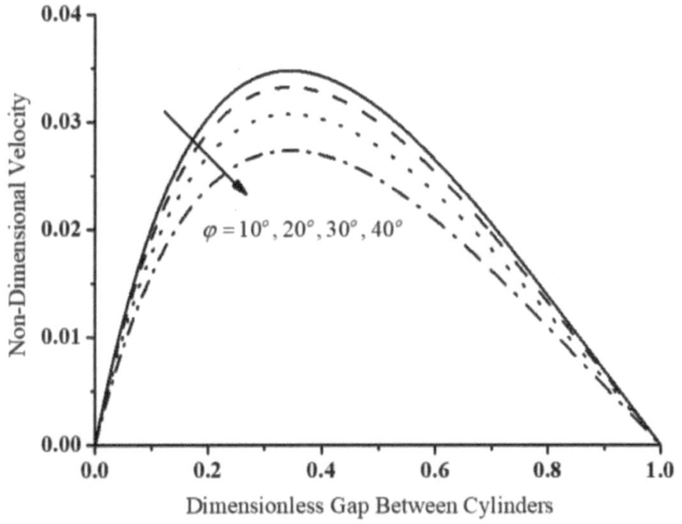

FIGURE 12.6 Change in velocity due to φ.

number Gr; this is because the Grashof number is the ratio of buoyancy to viscous force. An increase in Gr suggests an increase in buoyancy force is more relative to viscous force, and as a consequence, progress in the fluid movement is observed.

Figures 12.6 and 12.7 show the response of increasing φ and $\phi(=\phi_1+\phi_2)$ on the velocity field, and these figures illustrate that the velocity profile increases for

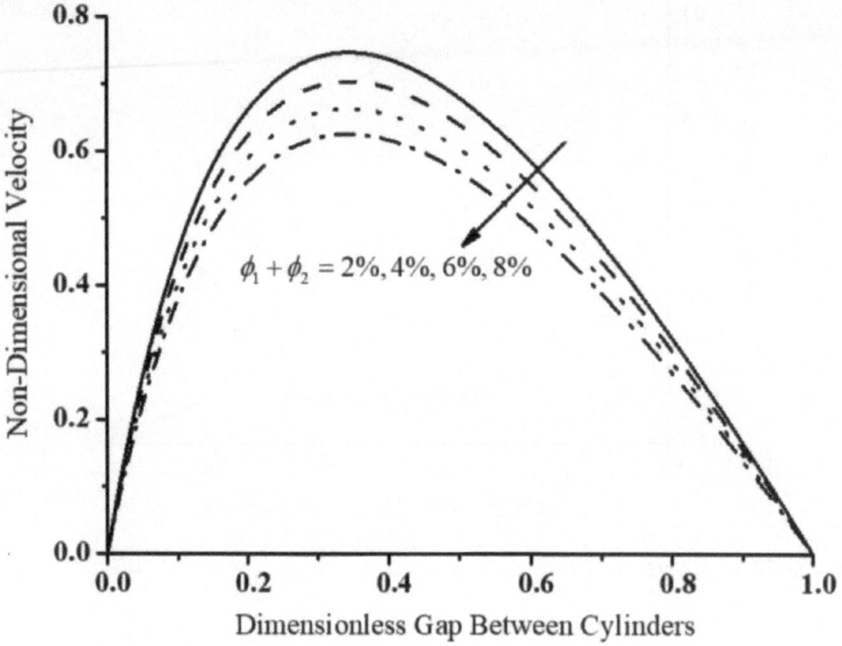

FIGURE 12.7 Change in velocity due to $\phi_1 + \phi_2$.

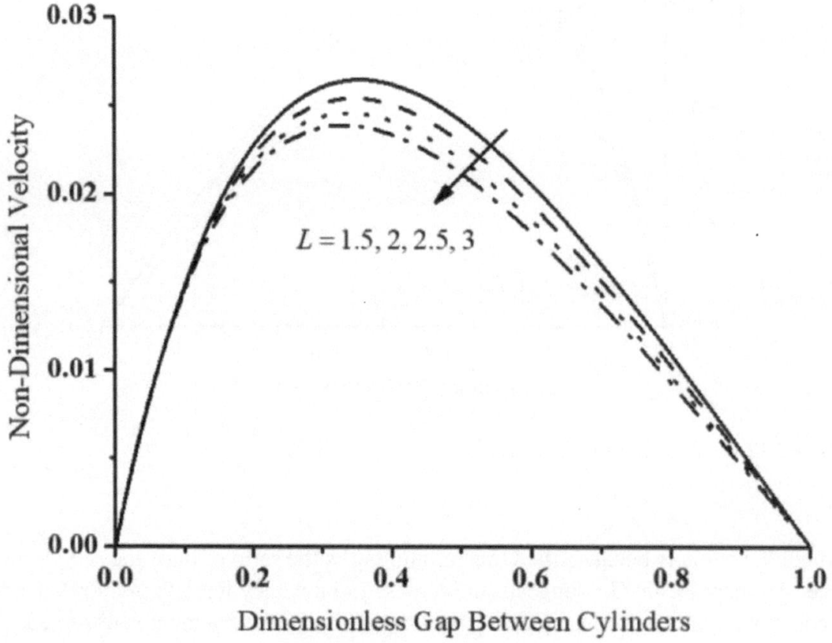

FIGURE 12.8 Change in velocity due to L.

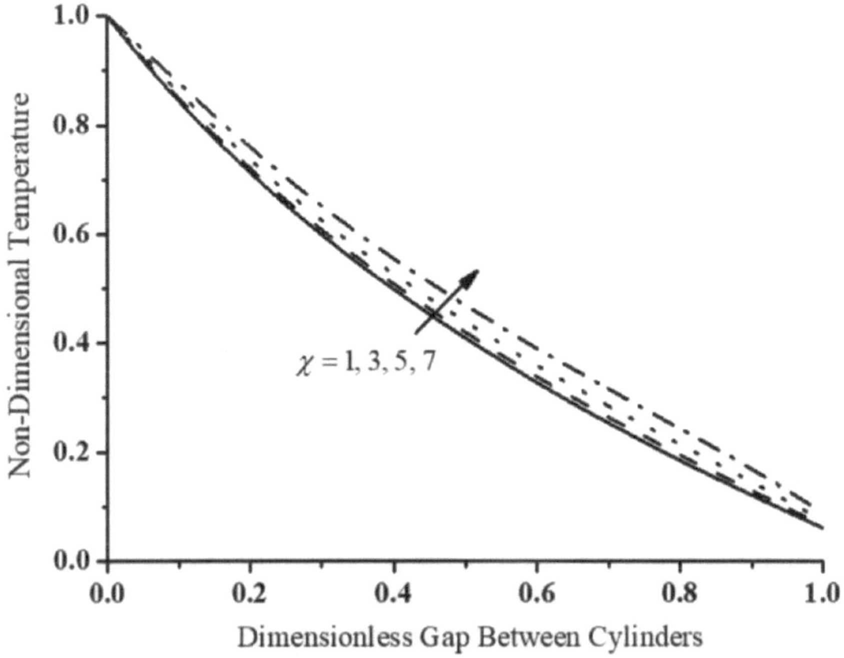

FIGURE 12.9 Change in temperature due to χ.

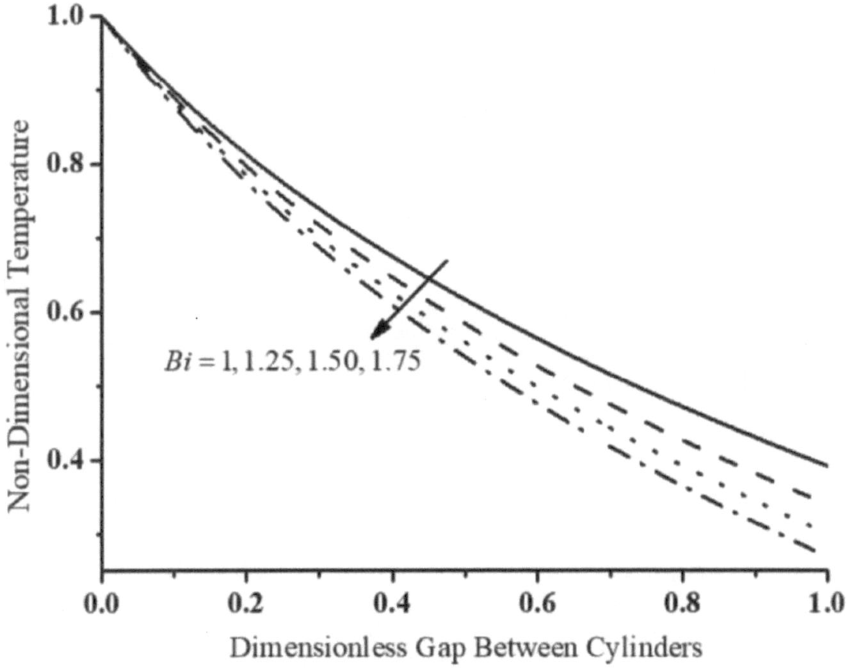

FIGURE 12.10 Change in temperature due to Bi.

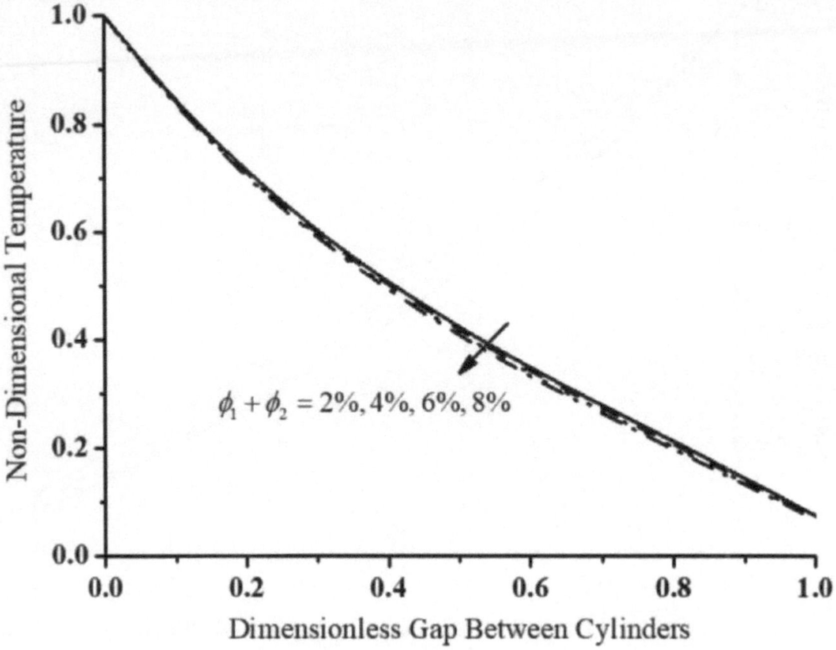

FIGURE 12.11 Change in temperature due to $\phi_1 + \phi_2$.

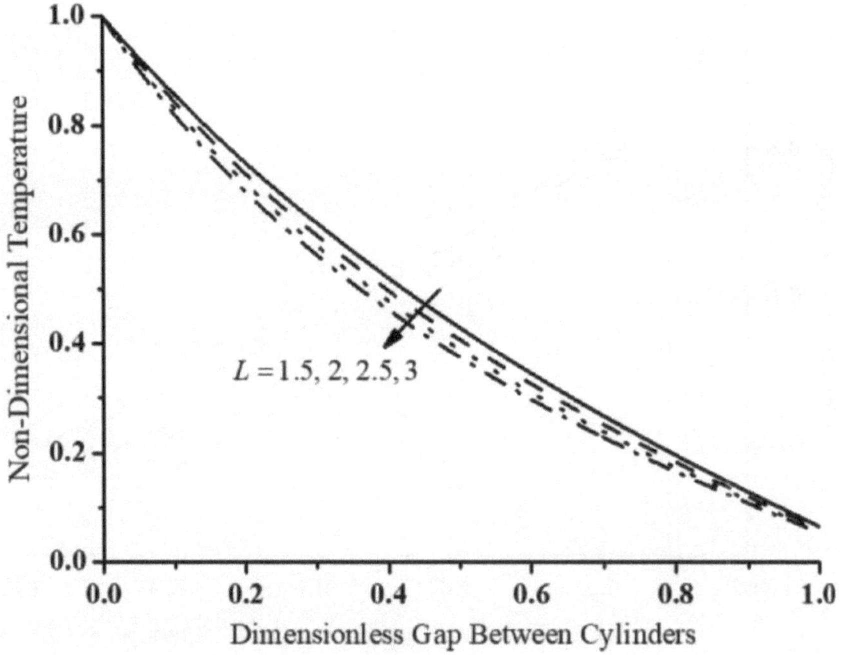

FIGURE 12.12 Change in temperature due to L.

increasing values of both φ and ϕ. The effect of L on fluid velocity is shown in Figure 12.8, which shows that fluid velocity near the surface has an insignificant variation for increasing L from 1.5 to 3, whereas for $\eta > 0.2$, velocity profiles show significant variations and it reduces with increasing L.

The effect of the active parameters χ, Bi, ϕ, and L on the temperature profile is presented in Figures 12.9–12.12, respectively. The graphs are plotted for the aforesaid parameters against the distance between the pipes, and it is illustrated from these graphs that with an increase in the magnitude of the aforesaid parameters, the temperature of the fluid increases for χ, whereas it reduces for Bi, ϕ, and L.

12.5 CONCLUSIONS

This study deals with the impact of mixed convection on hybrid NF flow between two concentric pipes. The main outcomes of this work are as follows:

- The velocity field increases with an increase in the pressure gradient parameter.
- The temperature profiles decrease with an increase in the Biot number.
- The heat transfer performance of the working fluid decreases as the value of L increases.
- On increasing the volume fraction of the solid particle, the heat transfer rate increases.

REFERENCES

1. RamReddy C., Murthy P. V., Chamkha A. J., & Rashad A. M. (2013). Soret effect on mixed convection flow in a nanofluid under convective boundary condition. International Journal of Heat and Mass Transfer, 64, 384–392.
2. Malvandi, A., & Ganji, D. D. (2014). Magnetohydrodynamic mixed convective flow of Al_2O_3–water nanofluid inside a vertical microtube. Journal of Magnetism and Magnetic Materials, 369, 132–141.
3. Das, S., Chakraborty, S., Jana, R. N., & Makinde, O. D. (2015). Mixed convective Couette flow of reactive nanofluids between concentric vertical cylindrical pipes. Journal of Nanofluids, 4(4), 485-493.
4. Hayat, T., Ullah, S., Khan, M. I., & Alsaedi, A. (2018). On framing potential features of SWCNTs and MWCNTs in mixed convective flow. Results in Physics, 8, 357–364.
5. Joshi, N., Pandey, A. K., Upreti, H., & Kumar, M. (2021). Mixed convection flow of magnetic hybrid nanofluid over a bidirectional porous surface with internal heat generation and a higher-order chemical reaction. Heat Transfer, 50(4), 3661–3682.
6. Pandey, A. K., & Upreti, H. (2021). Mixed convective flow of Ag–H_2O magnetic nanofluid over a curved surface with volumetric heat generation and temperature-dependent viscosity. Heat Transfer, 50(7), 7251–7270.
7. Rasool, G., & Wakif, A. (2021). Numerical spectral examination of EMHD mixed convective flow of second-grade nanofluid towards a vertical Riga plate using an advanced version of the revised Buongiorno's nanofluid model. Journal of Thermal Analysis and Calorimetry, 143, 2379–2393.
8. Hussain, Z., Hayat, T., Alsaedi, A., & Anwar, M. S. (2021). Mixed convective flow of CNTs nanofluid subject to varying viscosity and reactions. Scientific Reports, 11(1), 22838.

9. Tshehla, M. S., & Makinde, O. D. (2011). Analysis of entropy generation in a variable viscosity fluid flow between two concentric pipes with a convective cooling at the surface. International Journal of the Physical Sciences, 6(25), 6053–6060.
10. Bhatti, M. M., & Zeeshan, A. (2016). Analytic study of heat transfer with variable viscosity on solid particle motion in dusty Jeffery fluid. Modern Physics Letters B, 30(16), 1650196.
11. Mehmood, R., Tabassum, R., & Akbar, N. S. (2018). Oblique stagnation-point flow of non-Newtonian fluid with variable viscosity. Heat Transfer Research, 49(16), 1587–1603.
12. Vaidya, H., Rajashekhar, C., Gudekote, M., Prasad, K. V., Makinde, O. D., & Sreenadh, S. (2019). Peristaltic motion of non-Newtonian fluid with variable liquid properties in a convectively heated nonuniform tube: Rabinowitsch fluid model. Journal of Enhanced Heat Transfer, 26(3), 277–294.
13. Makinde O. D., & Eegunjobi A. S. (2020). Entropy analysis of a variable viscosity MHD Couette flow between two concentric pipes with convective cooling. Engineering Transactions, 68(4), 317–334.
14. Idowu, A. S., & Falodun, B. O. (2020). Variable thermal conductivity and viscosity effects on non-Newtonian fluids flow through a vertical porous plate under Soret-Dufour influence. Mathematics and Computers in Simulation, 177, 358–384.
15. Ryltseva, K. E., Borzenko, E. I., & Shrager, G. R. (2020). Non-Newtonian fluid flow through a sudden pipe contraction under non-isothermal conditions. Journal of Non-Newtonian Fluid Mechanics, 286, 104445.
16. Makinde, O. D. (2014). Thermal analysis of a reactive generalized Couette flow of power law fluids between concentric cylindrical pipes. The European Physical Journal Plus, 129, 1–9.
17. Makinde, O. D., & Eegunjobi, A. S. (2021). Inherent irreversibility of mixed convection within concentric pipes in a porous medium with thermal radiation. Journal of Mathematical and Fundamental Sciences, 53(3), 395–414.

13 Computational Study of Heat Transfer in Sisko Hybrid Nanofluid Flowing over a Radially Stretching Disc

Ankita Bisht, Sanjalee Maheshwari, and Arvind Singh Bisht

Nomenclature

A^*	Material parameter of Sisko fluid
b^*	Material constant of Sisko fluid
Bi^*	Thermal Biot number
c^*	Constant
C_f	Local skin friction coefficient
f	Dimensionless stream function
h_f	Heat transfer coefficient
M^*	Magnetic parameter
n^*	Power law index
Nu	Local Nusselt number
Pr^*	Prandtl number
s^*	Stretching parameter
T	Temperature inside the boundary layer
$U_w(x)$	Stretching sheet velocity
$\tilde{u}$	Axial velocity component
$\tilde{v}$	Radial velocity component
(r, z)	Space coordinates (m)
α	Thermal diffusivity

DOI: 10.1201/9781003595786-13

η_∞	Edge of the boundary
ν	Kinematic viscosity of nanofluid
κ	Thermal conductivity
ρ	Density of fluid
σ	Electrical conductivity
ψ	Dimensional stream function
θ	Dimensionless temperature

Subscripts

hnf	Hybrid nanofluid
f	Base fluid
np	Nanoparticles
1	Single-walled carbon nanotubes
2	Iron oxide

13.1 INTRODUCTION

In the world of rapidly growing technologies, heat transfer plays a critical role in several engineering and mechanical processes, for instance in heating and cooling of vehicles, electronic devices, and buildings; heat removal in nuclear reactors; controlling the chemical reaction rate; heat exchangers; and manufacturing of materials. Conventional fluids such as water, alcohol, and air have served as coolants for a long time; however, in some instances, they fail to efficiently transfer heat when the heat transfer process is not optimized or when the fluid properties are not well-suited for the specific application. To overcome this issue, Choi and Eastman [1] suggested that the addition of nanoscale particles (typically between 1 and 100 nm in size) to the base fluid can significantly enhance the thermal conductivity of the base fluid and named the mixture as nanofluid. In nanofluids, nanoparticles can be metallic, metallic oxide, carbon nanotubes, etc.

Despite the improved thermal properties of nanofluids, there are still limitations to their effectiveness. For example, nanoparticles can aggregate and settle, which can reduce their ability to improve thermal conductivity. Additionally, the cost of nanoparticles can be prohibitive, making it difficult to produce large quantities of nanofluids for practical applications. To address these limitations, researchers developed hybrid nanofluids. These fluids contain a combination of two or more types of nanoparticles or a mixture of nanoparticles and micro-sized particles. By combining different types of particles, hybrid nanofluids can enhance the thermal conductivity of the base fluid while also minimizing issues with aggregation and settling.

In addition to improving thermal conductivity, hybrid nanofluids may also offer other advantages, such as improved mechanical properties and stability. Studies suggest that hybrid nanofluids have potential applications in electronics cooling, heat exchangers, and solar thermal systems, and in clinical therapy [2, 3]. Madhesh and Kalaiselvam [4] performed experiments to calculate the heat transfer characteristic of a hybrid nanofluid as the coolant. Devi and Devi [5] numerically inspected the flow of water-based copper-alumina hybrid nanofluid past an expanding surface. Afridi et al. [6] compared the heat transfer rate of a water-based alumina nanofluid and a water-based copper-alumina hybrid nanofluid. Waini et al. [7] investigated the effect of the magnetic field on the flow dynamics of hybrid nanofluids over an expanding and a shrinking wedge. Zainal et al. [8] incorporated the effect of viscous dissipation and transversal magnetic field in their study of hybrid nanofluid flow over an expanding and a shrinking sheet. They established that the accumulation of nanoparticles significantly enhances the skin friction and heat transmission rate. Hayat and Nadeem [9] developed the three-dimensional mathematical model considering the effect of thermal radiation, chemical reaction, rotation, and heat source for an Ag-Cu nanoparticle-suspended water-based hybrid nanofluid flow over an expanding surface. Waini et al. [10] investigated the assisting/opposing flow of a hybrid nanofluid along a vertical surface using a single-phase nanofluid model. Hussain et al. [11] examined the effect of convective boundary conditions on the heat transmission rate of hybrid nanofluids. They considered the flow over a stretching rotating surface. Upreti et al. [12] presented a study of hybrid nanofluid flow under the influence of convective boundary conditions, viscous dissipation, and thermal radiation. They quantified the amount of heat transfer and entropy generated within the system. Joshi et al. [13] considered the single-walled carbon nanotube (SWCNT) and silver nanoparticle-suspended water-based hybrid nanofluid as the working fluid and studied its flow past a bidirectional porous medium under the effect of a higher-order chemical reaction and volumetric heat generation. Sanjalee et al. [14] discussed the magnetohydrodynamic flow of a hybrid nanofluid through a porous channel using the Lie group analysis. As of late, Pandey et al. [15] examined the flow of a Ag-SWCNT/water hybrid nanofluid past a cone and discussed the effect of variable viscosity and magnetic field on the velocity profile and the heat transmission rate.

The investigation of the boundary layer flow of non-Newtonian fluids over an expanding or stretching sheet has been a topic of research for many years. Due to the wide range of modern applications of non-Newtonian fluids in various industries such as food processing, cosmetics, oil drilling, pharmaceuticals, and biomedical devices, as well as their involvement in natural phenomena like glacier dynamics, researchers have devoted significant attention to understanding the flow dynamics of these fluids. The complexity of non-Newtonian fluids cannot be fully explained by a single-fluid model. Among the different types of non-Newtonian fluids, the Sisko fluid model stands out as a relatively simple and straightforward model. Named after its proposer, Sisko [16], this model combines the characteristic of both Newtonian and non-Newtonian fluids. Sisko fluid model is an extension of the power law fluid model. Sisko fluids can be commonly found in nature and have diverse implications in areas such as drug retention, tissue adhesion, coating process, and metal casting. A notable example of such fluids is the flow of greases and blood in small arteries, which exemplifies the relevance and significance of studying these types of fluids. Khan et al. [17] developed a mathematical model to study the flow dynamics and convective heat transfer of a Sisko fluid past a radially stretching surface under the effect of thermal radiation. Later on, Khan et al. [18] extended their work [17] for investigating the influence of transversal magnetic field on the convective heat transfer rate. Sharma and Bisht [19] numerically examined the effect of viscous dissipation on the magnetohydrodynamic boundary layer flow of a Sisko nanofluid over a vertically expanding surface. Bisht and Sharma [20] established that Joule heating decreases the convective heat transfer rate in the magnetohydrodynamic flow of a Sisko nanofluid past the stretching surface. Some other pioneering works on the boundary layer flow of Sisko nanofluids are well documented in [21–24] and several references therein. Khan et al. [25] discussed the flow dynamics of a Sisko-based hybrid nanofluid numerically. They obtained the dual solution for a radially shrinking porous surface. Almaneea [26] studied the thermal performance of a hybrid Sisko fluid over a stretching cylindrical surface and concluded that heat generation affects the thermal performance of the system adversely. Recently, Bilal and Saeed [27] explained the flow behaviour a Sisko hybrid nanofluid flowing over a non-linearly stretching or shrinking surface under the effect of the magnetic field.

A thorough review of the literature revealed that no attention yet has been paid to the study of the magnetohydrodynamic flow of hybrid Sisko nanofluids over a radially stretching surface considering the viscous dissipation effects. A mathematical model is developed for the Sisko hybrid nanofluid flow over a radially stretching surface considering the effect of transversal magnetic field. The flow governing dynamics are solved using the MATLAB bvp4c routine with convective boundary conditions. The outcomes of the study are graphically presented and discussed in detail. The key objective of the present study is to examine the thermal performance of the hybrid Sisko nanofluidic system.

13.2 MATHEMATICAL MODELLING

The steady axisymmetric Sisko hybrid nanofluid flow over a permeable radially stretching disc is investigated. Figure 13.1 shows a physical representation of the problem, where z and r represent the corresponding cylindrical polar coordinates. The permeable radially stretching disc is held in place at the region $z = 0$, and the motion of the Sisko hybrid nanofluid flow occurs at the region $z \geq 0$. The coordinate z-axis is occupied normal to the disc direction, while the radial coordinate r-axis is still engaged along the disc's surface. The power law velocity at the disc surface is taken as $U_w = c^* r^{s^*}$, where c^* is the stretching rate of the sheet (a positive constant) and s^* is the stretching sheet parameter. The variable magnetic field with intensity B_0 is implemented normal to the surface. Additionally, the lower surface of the disc is heated via convection from a hot fluid with temperature T_w, providing a heat transfer coefficient h_f. Moreover, T_∞ represents the free stream fluid temperature when $z \to \infty$, such that $T_w > T_\infty$. The physical sketch of the problem under consideration is shown in Figure 13.1.

Following the aforementioned assumptions and employing the hybrid nanofluid model proposed by Talebi and Salehi [28], the governing conservation equations of the model under consideration in terms of cylindrical polar coordinates (r, z) are as follows [25, 29]:

$$\frac{\partial(r\tilde{u})}{\partial r} + \frac{\partial(r\tilde{w})}{\partial z} = 0, \tag{13.1}$$

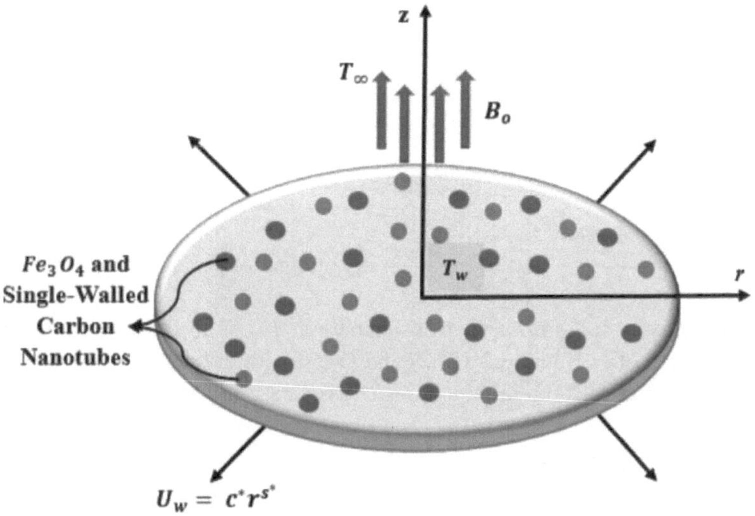

FIGURE 13.1 Physical sketch of the problem.

$$\tilde{u}\frac{\partial \tilde{u}}{\partial r}+\tilde{w}\frac{\partial \tilde{u}}{\partial z}+\frac{b^*}{r\rho_{hnf}}\left(-\frac{\partial \tilde{u}}{\partial z}\right)^{n^*}=\frac{\mu_{hnf}}{\rho_{hnf}}\frac{\partial^2 u}{\partial z^2}-\frac{\sigma_{hnf}B_o^{\ 2}}{\rho_{hnf}}\tilde{u}, \tag{13.2}$$

$$\tilde{u}\frac{\partial T}{\partial r}+\tilde{w}\frac{\partial T}{\partial z}=\frac{\kappa_{hnf}}{(\rho Cp)_{hnf}}\frac{\partial^2 T}{\partial z^2}+\frac{1}{(\rho Cp)_{hnf}}\left[\mu_{hnf}\left(-\frac{\partial u}{\partial z}\right)^2+b^*\left(-\frac{\partial u}{\partial z}\right)^{n^*+1}\right]. \tag{13.3}$$

The physical boundary conditions are

$$\text{At } z=0\ :\tilde{u}=U_w,\ \tilde{w}=0,\ \kappa_{hnf}\frac{\partial T}{\partial y}=-h_f(T_w-T),$$

$$\text{and as } z\to\infty:\ \tilde{u}\to 0, T\to T_\infty, \tag{13.4}$$

where $\tilde{u}$ and $\tilde{w}$ denote the nanofluid velocity in the axial (x) and radial (r) directions, respectively.

In accordance with Khan et al. [29], the following dimensionless variables are used:

$$\psi(r,z)=U_w\,\mathrm{r}^2\,\mathrm{Re}_{b^*}^{-1/(n^*+1)}f(\eta),\ \eta=\frac{z}{r}\mathrm{Re}_{b^*}^{1/(n^*+1)},\theta=\frac{T-T_\infty}{T_w-T_\infty}, \tag{13.5}$$

where the stream function ψ for velocity is given by

$$\tilde{u}=\frac{1}{r}\frac{\partial \psi}{\partial z},\tilde{w}=-\frac{1}{r}\frac{\partial \psi}{\partial r}. \tag{13.6}$$

Now, on using Equation (13.5), in Equation (13.6) we get

$$\left.\begin{array}{l} u = U_w f_\eta(\eta), \\ v = -U_w \mathrm{Re}_{b^*}^{\frac{-1}{(n^*+1)}} \left[\left(\dfrac{s^*(2n^*-1)+n^*+2}{n^*+1}\right) f(\eta) + \left(\dfrac{s^*(2-n^*)-1}{n^*+1}\right)\eta f_\eta(\eta)\right] \end{array}\right\}, \tag{13.7}$$

where $\mathrm{Re}_{b^*} = \dfrac{\rho_f r^{n^*} U_w^{2-n^*}}{b^*}$ and $\mathrm{Re} = \dfrac{U_w r}{\upsilon_f}$ denote the local Reynolds number.

On applying the self-similarity transformations (13.5) in the leading governing non-linear differential Equations (13.2) and (13.3) with boundary conditions (13.4), the required partial differential equations (PDEs) ease to the following dimensionless form of ODEs:

$$\begin{array}{l} \dfrac{n^*(-f_{\eta\eta})^{n^*-1}}{\rho_{hnf}/\rho_f} f_{\eta\eta\eta} + \dfrac{\mu_{hnf}/\mu_f}{\rho_{hnf}/\rho_f} A^* f_{\eta\eta\eta} + \left(\dfrac{s^*(2n^*-1)+1}{n^*+1}\right) \\ ff_{\eta\eta} - s^* f_\eta^{\,2} - \dfrac{\sigma_{hnf}/\sigma_f}{\rho_{hnf}/\rho_f} Ma^* f_\eta = 0, \end{array} \tag{13.8}$$

$$\begin{array}{l} \theta_{\eta\eta} + \mathrm{Pr}^*\left(\dfrac{s^*(2n^*-1)+1}{n^*+1}\right) \dfrac{(\rho\,\mathrm{Cp})_{hnf}/(\rho\,\mathrm{Cp})_f}{\kappa_{hnf}/\kappa_f} f\theta_\eta \\ + \dfrac{\kappa_f}{\kappa_{hnf}} \mathrm{Ec}^*\mathrm{Pr}^*[A^*(-f_{\eta\eta})^2 + (-f_{\eta\eta})^{n^*+1}] = 0 \end{array} \tag{13.9}$$

subjected to boundary conditions:

$$\left.\begin{array}{l} \text{At } \eta \to 0: f(0)=0, f_\eta(0)=1, \theta_\eta(0) = -B_i^*(1-\theta(0)), \\ \text{as } \eta \to \infty: f_\eta(\eta) \to 0, \theta(\eta) \to 0 \end{array}\right\}. \tag{13.10}$$

The physical parameters involved in Equations (13.8)–(13.10) are defined as Sisko material parameter (A^*), Prandtl number (Pr^*), magnetic parameter (Ma^*), Eckert number (Ec^*), and thermal Biot number (Bi^*), where

$$\left.\begin{array}{l} A^* = \dfrac{\mathrm{Re}_{b^*}^{2/(n^*+1)}}{\mathrm{Re}_{a^*}}, \mathrm{Pr}^* = \dfrac{\mathrm{r}U_w}{\alpha_f}\mathrm{Re}_{b^*}^{-2/(n^*+1)}, Ma^* = \dfrac{\sigma_f B_0^2 x}{\rho_f U_w}, \\ \mathrm{Ec}^* = \dfrac{U_w^2}{Cp_f(T_w - T_\infty)}, \mathrm{Bi}^* = \dfrac{\mathrm{r}h_f}{\kappa_f}\mathrm{Re}_{b^*}^{-1/(n^*+1)} \end{array}\right\}.$$

The physical quantities for the flow problem, i.e., skin friction coefficient (C_f) and Nusselt number (Nu_x) (or heat transfer rate), are, respectively, defined as

$$C_f = \frac{2\tau}{\rho_f U_w^2};\ \tau = \left[\mu_{hnf}\frac{\partial \tilde{u}}{\partial z} - b^*\left(-\frac{\partial \tilde{u}}{\partial z}\right)^{n^*}\right]\Bigg|_{z=0},\ \mathrm{Nu} = \frac{\kappa_{hnf}\, r}{\kappa_f (T_w - T_\infty)}\frac{\partial T}{\partial z}\bigg|_{z=0}.$$

After applying the similarity transformation to the above expression, we get

$$\frac{1}{2}C_f \mathrm{Re}_{b^*}^{1/(n^*+1)} = A^*\frac{\mu_{hnf}}{\mu_f}F_\eta(0) - (-F_\eta(0))^{n^*},$$

$$\mathrm{Re}_{b^*}^{-1/(n^*+1)}\mathrm{Nu} = -\frac{\kappa_{hnf}}{\kappa_f}\theta_\eta(0).$$

13.3 NUMERICAL METHODOLOGY

The numerical solutions are obtained for the non-linear Equations (13.8) and (13.9) together with the boundary conditions (13.10) by utilizing the finite difference technique. These non-linear differential equations are firstly converted into a set of first-order ODEs and then solved by a numerical technique based on a finite difference method called bvp4c routine. The thermophysical properties of the hybrid nanofluid are computed using the thermophysical laws given in Table 13.1 using the

TABLE 13.1
Thermophysical Laws for the Hybrid Nanofluid [30]

Property	Hybrid nanofluid model
Viscosity	$\mu_{hnf} = \frac{\mu_f}{(1-\phi_1-\phi_2)^{2.5}}$
Density	$\rho_{hnf} = \rho_f(1-\phi_1-\phi_2) + \rho_1\phi_1 + \rho_2\phi_2$
Heat capacity	$(\rho Cp)_{hnf} = (\rho Cp)_f(1-\phi_1-\phi_2) + (\rho Cp)_1\phi_1 + (\rho Cp)_2\phi_2$
Thermal conductivity	$\frac{\kappa_{hnf}}{\kappa_{bf}} = \frac{\kappa_2 + 2\kappa_f - 2\phi_2(\kappa_{bf} - \kappa_2)}{\kappa_2 + 2\kappa_f + \phi_2(\kappa_{bf} - \kappa_2)}$, where $\frac{\kappa_{bf}}{\kappa_f} = \frac{\kappa_1 + 2\kappa_f - 2\phi_1(\kappa_f - \kappa_1)}{\kappa_1 + 2\kappa_f + \phi_1(\kappa_f - \kappa_1)}$
Electrical conductivity	$\frac{\sigma_{hnf}}{\sigma_{bf}} = \frac{\sigma_2 + 2\sigma_f - 2\phi_2(\sigma_{bf} - \sigma_2)}{\sigma_2 + 2\sigma_f + \phi_2(\sigma_{bf} - \sigma_2)}$, where $\frac{\sigma_{bf}}{\sigma_f} = \frac{\sigma_1 + 2\sigma_f - 2\sigma_1(\sigma_f - \sigma_1)}{\sigma_1 + 2\sigma_f + \sigma_1(\sigma_f - \sigma_1)}$

thermophysical quantities of the base fluid and nanoparticles given in Table 13.2. For numerical computations, the range of numerical integration ($\eta_{max} = 6$) is considered to be the finite dimensions that are enough for the velocity and temperature profiles to attain the wide field asymptotically under the given boundary condition. The process is iteratively repeated until the solutions with a desired level of accuracy (i.e., up to 10^{-6}) is attained to fulfil the criterion of convergence.

13.4 VALIDATION OF THE RESULT

This numerical study is carried out to investigate the effect of heat transfer coefficients on radially stretching surfaces. To validate the numerical results obtained in this study, the value of the Nusselt number for different values of the Prandtl number and thermal Biot number is computed and compared with the previously published results, and the comparison is shown in Table 13.3. The table values show an excellent agreement with the results published by Khan et al. [18] and validate the accuracy of our numerical simulation.

TABLE 13.2
Thermophysical Quantities of Base Fluid Blood and Nanoparticles at 300 K [14]

	Viscosity (μ pas)	Density (ρ) kg/m³	Specific heat (Cp) J/kgK	Thermal conductivity (κ) W/mK	Electrical conductivity (σ) S/m
Blood	3.5×10^{-3}	1050	3900	0.492	0.8
SWCNTs	-	2600	425	6600	10^{6}
Iron oxide	-	5180	670	9.7	$10^{-5.99}$

TABLE 13.3
Comparison of the Limiting Case of the Present Study with the Published Work

A^*	Ma^*	Pr^*	Bi^*	Nu $(n^* = 1, s^* = 3)$ Khan et al. [18]	Nu $(n^* = 1, s^* = 3)$ Current work
1	1	1	0.1	0.091490	0.0914908
1	1	1	0.2	0.168632	0.1686325
1	1	1	0.5	0.341291	0.3412909
1	1	1	1	0.518121	0.5181208
1	1	2	0.1	0.094278	0.0942788
1	1	2	0.2	0.178353	0.1783538
1	1	2	0.5	0.383607	0.3836072
1	1	2	1	0.622343	0.6223423

13.5 RESULTS AND DISCUSSION

In this section, the impact of various significant flow parameters, including the Sisko material parameter, magnetic field parameter, and Eckert number, on velocity, temperature, and important engineering quantities is explored. The results are presented in the graphical form, comparing the outcomes for both Sisko fluids and Sisko hybrid nanofluids. This section is divided into two subsections for clarity:

(i) Results on velocity and temperature profiles
(ii) Results on engineering quantities of interest

13.5.1 Results on Velocity and Temperature Profiles

Figures 13.2 and 13.3 are presented to examine the influence of the Sisko material parameter on velocity and temperature profiles for both the Sisko fluid and Sisko hybrid nanofluid. The Sisko material parameter (A^*) inversely affects the fluid viscosity, meaning that an increase in the Sisko material parameter results in a decrease in fluid viscosity, allowing for freer movement of fluid particles. Consequently, as the Sisko material parameter increases, the velocity profile also increases, as depicted in Figure 13.2. The decrease in the fluid viscosity corresponds to an increase in the kinetic energy of the fluid particles, leading to an increase in fluid temperature, as demonstrated in Figure 13.3. Additionally, these plots indicate that the addition of nanoparticles significantly enhances the thickness of the momentum and thermal boundary layers. This outcome is expected because introduction of nanoparticles supports the Brownian motion and thermophoretic diffusion.

The influence of the magnetic field parameter on the velocity and temperature profiles is illustrated in Figures 13.4 and 13.5, respectively, and the results are compared for both the Sisko fluid and Sisko hybrid nanofluid. As the magnetic field parameter

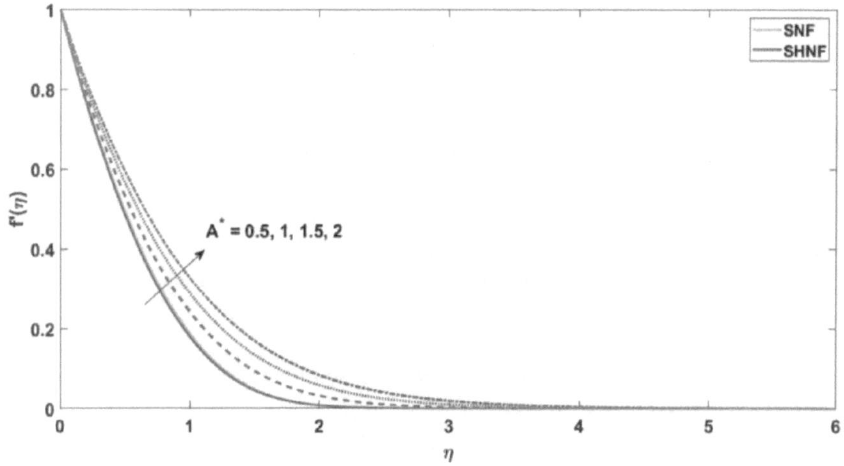

FIGURE 13.2 Variation in velocity profile with Sisko material parameter.

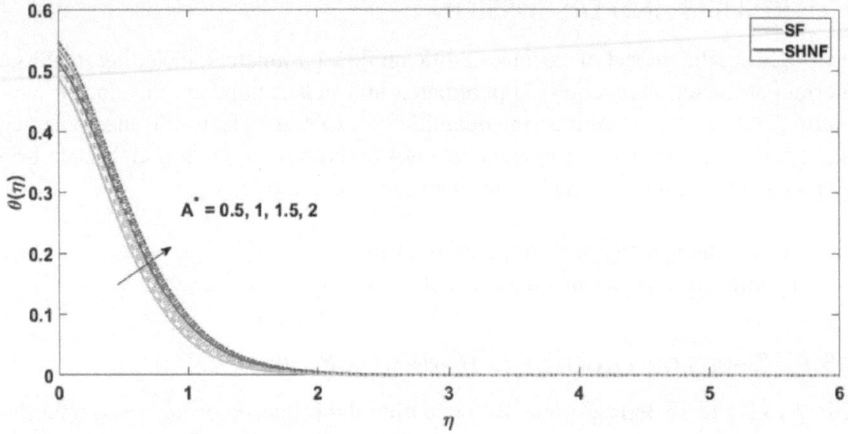

FIGURE 13.3 Variation in temperature profile with Sisko material parameter.

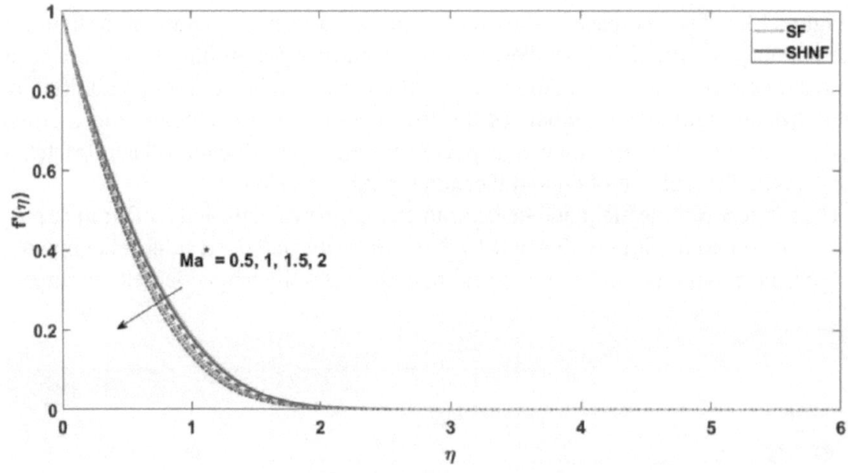

FIGURE 13.4 Variation in the velocity profile with magnetic parameter.

increases, the Lorentz force that acts against the flow intensifies. Consequently, the velocity profile decreases, as depicted in Figure 13.4. The resistance caused by the increased Lorentz force generates heat, resulting in an increase in the temperature profile, as shown in Figure 13.5.

The Sisko fluid belongs to the family of non-Newtonian fluids and obeys the power law model. The power law index $n^* > 1$ refers to the shear-thickening fluids, $n^* < 1$ corresponds to the shear-thinning fluids, and $n^* = 1$ signifies the Newtonian fluids. Figures 13.6 and 13.7 present the variations in the velocity and temperature profiles for different power law index values. Based on the observations from these, it can be

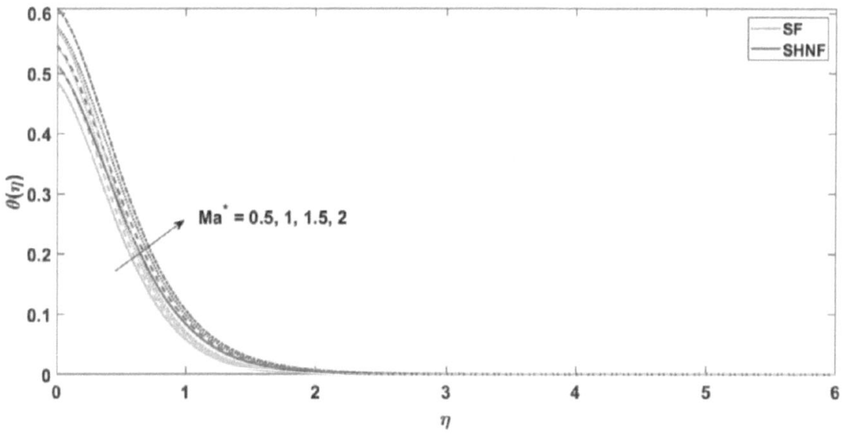

FIGURE 13.5 Variation in the temperature profile with magnetic parameter.

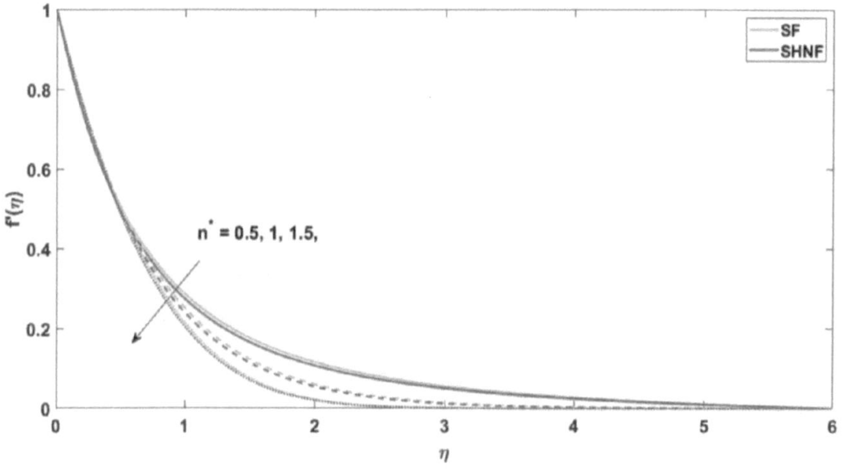

FIGURE 13.6 Variation in the velocity profile with the power law index.

concluded that both shear-thinning and shear-thickening fluids exhibit a decrease in momentum and thermal boundary layer thickness. Additionally, the boundary layer thickness for shear-thinning fluids is greater than that for shear-thickening fluids. This outcome aligns with the physical behaviour of shear-thinning fluids, as their effective viscosity decreases with an increasing shear rate. This decrease in viscosity reduces flow resistance, resulting in the thickening of both thermal and momentum boundary layers.

Figures 13.8 and 13.9 are presented to investigate the impact of the stretching parameter on the velocity and temperature profiles. Figure 13.8 illustrates that as the disc stretching increases, the radial velocity decreases, consequently reducing

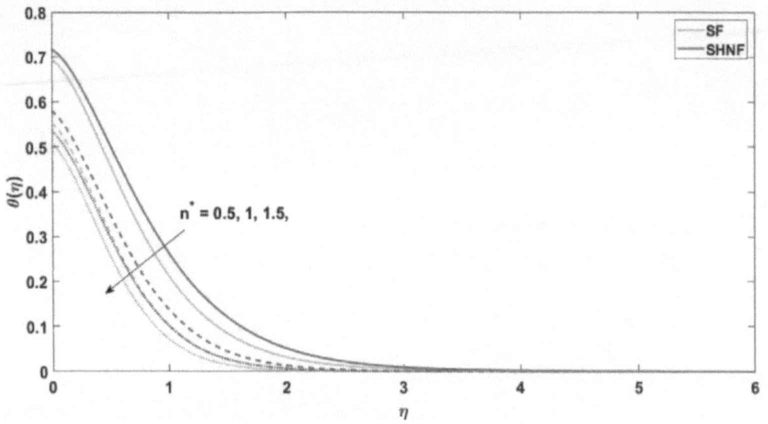

FIGURE 13.7 Variation in the temperature profile with the power law index.

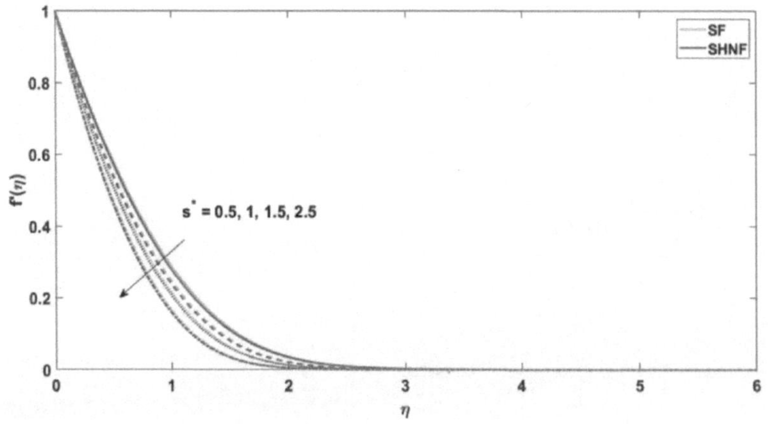

FIGURE 13.8 Variation in the velocity profile with the stretching parameter.

the thickness of the momentum boundary layer. Figure 13.9 demonstrates that an increase in the stretching parameter leads to a decrease in the temperature profile. Both the velocity and thermal profiles exhibit a monotonic decrease up to a certain distance from the disk, beyond which they asymptotically approach zero. This characteristic is particularly useful in applications involving the cooling of electronic equipment.

13.5.2 Results on the Quantities of Physical Interest

The influence of the Eckert number and Biot number on the heat transfer rate is illustrated in Figure 13.10 using the line plots. It is observed that an increase in the

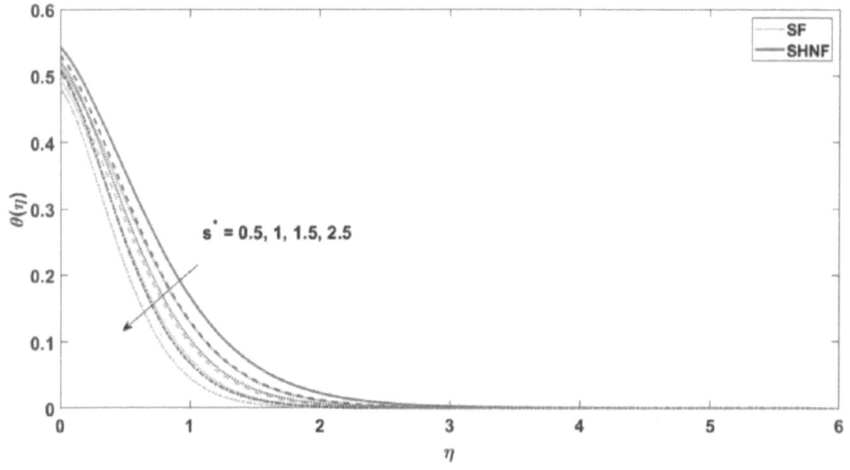

FIGURE 13.9 Variation in the temperature profile with the stretching parameter.

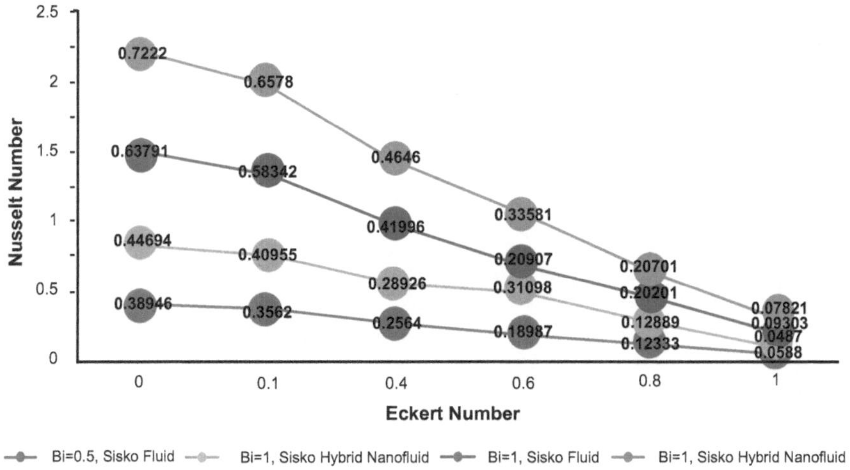

FIGURE 13.10 Variation in the Nusselt number with the Eckert number and Biot number.

Eckert number leads to a decrease in the heat transfer rate, while an increase in the Biot number results in an increase in the heat transfer rate. The Eckert number represents the conversion of mechanical energy into thermal energy, which corresponds to heat dissipation through internal friction. Consequently, an elevated Eckert number enhances fluid temperature (Figure 13.11) while reducing the heat transfer rate. On the other hand, an increase in the Biot number reduces thermal resistance near the surface of the disc, leading to an increase in convective heat transmission (Figure 13.12). The line plot suggests that the addition of nanoparticles to the Sisko base fluid significantly enhances the heat transmission rate. This can be attributed to the higher thermal conductivity of nanoparticles compared to the base fluid particles.

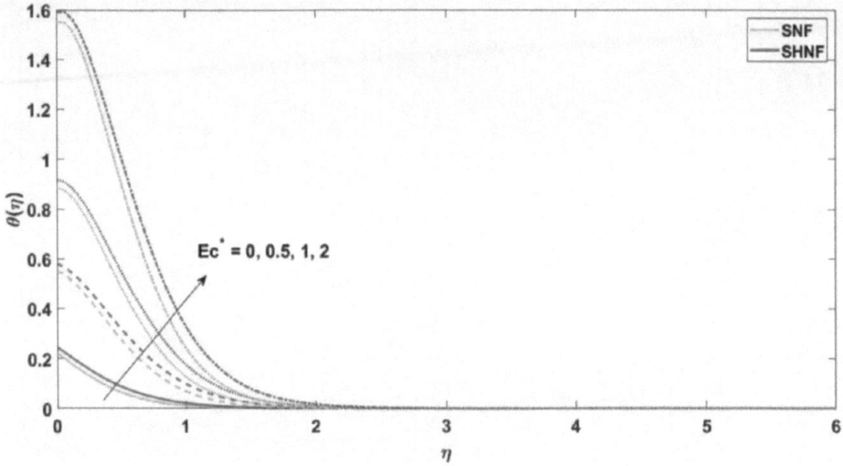

FIGURE 13.11 Variation in the temperature profile with the Eckert number.

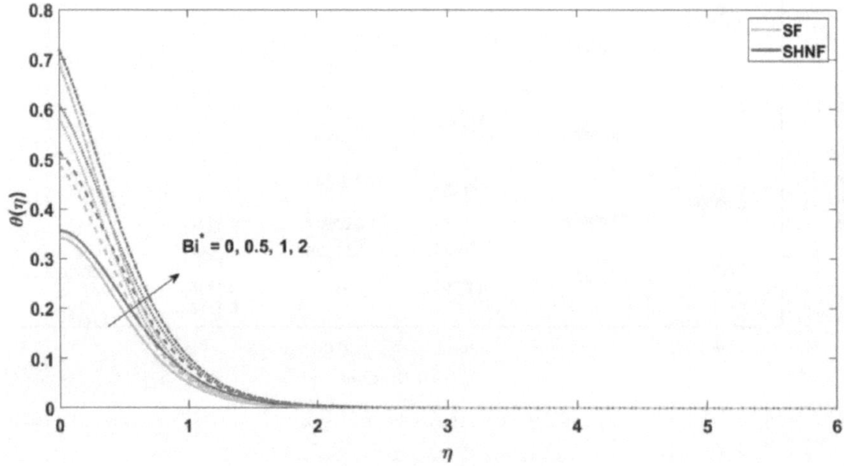

FIGURE 13.12 Variation in the temperature profile with the thermal Biot number.

This finding highlights the potential of using hybrid nanofluids to achieve the desired heat transfer rate.

Figure 13.13 presents the changes in the skin friction coefficient as the Sisko material parameter increases for both shear-thinning and shear-thickening fluids. The line plot clearly demonstrates that an increase in the Sisko material parameter corresponds to an increase in the skin friction coefficient. This observation can be explained by considering the definition of the Sisko material parameter, which is the ratio of the shear viscosity to the consistency index. As A^* increases, the viscous force decreases, which leads to the thinning of momentum boundary layer thickness

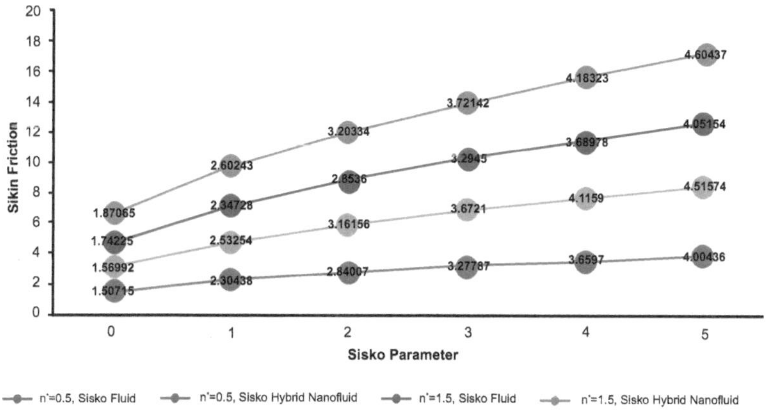

FIGURE 13.13 Variation in skin friction coefficient with the Sisko material parameter and power law index.

and strengthening of the skin friction coefficient. The skin friction coefficient for the shear-thickening and shear-thinning Sisko hybrid nanofluid is higher than that for the shear-thickening and shear-thinning Sisko fluid.

13.6 CONCLUSIONS

The flow dynamics and thermal performance of the magnetohydrodynamic flow of the Sisko hybrid nanofluid over a radially stretching disc with convective boundary conditions and viscous dissipation were numerically investigated in the present study. The major findings of the study are summarized as follows:

- The magnitude of the velocity and temperature can be monitored by the magnetic field. By accelerating the magnetic field intensity, the magnitude of velocity decreases while temperature increases.
- Sisko material parameter increases the thermal and momentum boundary layer thickness.
- With the addition of nanoparticles in the base fluid, the heat transmission rate significantly increases; therefore, addition of nanoparticles in the base fluid is recommended.
- Viscous dissipation decreases the heat transfer rate, and Biot number increases the heat transfer rate.
- Shear-thickening Sisko hybrid nanofluids exert the maximum the drag force, whereas shear-thinning Sisko fluids exert the minimum drag force on the disc surface.

REFERENCES

1. Choi, S. U., & Eastman, J. A. (2001). *Enhanced Heat Transfer Using Nanofluids* (No. US 6221275). Argonne, IL: Argonne National Lab (ANL).

2. Sidik, N. A. C., Jamil, M. M., Japar, W. M. A. A., & Adamu, I. M. (2017). A review on preparation methods, stability and applications of hybrid nanofluids. *Renewable and Sustainable Energy Reviews*, 80, 1112–1122.
3. Sarkar, J., Ghosh, P., & Adil, A. (2015). A review on hybrid nanofluids: Recent research, development and applications. *Renewable and Sustainable Energy Reviews*, 43, 164–177.
4. Madhesh, D., & Kalaiselvam, S. (2014). Experimental analysis of hybrid nanofluid as a coolant. *Procedia Engineering*, 97, 1667–1675.
5. Devi, S. A., & Devi, S. S. U. (2016). Numerical investigation of hydromagnetic hybrid Cu–Al_2O_3/water nanofluid flow over a permeable stretching sheet with suction. *International Journal of Nonlinear Sciences and Numerical Simulation*, 17(5), 249–257.
6. Afridi, M. I., Qasim, M., & Saleem, S. (2018). Second law analysis of three-dimensional dissipative flow of hybrid nanofluid. *Journal of Nanofluids*, 7(6), 1272–1280.
7. Waini, I., Ishak, A., & Pop, I. (2020). MHD flow and heat transfer of a hybrid nanofluid past a permeable stretching/shrinking wedge. *Applied Mathematics and Mechanics*, 41(3), 507–520.
8. Zainal, N. A., Nazar, R., Naganthran, K., & Pop, I. (2021). Viscous dissipation and MHD hybrid nanofluid flow towards an exponentially stretching/shrinking surface. *Neural Computing and Applications*, 1–11.
9. Hayat, T., & Nadeem, S. (2017). Heat transfer enhancement with Ag–CuO/water hybrid nanofluid. *Results in Physics*, 7, 2317–2324.
10. Waini, I., Ishak, A., Groşan, T., & Pop, I. (2020). Mixed convection of a hybrid nanofluid flow along a vertical surface embedded in a porous medium. *International Communications in Heat and Mass Transfer*, 114, 104565.
11. Hussain, A., Alshbool, M. H., Abdussattar, A., Rehman, A., Ahmad, H., Nofal, T. A., & Khan, M. R. (2021). A computational model for hybrid nanofluid flow on a rotating surface in the existence of convective condition. *Case Studies in Thermal Engineering*, 26, 101089.
12. Upreti, H., Pandey, A. K., & Kumar, M. (2021). Assessment of entropy generation and heat transfer in three-dimensional hybrid nanofluids flow due to convective surface and base fluids. *Journal of Porous Media*, 24(3).
13. Joshi, N., Upreti, H., Pandey, A. K., & Kumar, M. (2021). Heat and mass transfer assessment of magnetic hybrid nanofluid flow via bidirectional porous surface with volumetric heat generation. *International Journal of Applied and Computational Mathematics*, 7(3), 64.
14. Maheshwari, S., Sharma, Y. D., & Yadav, O. P. (2022). Entropy generation in magnetohydrodynamics flow of hybrid Casson nanofluid in porous channel: Lie group analysis. *International Journal of Applied and Computational Mathematics*, 8(5), 247.
15. Pandey, A. K., Upreti, H., & Uddin, Z. (2023). Magnetic SWCNT–Ag/H_2O nanofluid flow over cone with volumetric heat generation. *International Journal of Modern Physics B*, 2350253.
16. Sisko, A. W. (1958). The flow of lubricating greases. *Industrial & Engineering Chemistry*, 50(12), 1789–1792.
17. Khan, M., Malik, R., & Munir, A. (2015). Mixed convective heat transfer to Sisko fluid over a radially stretching sheet in the presence of convective boundary conditions. *AIP Advances*, 5(8), 087178.
18. Khan, M., Malik, R., Munir, A., & Shahzad, A. (2016). MHD flow and heat transfer of Sisko fluid over a radially stretching sheet with convective boundary conditions. *Journal of the Brazilian Society of Mechanical Sciences and Engineering*, 38, 1279–1289.

19. Sharma, R. K., & Bisht, A. (2020). Effect of buoyancy and suction on Sisko nanofluid over a vertical stretching sheet in a porous medium with mass flux condition. *Indian Journal of Pure & Applied Physics*, 58(3), 178–188.
20. Sharma, R., & Bisht, A. (2019). MHD flow of Sisko nanofluid over a stretching sheet with Joule heating. In *AIP Conference Proceedings* (Vol. 2134, No. 1, p. 030002). AIP Publishing LLC.
21. Al-Mamun, A., Arifuzzaman, S. M., Reza-E-Rabbi, S., Biswas, P., & Khan, M. S. (2019). Computational modelling on MHD radiative Sisko nanofluids flow through a nonlinearly stretching sheet. *International Journal of Heat and Technology*, 37(1), 285–295.
22. Bisht, A., & Sharma, R. (2020). Comparative analysis of a Sisko nanofluid over a stretching cylinder through a porous medium. *Heat Transfer*, 49(6), 3477–3488.
23. Bisht, A., & Sharma, R. (2021). Entropy generation analysis in magnetohydrodynamic Sisko nanofluid flow with chemical reaction and convective boundary conditions. *Mathematical Methods in the Applied Sciences*, 44(5), 3396–3417.
24. Hayat, U., & Shahzad, A. (2023). Analysis of heat transfer and thin film flow of Au–Np over an unsteady radial stretching sheet. *Numerical Heat Transfer, Part A: Applications*, 1–14.
25. Khan, U., Zaib, A., Ishak, A., Al-Mubaddel, F. S., Bakar, S. A., Alotaibi, H., & Aljohani, H. M. (2021). Computational modeling of hybrid Sisko nanofluid flow over a porous radially heated shrinking/stretching disc. *Coatings*, 11(10), 1242.
26. Almaneea, A. (2022). Numerical study on thermal performance of Sisko fluid with hybrid nano-structures. *Case Studies in Thermal Engineering*, 30, 101754.
27. Bilal, M., & Saeed, A. (2022). Numerical computation for the dual solution of Sisko hybrid nanofluid flow through a heated shrinking/stretching porous disk. *International Journal of Ambient Energy*, 43(1), 8802–8811.
28. Takabi, B., & Salehi, S. (2014). Augmentation of the heat transfer performance of a sinusoidal corrugated enclosure by employing hybrid nanofluid. *Advances in Mechanical Engineering*, 6, 147059.
29. Khan, U., Zaib, A., Shah, Z., Baleanu, D., & Sherif, E. S. M. (2020). Impact of magnetic field on boundary-layer flow of Sisko liquid comprising nanomaterials migration through radially shrinking/stretching surface with zero mass flux. *Journal of Materials Research and Technology*, 9(3), 3699–3709.
30. Kumar, S., & Sharma, K. (2022). Entropy optimized radiative heat transfer of hybrid nanofluid over vertical moving rotating disk with partial slip. *Chinese Journal of Physics*, 77, 861–873.

14 Influence of Slip Mechanisms and Suction on Hybrid Nanofluid Flow via a Stretching Cylinder with Heat Generation

Himanshu Upreti, Alok Kumar Pandey, and Oluwole Daniel Makinde

14.1 INTRODUCTION

In fluid mechanics, boundary layer flow (BLF) over a stretching cylinder is an important research topic, especially for comprehending the behaviour of viscous fluids surrounding deforming solid surfaces. Numerous engineering domains use this phenomenon, such as polymer processing, fibre spinning, and coating procedures. Continuous stretching or elongation of a cylindrical surface in a fluid flow creates a boundary layer close to the surface where there are noticeable fluid velocity gradients. The knowledge acquired by researching BLF across a stretching cylinder can be applied to the optimization of fluid–solid interactions in industrial processes. Controlling the BLF across the spinning cylinder, for instance, can have a major impact on the morphology and properties of the fibre throughout the spinning process. Overall, studying BLF over a stretching cylinder provides valuable insights into fluid dynamics and contributes to the development of innovative technologies in various industrial applications. The BLF of a nanofluid over a cylinder under the influence of a magnetic field was examined by Ashorynejad et al. [1]. Ahmed et al. [2] examined the effects of BLF caused by a permeable stretching tube with a heat source/sink in the context of uncertainties in the thermal conductivity (TC) and dynamic viscosity (DV) of the nanofluid. Hayat et al. [3] examined the axisymmetric flow of a third-grade fluid across a stretching cylinder and discovered that the third-grade fluid has superior velocity profiles to both Newtonian and second-grade fluids. Furthermore, entropy generation in viscous flow across a cylinder with the aid of a magnetic field was studied by Butt et al. [4]. Later, Ganesh et al. [5] theoretically analysed heat transfer flow over a vertical stretching cylinder and revealed the increment in boundary layer thickness due to velocity slip. Khan et al. [6] studied

 DOI: 10.1201/9781003595786-14

the thermal slip magnetohydrodynamic (MHD) flow by a non-linear cylinder with radiation. Many more works of the literature on the study of BLF over a stretching cylinder can be found in [7–13]. By gaining insights into the behaviour of the boundary layer and its influence on overall flow patterns, engineers and researchers can develop more efficient and effective processes in various engineering applications.

Hybrid nanofluids (HNFs) represent a promising class of fluids with diverse applications across various fields, ranging from thermal management and energy conservation to biomedical and manufacturing applications. Hayat and Nadeem [14] analysed the heat transfer enhancement by considering water-based HNFs. In comparison to nanofluids, they achieved a higher rate of heat transmission. Using Xue and Yamada–Ota models, Abbas et al. [15] examined the stagnation point flow of an HNF across a moving cylinder in order to investigate the rate of heat transfer. They concluded a better heat transfer rate through the Yamada–Ota model than the Xue model. Later, Waqas et al. [16] examined the MHD flow of an engine oil-based HNF over a vertical stretching cylinder under the impact of thermal radiation. They disclosed a better temperature field due to the temperature ratio parameter. Recently, a comparison study between a nanofluid and an HNF over a stretching/shrinking cylinder was carried out by Yasir et al. [17], and the results showed that HNFs are better than nanofluids. Huminic and Huminic [18] studied the entropy generation of an HNF in thermal systems. More pieces of the literature can be seen to comprehend the characteristics of HNFs [19–25].

14.2 MATHEMATICAL FORMULATION

Let us assume that the two-dimensional laminar motion of nanofluid flow over a stretching cylinder is steady, incompressible, and of axisymmetric radius b. The axes (z,r) represent the horizontal and vertical directions of the cylinder (see Figure 14.1). It is assumed that (T_w, T_∞) are the surface and ambient temperature of the cylinder, respectively, where $(T_w - T_\infty) > 0$.

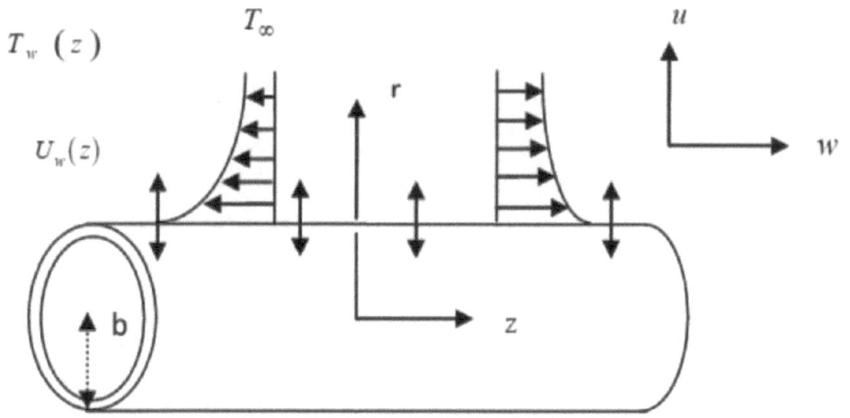

FIGURE 14.1 Geometry of the flow model.

The flow governing equations are [9]

$$\frac{\partial(rw)}{\partial z}+\frac{\partial(ru)}{\partial r}=0, \tag{14.1}$$

$$\frac{\rho_{hnf}}{\mu_{hnf}}\left(ww_z+uw_r\right)=w_{zz}+\frac{1}{r}w_r, \tag{14.2}$$

$$\frac{\rho_{hnf}}{\mu_{hnf}}\left(wu_z+uu_r\right)=-\frac{1}{\mu_{hnf}}P_r+u_{rr}+\frac{1}{r}u_r-\frac{1}{r^2}u, \tag{14.3}$$

$$\left(\rho C_p\right)_{hnf}\left(wT_z+uT_r\right)=k_{hnf}\left(T_{rr}+\frac{1}{r}T_r\right)+Q_0\left(T-T_\infty\right)+\mu_{hnf}\left(w_r\right)^2. \tag{14.4}$$

The boundary conditions (BCs) are

$$\left.\begin{array}{l} u=U_w,\, w-W_w=l_1\dfrac{\partial w}{\partial r},\, T-T_w=l_2\dfrac{\partial T}{\partial r} \text{ when } r=b \\ u=0,\, T=T_\infty \text{ when } r\to\infty \end{array}\right\}, \tag{14.5}$$

where (w,u) are the velocity components in the (z,r) directions, and (l_1,l_2) are the velocity slip and thermal slip factors. For no slip condition, both are zero.

The correlations of the thermophysical properties of an HNF are

$$\left.\begin{array}{l}
\dfrac{\mu_{hnf}}{\mu_f}=\left[\left(1-\phi_{Al_2O_3}\right)\left(1-\phi_{Cu}\right)\right]^{-2.5}, \\
\rho_{hnf}=\left[\left(1-\phi_{Cu}\right)\left\{\left(1-\phi_{Al_2O_3}\right)\rho_f+\phi_{Al_2O_3}\rho_{Al_2O_3}\right\}\right]+\phi_{Cu}\rho_{Cu}, \\
\left(\rho C_P\right)_{hnf}=\left[\left(1-\phi_{Cu}\right)\left\{\left(1-\phi_{Al_2O_3}\right)\left(\rho C_P\right)_f+\phi_{Al_2O_3}\left(\rho C_P\right)_{Al_2O_3}\right\}\right]+\phi_{Cu}\left(\rho C_P\right)_{Cu}, \\
\dfrac{\kappa_{hnf}}{\kappa_f}=\dfrac{\kappa_{hnf}}{\kappa_{bf}}\times\dfrac{\kappa_{bf}}{\kappa_f}, \text{ where } \dfrac{\kappa_{hnf}}{\kappa_{bf}}=\dfrac{\left(\kappa_{Cu}+2\kappa_{bf}\right)-2\phi_{Cu}\left(\kappa_{bf}-\kappa_{Cu}\right)}{\left(\kappa_{Cu}+2\kappa_{bf}\right)+\phi_{Cu}\left(\kappa_{bf}-\kappa_{Cu}\right)} \text{ and} \\
\qquad\qquad\qquad\qquad \dfrac{\kappa_{bf}}{\kappa_f}=\dfrac{\left(\kappa_{Al_2O_3}+2\kappa_f\right)-2\phi_{Al_2O_3}\left(\kappa_f-\kappa_{Al_2O_3}\right)}{\left(\kappa_{Al_2O_3}+2\kappa_f\right)+\phi_{Al_2O_3}\left(\kappa_f-\kappa_{Al_2O_3}\right)}
\end{array}\right\} \tag{14.6}$$

The similarity variables are expressed as

$$u=-\frac{ca}{\sqrt{\eta}}f(\eta),\, w=2zcf'(\eta),\, \eta=\left(\frac{r}{a}\right)^2,\, \theta(\eta)=\frac{T-T_\infty}{T_w-T_\infty}. \tag{14.7}$$

Substituting Equations (14.6) and (14.7) into Equations (14.2)–(14.4), we get the system of non-linear ODEs:

$$\frac{\mu_{hnf}}{\mu_f}\left[\eta f''' + f''\right] + \frac{\rho_{hnf}}{\rho_f} Re\left[ff'' - f'^2\right] = 0, \tag{14.8}$$

$$RePr\frac{\left(\rho C_p\right)_{hnf}}{\left(\rho C_p\right)_f} f\theta' + \frac{\mu_{hnf}}{\mu_f}\left(\eta PrEc\right)f''^2 + \frac{\kappa_{hnf}}{\kappa_f}\left(\eta\theta'' + \theta'\right) + \mathrm{Q}\theta = 0. \tag{14.9}$$

Now, Equation (14.5) is transformed into non-dimensional terms of BCs:

$$\left.\begin{array}{l} f(1) = \mathrm{F_w}, f'(1) = 1 + pf''(1), \theta(1) = 1 + q\theta'(1) \text{ when } \eta = 1, \\ f'(\infty) \to 0, \theta(\infty) \to 0 \text{ when } \eta \to \infty \end{array}\right\}, \tag{14.10}$$

where dimensionless parameters are defined as

$$\left.\begin{array}{l} Re = \dfrac{cb^2}{2\alpha_{bf}}, \mathrm{Pr} = \dfrac{\nu_{bf}}{\alpha_{bf}}, \mathrm{Q} = \dfrac{Q_0 b^2}{4\kappa_{bf}}, \mathrm{F_w} = -\dfrac{\sqrt{\eta}U_w}{cb}, \\ Ec = \dfrac{4\rho_{bf}\left(cz\right)^2}{\left(\rho C_p\right)_{bf}\left(T_w - T_\infty\right)}, p = \dfrac{2l_1}{b}, q = \dfrac{2l_2}{b} \end{array}\right\}. \tag{14.11}$$

The local Nusselt number can be stated as

$$Nu = -\frac{b\kappa_{hnf}}{\kappa_f\left(T_w - T_\infty\right)}\left(T_r\right)_{r=a}. \tag{14.12}$$

Now, using Equation (14.7) in Equation (14.12), we get

$$Nu = -\frac{2\kappa_{hnf}}{\kappa_{bf}}\theta'(1). \tag{14.13}$$

14.3 NUMERICAL SCHEME

Equations (14.8) and (14.9) with supportive Equation (14.10) are handled numerically by using the fourth- and fifth-order Runge–Kutta–Fehlberg (RKF) technique. Initially change the differential equations into first-order ordinary differential equations (ODEs).

Let $y_1 = \eta$, $y_2 = f$, $y_3 = f'$, $y_4 = f''$, $y_5 = \theta$, $y_6 = \theta'$.

Using the above substitutions, the existing ODEs are converted into non-linear first-order ODEs, which are written as

$$\begin{pmatrix} y_1' \\ y_2' \\ y_3' \\ y_4' \\ y_5' \\ y_6' \end{pmatrix} = \begin{pmatrix} 1 \\ y_3 \\ y_4 \\ \left(A_1 Re\left(y_3 y_3 - y_2 y_4\right) - A_3 y_4\right) / \left(y_1 A_3\right) \\ y_6 \\ \left(-A_4 y_6 - A_2 PrRe y_2 y_6 - Q y_5 - A_3 y_1 PrEc y_4 y_4\right) / \left(y_1 A_4\right) \end{pmatrix}, \quad (14.14)$$

and the initial conditions are

$$\begin{pmatrix} y_1 \\ y_2 \\ y_3 \\ y_4 \\ y_5 \\ y_6 \end{pmatrix} = \begin{pmatrix} 0 \\ F_w \\ 1 + p\alpha \\ \alpha \\ 1 + q\beta \\ \beta \end{pmatrix}. \quad (14.15)$$

The ODEs (14.14) with BCs (14.15) are solved using the fourth- and fifth-order RKF scheme. Here, we regulate α and β until the conditions are satisfied, i.e., $f'(\infty) = 0$ and $\theta(\infty) = 0$.

14.4 RESULTS AND DISCUSSION

The outcomes of non-dimensional terms, i.e., $f'(\eta)$ and $\theta(\eta)$, are revealed by graphs for various values of acting parameters. In the entire computation, $Pr = 6.2$ is fixed. The thermophysical properties of the solid particles and the base fluid are shown in Table 14.1.

Figure 14.2 and 14.3 show the influence of the Reynolds number (Re) on the velocity and temperature profiles. It is observed from these figures that both velocity

TABLE 14.1
Thermophysical Properties of Nanoparticles and the Base Fluid [26]

	Base fluid	Nanoparticle	
Physical property	**Water**	**Cu**	**Al_2O_3**
ρ (kg/m³)	997.1	8933	3970
C_p (J/kgK)	4179	385	765
κ (W/mK)	0.613	400	40

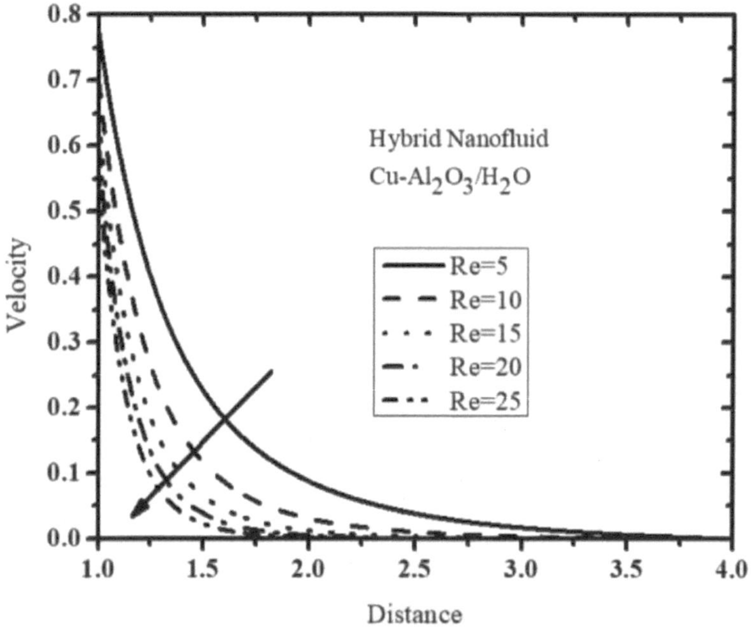

FIGURE 14.2 Variation in velocity due to Re.

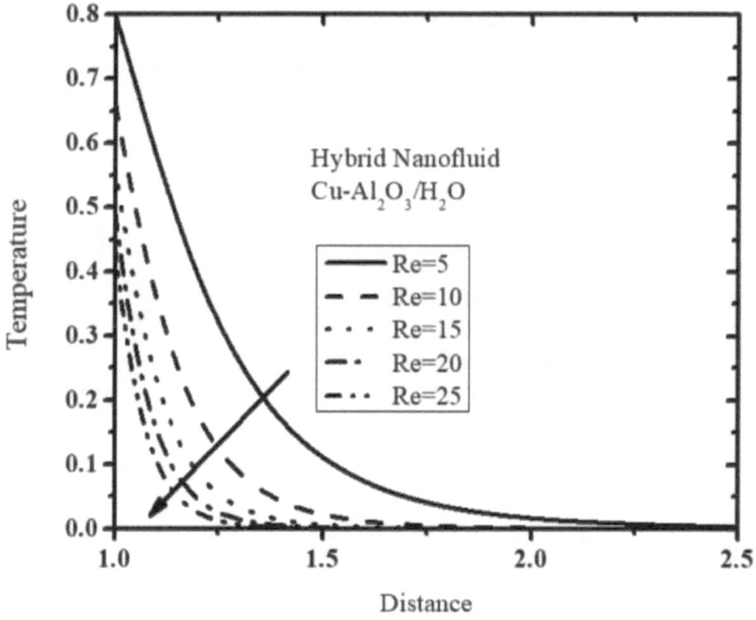

FIGURE 14.3 Variation in temperature due to Re.

and temperature of the fluid (HNF) decrease gradually with an increase in the values of the Reynolds number. Also, the associated boundary layers become thinner with an increase of the Reynolds number.

Figures 14.4 and 14.5 display the effect of increasing suction parameter $(F_w > 0)$ on the velocity and temperature profiles. The figure reveals that an increase in F_w

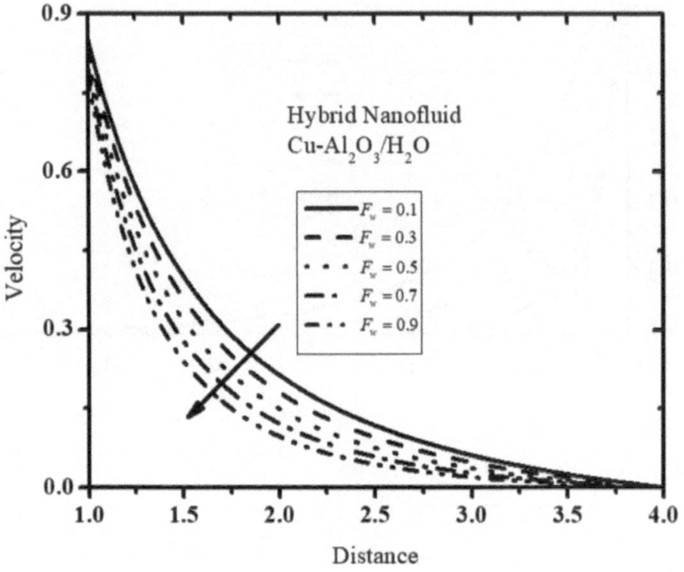

FIGURE 14.4 Variation in velocity due to F_w.

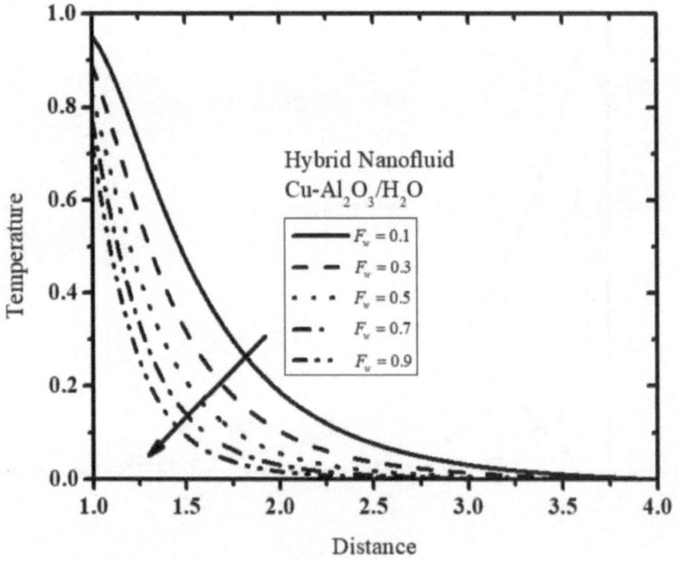

FIGURE 14.5 Variation in temperature due to F_w.

from 0.1 to 0.9 results in the deceleration of the fluid movement and also reduces the temperature of the fluid.

The effect of velocity (p) and temperature (q) slips on flow characteristics (velocity and temperature) is elucidated through Figures 14.6–14.8. The figures illustrate that increasing the velocity slip parameter (p) from 0 to 0.5 results in the decline of the velocity profile (see Figure 14.6), whereas the temperature profile shows dual nature, i.e., for $\eta < 2.5$ it decreases with an increase in p, while for $\eta > 3.5$ it increases (see Figure 14.7). However, the variation in the temperature profile for distinct p for $\eta > 3.5$ is very less. In addition, the fluid temperature decreases with an increasing value of the thermal slip parameter (see Figure 14.8).

Figures 14.9–14.11 illustrate the influence of dissipation parameter (Ec), heat generation parameter (Q), and nanoparticle concentration ($\phi_{Al_2O_3}$ and ϕ_{Cu}) on the temperature profile of an HNF. It can be perceived from Figures 14.9 and 14.10 that the temperature of the moving fluid increases with an increase in Ec and in Q. A similar pattern was observed for increasing nanoparticle concentration in the base fluid (see Figure 14.11). From this figure, it was noticed that increasing the concentration of either dispersed nanoparticle results in the increase in temperature. Also, the associated boundary layer becomes thicker with increasing $\phi_{Al_2O_3}$ and ϕ_{Cu}.

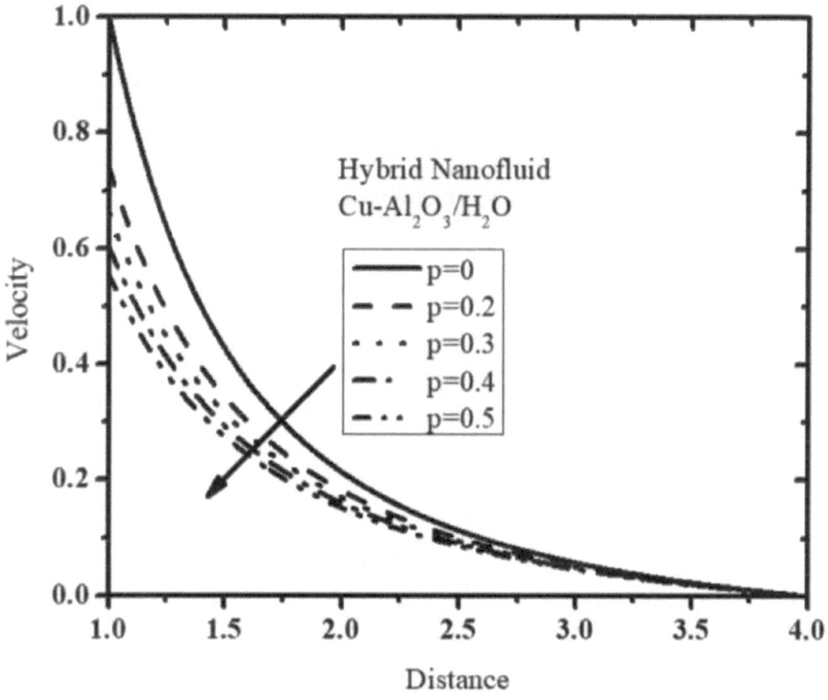

FIGURE 14.6 Variation in velocity due to p.

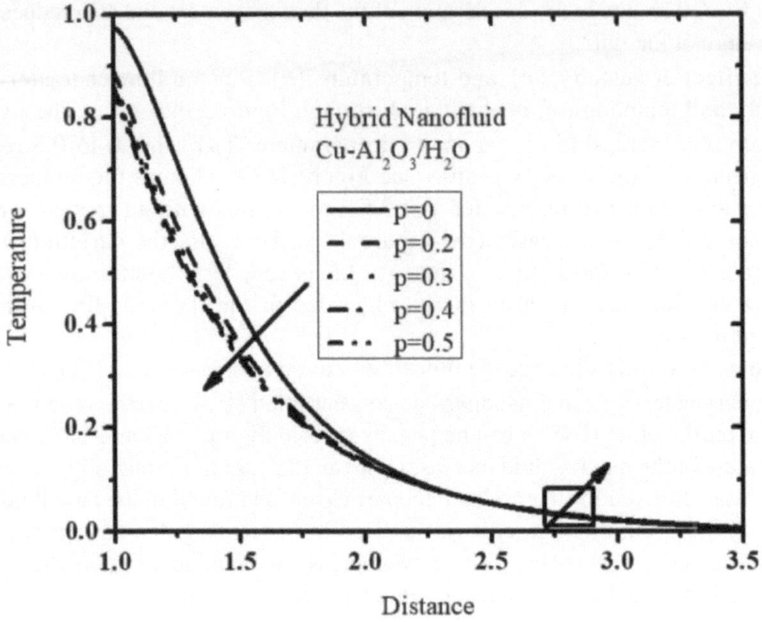

FIGURE 14.7 Variation in temperature due to p.

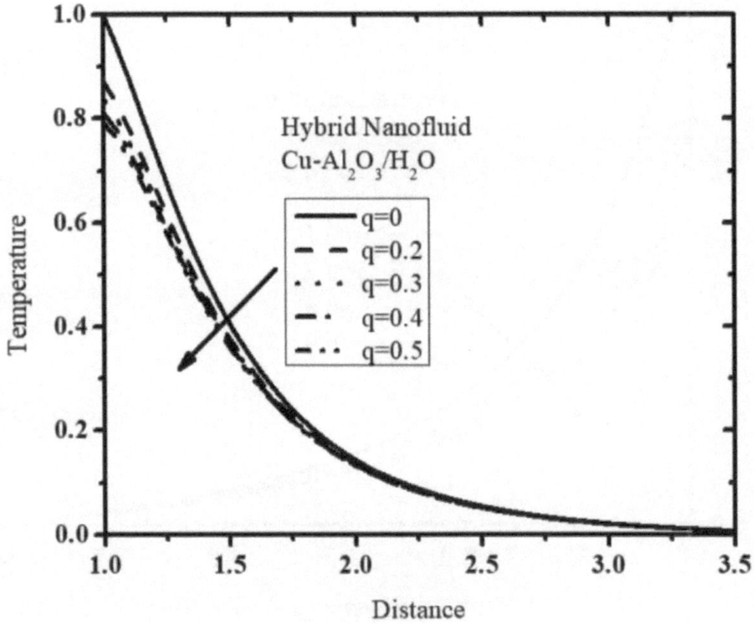

FIGURE 14.8 Variation in temperature due to q.

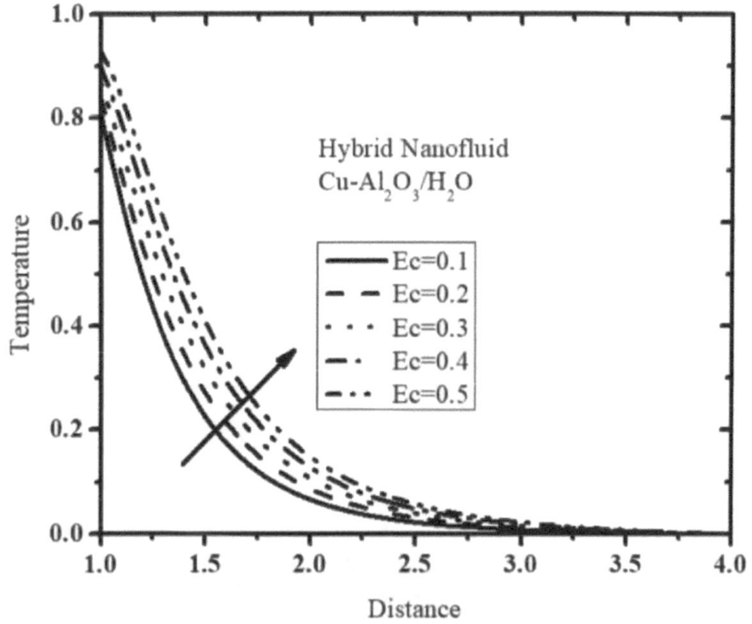

FIGURE 14.9 Variation in temperature due to Ec.

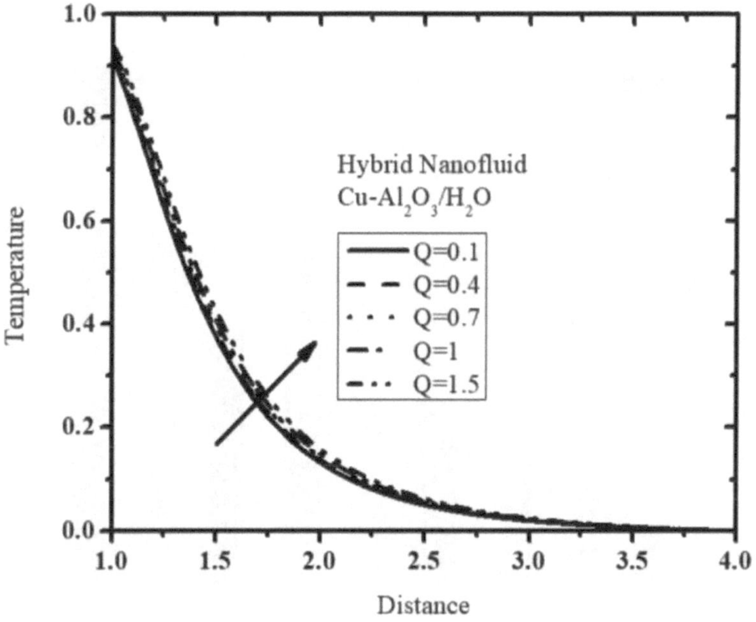

FIGURE 14.10 Variation in temperature due to Q.

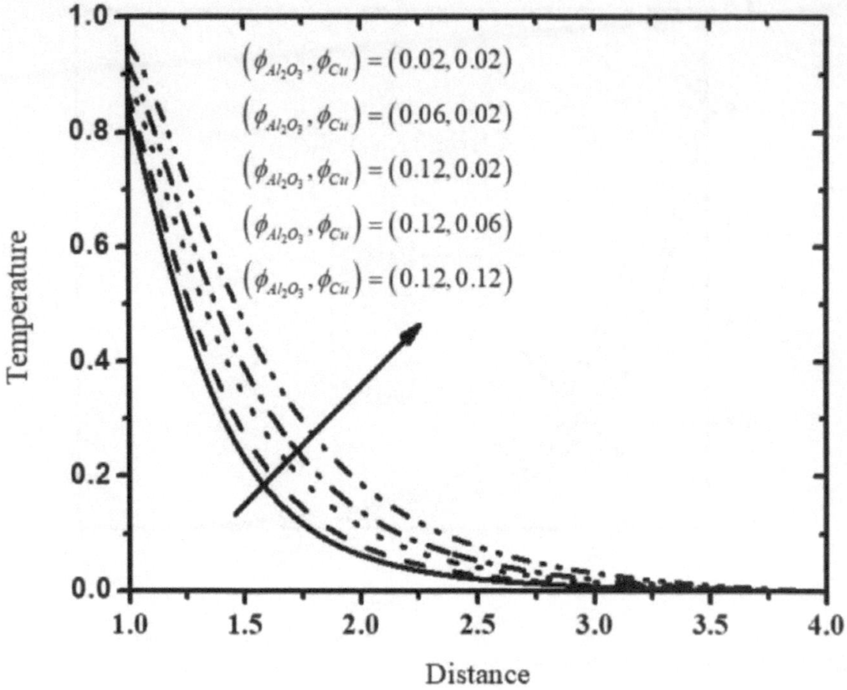

FIGURE 14.11 Variation in temperature due to $\phi_{Al_2O_3}$ and ϕ_{Cu}

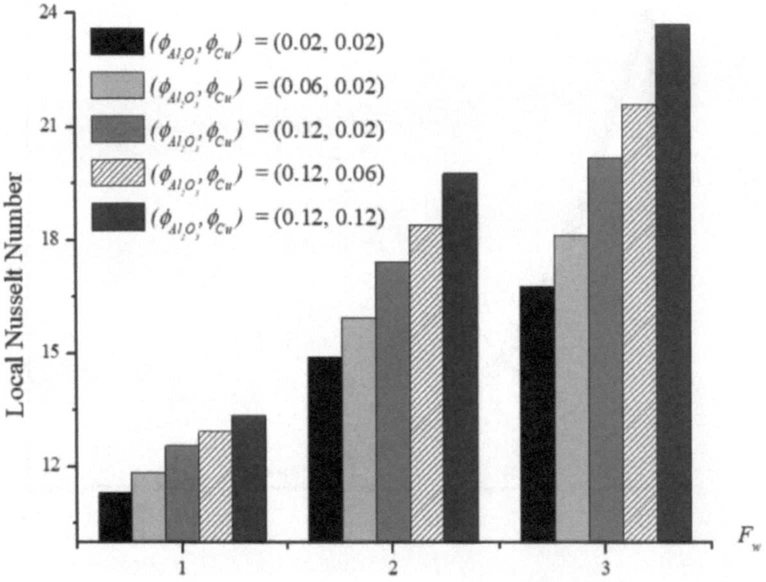

FIGURE 14.12(a) Variation in LNN due to $\phi_{Al_2O_3}$, ϕ_{Cu} and F_w

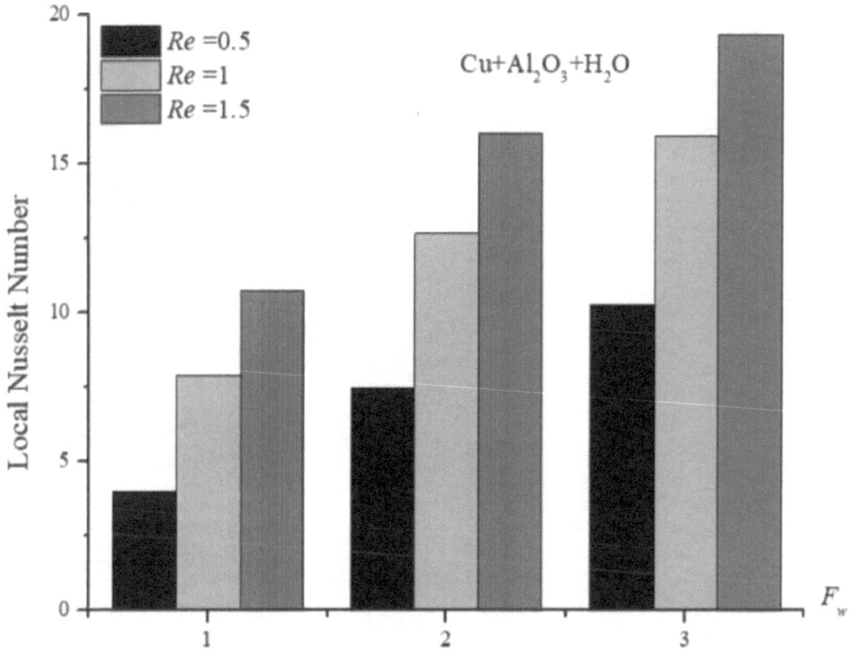

FIGURE 14.12(b) Variation in LNN due to Re and F_w.

Local flux parameter (LNN) is plotted against F_w to show the influence of increasing nanoparticle concentration and Reynolds number (see Figures 14.12 (a, b)). The figures show that with increasing values of the concerned parameter, the LNN increases. Figures 14.13 (a, b) present a comparison of the LNN for different working fluids (H_2O, Cu, Ag, and hybrid) when plotted against F_w and Re. These plots show that HNFs have superior LNN, followed by mono-nanofluids and water.

14.5 CONCLUSIONS

The aim of this work was to investigate the impact of velocity and thermal slip parameters on Cu-Al_2O_3/H_2O flow over a stretching cylinder with heat generation, suction, and viscous dissipation. The numerical outcomes were obtained using the fourth- to fifth-order RKF technique. The key results of this study are as follows:

- The heat transfer performance of an HNF (Cu-Al_2O_3/H_2O) is better than that of mono-nanofluids and conventional fluids.
- The velocity outlines declined as the velocity slip parameter increased.
- Both velocity and temperature outlines declined with an increase in the Reynolds number and suction parameter values.
- On increasing the values of the heat generation parameter and Eckert number, the temperature profile accelerated.

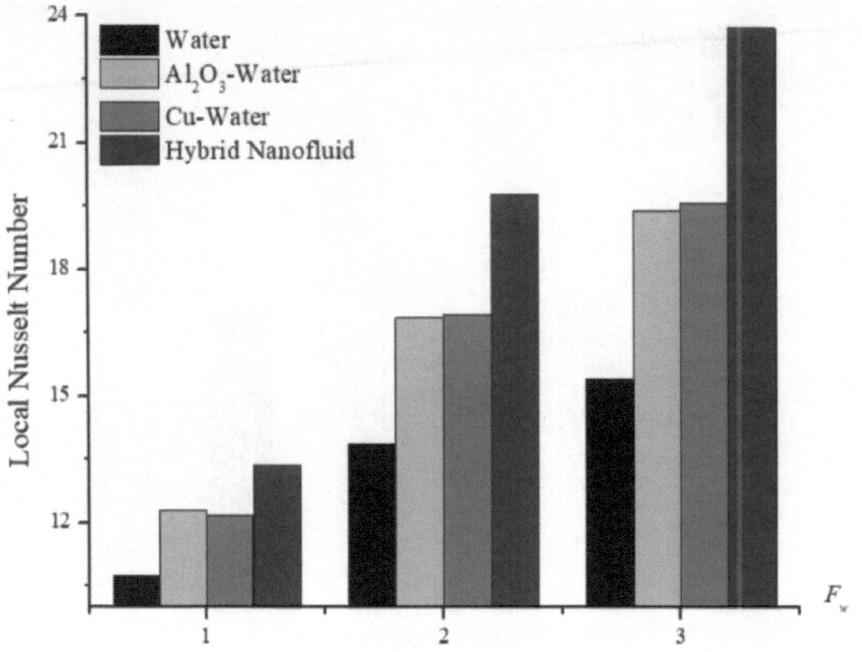

FIGURE 14.13(a) Variation in LNN due to F_w for different fluids.

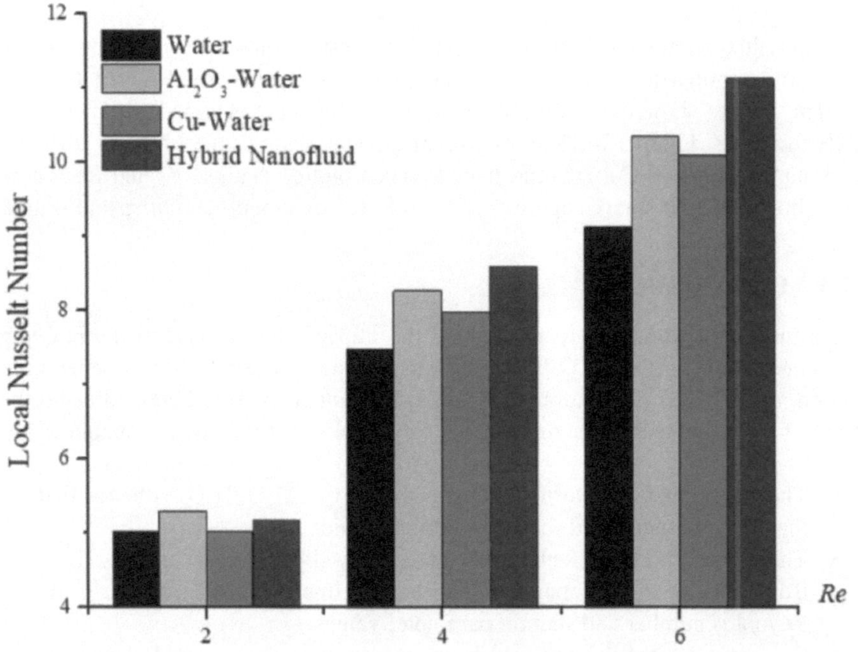

FIGURE 14.13(b) Variation in LNN due to Re for different fluids.

- The temperature profile enhanced due to an increase in the volume fraction of solid particles, i.e., Cu and Al_2O_3.

REFERENCES

1. Ashorynejad, H. R., Sheikholeslami, M., Pop, I. and Ganji, D. D. (2013). Nanofluid flow and heat transfer due to a stretching cylinder in the presence of magnetic field. Heat and Mass Transfer, 49(3), 427–436.
2. Ahmed, S. E., Hussein, A. K., Mohammed, H. A. and Sivasankaran, S. (2014). Boundary layer flow and heat transfer due to permeable stretching tube in the presence of heat source/sink utilizing nanofluids. Applied Mathematics and Computation, 238, 149–162.
3. Hayat, T., Shafiq, A. and Alsaedi, A. (2015). MHD axisymmetric flow of third-grade fluid by a stretching cylinder. Alexandria Engineering Journal, 54(2), 205–212.
4. Butt, A. S., Ali, A. and Mehmood, A. (2016). Numerical investigation of magnetic field effects on entropy generation in viscous flow over a stretching cylinder embedded in a porous medium. Energy, 99, 237–249.
5. Ganesh, N. V., Ganga, B., Hakeem, A. A., Saranya, S. and Kalaivanan, R. (2016). Hydromagnetic axisymmetric slip flow along a vertical stretching cylinder with convective boundary condition. St. Petersburg Polytechnical University Journal: Physics and Mathematics, 2(4), 273–280.
6. Khan, M. I., Tamoor, M., Hayat, T. and Alsaedi, A. (2017). MHD boundary layer thermal slip flow by nonlinearly stretching cylinder with suction/blowing and radiation. Results in Physics, 7, 1207–1211.
7. Hussain, A., Malik, M. Y., Salahuddin, T., Bilal, S. and Awais, M. (2017). Combined effects of viscous dissipation and Joule heating on MHD Sisko nanofluid over a stretching cylinder. Journal of Molecular Liquids, 231, 341–352.
8. Ishak, A. and Nazar, R. (2009). Laminar boundary layer flow along a stretching cylinder. European Journal of Scientific Research, 36(1), 22–29.
9. Pandey, A. K. and Kumar, M. (2017). Natural convection and thermal radiation influence on nanofluid flow over a stretching cylinder in a porous medium with viscous dissipation. Alexandria Engineering Journal, 56(1), 55–62.
10. Pandey, A. K. and Kumar, M. (2017). Boundary layer flow and heat transfer analysis on Cu-water nanofluid flow over a stretching cylinder with slip. Alexandria Engineering Journal, 56(4), 671–677.
11. Wang, C. Y. (1988). Fluid flow due to a stretching cylinder. The Physics of Fluids, 31(3), 466–468.
12. Sheikholeslami, M. (2015). Effect of uniform suction on nanofluid flow and heat transfer over a cylinder. Journal of the Brazilian Society of Mechanical Sciences and Engineering, 37(6), 1623–1633.
13. Ishak, A., Nazar, R. and Pop, I. (2008). Uniform suction/blowing effect on flow and heat transfer due to a stretching cylinder. Applied Mathematical Modelling, 32(10), 2059–2066.
14. Hayat, T. and Nadeem, S. (2017). Heat transfer enhancement with Ag–CuO/water hybrid nanofluid. Results in Physics, 7, 2317–2324.
15. Abbas, N., Nadeem, S., Saleem, A., Malik, M. Y., Issakhov, A. and Alharbi, F. M. (2021). Models base study of inclined MHD of hybrid nanofluid flow over nonlinear stretching cylinder. Chinese Journal of Physics, 69, 109–117.
16. Waqas, H., Naqvi, S. M. R. S., Alqarni, M. S. and Muhammad, T. (2021). Thermal transport in magnetized flow of hybrid nanofluids over a vertical stretching cylinder. Case Studies in Thermal Engineering, 27, 101219.

17. Yasir, M., Malik, Z. U., Alzahrani, A. K. and Khan, M. (2023). Study of hybrid Al_2O_3-Cu nanomaterials on radiative flow over a stretching/shrinking cylinder: Comparative analysis. Ain Shams Engineering Journal, 14(9), 102070.
18. Huminic, G. and Huminic, A. (2020). Entropy generation of nanofluid and hybrid nanofluid flow in thermal systems: A review. Journal of Molecular Liquids, 302, 112533.
19. Sajid, M. U. and Ali, H. M. (2018). Thermal conductivity of hybrid nanofluids: A critical review. International Journal of Heat and Mass Transfer, 126, 211–234.
20. Suresh, S., Venkitaraj, K. P., Selvakumar, P. and Chandrasekar, M. (2012). Effect of Al_2O_3–Cu/water hybrid nanofluid in heat transfer. Experimental Thermal and Fluid Science, 38, 54–60.
21. Waini, I., Ishak, A., Groşan, T. and Pop, I. (2020). Mixed convection of a hybrid nanofluid flow along a vertical surface embedded in a porous medium. International Communications in Heat and Mass Transfer, 114, 104565.
22. Rafique, K., Mahmood, Z. and Khan, U. (2023). Mathematical analysis of MHD hybrid nanofluid flow with variable viscosity and slip conditions over a stretching surface. Materials Today Communications, 36, 106692.
23. Ekiciler, R., Arslan, K., Turgut, O. and Kurşun, B. (2021). Effect of hybrid nanofluid on heat transfer performance of parabolic trough solar collector receiver. Journal of Thermal Analysis and Calorimetry, 143, 1637–1654.
24. Alsaedi, A., Muhammad, K. and Hayat, T. (2022). Numerical study of MHD hybrid nanofluid flow between two coaxial cylinders. Alexandria Engineering Journal, 61(11), 8355–8362.
25. Maskeen, M. M., Zeeshan, A., Mehmood, O. U. and Hassan, M. (2019). Heat transfer enhancement in hydromagnetic alumina–copper/water hybrid nanofluid flow over a stretching cylinder. Journal of Thermal Analysis and Calorimetry, 138, 1127–1136.
26. Aladdin, N. A. L., Bachok, N. and Pop, I. (2020). Cu-Al_2O_3/water hybrid nanofluid flow over a permeable moving surface in presence of hydromagnetic and suction effects. Alexandria Engineering Journal, 59(2), 657–666.

15 Study on Effects of Uncertain Volume Fraction on Hybrid Nanofluid Flow Using the Homotopy Analysis Method

Ramakanta Meher, Lalchand Verma, and Parthkumar P. Sartanpara

Nomenclature

r	Radial coordinate
θ	Angular coordinate
v	Radial velocity
P	Pressure
μ_{nf}	Effective dynamic viscosity
ρ_{nf}	Effective density
ν_{nf}	Kinematic viscosity
Re	Reynolds number
ϕ	Nanoparticle volume fraction

Subscripts

f	Fluid
nf	Nanofluid
s	Solid

15.1 INTRODUCTION

In recent years, there has been a notable surge in scholarly attention towards the investigation of non-linear phenomena, specifically the magnetohydrodynamic

DOI: 10.1201/9781003595786-15

(MHD) Jeffery–Hamel flow problems. Consequently, the field of nanofluidics has gained considerable importance in the field of computational fluid dynamics applications [1, 2]. The phenomenon of incompressible fluid flow in a convergent–divergent channel has garnered significant attention from numerous scientists in recent times. The transportation of fluid across two plates is a commonly employed example and application in the fields of industrial and mechanical engineering. For example, in the production of polymers using hot and cold procedures, converging dies, and similar methods, pressure-driven particles are transferred along channels that exhibit symmetrical convergence and divergence. Initially, research was done on the fluid flow between two sloped planes. Jeffery and Hamel first discussed the Jeffery–Hamel problem [3, 4], which introduced the issue of fluid flow across several diverging channels. MHD Jeffery–Hamel nanofluids have many applications in engineering domains such as aircrafts, thermal management systems, and microfluidics. In the aerospace industry, the utilization of MHD Jeffery–Hamel nanofluids has the potential to optimize cooling systems within propulsion systems and improve thermal management in spacecrafts. In microfluidic devices, the precise control of nanofluids under MHD conditions in Jeffery–Hamel flows can enhance mixing efficiency or ease particle separation procedures. In the literature, a number of numerical and analytical methods were put out to address these challenging computational fluid dynamical problems [5, 6]. Numerical solutions have the potential to tackle nonlinear fluid dynamics problems in the presence of high magnetic fields, including nanomaterials [7, 8]. These current well-defined methodologies have benefits, uses, and limitations. A MHD field can influence this fluid flow [9, 10]. The term MHD originally appeared in the work by Alfvên [11]. According to the MHD theory, an electrical flow that is generated in a high magnetic field exerts pressure on the ions of the conductive fluid [12, 13]. Accelerators, liquid metal cooling system design, flow meters, MHD power production, and pumps are just a few of the applications that have benefited from studies based on MHD channels.

As a novel category of working fluids, hybrid nanofluids are distinguished by the existence of two distinct types of nanoparticles, each of which has a diameter of less than 100 nm. This is the defining characteristic of hybrid nanofluids. Conventional fluids, including water, lubricants, ethylene glycol, and biological fluids alike, contain nanoparticles that are suspended within them. In comparison to the thermophysical properties and thermal conductivity of traditional fluids, this cutting-edge nanofluid possesses superior qualities in both of these areas. In recent years, it has been revealed that hybrid nanofluids can be useful in a wide variety of applications that include heat transfer. Heat pipe technology, solar energy conversion, refrigeration, heating systems, heat exchangers, ventilation systems, air conditioning systems, coolant systems in machining and manufacturing, biomedical applications, space exploration, maritime transportation, military systems, and a great number of other applications are included in this category of applications.

The investigation of hybrid nanofluids, an enhanced variant of nanofluids, has been stimulated by the increasing fascination with creating nanofluids when various nanoparticles are dissolved in a base fluid. The production of nanofluids can be achieved by either a one-step or a two-step method, resulting in hybrid nanofluids. The approach for the one-step process was devised by Ali et al. [14] by the utilization of vacuum-induced evaporation of a metal container. The minimum density of nanoparticles was achieved with the repetition of this method. In contrast to the one-step process, the two-step procedure involved distributing nanoparticles into dry powders. According to the study conducted by Jamaludin et al. [15], the utilization of a hybrid nanofluid resulted in enhanced thermal conductivity and facilitated effective heat transfer within a heat exchanger. According to Mishra and Upreti [16], hybrid nanofluid applications have the potential to be utilized in the development of electronic coolants, refrigeration systems, cars, and heat conversion technologies. According to Ahmad et al. [17], titanium oxide (TiO_2) exhibits favourable thermal characteristics and demonstrates rapid dissolution in host fluids. On the other hand, copper (Cu) nanoparticles are metallic nanostructures that possess exceptional electrical conductivity. It is possible that the combination of the nanoparticles discussed above may improve flow stability and attain exceptional thermal conductivity properties. An analysis of the heat transfer rate of a hybrid nanofluid conducted by Gupta and Misra [18] revealed that the combination of copper-titanium oxide ($Cu\text{-}TiO_2$/water) with a water-based fluid offers enhanced stability, improved thermal conductivity, and reduced thermal resistance. This makes it advantageous for the microelectronics cooling devices market due to its small size and exceptional surface stability properties. The widespread utilization of hybrid nanofluids in real-world applications necessitates the removal of many hurdles and limits. These concerns encompass stability, dispersion of particles, aggregation, and sedimentation. Additionally, they emphasize the necessity for dependable and cost-effective synthesis techniques. Hence, further investigation is warranted to enhance the development and refinement of hybrid nanofluids, and to evaluate their efficacy across diverse scenarios and applications. Recent research has focused on several hybrid nanofluids, including Al_2O_{3}-Cu, Cu-Ag, and Ag-Al_2O_3.

This chapter comprises multiple sections. Section 15.2 provides an in-depth analysis of the fundamental concept of a fuzzy number, along with the β-cut and double parametric formulation (DPF). On the other hand, Section 15.3 delves into the fundamental mathematical structure of the Jeffery–Hamel nanofluid flow. The fuzzy controlling system of differential Equation 15.4 is solved using the double-parametric method based on fuzzy homotopy analysis. Section 15.6 examines the findings and analysis. Section 15.7 provides an overview of the primary findings.

15.2 BASIC CONCEPTS

This section provides an overview of the fundamental notion of fuzzy sets and explores its properties that can be utilized for subsequent computational purposes.

Definition 1. [19] Let X be a universal set and $Z : X \to [0,1]$, then A fuzzy set $\tilde{B}$ can be defined as a collection of a set of ordered pairs:

$$\tilde{B} = \{(p_0, z(p_0)) : p_0 \in X, z(p_0) \in [0,1]\},$$

where z is the membership function.

Definition 2. [20] A fuzzy number is a fuzzy set $\tilde{B}$ if its membership function $z: R \to [0,1]$, satisfying the following conditions:

i. It is fuzzy convex, i.e., $z(kp_0 + (1-k)q_0) \geq \{z(p_0), z(q_0)\}$, where $k \in [0,1]$.
ii. It is a normal, i.e., $\exists z(x_0) = 1$.
iii. The membership function is piece-wise continuous.
iv. The closure of Supp $(\tilde{B}) = \{p \in R, z(p) > 0\}$ is compact.

Definition 3. [21, 22] Triangular Fuzzy number (TFN) $\tilde{B} = [e, s, r]$ is a fuzzy set on set of real numbers in which its membership function is defined as follows:

$$z(x) = \begin{cases} 0 & x \leq e \\ \dfrac{x-e}{s-e} & e \leq x \leq s \\ \dfrac{r-x}{r-s} & s \leq x \leq r \\ 0 & x \geq r \end{cases}.$$

Definition 4. [23, 24] Let $z \in \tilde{B}$ and $\beta \in [0,1]$, where β is a crisp and defined as

$$\widetilde{B_\beta} = \{x \in X : z(x) \geq \beta, \beta \in [0,1]\}.$$

A given TNF $\tilde{B} = [e, s, r]$ can be written in an interval form by applying the idea of "β-cut," i.e., $B_\beta = [\underline{B}(\beta), \overline{B}(\beta)] = [(s-e)\beta + e, r - (r-s)\beta]$.

Definition 5. [25, 26] In a fuzzy, β-cuts can be expressed as an interval form $J = [\underline{J}, \overline{J}]$. The J can be written as

$$J = \delta(\overline{J} - \underline{J}) + \underline{J},\ \delta \in [0,1].$$

The TFN may also be expressed as a discretized parametric form with two parameters, β and δ, which is referred to as the double-parametric form concept.

Definition 6. [27, 28] Let us consider an n-th order fuzzy initial value problem of the form

$$\begin{cases} X^n(t) = F(t, X(t), X'(t), \ldots, X^{n-1}(t)) \\ X(t_0) = X_n \end{cases},$$

where $F : [t_0, T] \times E^n \to E$ is continuous and $X_n \in E; E$ is a fuzzy number, $n = 1, 2, .., i$. This fuzzy IVP has a unique solution if it satisfies the following properties:

i. Let $F : [t_0, T] \times E^n \to E$ is continuous and bounded. Then there exists at least one solution to the fuzzy initial value problem on the interval $[t_0, T]$.

ii. The fuzzy IVP has a unique solution if the function $F:[t_0,T]\times E^n \to E$ satisfies the Lipschitz condition, i.e.,

$$d_\infty\left(F\left(t,X,X',..X^{n-1}\right),F\left(t,Y,Y',...,Y^{n-1}\right)\right)$$
$$\leq k_1 d_\infty\left(X,Y\right)+k_2 d_\infty\left(X',Y'\right)+...+k_n d_\infty\left(X^{n-1},Y^{n-1}\right)$$

15.3 FORMULATION OF THE PROBLEM

This study considers the case where a source or sink is situated between two walls that are not perpendicular to one another, resulting in a continuous laminar flow of Cu-water and hybrid Cu-Al_2O_3 nanofluid in two dimensions. The angle between the walls is 2ω, as indicated in Figure 15.1. In this instance, it is presumed that the velocity is entirely radial. This radial velocity is dependent on the angle θ and r. This demonstrates that the flow parameter does not experience any changes in the z direction. It is also considered that these channels are convergent and divergent, respectively, as per the sign of "$\omega < 0$ (convergent channel) or $\omega > 0$ (divergent channel)." Here, the fluid flow with nanoparticles is considered to be "purely radial" in nature and relies only on "r and θ" so that "$v=\left(\vartheta(r,\theta),0\right)$." Thus, the reduced equation of continuity and the Navier–Stokes equations in polar coordinates [29] can be expressed as

$$\frac{\rho_{nf}}{r}\frac{\partial\left(r\vartheta(r,\theta)\right)}{\partial r}=0, \tag{15.1}$$

$$\vartheta(r,\theta)\frac{\partial\vartheta}{\partial r}=-\frac{1}{\rho_{nf}}\frac{\partial P}{\partial r}+\mu_{nf}\left[\frac{\partial^2\vartheta}{\partial r^2}+\frac{1}{r}\frac{\partial\vartheta}{\partial r}+\frac{1}{r^2}\frac{\partial^2\vartheta}{\partial\theta^2}-\frac{\vartheta}{r^2}\right]-\frac{\sigma_{nf}B_0^2}{\rho_{nf}r^2}\vartheta, \tag{15.2}$$

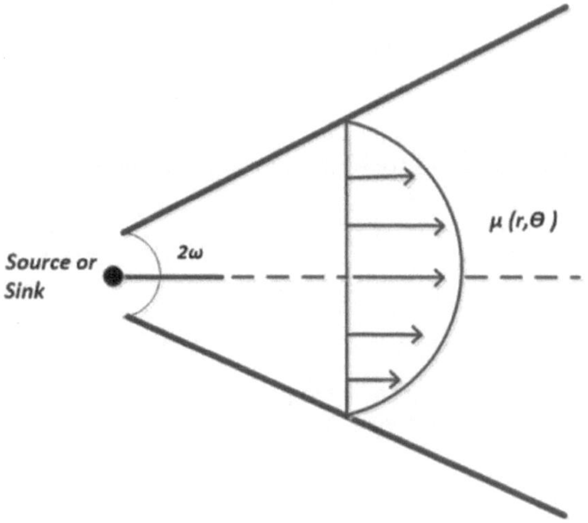

FIGURE 15.1 Diagram of the Jeffery–Hamel flow.

$$\frac{1}{r\rho_{nf}}\frac{\partial P}{\partial \theta}-\frac{2\nu_{nf}}{r^2}\frac{\partial \vartheta}{\partial \theta}=0, \tag{15.3}$$

$$\begin{aligned}\vartheta(r,\theta)\frac{\partial T}{\partial r}&=\frac{k_{nf}}{(\rho C_P)_{nf}}\left(\frac{\partial^2 T}{\partial r^2}+\frac{1}{r}\frac{\partial T}{\partial r}+\frac{1}{r^2}\frac{\partial^2 T}{\partial \theta^2}\right)\\&+\frac{k_{nf}}{(\rho C_P)_{nf}}\left[4\left\{\left(\frac{\partial (r\vartheta)}{\partial r}\right)^2+\frac{1}{r^2}\left(\frac{\partial (r\vartheta)}{\partial \theta}\right)^2\right\}\right]\end{aligned} \tag{15.4}$$

having boundary conditions

$$\begin{aligned}\vartheta=U_{\max},\frac{\partial \vartheta}{\partial \theta}&=0,\frac{\partial T}{\partial \theta}=0,\text{at } \theta=0,\\ \vartheta&=0,T=T_w,\text{ at } \theta=\omega\end{aligned}. \tag{15.5}$$

The physical variation in parameters is given by

$$\begin{cases}\rho_{nf}=\rho_f(1-\phi)+\rho_s\phi\\ \mu_{nf}=\dfrac{\mu_f}{(1-\phi)^{2.5}}\\ (\rho C_P)_{nf}=(1-\phi)(\rho C_P)_f+\phi(\rho C_P)_s\\ \dfrac{\sigma_{nf}}{\sigma_f}=\dfrac{(\sigma_s)+2\sigma_f+2(\sigma_s-\sigma_f)\phi}{(\sigma_s)+2\sigma_f-(\sigma_s-\sigma_f)\phi}\end{cases}. \tag{15.6}$$

The "effective thermal conductivity of the nanofluid" can be determined by

$$\frac{k_{nf}}{k_f}=\left(\frac{k_s+2k_f-2\phi(k_f-k_s)}{k_s+2k_f+\phi(k_f-k_s)}\right). \tag{15.7}$$

From Equation (15.1) and using non-dimensional parameters, we get

$$f(\theta)=r\vartheta, \tag{15.8}$$

$$\xi=\frac{\theta}{\omega},\eta(\xi)=\frac{f(\theta)}{f_{\max}},\kappa=\frac{T}{T_w}. \tag{15.9}$$

Now, the dimensionless velocity and temperature profiles are obtained by substituting Eqluations (15.1) and (15.2) in Equations (15.8) and (15.9), respectively, by removing the pressure *P*.

$$\eta'''(\xi)+2\omega\mathrm{Re}\Omega_f(1-\phi)^{2.5}\eta(\xi)\eta'(\xi)+\left(4-(1-\phi)^{2.5}Ha\right)\omega^2\eta'(\xi)=0, \tag{15.10}$$

$$\Omega_k \kappa''(\xi) + \frac{E_c P_r}{(1-\phi)^{2.5}} \Omega_c \left[4\omega^2 \eta^2 + (\eta')^2\right] = 0, \tag{15.11}$$

where the parameters $\Omega_f, \Omega_c, \Omega_\sigma$, and Ω_k, *Re* and *Ha* are

$$\Omega_f = (1-\phi) + \frac{\rho_s}{\rho_f}\phi, \tag{15.12}$$

$$\Omega_c = (1-\phi) + \frac{(\rho C_P)_s}{(\rho C_P)_f}\phi, \tag{15.13}$$

$$\Omega_\sigma = \frac{\sigma_{nf}}{\sigma_f} = \frac{\sigma_s + 2\sigma_f + 2\phi(\sigma_s - \sigma_f)}{\sigma_s + 2\sigma_f - \phi(\sigma_s - \sigma_f)}, \tag{15.14}$$

$$\Omega_k = \frac{k_{nf}}{k_f} = \left(\frac{(k_s + 2k_f) - \phi 2(k_f - k_s)\phi}{k_s + 2k_f + \phi(k_f - k_s)}\right), \tag{15.15}$$

$$Re = \frac{f_{\max}\omega}{\nu_f} = \frac{U_{\max} r\omega}{\nu_f}, \tag{15.16}$$

$$Ha = B_0 \sqrt{\frac{\sigma}{\rho_f \nu_f}}, \tag{15.17}$$

subject to the condition

$$\begin{aligned} &\eta(0) = 1, \eta'(0) = 0, \kappa'(0) = 0 \\ &\eta(1) = 0, \tilde{\kappa}(1) = 1. \end{aligned} \tag{15.18}$$

An alteration in the volume fraction is expected to have an impact on the velocity and temperature distributions of the fluid flow problems. In order to gain a comprehensive understanding of the behaviour of the magnetohydrodynamic (MHD) nanofluid flow, we examine the nanoparticle's volume fraction as an unforeseeable parameter in this context. The term TNF is employed to represent this parameter that is subject to uncertainty."

Equations (15.10) and (15.11) are derived in the fuzzy form as follows:

$$\tilde{\eta}'''(\xi) + 2\omega \mathrm{Re}\Omega_f (1-\phi)^{2.5} \tilde{\eta}(\xi)\tilde{\eta}'(\xi) + \left(4 - (1-\phi)^{2.5} Ha\right)\omega^2 \tilde{\eta}'(\xi) = 0, \tag{15.19}$$

$$\frac{k_{nf}}{k_f}\tilde{\kappa}''(\xi) + \frac{E_c P_r}{(1-\phi)^{2.5}} \Omega_c \left[4\omega^2 \tilde{\eta}^2 + (\tilde{\eta}')^2\right] = 0 \tag{15.20}$$

having subject to conditions

$$\begin{aligned} &\tilde{\eta}(0) = 1,\ \tilde{\eta}'(0) = 0, \tilde{\kappa}'(0) = 0 \\ &\tilde{\eta}(1) = 0,\ \tilde{\kappa}(1) = 1. \end{aligned} \tag{15.21}$$

15.4 DOUBLE-PARAMETRIC FORM-BASED HOMOTOPY ANALYSIS METHOD

This section examines a new analytical technique, homotopy analysis method (HAM) [30, 31], which uses a double-parametric approach to obtain the solution of the fuzzy differential equations. The interval form of Equations (15.10) and (15.11) can be derived by employing the β-cut notion.

$$\begin{aligned}&\left\{\underline{\eta}'''(\xi,\beta),\overline{\eta'''}(\xi,\beta)\right\}\\&+2\omega Re\left[\underline{\Omega}_f(\beta),\overline{\Omega}_f(\beta)\right]\left\{1-[\underline{\phi}(\beta),\overline{\phi}(\beta)\}^{2.5}\left[\underline{\eta}'(\xi,\beta),\overline{\eta}'(\xi,\beta)\right]\left[\underline{\eta}(\xi,\beta),\overline{\eta}(\xi,\beta)\right]\right.\\&+\left(4-\left(1-\left[\underline{\phi}(\beta),\overline{\phi}(\beta)\right]\right)^{2.5}Ha\left[\underline{\Omega}_\sigma(\beta),\overline{\Omega}_\sigma(\beta)\right]\right)\omega^2\left[\underline{\eta}'(\xi,\beta),\overline{\eta}'(\xi,\beta)\right]=0\end{aligned}\tag{15.22}$$

$$\begin{aligned}&\left\{\underline{\kappa}''(\xi,\beta),\overline{\kappa}''(\xi,\beta)\right\}\\&+\frac{E_cP_r}{\left(\underline{P}_k(\beta),\overline{\Omega}_k(\beta)\right)(1-\left[\underline{\phi}(\beta),\overline{\phi}(\beta)\right])^{2.5}}\left[\underline{\Omega}_c(\beta),\overline{\Omega}_c(\beta)\right]\\&\left[4\omega^2\left\{(\underline{\eta}(\xi,\beta),\overline{\eta}(\xi,\beta)\right\}^2+\left\{\underline{\eta}'(\xi,\beta),\overline{\eta}'(\xi,\beta)\right\}^2\right]=0\end{aligned}\tag{15.23}$$

Using parameter δ in Equations (15. 22) and (15. 23), we find

$$\begin{aligned}&\left[\delta\left(\overline{\eta'''}(\xi,\beta)-\underline{\eta}'''(\xi,\beta)\right)+\underline{\eta}'''(\xi,\beta)\right]\\&+\left[2\omega Re\left[\left\{\delta\left(\overline{\Omega_f}(\beta)-\underline{\Omega}_f(\beta)\right)+\underline{\Omega}_f(\beta)\right\}\{1-\{\delta\left(\overline{\phi}(\beta)-\underline{\phi}(\beta)\right)\right.\right.\\&\left.+\underline{\phi}(\beta)\}\}^{2.5}\times\left\{\delta\left(\overline{\eta}'(\xi,\beta)-\underline{\eta}'(\xi,\beta)\right)+\underline{\eta}'(\xi,\beta)\right\}\left\{\delta\left(\overline{\eta}(\xi,\beta)-\underline{\eta}(\xi,\beta)\right)+\underline{\eta}(\xi,\beta)\right\}\right]\\&+\left(4-\left(1-\left[\delta\left\{\overline{\phi}(\beta)-\underline{\phi}(\beta)\right\}+\underline{\phi}(\beta)\right]\right)^{2.5}Ha\left[\delta\left\{\overline{P}_\sigma(\beta)-\underline{\Omega}_\sigma(\beta)\right\}+\underline{P}_\sigma(\beta)\right]\right)\omega^2\\&\times\left\{\delta\left(\overline{\eta'}(\xi,\beta)-\underline{\eta}'(\xi,\beta)\right)+\underline{\eta}'(\xi,\beta)\right\}=0,\end{aligned}\tag{15.24}$$

$$\begin{aligned}&\left[\overline{\kappa''}\delta\left((\xi,\beta)-\underline{\kappa}''(\xi,\beta)\right)+\underline{\kappa}''(\xi,\beta)\right]\\&+\frac{E_cP_r}{\left[\delta\left(\overline{\Omega_k}(\beta)-\underline{\Omega}_k(\beta)\right)+\underline{\Omega}_k(\beta)\right]\left(1-\left[\delta\left(\overline{\phi}(\beta)-\underline{\phi}(\beta)\right)+\underline{\phi}(\beta)\right]^{2.5}\right)}\times\\&\left[\left[\delta\left(\overline{\Omega}_c(\beta)-\underline{\Omega}_c(\beta)\right)+\underline{\Omega}_c(\beta)\right]\left[4\omega^2\{\delta\left(\overline{\eta}(\xi,\beta)-\underline{\eta}(\xi,\beta)\right)+\underline{\eta}(\xi,\beta)\}^2\times\right.\right.\\&\left.+\left\{\delta\left(\overline{\eta}'(\xi,\beta)-\underline{\eta}'(\xi,\beta)\right)+\underline{\eta}'(\xi,\beta)\right\}^2\right]=0.\end{aligned}\tag{15.25}$$

Here, $\beta,\delta\in[0,1]$,"where β and δ are the uncertain parameters that regulate the fuzziness of the governing fuzzy differential Equations (15.19) and (15.20), which are expressed in the double-parametric form."

Now, denoting

$$\begin{aligned}
\delta\left(\overline{\eta}(\xi,\beta)-\underline{\eta}(\xi,\beta)\right)+\underline{\eta}(\xi,\beta)&=\eta(\xi,\beta,\delta),\\
\delta\left(\overline{\eta}'(\xi,\beta)-\underline{\eta}'(\xi,\beta)\right)+\underline{\eta}'(\xi,\beta)&=\eta'(\xi,\beta,\delta),\\
\delta\left(\overline{\eta'''}(\xi,\beta)-\underline{\eta}'''(\xi,\beta)\right)+\underline{\eta}'''(\xi,\beta)&=\eta(\xi,\beta,\delta),\\
\delta\left(\Omega_f(\beta)-\underline{\Omega}_f(\beta)\right)+\underline{\Omega}_f(\beta)&=\Omega_f(\beta,\delta),\\
\delta\left(\overline{\Omega}_c(\beta)-\underline{\Omega}_c(\beta)\right)+\underline{\Omega}_c(\xi,\beta)&=\Omega_c(\beta,\delta),\\
\delta\left(\overline{\Omega}_\sigma(\beta)-\underline{\Omega}_\sigma(\beta)\right)+\underline{\Omega}_\sigma(\beta)&=\Omega_\sigma(\beta,\delta),\\
\delta\left(\overline{\Omega}_k(\beta)-\underline{\Omega}_k(\beta)\right)+\underline{\Omega}_k(\beta)&=\Omega_k(\beta,\delta),\\
\delta\left(\overline{\phi}(\beta)-\underline{\phi}(\beta)\right)+\underline{\phi}(\beta)&=\phi(\beta,\delta),
\end{aligned}\tag{15.26}$$

We can write Equations (15.24) and (15.25) with the help of Equation (15.26), and we obtain

$$\begin{aligned}
&\eta'''(\beta,\delta,\xi)+2\omega\mathrm{Re}\Omega_f(\beta,\delta)(1-\phi(\beta,\delta))^{2.5}\eta(\xi,\beta,\delta)\eta'(\xi,\beta,\delta)\\
&\quad+\left(4-(1-\phi)^{2.5}Ha\right)\omega^2\eta'(\xi,\beta,\delta)=0,
\end{aligned}\tag{15.27}$$

$$\begin{aligned}
&\kappa''(\xi,\beta,\delta)+\frac{E_cP_r}{\Omega_k(\beta,\delta)(1-\phi(\beta,\delta))^{2.5}}\Omega_c(\beta,\delta)\\
&\quad\times\left[4\omega^2\left(\eta(\beta,\delta,\xi)\right)^2+\left(\eta'(\xi,\beta,\delta)\right)^2\right]=0.
\end{aligned}\tag{15.28}$$

This section presents the fuzzy HAM as a means to assess the results of the nanofluid flow simulation. The linear and non-linear operators of the governing equations are represented by the following expressions:

$$\begin{aligned}
N1\left[\tilde{\eta}(\beta,\delta,\xi,q)\right]&=\eta'''(\beta,\delta,\xi,q)\\
&\quad+2\omega\mathrm{Re}\Omega_f(\beta,\delta)\left(1-\phi(\beta,\delta)\right)^{2.5}\eta(\beta,\delta,\xi,q)\eta'(\beta,\delta,\xi,q)\\
&\quad+\left(4-(1-\phi)^{2.5}Ha\right)\omega^2\eta'(\beta,\delta,\xi,q),
\end{aligned}\tag{15.29}$$

$$\begin{aligned}
N2\left[\tilde{\kappa}(\beta,\delta,\xi,q)\right]&=\kappa''(\beta,\delta,\xi,q)\\
&\quad+\frac{E_cP_r}{\Omega_k(\beta,\delta)(1-\phi(\beta,\delta))^{2.5}}\Omega_c(\beta,\delta)\left[4\omega^2(\eta(\beta,\delta,\xi))^2+\left(\eta'(\beta,\delta,\xi,q)\right)^2\right],
\end{aligned}\tag{15.30}$$

and

$$L1\left[\tilde{\psi}(\xi,\beta,\delta)\right] = \tilde{\eta}'''(\xi,\beta,\delta). \tag{15.31}$$

$$L2\left[\tilde{\psi}(\xi,\beta,\delta)\right] = \tilde{\kappa}''(\xi,\beta,\delta) \tag{15.32}$$

having the property

$$L1\left[b_1 + b_2\xi + b_3\xi^2\right] = 0,\ L2\left[b_4 + b_5\xi\right] = 0, \tag{15.33}$$

where b_1, b_2, b_3, b_4, and b_5 are constants.

The initial approximations are

$$\tilde{\eta}_0(\xi) = 1 - \xi^2,\ \tilde{\kappa}_0(\xi) = 1. \tag{15.34}$$

The n-th order deformation equation for $n \geqslant 1$ [31] may be written as

$$\begin{aligned}
\tilde{\eta}_n(\xi,\beta,\delta) &= \chi_n \tilde{\eta}_{n-1}(\xi,\beta,\delta) + h\int_0^{\xi}\int_0^{\xi}\int_0^{\xi} R_n\left(\overline{\tilde{\eta}_{n-1}}\right) d\xi d\xi d\xi + b_1 + b_2\xi + b_3\xi^2 \\
\tilde{\kappa}_n(\xi,\beta,\delta) &= \chi_n \tilde{\kappa}_{n-1}(\xi,\beta,\delta) + h\int_0^{\xi}\int_0^{\xi} R_n\left(\overline{\tilde{\kappa}_{n-1}}\right) d\xi d\xi + b_3 + b_4\xi
\end{aligned} \tag{15.35}$$

with

$$\tilde{\eta}_n(0) = \tilde{\eta}_n(0) = \tilde{\eta}_n(1) = \tilde{\kappa}_n(1) = \tilde{\kappa}_n(0) = 0, \tag{15.36}$$

where

$$\begin{aligned}
R_n\left(\vec{\eta}_{n-1}(\xi,\beta,\delta)\right) &= \tilde{\eta}_{n-1}'''(\xi,\beta,\delta) \\
&+ 2\omega Re\Omega_f(\beta,\delta)\left(1-\phi(\beta,\delta)\right)^{2.5} \sum_{i=0}^{n-1} \tilde{\eta}_i(\xi,\beta,\delta)\tilde{\eta}_{n-1-i}'(\xi,\beta,\delta) \\
&+ \left(4 - (1-\phi)^{2.5} Ha\Omega_\sigma(\beta,\delta)\right)\omega^2 \tilde{\eta}_{n-1-i}'(\xi,\beta,\delta),
\end{aligned} \tag{15.37}$$

$$\begin{aligned}
R_n\left(\vec{\kappa}_{n-1}(\xi,\beta,\delta)\right) &= \kappa_{n-1}''(\xi,\beta,\delta) \\
&+ \frac{E_c P_r}{\Omega_k(\beta,\delta)(1-\phi(\beta,\delta))^{2.5}} \Omega_c(\beta,\delta) \sum_{i=0}^{n-1} \begin{bmatrix} 4\omega^2 \eta_i(\beta,\delta,\xi)\eta_{n-i-1}(\beta,\delta,\xi) \\ +\eta_i'(\xi,\beta,\delta)\eta_{n-i-1}'(\xi,\beta,\delta) \end{bmatrix},
\end{aligned} \tag{15.38}$$

And

$$\chi_n = \begin{cases} 0 & n <= 1 \\ 1 & n > 1 \end{cases}. \tag{15.39}$$

Solving Equations (15.35)–(15.39) using Equation (15.34), we get a series of solutions:

$$\tilde{\eta}(\xi,\beta,\delta)=1-\xi^2+h\left(\frac{1}{60}A_1\xi^2-\frac{1}{12}(A_1+B_1)\xi^4\right)+h\left(\frac{1}{15}A_1+\frac{1}{12}B_1\right)\xi^2$$
$$+h\left(\frac{1}{60}A_1\xi^6-\frac{1}{12}(A_1+B_1)\xi^4\right)+h\left(\frac{1}{15}A_1+\frac{1}{12}B_1\right)\xi^2$$
$$+h\left(-\frac{1}{5400}A_1^2h\xi^{10}+\frac{1}{560}\left(A_1^2h+A_1B_1h\right)\xi^8+\frac{1}{6}\left(\frac{1}{10}A_1h-\frac{3}{100}A_1^2h-\frac{1}{20}A_1Bh-\frac{1}{60}B_1^2h\right)\xi^6\right)$$
$$+h\frac{1}{4}\left(-\frac{1}{3}A_1h-\frac{1}{3}B_1h+\frac{1}{45}A_1^2h+\frac{1}{20}A_1B_1h+\frac{1}{36}B_1^2h\right)\xi^4$$
$$-h\left(\frac{163}{75600}A_1^2h+\frac{1}{168}A_1B_1h-\frac{1}{15}+\frac{1}{240}B_1^2h-\frac{1}{12}B_1h\right)\xi^2+\ldots\ldots \quad (15.40)$$

and

$$\tilde{\kappa}(\xi,\beta,\delta)$$
$$=1+\mathrm{h}\left(\frac{1}{163800}\omega^2h^2A_1^2+\frac{1}{990}\omega^2h^2\left(-\frac{1}{12}A_1-\frac{1}{12}B_1\right)A_1+\frac{1}{13200}h^2A_1^2\right.$$
$$+\frac{2}{45}\omega^2\left(\left(-\frac{3}{30}+\frac{1}{30}h\left(\frac{1}{15}A_1+\frac{1}{12}B_1\right)hA_1+h^2\left(-\frac{1}{12}A_1-\frac{1}{12}B_1\right)^2\right)\right)$$
$$+\frac{1}{450}h^2\left(-\frac{1}{3}A_1-\frac{1}{3}B_1\right)A_1$$
$$+\frac{1}{14}\omega^2\left(\frac{1}{30}hA_1+\left(-2+2h\left(\frac{1}{15}A_1+\frac{1}{12}B_1\right)\right)h\left(-\frac{1}{12}A_1-\frac{1}{12}B_1\right)\right),$$
$$+\left(-\frac{1}{40}+\frac{1}{40}h\left(\frac{1}{15}A_1+\frac{1}{12}B_1\right)\right)hA_1 \quad (15.41)$$
$$+\frac{1}{56}h^2\left(-\frac{1}{3}A_1-\frac{1}{3}B_1\right)^2$$
$$\left.+\frac{2}{15}\omega^2\left(2h\left(-\frac{1}{12}A_1-\frac{1}{12}B_1+\left(-1+h\left(\frac{1}{15}A_1+\frac{1}{12}B_1\right)\right)^2\right)+\ldots\right)\right)$$

where $A_1=2\omega\mathrm{Re}\Omega_f(\beta,\delta)\left(1-\phi(\beta,\delta)\right)^{2.5}$, $B_1=\left(4-\left(1-\phi(\beta,\delta)\right)^{2.5}Ha\Omega\sigma(\beta,\delta)\right)\omega^2$, and

$$Y=\frac{E_cP_r}{V_k(\beta,\delta)(1-\phi(\beta,\delta))^{2.5}}V_c(\beta,\delta).$$

15.5 CONVERGENCE ANALYSIS OF PROBLEM

Theorem 1. [30, 32] "Let $\sum_{n=0}^{\infty}\eta_n(\xi,\beta,\delta)$ and $\sum_{n=0}^{\infty}\kappa_n(\xi,\beta,\delta)$ uniformly converge to $\eta(\xi,\beta,\delta)$ and $\kappa(\xi,\beta,\delta)$, respectively, where $\eta(\xi,\beta,\delta)\in L(R^+)$ and $\kappa(\xi,\beta,\delta)\in L(R^+)$ are generated by n-th order deformation equations. Then $\eta(\xi,\beta,\delta)$ and $\kappa(\xi,\beta,\delta)$ are the exact solutions of Equation (15.19) and (15.20)."

Proof. Let us represent the system as a series of expressions:

$$\begin{aligned}\eta(\xi,\beta,\delta)&=\eta_0(\xi,\beta,\delta)+\sum_{n=1}^{\infty}\eta_n(\xi,\beta,\delta),\\ \kappa(\xi,\beta,\delta)&=\kappa_0(\xi,\beta,\delta)+\sum_{n=1}^{\infty}\kappa_n(\xi,\beta,\delta)\end{aligned}\tag{15.42}$$

and they are uniformly convergent. Then

$$\begin{aligned}&\lim_{n\to\infty}\eta_n(\xi,\beta,\delta)=0\\ &\lim_{n\to\infty}\kappa_n(\xi,\beta,\delta)=0.\end{aligned}\tag{15.43}$$

Now we have

$$\begin{aligned}\sum_{n=1}^{j}\left[\eta_n(\xi,\beta,\delta)-\chi\eta_{n-1}(\xi,\beta,\delta)\right]&=\eta_1(\xi,\beta,\delta)+\left(\eta_2(\xi,\beta,\delta)-\eta_1(\xi,\beta,\delta)\right)\\ &+\left(\eta_3(\xi,\beta,\delta)-\eta_2(\xi,\beta,\delta)\right)+\ldots+\left(\eta_j(\xi,\beta,\delta)-\eta_{j-1}(\xi,\beta,\delta)\right)\\ \sum_{n=1}^{j}\left[\kappa_n(\xi,\beta,\delta)-\chi\kappa_{n-1}(\xi,\beta,\delta)\right]&=\kappa_1(\xi,\beta,\delta)+\left(\kappa_2(\xi,\beta,\delta)-\kappa_1(\xi,\beta,\delta)\right)\\ &+\left(\kappa_3(\xi,\beta,\delta)-\kappa_2(\xi,\beta,\delta)\right)+\ldots+\left(\kappa_j(\xi,\beta,\delta)-\kappa_{j-1}(\xi,\beta,\delta)\right).\end{aligned}\tag{15.44}$$

From Equations (15.42)–(15.44), we get

$$\begin{aligned}\sum_{n=1}^{\infty}\left[\eta_n(\xi,\beta,\delta)-\chi\eta_{n-1}(\xi,\beta,\delta)\right]&=\lim_{n\to\infty}\eta_n(\xi,\beta,\delta)=0\\ \sum_{n=1}^{\infty}\left[\kappa_n(\xi,\beta,\delta)-\chi\kappa_{n-1}(\xi,\beta,\delta)\right]&=\lim_{n\to\infty}\kappa_n(\xi,\beta,\delta)=0.\end{aligned}\tag{15.45}$$

According to the definition of the operators $L1, L2$, we have

$$\begin{aligned}L1\left(\sum_{n=1}^{\infty}\left[\eta_n(\xi,\beta,\delta)-\chi\eta_{n-1}(\xi,\beta,\delta)\right]\right)&=\sum_{n=1}^{\infty}L1\left[\eta_n(\xi,\beta,\delta)-\chi\eta_{n-1}(\xi,\beta,\delta)\right]\\ L2\left(\sum_{n=1}^{\infty}\left[\kappa_n(\xi,\beta,\delta)-\chi\kappa_{n-1}(\xi,\beta,\delta)\right]\right)&=\sum_{n=1}^{\infty}L2\left[\kappa_n(\xi,\beta,\delta)-\chi\kappa_{n-1}(\xi,\beta,\delta)\right].\end{aligned}\tag{15.46}$$

Using the nth-ordered deformation equation, we have

$$\begin{aligned} h\sum_{n=1}^{\infty}R_n\left(\vec{\eta}_{n-1}(\xi,\beta,\delta)\right) &= \sum_{n=1}^{\infty}L1\left[\eta_n(\xi,\beta,\delta)-\chi\eta_{n-1}(\xi,\beta,\delta)\right]=0 \\ h\sum_{n=1}^{\infty}R_n\left(\vec{\kappa}_{n-1}(\xi,\beta,\delta)\right) &= \sum_{n=1}^{\infty}L2\left[\kappa_n(\xi,\beta,\delta)-\chi\kappa_{n-1}(\xi,\beta,\delta)\right]=0. \end{aligned} \tag{15.47}$$

Since h cannot vanish, we obtain

$$\begin{aligned} \sum_{n=1}^{\infty}R_n\left(\vec{\eta}_{n-1}(\xi,\beta,\delta)\right) &= 0 \\ \sum_{n=1}^{\infty}R_n\left(\vec{\kappa}_{n-1}(\xi,\beta,\delta)\right) &= 0. \end{aligned} \tag{15.48}$$

From Equations (15.37) and (15.38), we get

$$\begin{aligned} \sum_{n=1}^{\infty}R_n\left(\vec{\eta}_{n-1}(\xi,\beta,\delta)\right) &= \sum_{n=1}^{\infty}\eta'''_{n-1}(\xi,\beta,\delta) \\ &+2\omega Re\Omega_f(\Omega,\delta)(J-\psi(J,\delta))^{2.5}\sum_{n=1}^{\infty}\sum_{i=0}^{n-1}\eta_i(\xi,\beta,\delta)\eta'_{n-1-i}(J,\delta,\zeta) \\ &+\left(4-(1-\psi)^{2.5}Ha\right)\omega^2\sum_{n=1}^{\infty}\eta'_{n-1-i}(\xi,\beta,\delta) \\ \sum_{n=1}^{\infty}R_n\left(\vec{r}_{n-1}(\xi,\beta,\delta)\right) &= \sum_{n=1}^{\infty}\kappa''_{n-1}(\xi,\beta,\delta)+\frac{E_cP_r}{\Omega_k(J,\delta)\left(1-\psi(\beta,\delta)\right)^{2.5}}(J,\delta)\times \\ &\sum_{n=1}^{\infty}\sum_{i=0}^{n-1}\left[4\omega^2\eta_i(\xi,\beta,\delta)\eta_{n-1-i}(\xi,\beta,\delta)+\eta'_i(\xi,\beta,\delta)\eta'_{n-1-i}(\xi,\beta,\delta)\right]. \end{aligned} \tag{15.49}$$

Simplifying Equation (15.49), we get

$$\begin{aligned} &\eta'''(\xi,\beta,\delta)+2\omega\mathrm{Re}_f(J,\delta)\left(1-\psi(J,\delta)\right)^{2.5}\eta(\xi,\beta,\delta)\eta'(\xi,\beta,\delta) \\ &+\left(4-(1-\psi)^{2.5}Ha\right)\omega^2\eta'(\xi,\beta,\delta)=0 \\ &\kappa''(\xi,\beta,\delta)+\frac{E_cP_r}{\Omega_k(J,\delta)(1-\psi(\beta,\delta))^{2.5}}\Omega_c(J,\delta)\times \\ &\left[4\omega^2\left(\eta(\xi,\beta,\delta)\right)^2+\left(\eta'(\xi,\beta,\delta)\right)^2\right]=0. \end{aligned} \tag{15.50}$$

This is the primary problem being studied and represented by the system of controlling differential equations. Thus, the series solution $\sum_{n=0}^{\infty}\eta_n(\xi,\beta,\delta)$ and $\sum_{n=0}^{\infty}\kappa_n(\xi,\beta,\delta)$ are converging to the precise solution of the problem.

15.6 RESULTS AND DISCUSSION

In this study, HAM is used to model the Jeffery–Hamel hybrid nanofluid to simulate the flow along with heat transfer within a channel having non-parallel walls. Here, we use H_2O as the base fluid, with Cu and Cu-Al_2O_3 nanoparticles and hybrid nanoparticles as the other components. The thermophysical properties of the particles and water are given in Table 15.1. Table 15.2 presents the effective model in the case of a hybrid nanofluid. For the validation of the results, Table 15.3 shows that HAM provides better results in comparison with adomian decomposition method (ADM) as the results of HAM are very effective and more accurate than numerical method results. In order to determine the outcome in fuzzy nature, we suppose that ϕ is taken as $[0\%,1\%,2\%]$, a fuzzy triangular number as shown in Figure 15.2. Tables 15.4 and 15.5 show the series solution is more stable and the convergent nature as the term approximation is increasing for both velocity and temperature in the convergent and divergent channels.

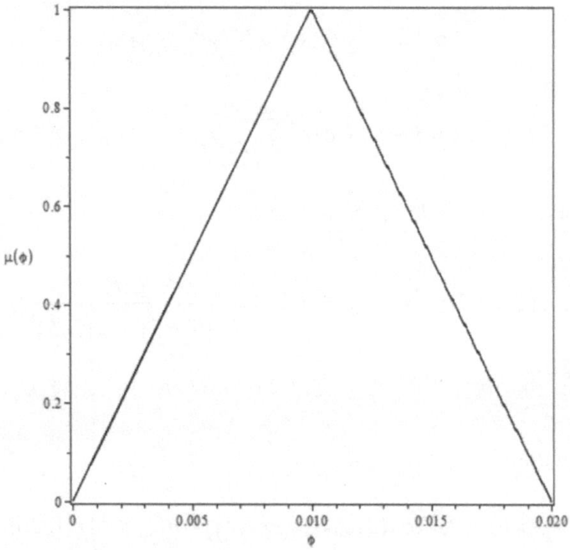

FIGURE 15.2 TFN $\tilde{\phi} = [0,0.01,0.02]$

TABLE 15.1
Thermophysical Properties of the Base Fluid and Nanoparticles

	Fluid phase	Nanoparticles	
Property	H_2O	Cu	Al_2O_3
Density	997.1	8933	3970
Thermal conductivity	0.613	400	40
Specific heat	4179	385	765

TABLE 15.2
Effective Representations of the Properties of Hybrid Nanofluid

Property	Nanofluid	Hybrid nanofluid
Density	$\rho_{nf}=\rho_f(1-\phi)+\rho_s\phi$	$\rho_{hnf}=\left[(1-\phi_2)\left(1-\phi_1+\frac{\rho_{s_1}}{\rho_f}\phi_1\right)+\frac{\rho_{s_2}}{\rho_f}\phi_2\right]\rho_f$
Viscosity	$\mu_{nf}=\frac{\mu_f}{(1-\phi)^{2.5}}$	$\mu_{hnf}=\frac{\mu_f}{(1-\phi_1)^{2.5}(1-\phi_2)^{2.5}}$
Heat capacity	$(\rho C_P)_{nf}=(1-\phi)(\rho C_P)_f+\phi(\rho C_P)_s$	$(\rho C_P)_{hnf}=(1-\phi_2)\left[(1-\phi_1)(\rho C_P)_f+\phi_1(\rho C_P)_{s_1}\right]+\phi_2(\rho C_P)_{s_2}$
Thermal conductivity	$\frac{k_{nf}}{k_f}=\left(\frac{(k_s+2k_f)-\phi 2(k_f-k_s)\phi}{k_s+2k_f+\phi(k_f-k_s)}\right)$	$\frac{k_{hnf}}{k_{bf}}=\frac{(k_{s_2}+(m-1)k_{bf})-(m-1)(k_{bf}-k_{s_2})\phi_2}{k_{s_2}+(m-1)k_f+(k_{bf}-k_{s_2})\phi_2}$, where $\frac{k_{bf}}{k_f}=\frac{(k_{s_1}+(m-1)k_f)-(m-1)(k_f-k_{s_1})\phi_1}{k_{s_1}+(m-1)k_f+(k_f-k_{s_1})\phi_1}$

TABLE 15.3
Comparison of the Velocity at $\tilde{\phi}=0(\beta=0,\delta=0)$, Re $=25,\omega=5^0$, and Ha = 250 with ADM [9] and NM [9]

ξ	HAM	ADM	NM	$\lVert NM-HAM\rVert$	$\lVert NM-ADM\rVert$
0	1.0	1.0	1.0	0.00000	0.00000
0.1	0.9886181310	0.990196	0.988606	$1.213E-05$	0.00159
0.2	0.9547007589	0.960841	0.954700	$7.589E-07$	0.006141
0.3	0.8990772239	0.912079	0.899076	$1.224E-06$	0.013003
0.4	0.8229277514	0.811225	0.822925	$2.751E-06$	0.0117
0.5	0.7276679619	0.713799	0.727664	$3.962E-06$	0.01387
0.6	0.6147150445	0.604866	0.614709	$6.045E-06$	0.00984
0.7	0.4852389737	0.476625	0.485232	$6.974E-06$	0.00861
0.8	0.3398981712	0.325834	0.339890	$8.171E-06$	0.01406
0.9	0.1785610105	0.171160	0.178555	$6.011E-06$	0.00739
1.0	0.0	0.0	0.0	0.00000	0.00000

TABLE 15.4

Velocity at $\tilde{\phi} = 0.01(\beta = 0, \delta = 1)$, Re = 50, and Ha = 150

	$\omega = -5°$				$\omega = 5°$			
ξ	**Two-term**	**Three-term**	**Four-term**	**Five-term**	**Two-term**	**Three-term**	**Four-term**	**Five-term**
0.0	1.0	1.0	1.0	1.0	1.0	1.0	1.0	1.0
0.1	0.9835085524	0.9834252467	0.9834528948	0.9834600832	0.9944410201	0.9949965519	0.9949813408	0.9949458569
0.2	0.9352626931	0.9350139411	0.9351279256	0.9351541425	0.9774645032	0.9794138814	0.9794097017	0.9792668995
0.3	0.8587703820	0.8584725087	0.8587303174	0.8587793844	0.9480458240	0.9515369879	0.9516892573	0.9513768097
0.4	0.7593136385	0.7592604485	0.7596919162	0.7597553060	0.9040919530	0.9084925513	0.9090137422	0.9085099354
0.5	0.6431933994	0.6436964470	0.6442711069	0.6443296412	0.8419806463	0.8461687485	0.8471956354	0.8465494895
0.6	0.5168451324	0.5180183799	0.5186382648	0.5186723321	0.7561390164	0.7590464844	0.7605069612	0.7598338735
0.7	0.3859731913	0.3875961695	0.3881329170	0.3881354884	0.6388603532	0.6400042189	0.6415995692	0.6410328448
0.8	0.2548586977	0.2564275955	0.2567858636	0.2567674622	0.4805794133	0.4803000655	0.4816295327	0.4812579870
0.9	0.1259759465	0.1269385806	0.1270978440	0.1270790147	0.2708211413	0.2700954644	0.2708363773	0.2706733603
1.0	0.0	0.0	0.0	0.0	0.0	0.0	0.0	0.0

TABLE 15.5

Velocity at $\tilde{\phi} = 0.01(\beta = 0, \delta = 1)$, Re = 50, Ec= 0.01, Pr = 6.2, and Ha = 150

	$\omega = -5^{\circ}$				$\omega = 5^{\circ}$			
ξ	Two-term	Three-term	Four-term	Five-term	Two-term	Three-term	Four-term	Five-term
0.0	1.027052973	1.028668186	1.02900012	1.02900012	1.018635368	1.021232933	1.020402033	1.020540853
0.1	1.027038047	1.028651932	1.028983463	1.028983463	1.018625684	1.021221217	1.020391207	1.020529765
0.2	1.026940468	1.028536217	1.028862361	1.028862361	1.018604215	1.021171722	1.020353806	1.020489020
0.3	1.026610037	1.028136878	1.028444175	1.028444175	1.018583259	1.021044164	1.020269239	1.020393890
0.4	1.025822922	1.027197912	1.027468962	1.027468962	1.018549347	1.020775664	1.020085474	1.020193161
0.5	1.024318427	1.025452633	1.025676567	1.025676567	1.018415271	1.020271483	1.019695585	1.019787188
0.6	1.021844493	1.022682785	1.022862505	1.022862505	1.017958889	1.019365842	1.018903315	1.018986604
0.7	1.018205923	1.018757852	1.018905494	1.018905494	1.016754655	1.017738093	1.017363941	1.017444223
0.8	1.013307663	1.013642460	1.013763565	1.013763565	1.014105601	1.014779924	1.014479418	1.014549760
0.9	1.007183642	1.007370680	1.007449281	1.007449281	1.008985159	1.009427790	1.009234978	1.009277649
1.0	1.0	1.0	1.0	1.0	1.0	1.0	1.0	1.0

Cu/H_2O and $Cu\text{-}Al_2O_3/H_2O$ nanofluids are shown in Figures 15.3(a, b) as having different effects on fluid velocity depending on the Reynolds number at $\omega > 0$ and $\omega < 0$, respectively. As the Reynolds number increases, the velocity profile of the Cu/H_2O nanofluid in $\omega < 0$ increases (see Figure 15.3(a)). Since the Re in physics represents the proportion of inertial and viscous forces, this also improves the momentum barrier layer and fluid velocity. The fluid's speed in the diverging channel slows down as the Re increases, as seen in Figure 15.3(b). Similar to this, Figures 15.4(a, b) show how Re temperature affects the convergent and divergent channels. As the Re rises, the temperature profile in the $\omega > 0$ and $\omega < 0$ channels becomes higher.

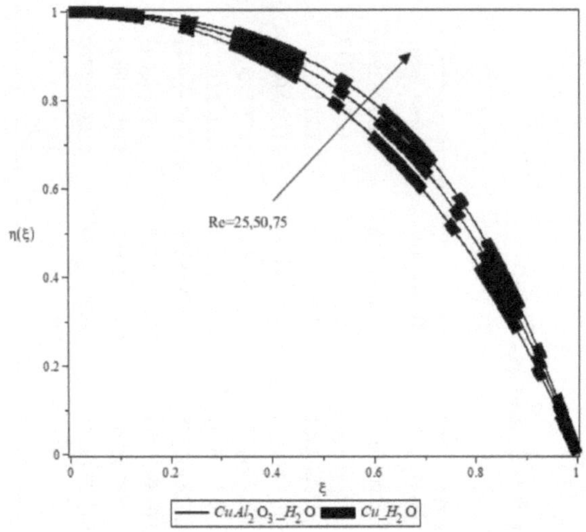

FIGURE 15.3(a) Velocity with the change in Reynolds parameter Re ($\omega < 0$).

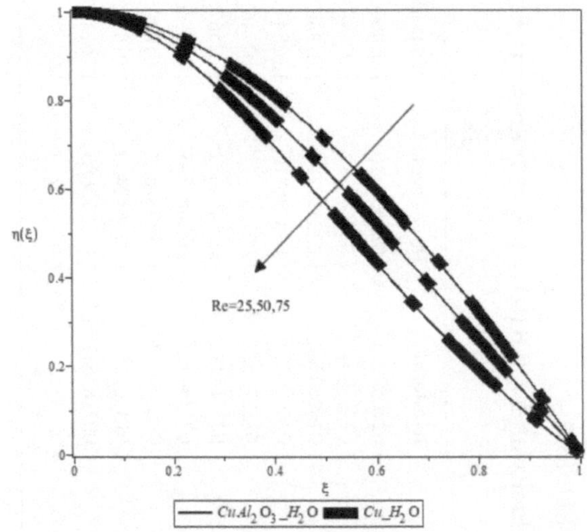

FIGURE 15.3(b) Velocity with the change in Reynolds parameter Re ($\omega > 0$).

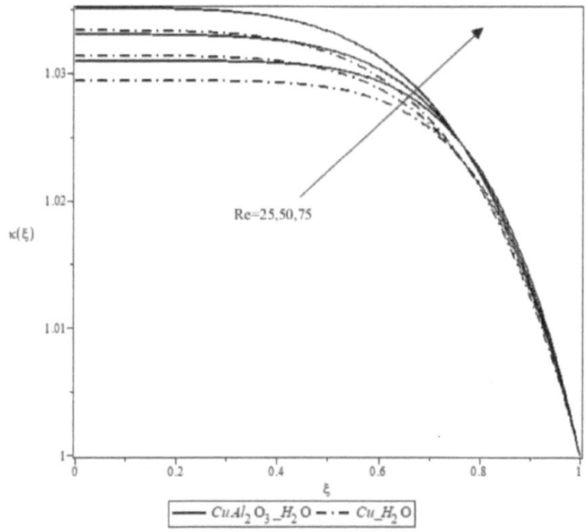

FIGURE 15.4(a) Temperature with the change in Reynolds parameter Re ($\omega < 0$).

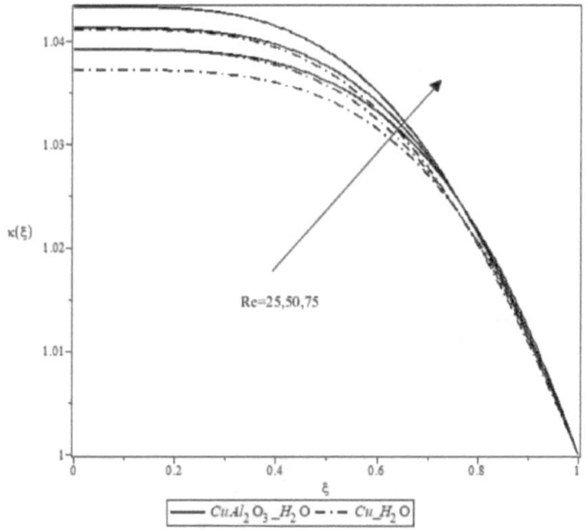

FIGURE 15.4(b) Temperature with the change in Reynolds parameter Re ($\omega > 0$).

It is obvious that the velocity profiles are affected by the inclination angle with nanoparticles Cu/H_2O and Cu-Al_2O_3/H_2O, respectively, which corresponds to convergent and divergent channels. Figures 15.5(a, b) demonstrate that when the inclination angle increases, the thickness of the boundary layer for both convergent and divergent channels decreases. Physically, the stream's velocity decreases under diverging conditions as the wall's physical influence on the fluid flow lowers as it expands. In Figures 15.6(a, b), the fluctuations in fluid temperature for the channels Cu/H_2O and Cu-Al_2O_3/H_2O are depicted. It is evident that the temperature of the fluid would rise in both situations, i.e., Cu/H_2O and $Cu\text{-}Al_2O_3/H_2O$, via the convergent and divergent channels, respectively. Additionally, it should be observed that both the $\omega > 0$ and $\omega < 0$ channels' velocity profiles for the Cu/H_2O and $Cu\text{-}Al_2O_3/H_2O$ nanofluids are similar. In contrast to the Cu/H_2O nanofluid, the hybrid $Cu\text{-}Al_2O_3/H_2O$ nanofluid shows a greater temperature profile in both channels.

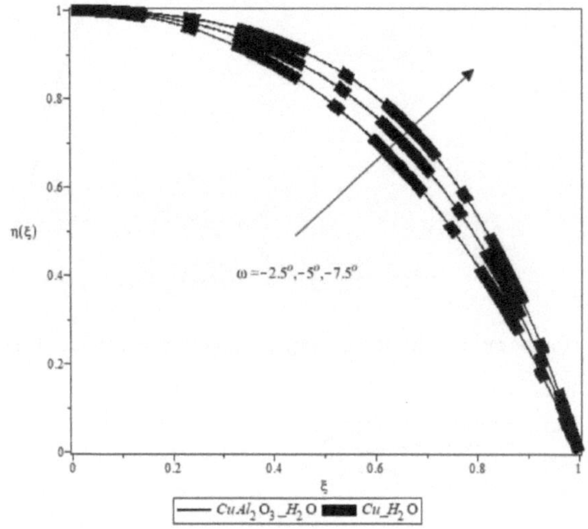

FIGURE 15.5(a) Velocity with the change in ω $(\omega < 0)$.

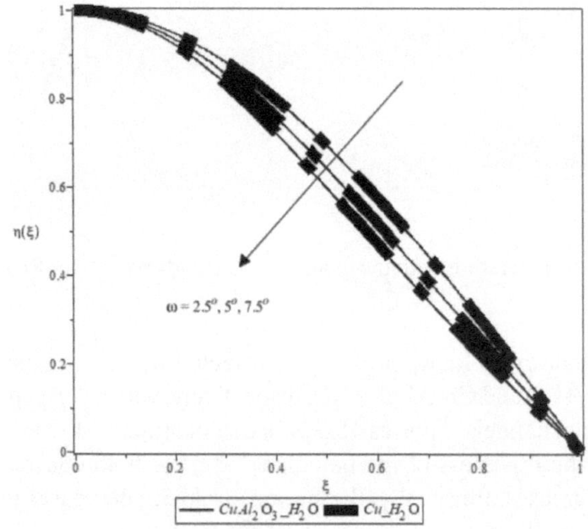

FIGURE 15.5(b) Velocity with the change in ω $(\omega > 0)$.

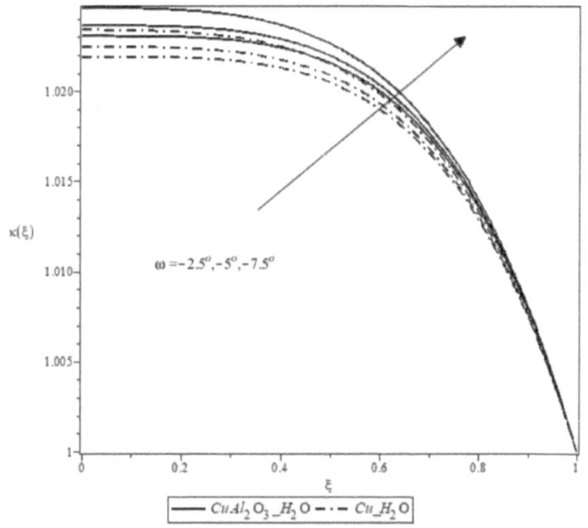

FIGURE 15.6(a) Temperature with the change of ω ($\omega < 0$).

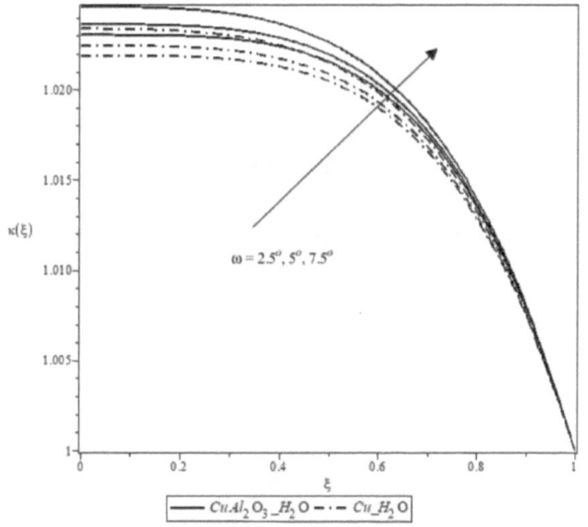

FIGURE 15.6(b) Temperature with the change of ω ($\omega > 0$).

The velocity profiles of Cu/H_2O and Cu-Al_2O_3/H_2O nanofluids in $\omega > 0$ and $\omega < 0$ channels are discussed in Figures 15.7(a, b) for various volume fraction values and with other constant parameters. As the volume percentage increases, the fluid's velocity profile is seen to increase in convergent channels, whereas the opposite is seen for divergent channels. The capacity of nanoparticles to move quickly inside the fluid accounts for the enhanced velocity following the introduction of the nanoparticles. Figures 15.8(a, b) depict the effect of nanoparticles on fluid temperature in convergent and divergent channels for Cu/H_2O and $Cu\text{-}Al_2O_3/H_2O$ nanofluids. Additionally, the velocity profiles for the nanofluids of Cu/H_2O and $Cu\text{-}Al_2O_3/H_2O$ can be found to be comparable in both channels. In both channels, the increase in volume fraction shows the increasing heat.

The relationship between the fluid velocity for Cu/H_2O and $Cu\text{-}Al_2O_3/H_2O$ nanoparticles in $\omega > 0$ and $\omega < 0$ channels and the Hartmann number is shown in Figures 15.9(a, b). As can be observed, increasing the Hartmann number increases

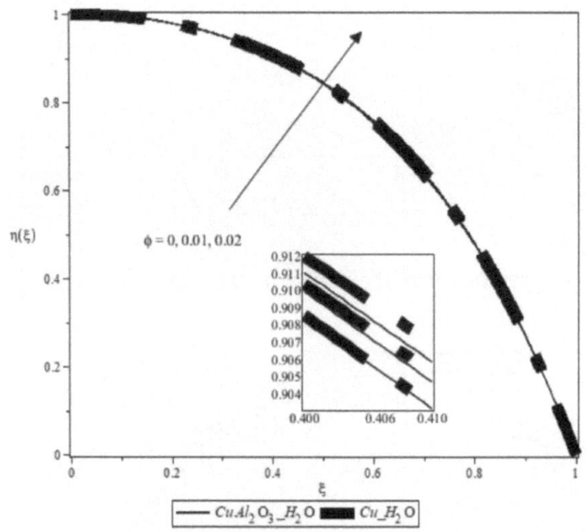

FIGURE 15.7(a) Velocity with the change in ϕ ($\omega < 0$).

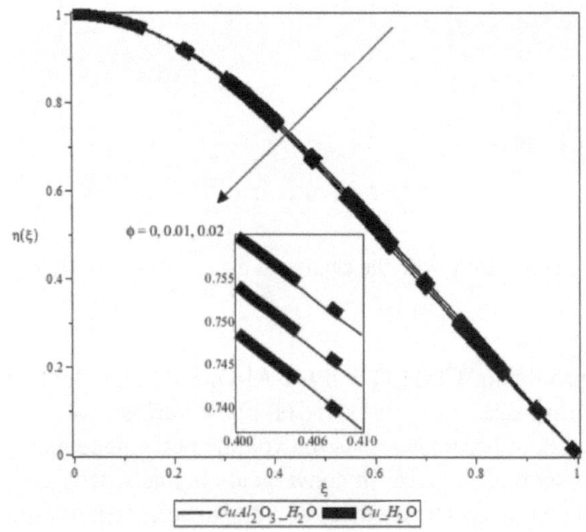

FIGURE 15.7(b) Velocity with the change in ϕ ($\omega > 0$).

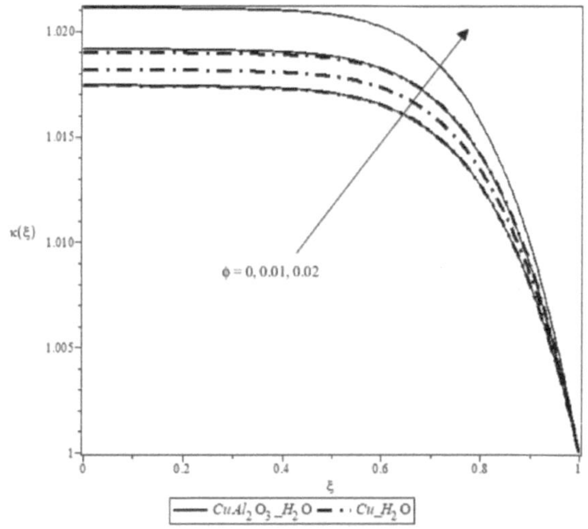

FIGURE 15.8(a) Temperature with the change in ϕ ($\omega < 0$).

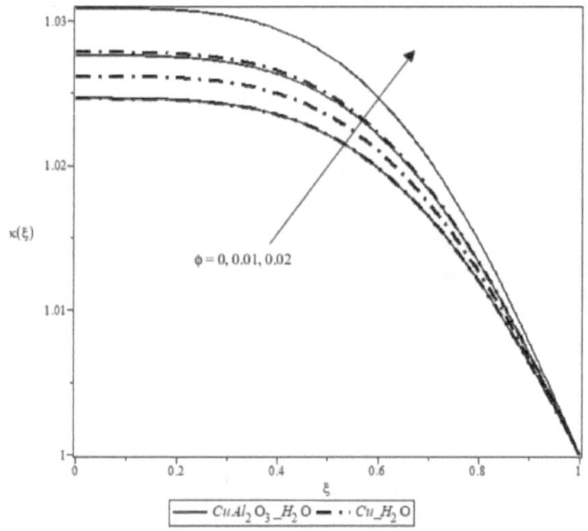

FIGURE 15.8(b) Temperature with the change in ϕ ($\omega > 0$).

the fluid's velocity in both the $\omega > 0$ and $\omega < 0$ channels. Figures 15.10(a, b) show the fluctuation in fluid temperature for both channels while maintaining constant values for other parameters and the Hartmann number. It can also be shown that for both the $\omega > 0$ and $\omega < 0$ channels, a higher Harmann number results in a higher fluid temperature. Physically speaking, the dominating magnetic field increases the viscous forces, which leads to an increase in resistance and, ultimately, an increase in the fluid's heat production. The fluid temperature increases as a result in both the $\omega > 0$ and $\omega < 0$ channels. It is also important to note that the velocity profiles on the two channels are very similar. Although the hybrid nanofluid Cu-Al_2O_3/H_2O has a higher temperature profile in both channels, Cu/H_2O and Cu-Al_2O_3/H_2O nanofluids have $\omega > 0$ and $\omega < 0$ channels, respectively.

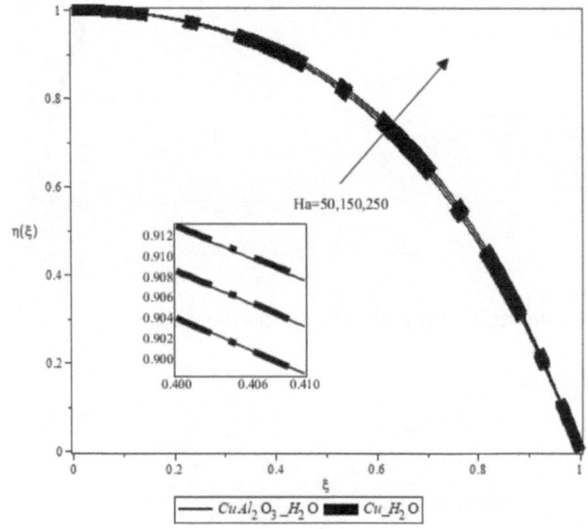

FIGURE 15.9(a) Velocity with the change in Ha ($\omega < 0$).

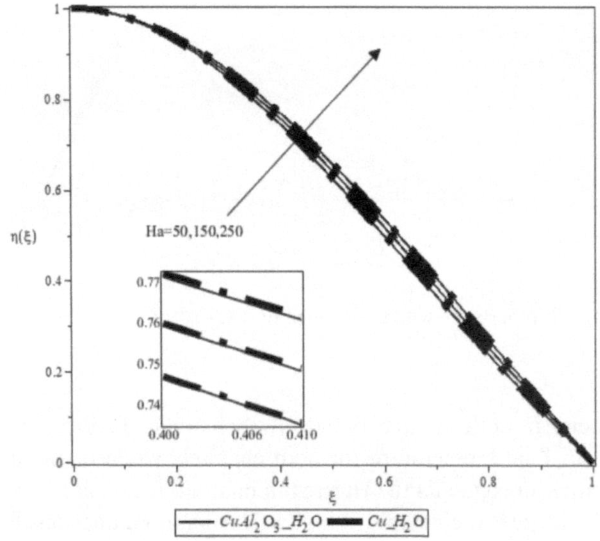

FIGURE 15.9(b) Velocity with the change in Ha ($\omega > 0$).

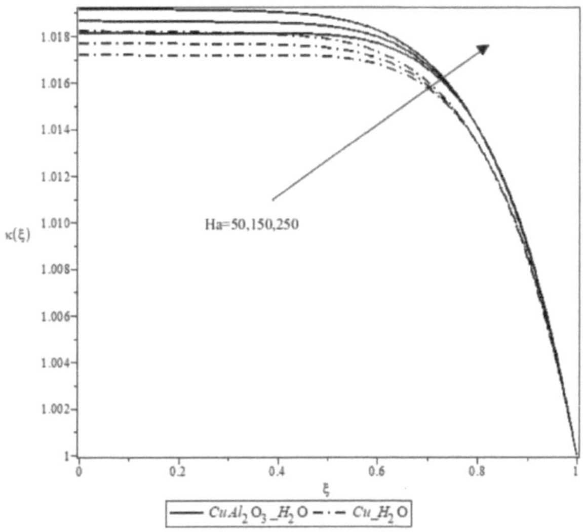

FIGURE 15.10(a) Temperature with the change in Ha ($\omega < 0$).

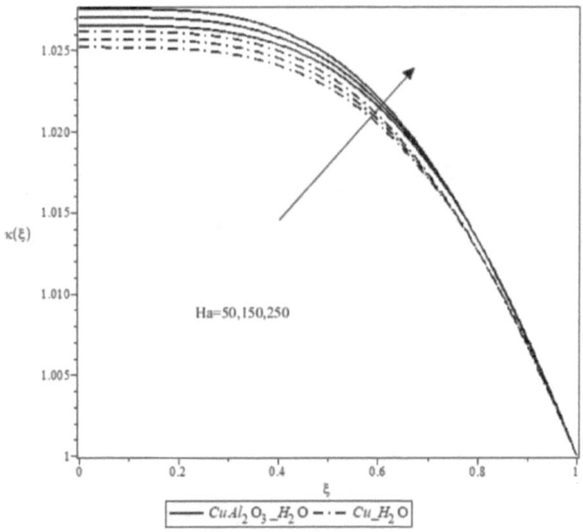

FIGURE 15.10(b) Temperature with the change in Ha ($\omega > 0$).

With all parameters held constant, Figures 15.11(a, b) illustrate the variation in temperature for nanofluid Cu/H_2O and $Cu\text{-}Al_2O_3/H_2O$ for different Eckert numbers in both convergent and divergent channels. The temperature profile is shown to increase as the Eckert number increases, and it can be seen that the $Cu\text{-}Al_2O_3/H_2O$ nanofluid fluid temperature is greater than the Cu/H_2O nanofluid temperature in both channels.

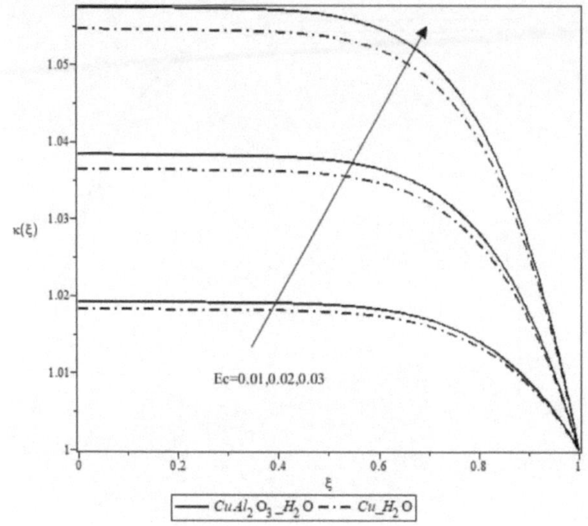

FIGURE 15.11(a) Temperature with the change in Ec ($\omega < 0$).

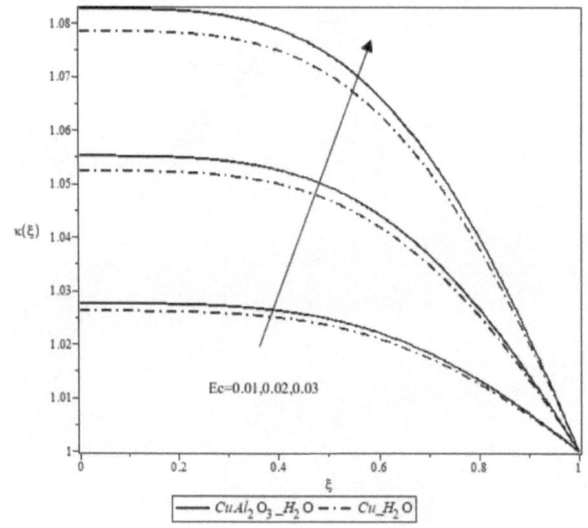

FIGURE 15.11(b) Temperature with the change in Ec ($\omega > 0$).

Figures 15.12(a, b) and 15.13(a, b) show the fuzzy velocity and temperature profiles of Cu/H_2O and Cu-Al_2O_3/H_2O when the volume fraction is treated as a TFN, i.e., $\phi = [0\%, 1\%, 2\%]$ at $\xi = 0.3$. In both converging and diverging channels, the graphs show the fuzzy non-dimensional velocity and temperature profiles for $\xi = 0.3$. The results shown at various positions in the diagram are the fuzzy velocities. At

TABLE 15.6
Fuzzy Velocity at $\xi = 0.3$

	$\omega = -5^{\circ}$				$\omega = 5^{\circ}$			
	Cu/H_2O		$Cu\text{-}Al_2O_3/H_2O$		Cu/H_2O		$Cu\text{-}Al_2O_3/H_2O$	
β	$\underline{\eta}$	$\overline{\eta}$	$\underline{\eta}$	$\overline{\eta}$	$\underline{\eta}$	$\overline{\eta}$	$\underline{\eta}$	$\overline{\eta}$
0.0	0.9503772090	0.9526724791	0.9503772089	0.9524083630	0.8628081164	0.8551436232	0.8628081164	0.8547828682
0.1	0.9505031495	0.9525683570	0.9505040256	0.9523300218	0.8624103833	0.8555117582	0.8623502721	0.8551278150
0.2	0.9506278544	0.9524631683	0.9506279501	0.9522494346	0.8620141132	0.8558815397	0.8618983857	0.8554786370
0.3	0.9507513340	0.9523569050	0.9507490244	0.9521665710	0.8616193174	0.8562529590	0.8614524620	0.8558353472
0.4	0.9508735986	0.9522495582	0.9508672900	0.9520814002	0.8612260070	0.8566260060	0.8610125034	0.8561979573
0.5	0.9509946584	0.9521411188	0.9509827874	0.9519938908	0.8608341929	0.8570006720	0.8605785124	0.8565664782
0.6	0.9511145228	0.9520315786	0.9510955565	0.9519040105	0.8604438858	0.8573769468	0.8601504896	0.8569409195
0.7	0.9512332021	0.9519209284	0.9512056362	0.9518117267	0.8600550967	0.8577548211	0.8597284351	0.8573212911
0.8	0.9513507061	0.9518091588	0.9513130647	0.9517170061	0.8596678361	0.8581342850	0.8593123469	0.8577076003
0.9	0.9514670442	0.9516962612	0.9514178796	0.9516198145	0.8592821142	0.8585153284	0.8589022232	0.8580998546
1.0	0.9515822262	0.9515822262	0.9515201175	0.9515201175	0.8588979414	0.8588979414	0.8584980606	0.8584980606

the very last, the velocity and temperature limits of the Cu/H_2O and Cu-Al_2O_3/H_2O nanofluids over a range of beta values and at $\xi = 0.3$ in the convergent and divergent channels are reported in Tables 15.6 and 15.7. The tables show the variations in nanofluid velocity and temperature for a volume percentage of [0%, 1%, and 2%] that is unknown. It is important to keep in mind that when the volume fraction decreases, the issue is solely one of fluid flow.

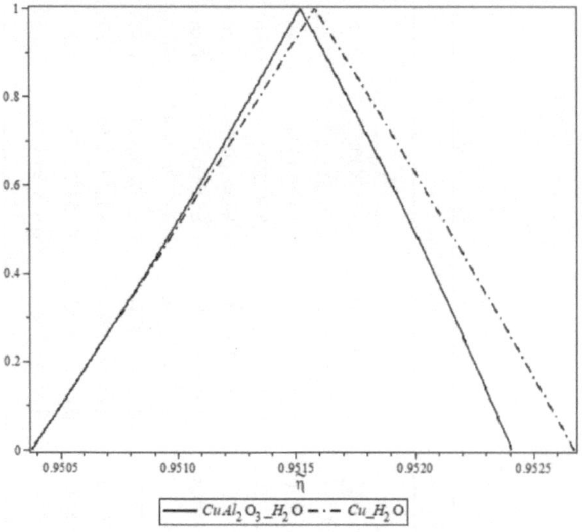

FIGURE 15.12(a) Fuzzy velocity at $\xi = 0.3$ $(\omega < 0)$.

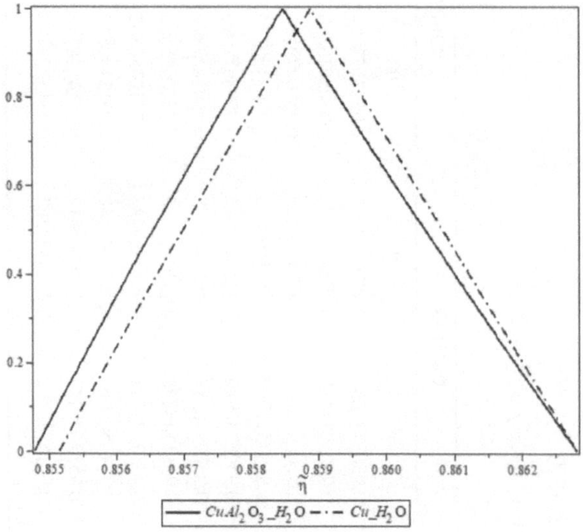

FIGURE 15.12(b) Fuzzy velocity at $\xi = 0.3$ $(\omega > 0)$.

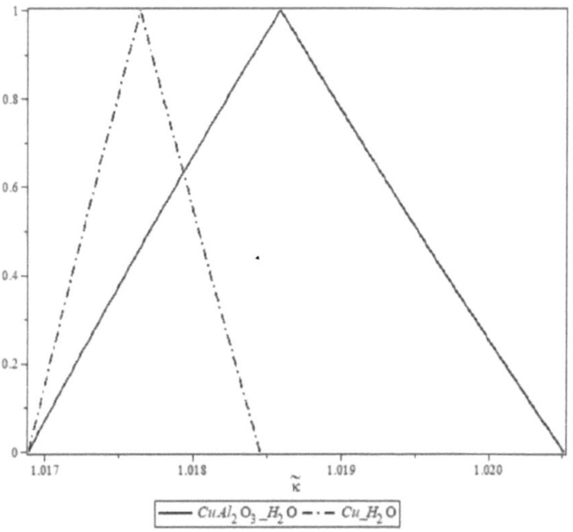

FIGURE 15.13(a) Fuzzy temperature at $\xi = 0.3 \ (\omega < 0)$.

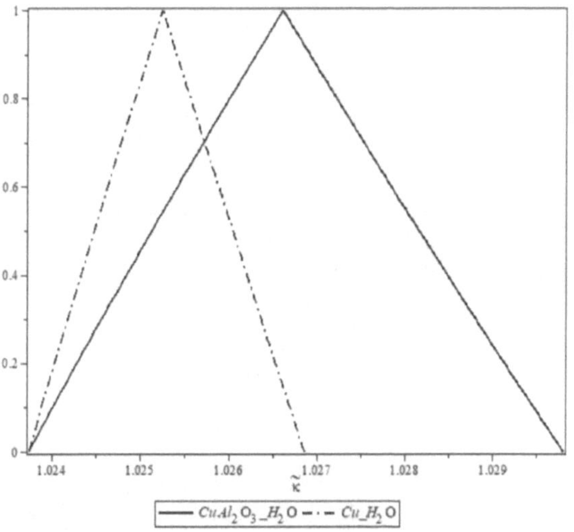

FIGURE 15.13(b) Fuzzy temperature at $\xi = 0.3 \ (\omega > 0)$.

15.7 CONCLUSIONS

Here, a double-parametric form-based HAM is proposed to study the fuzzy differential equation through the β-cut concept. Convergence analysis of the technique with an uncertain volume fraction for both crisp and uncertain conditions has been checked through theorems, corollaries, and numerically using term approximations for both divergent and convergent channels. The existing result is available only for

TABLE 15.7
Fuzzy Temperature at $\xi = 0.3$

	$\omega = -5^{\circ}$				$\omega = 5^{\circ}$			
	Cu/H_2O		$Cu\text{-}Al_2O_3/H_2O$		Cu/H_2O		$Cu\text{-}Al_2O_3/H_2O$	
β	$\underline{\kappa}$	$\overline{\kappa}$	$\underline{\kappa}$	$\overline{\kappa}$	$\underline{\kappa}$	$\overline{\kappa}$	$\underline{\kappa}$	$\overline{\kappa}$
0.0	1.012685751	1.014161835	1.012685751	1.015697089	1.023744444	1.026863487	1.023744444	1.029805806
0.1	1.012755198	1.014083467	1.012820529	1.015529481	1.023891484	1.026698215	1.024016914	1.029469803
0.2	1.012825083	1.014005602	1.012956846	1.015363807	1.024039421	1.026533968	1.024292371	1.029137527
0.3	1.012895409	1.013928236	1.013094720	1.015200042	1.024188260	1.026370740	1.024570854	1.028808933
0.4	1.012966180	1.013851365	1.013234171	1.015038161	1.024338010	1.026208522	1.024852400	1.028483975
0.5	1.013037398	1.013774986	1.013375218	1.014878141	1.024488676	1.026047306	1.025137046	1.028162608
0.6	1.013109068	1.013699095	1.013517883	1.014719957	1.024640268	1.025887085	1.025424829	1.027844788
0.7	1.013181192	1.013623687	1.013662186	1.014563588	1.024792790	1.025727853	1.025715790	1.027530472
0.8	1.013253774	1.013548758	1.013808148	1.014409010	1.024946249	1.025569600	1.026009965	1.027219617
0.9	1.013326819	1.013474308	1.013955790	1.014256198	1.025100652	1.025412321	1.026307396	1.026912180
1.0	1.013400328	1.013400328	1.014105132	1.014105132	1.025256007	1.025256007	1.026608120	1.026608120

the divergent case. Finally, the obtained results were validated to check the efficiency of the proposed method, and we found that the obtained numerical results are in good agreement with the available results with the crisp and fuzzy conditions with the more minor error approximations.

The following observations were also made:

i. In both channels, the velocity profiles for the nanofluids Cu/H_2O and $Cu\text{-}Al_2O_3/H_2O$ are the same. However, in both channels, the temperature profile of the hybrid $Cu\text{-}Al_2O_3/H_2O$ nanofluid is greater than that of the Cu/H_2O nanofluid.
ii. In both channels, the temperature profile increases as the Reynolds number increases.
iii. The fluid temperature of $Cu\text{-}Al_2O_3/H_2O$ is greater than that of the Cu/H_2O nanofluid for both channels.
iv. $Cu\text{-}Al_2O_3/H_2O$ nanofluids have nearly comparable velocity profiles to Cu/H_2O; however, both channels have a wider temperature distribution.
v. The temperature profile is found to increase as the Eckert number increased. The $Cu\text{-}Al_2O_3/H_2O$ nanofluid temperature is greater in both channels than the Cu/H_2O nanofluid temperature.

Consequently, fuzzy triangular numbers can also be employed to describe the velocities at different places along both the convergent and divergent channels. The calculation of the velocity bound is performed at different positions along the convergent and divergent channels, given a specific set of values for the associated parameters.

REFERENCES

1. Sheikholeslami, M., Mollabasi, H., Ganji, D. (2015). Analytical investigation of MHD Jeffery–Hamel nanofluid flow in non-parallel walls. International Journal of Nanoscience and Nanotechnology, 11(4), 241–248.
2. Dogonchi, A., Ganji, D. (2016). Study of nanofluid flow and heat transfer between non-parallel stretching walls considering Brownian motion. Journal of the Taiwan Institute of Chemical Engineers, 69, 1–13.
3. Jeffery, G.B. (1915). The two-dimensional steady motion of a viscous fluid. The London, Edinburgh, and Dublin Philosophical Magazine and Journal of Science, 29(172), 455–465.
4. Hamel, G. (1917). Spiralförmige bewegungen zäher flüssigkeiten. Jahresbericht der deutschen mathematiker-vereinigung, 25, 34–60.
5. Nagler, J. (2017). Jeffery-Hamel flow of non-Newtonian fluid with nonlinear viscosity and wall friction. Applied Mathematics and Mechanics, 38, 815–830.
6. Rezaei, M., Azimian, A., Toghraie, D. (2015). Molecular dynamics study of an electro-kinetic fluid transport in a charged nanochannel based on the role of the stern layer. Physica A: Statistical Mechanics and its Applications, 426, 25–34.
7. Hosseinzadeh, K., Roghani, S., Mogharrebi, A., Asadi, A., Ganji, D. (2021). Optimization of hybrid nanoparticles with mixture fluid flow in an octagonal porous medium by effect of radiation and magnetic field. Journal of Thermal Analysis and Calorimetry, 143, 1413–1424.

8. Salehi, S., Nori, A., Hosseinzadeh, K., Ganji, D. (2020). Hydrothermal analysis of MHD squeezing mixture fluid suspended by hybrid nanoparticles between two parallel plates. Case Studies in Thermal Engineering, 21, 100650.
9. Sheikholeslami, M., Ganji, D., Ashorynejad, H., Rokni, H.B. (2012). Analytical investigation of Jeffery-Hamel flow with high magnetic field and nanoparticle by Adomian decomposition method. Applied Mathematics and Mechanics, 33, 25–36.
10. Motsa, S., Sibanda, P., Awad, F., Shateyi, S. (2010). A new spectral-homotopy analysis method for the MHD Jeffery–Hamel problem. Computers & Fluids, 39(7), 1219–1225.
11. Bansal, J. (1994). Magnetofluiddynamics of viscous fluids. Jaipur Publishing House.
12. Nijsing, R., Eifler, W. (1980). A computational analysis of transient heat transfer in fuel rod bundles with single phase liquid metal cooling. Nuclear Engineering and Design, 62(1–3), 39–68.
13. Cha, J.E., Ahn, Y.C., Kim, M.H. (2002). Flow measurement with an electromagnetic flowmeter in two-phase bubbly and slug flow regimes. Flow Measurement and Instrumentation, 12(5–6), 329–339.
14. Ali, H.M. (2020). Hybrid nanofluids for convection heat transfer. Academic Press.
15. Jamaludin, A., Nazar, R., Naganthran, K., Pop, I. (2021). Mixed convection hybrid nanofluid flow over an exponentially accelerating surface in a porous media. Neural Computing and Applications, 33(22), 15719–15729.
16. Mishra, A., Upreti, H. (2022). A comparative study of Ag–MgO/water and Fe_3O_4–$COFe_2O_4$/Eg–water hybrid nanofluid flow over a curved surface with chemical reaction using Buongiorno model. Partial Differential Equations in Applied Mathematics, 5, 100322.
17. Ahmad, S., Ali, K., Faridi, A.A., Ashraf, M. (2021). Novel thermal aspects of hybrid nanoparticles Cu-TiO_2 in the flow of ethylene glycol. International Communications in Heat and Mass Transfer, 129, 105708.
18. Gupta, S.K., Misra, R.D. (2020). Development of micro/nanostructured-Cu-TiO_2-nanocomposite surfaces to improve pool boiling heat transfer performance. Heat and Mass Transfer, 56(8), 2529–2544.
19. Bede, B., Gal, S.G. (2005). Generalizations of the differentiability of fuzzy-number-valued functions with applications to fuzzy differential equations. Fuzzy Sets and Systems, 151(3), 581–599.
20. Sartanpara, P.P., Meher, R. (2023). A robust fuzzy-fractional approach for the atmospheric internal wave model. Journal of Ocean Engineering and Science, 8(3), 308–322.
21. Verma, L., Meher, R. (2024). Solution for generalized fuzzy time-fractional fisher's equation using a robust fuzzy analytical approach. Journal of Ocean Engineering and Science, 9(5), 475–488.
22. Sartanpara, P.P., Meher, R. (2023). Solution of generalised fuzzy fractional Kaup–Kupershmidt equation using a robust multi parametric approach and a novel transform. Mathematics and Computers in Simulation, 205, 939–969.
23. Verma, L., Meher, R., Avazzadeh, Z., Nikan, O. (2023). Solution for generalized fuzzy fractional Kortewege-de varies equation using a robust fuzzy double parametric approach. Journal of Ocean Engineering and Science, 8(6), 602–622.
24. Meher, R., Verma, L., Avazzadeh, Z., Nikan, O. (2023). Study of MHD nanofluid flow with fuzzy volume fraction in thermal field-flow fractionation. AIP Advances, 13(1), 015204.
25. Chakraverty, S., Tapaswini, S., Behera, D. (2016). Fuzzy differential equations and applications for engineers and scientists. CRC Press.

26. Biswal, U., Chakraverty, S., Ojha, B. (2020). Natural convection of non-Newtonian nanofluid flow between two vertical parallel plates in uncertain environment. In: Recent Trends in Wave Mechanics and Vibrations. Springer, pp. 295–309.
27. Allahviranloo, T., Kiani, N.A., Motamedi, N. (2009). Solving fuzzy differential equations by differential transformation method. Information Sciences, 179(7), 956–966.
28. Ahmad, M., Hasan, M.K., Abbasbandy, S. (2013). Solving fuzzy fractional differential equations using Zadeh's extension principle. The Scientific World Journal, 2013(1), 454969.
29. Patel, N., Meher, R. (2018). Investigation of a Jeffery–Hamel flow between two rectangular inclined smooth walls using the differential transform method. Journal of Applied Mathematics and Computational Mechanics, 17(4), 47–57.
30. Verma, L., Meher, R. (2022). Effect of heat transfer on Jeffery–Hamel Cu/Ag–water nanofluid flow with uncertain volume fraction using the double parametric fuzzy homotopy analysis method. The European Physical Journal Plus, 137(3), 1–20.
31. Gohil, V., Meher, R. (2017). Homotopy analysis method for solving counter current imbibition phenomena of the time positive fractional type arising in heterogeneous porous media. International Journal of Mathematics & Computation, 28(2), 77–85.
32. Odibat, Z.M. (2010). A study on the convergence of homotopy analysis method. Applied Mathematics and Computation, 217(2), 782–789.

16 Marangoni Boundary Layer Flow of an Electrically Conducting Hybrid Nanofluid for the Impact of Particle Shape and Thermal Radiation

S.R. Mishra, Rupa Baithalu, and Subhajit Panda

16.1 INTRODUCTION

Over the past 20 years, there has been significant research into the basic principles of the transport mechanisms of nanofluids, primarily driven by their augmented thermal conductivity. Choi [1] coined the term "nanofluids" to describe fluids including nanoparticles. Their research suggests that the integration of solid particles, such as metals, oxides, and carbon nanotubes, into base fluids can greatly increase the conductivity and characteristics of the heat transfer fluid. As a result, nanofluids find extensive applications in a variety of industries, including the aerospace industry, microelectronics, automotive engineering, powder energy, and biomedical area.

Earlier investigation indicated that nanofluids demonstrate enhanced heat transport properties compared to traditional fluids. A contributing factor to this phenomenon is the significant development of conductivity for the presence of suspended particles within nanofluids. However, to date, the development of a comprehensive theory to accurately forecast the conductivity of nanofluids remains an unresolved challenge. In their investigation of the heat transport properties of nanofluids, Lomascolo and colleagues [2] employed an analysis of the experimental findings pertaining to convective, radiative, and conductive processes.

Their research focused on the results of experiments and the thermal properties of nanofluids. The analysis delved into various parameters affecting the thermal conductivity of nanofluids, including particle type, their size including shape, adequate temperature, and acidity. Furthermore, Pang et al. [3] conducted a comprehensive assessment focusing on the latest advancements concerning various nanofluids and their thermal properties. Their objective was to offer insights for future research directions in this field. The review highlighted a predominant focus on investigating the thermal conductivity, and the role of suspended particles shows greater performance within nanofluids.

 DOI: 10.1201/9781003595786-16

The Marangoni boundary layer is the layer of dissipation that forms at the interfaces between liquid and gas, or liquid and liquid. Marangoni flow is triggered by either the phenomenon governed by "*surface temperature gradient*" or a "*surface concentration gradient*." It is observed in many real-world applications, including chemical reaction processes [4, 5], crystal growth [6, 7], thin liquid films [8–9], and silicon melts [10]. The proposed effects can be characterized into the "*thermal Marangoni effect*" (EMT) and the "*solute Marangoni effect*" (EMS). Pearson [11] explored the mechanism of EMT, elucidating that heating a thin fluid layer from beneath generates a temperature gradient wherein slight alterations in surface temperature prompt surface tractions, inducing fluid movement that aims to sustain the actual heat differences. The application of the EMT is thus established.

16.2 MATHEMATICAL FORMULATION

The flow of a CNT-water-based hybrid nanofluid with Marangoni convection is analysed in the proposed study. The proposed coordinate system reveals that the main flow is towards the x-axis and the normal flow is along the y-axis. Additionally, the applied magnetic field is imposed along the transverse direction of the flow (Figure 16.1(a)).

The affirmation subject cited here leads to design of the governing equation for the several flow

$$U_X + V_Y = 0, \tag{16.1}$$

$$\rho_{hnf}\left(uU_X + vU_Y\right) = \mu_{hnf}U_{YY} - \sigma_{hnf}B_0^2U, \tag{16.2}$$

$$\left(\rho Cp\right)_{hnf}\left(UT_X + VT_Y\right) = k_{hnf}T_{YY} - \left(q_r\right)_r. \tag{16.3}$$

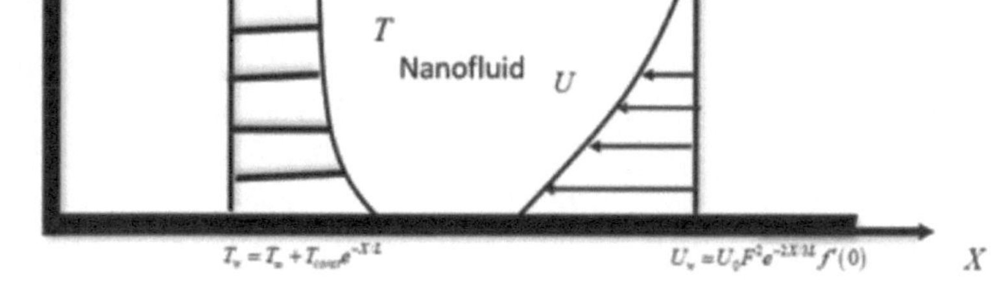

FIGURE 16.1 Flow diagram.

The boundary conditions are

$$\left. \begin{array}{l} Y = 0: \mu_{hnf} U_y\big|_{Y=0} = \sigma_X\big|_{Y=0}, V\big|_{Y=0} = 0, T\big|_{Y=0} = T_w = T_\infty + T_{const} e^{-X/L_0} \\ Y \to \infty: U\big|_{Y\to\infty} = 0, T\big|_{Y\to\infty} = T_\infty \end{array} \right\}. \quad (16.4)$$

Additionally, the viscosity, denoted as μ_{nf}, and base fluid viscosity, μ_f, that incorporates a sparse dispersion of fine spherical particles are described by Baithalu et al. [12]

By employing the Rosseland approximation, the radiative heat flux is given as

$$q_r = -\frac{4\sigma^*}{3k^*}\left(T^4\right)_Y. \quad (16.5)$$

Here, δ^* represents the Stefan–Boltzmann constant, while k^* denotes the average absorption.

$$T^4 \approx 4T_\infty^3 T - 3T_\infty^4$$

In Equation (16.4), σ represents the surface tension along with the surface tension gradient (interfacial).

$$(\sigma)_X = (\sigma)_T * (T)_X$$

It is generally presumed that the surface tension increases linearly with temperature, meaning that

$$\sigma = \sigma_0 - \gamma_T (T - T_\infty), \gamma_T = -(\sigma)_T.$$

The subsequent dimensionless variables are presented:

$$u = \frac{U}{U_0}, v = \frac{V}{U_0}\left(\frac{U_0 L_0}{\gamma_f}\right) = \frac{V}{U_0}\mathrm{Re}^{1/2}, x = \frac{X}{L_0}, y = \frac{Y}{L_0}\left(\frac{U_0 L_0}{\gamma_f}\right)^{1/2} = \frac{Y}{L_0}\mathrm{Re}^{1/2}, t = \frac{T}{T_\infty},$$

$$\mathrm{Re} = \frac{U_0 L_0}{\gamma_f}, \mathrm{Pr} = \frac{\upsilon_f}{\alpha_f}, \upsilon_f = \frac{\mu_f}{\rho_f}, Ma = \frac{\gamma_T T_{const} L_0}{\mu_f \alpha_f}, Nr = \frac{16\delta^* T_\infty}{3k^* k_{nf}}.$$

Here, Ma is the Marangoni number and Nr is the radiation parameter. Implementing these, the resulting Equations (16.1)–(16.3) are obtained as

$$u_x + v_y = 0, \quad (16.6)$$

$$uu_x + vu_y = \frac{A_1}{A_2}u_{yy} - \frac{A_3}{A_2}Mf', \quad (16.7)$$

$$ut_x + vt_y = \frac{1}{\Pr}(A_4 + Nr)t_{yy} - \frac{1}{A_5}(q_r)_r. \tag{16.8}$$

The boundary conditions (16.4) and (16.5) can be expressed as

$$\left.\begin{array}{l} y=0: \dfrac{T_{const}}{T_\infty} u_y\Big|_{y=0} = -\dfrac{Ma}{\Pr}\left(\dfrac{1}{\text{Re}}\right)^{3/2} ct_x\Big|_{y=0}, v=0, \\ t\big|_{y=0} = 1 + \dfrac{T_{const}}{T_\infty} e^{-x}, \\ y \to \infty : u\big|_{y\to\infty} = 0, t\big|_{y\to\infty} = 1 \end{array}\right\}. \tag{16.9}$$

Using the usual stream function, i.e., $u = \psi_y, v = -\psi_x$, the transformation variables are deployed as

$$\left.\begin{array}{l} \psi(x,y) = Ae^{-x/3} f(\eta), t(x,y) = 1 + \dfrac{T_{const}}{T_\infty} e^{-x}\theta(\eta) \\ A = \left(\dfrac{Ma}{\Pr}\right)^{1/3}\left(\dfrac{1}{\text{Re}}\right)^{1/2}, \eta = Ae^{-x/3}y, \end{array}\right\}, \tag{16.10}$$

Furthermore, the proposed Equations. (6)–(8) can be transformed into

$$f'''(\eta) - \frac{A_2}{A_1}\left[\frac{1}{3} f(\eta) f''(\eta) - \frac{2}{3} f'(\eta)^2\right] - \frac{A_3}{A_1} Mf' = 0, \tag{16.11}$$

$$\theta''(\eta) - \frac{\Pr}{(A_4 + Nr)}\left[\frac{1}{3} f(\eta)\theta'(\eta) - f'(\eta)\theta(\eta)\right] = 0, \tag{16.12}$$

and the surface conditions (16.9) are reduced to

$$\left.\begin{array}{l} f(0) = 0, f''(0) = c, f'(\infty) = 0, \\ \theta(0) = 1, \theta(\infty) = 0. \end{array}\right\}. \tag{16.13}$$

Furthermore, the velocity and temperature transformations are given as

$$\left.\begin{array}{l} U = U_0\left(\dfrac{Ma}{\Pr}\right)^{2/3} \dfrac{1}{\text{Re}} e^{\frac{-2X}{3L_0}} f'(\eta) = \left(\dfrac{\gamma_T T_{const}\gamma_f}{\rho_f L_0^2}\right)^{2/3} \dfrac{\text{Re}}{U_0} e^{\frac{-2X}{3L_0}} f'(\eta), \\ V = \dfrac{1}{3}\left(\dfrac{\gamma_T T_{const}\gamma_f}{\rho_f L_0^2}\right)^{1/3} e^{\frac{-2X}{3L_0}} \left[f(\eta) + \eta f'(\eta)\right], \\ T = T_\infty + T_{const} e^{\frac{-X}{L_0}} \theta(\eta) \end{array}\right\}. \tag{16.14}$$

The physical quantity of local Nusselt number (Nu_x) is presented as

$$Nu_x = \frac{Xq_w(X)}{k(T)\left[T(X,0) - T(X,\infty)\right]}, \tag{16.15}$$

where $q_w(X)$ is the heat flux of the nanofluid. Using Equations. (16.14) and (16.15), Nu_x becomes

$$Nu_x = -\frac{X}{L_0} e^{-\frac{X}{3L_0}} \left(\frac{\gamma_T^2 T_{const}^2}{\rho_f^{\,2} U_0^3 L_0 \gamma_f} \right)^{1/6} \theta'(0).$$

Here $-\theta'(0)$ represents the local Nu_x.

16.3 RESULTS AND DISCUSSION

The Marangoni forced convection of an electrically conducting hybrid nanofluid combined with CNT nanoparticles in the base liquid water is presented in this investigation. A transverse magnetic field was applied to conduct the effect of magnetization along with the radiating heat for the consideration of Rosseland approximation in the heat transport phenomenon. Based on the model of thermal conductivity considered here, Figure 16.2 presents various shapes of the nanoparticles.

Table 16.1 presents the physical properties such as viscosity, density, thermal and electrical conductivity models. Furthermore, Table 16.2 shows the numerical results obtained experimentally for the thermal attributes of both the CNT nanoparticles and the base fluid water at a standard temperature of 300 K. The transformed governing set of non-linear equations with the proposed boundary conditions is solved numerically, implementing the shooting-based Runge–Kutta fourth-order technique. The parametric analysis for each of the parameters is presented briefly through graphs, and the range of these factors is depicted in the corresponding figure.

Figure 16.2 illustrates the significant structural behaviour of the magnetized field effect due to the presence of the magnetic parameter on the fluid velocity distribution. The variation in the magnetic parameter is presented within the range of [0.1, 0.4], and the behaviour is described in three distinct cases, such as pure fluid $(\phi_{1SWCNT} = \phi_{1MWCNT} = 0)$, nanofluid $(\phi_{1SWCNT} = 0.01, \phi_{1MWCNT} = 0)$, and hybrid nanofluid $(\phi_{1SWCNT} = 0.01, \phi_{1MWCNT} = 0.02)$. The inclusion of an applied magnetic field, a resistive force that produces Lorentz force, resists the fluid motion. Therefore, although an increase in the magnetic parameter increases the magnitude of the

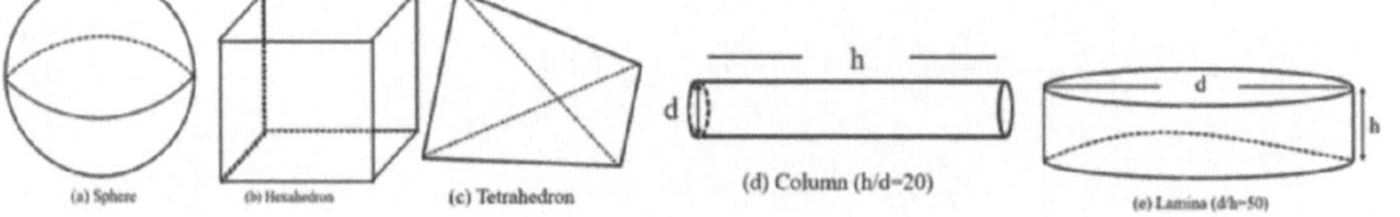

FIGURE 16.2 Shape of nanoparticles.

TABLE 16.1
Thermophysical Models for Hybrid Nanofluids

Attribute	Hybrid nanofluid
Density	$\rho_{hnf} = \rho_{nf}\left[(1-\phi_{SWCNT})\left[1-\phi_{MWCNT}+\phi_{MWCNT}\left(\frac{\rho_{MWCNT}}{\rho_{water}}\right)\right]+\phi_{SWCNT}\left(\frac{\rho_{SWCNT}}{\rho_{water}}\right)\right]$
viscosity	$\mu_{hnf} = \mu_{nf}(0.904)^2 e^{14.8(\phi_{SWCNT}+\phi_{MWCNT})}$
Heat capacity	$(\rho c_p)_{hnf} = (1-\phi_{SWCNT})\left[(1-\phi_{MWCNT})(\rho c_p)_{Water} + \phi_{MWCNT}(\rho c_p)_{MWCNT}\right] + \phi_{SWCNT}(\rho c_p)_{SWCNT}$
Thermal conductivity	$\frac{k_{hnf}}{k_{nf}} = \left[\frac{k_{MWCNT}+(n-1)k_{nf}-(n-1)\phi_{MWCNT}(k_{nf}-k_{MWCNT})}{k_{MWCNT}+(n-1)k_{nf}-\phi_{MWCNT}(k_{nf}-k_{MWCNT})}\right],$ where $\frac{k_{nf}}{k_f} = \left[\frac{k_{SWCNT}+(n-1)k_{Water}-(n-1)\phi_{SWCNT}(k_{Water}-k_{SWCNT})}{k_{SWCNT}+(n-1)k_{Water}-\phi_{SWCNT}(k_{Water}-k_{SWCNT})}\right]$
Electrical conductivity	$\frac{\sigma_{hnf}}{\sigma_{nf}} = \frac{\sigma_{MWCNT}+2\sigma_{nf}-2\phi_{MWCNT}(\sigma_{nf}-\sigma_{MWCNT})}{\sigma_{MWCNT}+2\sigma_{nf}+\phi_{MWCNT}(\sigma_{nf}-\sigma_{MWCNT})},$ where $\frac{\sigma_{nf}}{\sigma_f} = \frac{\sigma_{SWCNT}+2\sigma_{Water}-2\phi_{SWCNT}(\sigma_{Water}-\sigma_{SWCNT})}{\sigma_{SWCNT}+2\sigma_{Water}+\phi_{SWCNT}(\sigma_{Water}-\sigma_{SWCNT})}$

TABLE 16.2
Thermophysical Properties of CNT Nanoparticles and Water

	ρ	Cp	k	σ
Water (H_2O)	997.1	4179	0.613	0.05
SWCNTs	2600	425	6600	10^6
MWCNTs	1600	796	3000	10^6

velocity profile, it led to a decrease in the bounding surface thickness, which validates the fact of the magnetization. Furthermore, the comparative analysis presents that the magnitude of the profile augments for the case of pure fluids as the thickness of the bounding surface decreases greatly. The combination of SWCNTs as well as both SWCNTs and MWCNTs in the base fluid water retards the fluid velocity gradually. The fact is that infusion of nanoparticles of heavier density results in the clumping of the particles, which attenuates the profile significantly.

Figure 16.3 shows the structural behaviour of the particle concentration of SWCNTs depicted in the velocity profile of a hybrid nanofluid. Here, the variation in $\phi_1 \approx \phi_{SWCNT}$ is obtained keeping $\phi_2 \approx \phi_{MWCNT}$ as fixed. However, the result is depicted

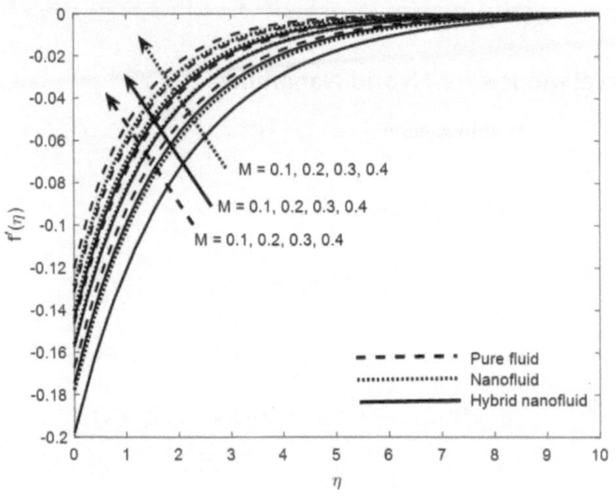

FIGURE 16.3 Variation in M based on $f'(\eta)$.

for the two different cases of magnetization, particularly considering the absence/presence of a magnetic field. As described in the earlier figure, it is seen that with increasing concentration of SWCNT nanoparticles, the profile strength decreases significantly, which causes the bounding surface thickness to gradually increase. However, the impact is reversed for increasing magnetization. The analysis reveals that the absence of a magnetic field produces greater deceleration in the velocity profile, which is beneficial for the production process of different products in industries. This also helps in achieving better shape and size without damage of the product.

Figure 16.4 shows the curvature parameter obtained due to the surface tension gradient at the surface. The variation is presented considering pure, nano, and hybrid nanofluid cases. With the increasing curvature constant, the fluid velocity decreases significantly, and the retardation gradually increases with the addition of various nanoparticles. That is, the hybrid nanofluid case dominates over the case of nanofluids and pure fluids. The variation in the contributing factors on the fluid temperature is also presented briefly.

Figure 16.5 explores the role of magnetization affecting the fluid temperature distribution, and the effect is deliberated for various fluids. In particular, the comparison is obtained by considering the pure fluid water, SWCNT-water nanofluid, and SWCNT + MWCNT~water hybrid nanofluid. The increasing magnetization that decelerates the fluid velocity bounding surface thickness due to its resistivity offered by the generation of Lorentz force, it adequately stored energy at the surface. Since particle clogging is obtained near the surface region with greater magnetization, the energy is stored there, and further it boosts up to enhance it throughout the domain. Therefore, the increase in the magnetic parameter suitably increases the fluid temperature. However, it is interesting to observe that the hybrid nanofluid favours controlling the fluid temperature throughout.

The proposed design is obtained for the combined effect of CNT nanoparticles that perform the role of hybrid nanofluid; the contribution of the physical properties

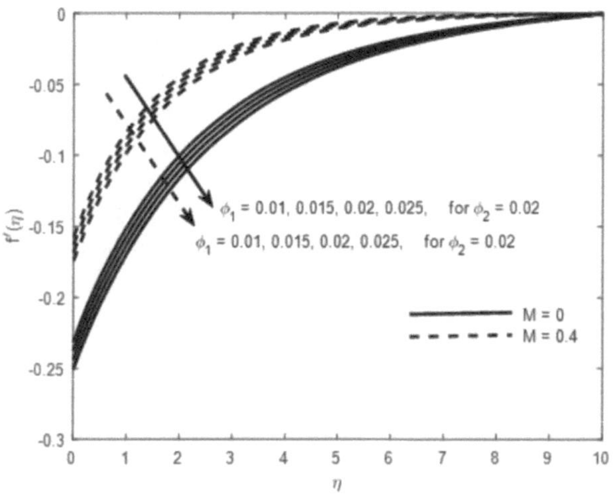

FIGURE 16.3 Variation of ϕ_1 on $f'(\eta)$.

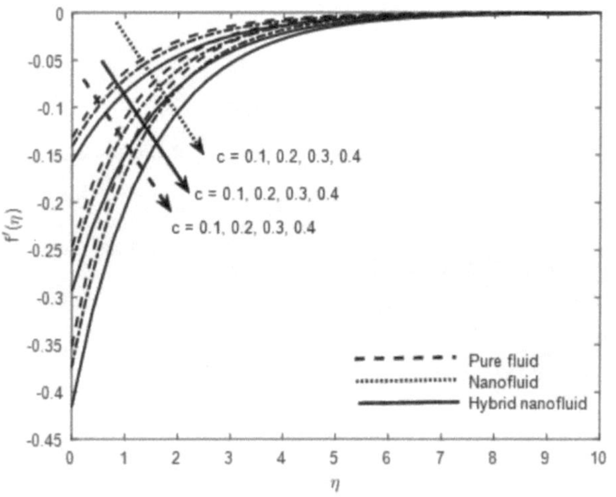

FIGURE 16.4 Variation of c on $f'(\eta)$.

is based on the role of volume fraction. Since each of the thermophysical properties are equipped with the particle concentration i.e. the enhanced properties of the viscosity, specific heat, thermal and electrical conductivity are obtained due to the presence of particle concentration and this overrides the fact of base fluid. Since Figure 16.6, the concentration of MWCNT is fixed and variation of SWCNT is obtained. This shows the effect of hybrid nanofluid. Although the variation is not so significant, it is clear to observe that for the case of hybrid nanofluid the profile shows greater enhancement in the fluid temperature. This is because of the inclusion of particle concentration in the base fluid, and the increase in conductivity of the base fluid increases the fluid temperature. Furthermore, with the increasing concentration of SWCNT nanoparticles, the profile also augments significantly.

Figure 16.7 shows the impact of thermal radiation on the fluid temperature. Also, the behaviour is projected for the consideration of pure, nano, and hybrid nanofluids Thermal radiation is the measure of electromagnetic wave that radiates from the surface of the fluid elements. This electromagnetic wave is then transformed into radiating heat, known as thermal radiation. The heat that radiates from the surface

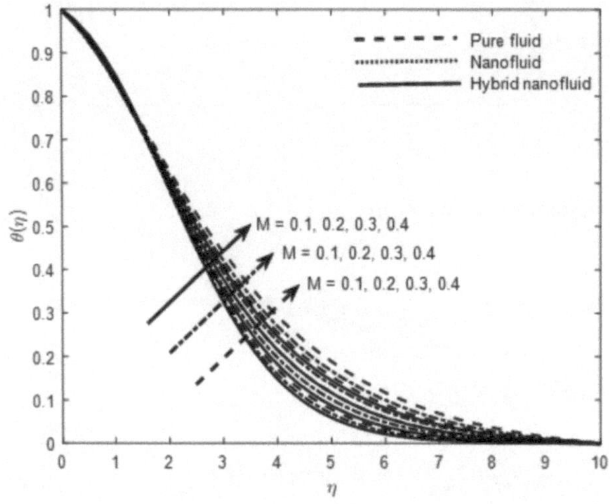

FIGURE 16.5 Variation of M on $\theta(\eta)$.

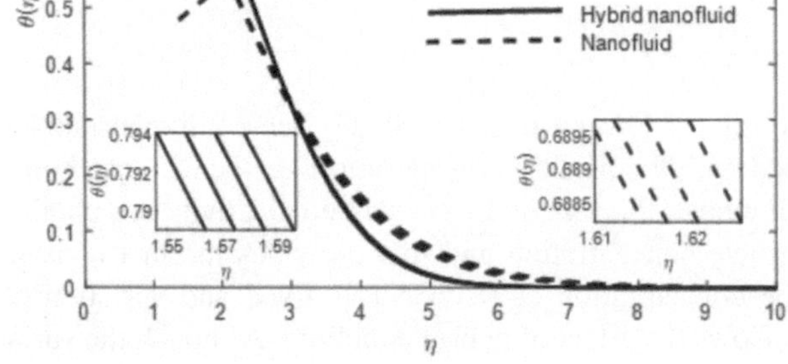

FIGURE 16.6 Variation of ϕ_1 on $\theta(\eta)$.

provides a cooling effect, i.e., the surface becomes cool. Therefore, the increase in thermal radiation enhances the fluid temperature as the radiation from the surface overshoots the profile temperature. The greater conductivity of the hybrid nanofluid significantly encourages the fluid temperature.

Finally, the numerical computation of the heat transfer rate for various parameters in the case of pure, nanofluid, and hybrid nanofluid is presented in Table 16.3. The result reveals that with enhanced magnetization and resistivity, the heat transfer rate increases, whereas with an increase in the Prandtl number, the heat transfer rate decreases significantly. Furthermore, the radiating heat also overshoots the heat transfer rate throughout. The comparative analysis reveals that for the increasing concentration in the case of hybrid nanofluid, the heat transfer rate decreases.

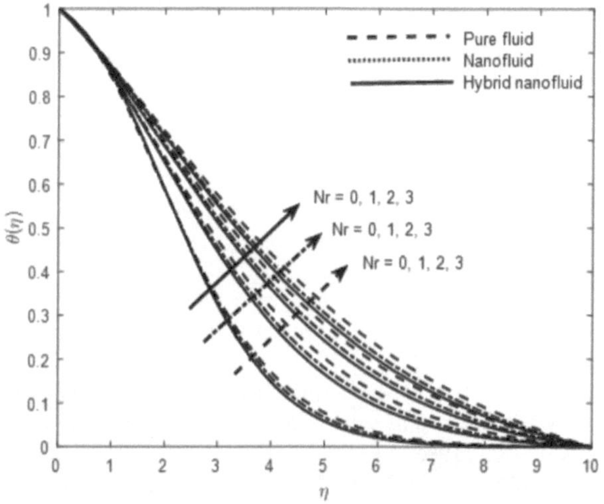

FIGURE 16.7 Variation of Nr on $\theta(\eta)$.

TABLE 16.3
Heat Transfer for Various Parameters

M	*Pr*	*Nr*	**Pure**	**Nano**	**Hybrid**
0	7	0	0.067939	0.063366	0.061392
0.1			0.091413	0.086043	0.084208
0.2			0.105886	0.101251	0.100819
0.3			0.113827	0.110419	0.111779
0.1	6		0.099262	0.094682	0.094159
	4		0.112274	0.109932	0.112926
	7	1	0.228719	0.225324	0.225178
		2	0.348164	0.347266	0.35104
		3	0.458984	0.459804	0.466722

16.4 CONCLUSIONS

The Marangoni boundary layer flow of a hybrid nanofluid comprising CNT nanoparticles in the base fluid water is exhibited in the present analysis. The electrically conducting hybrid nanofluid flow in association with thermal radiation and magnetization affects the flow phenomena. The transformed governing equations equipped with several characterizing parameters are solved numerically, and the behaviours of this parameter are presented briefly. The important outcomes of the results are presented as follows:

i. The resistivity offered by the magnetization due to the inclusion of the magnetic parameter reduces the thickness of the bounding surface.
ii. The higher density of the particle concentration diminishes the fluid velocity, whereas an increase in the temperature profile augments fluid velocity significantly.
iii. The enhanced magnetic field augments the fluid temperature irrespective of the choice of fluid considered in the study.
iv. Both the thermal radiation and the magnetic field show a greater impact on the heat transfer rate of the hybrid nanofluid.

REFERENCES

1. Choi S.U.S. (1995). Enhancing thermal conductivity of fluids with nanoparticles. ASME Publications Fed, 231, 99–106.
2. Lomascolo M., Colangelo G., Milanese M., Risi A. D. (2015). Review of heat transfer in nanofluids: Conductive, convective and radiative experimental results. Renew. Sust. Energ. Rev., 43, 1182–1198.
3. Pang C.W., Lee J.W., Kang Y.T. (2015). Review on combined heat and mass transfer characteristics in nanofluids. Int. J. Therm. Sci., 87, 49–67.
4. Zhang Y., Zheng L.C. (2012). Analysis of MHD thermosolutal Marangoni convection with the heat generation and a first-order chemical reaction. Chem. Eng. Sci., 69, 449–455.
5. Zhang Y., Zheng L.C. (2014). Similarity solutions of Marangoni convection boundary layer flow with gravity and external pressure. Chin. J. Chem. Eng., 22(4), 365–369.
6. Arafune K., Hirata A. (1999). Thermal and solutal Marangoni convection in In-Ga-Sb system. J. Cryst. Growth, 197, 811–817.
7. Ruyer-Quil C., Scheid B., Kalliadasis S., Velarde M.G., Zeytounian K. (2005). Thermocapillary long wave in a liquid film flow. Part 1. Low-dimensional formulation. J. Fluid Mech., 538, 199–222.
8. Scheid B., Ruyer-Quil C., Kalliadasis S., Velarde M.G., Zeytounian K. (2005). Thermocapillary long wave in a liquid film flow. Part 2. Linear stability and nonlinear waves. J. Fluid Mech., 538, 223–244.
9. Chen C. (2007). Marangoni effects on forced convection of power-law liquids in a thin film over a stretching surface. Phys. Lett. A, 370, 51–57.
10. Li J., Li M.W., Hu W.R., Zeng D.L. (2003). Suppression of Marangoni convection of silicon melt by a non-contaminating method. Int. J. Heat Mass Transf., 46, 4969–4973.

11. Pearson J.R.A. (1958). On convection cells induced by surface tension. J. Fluid Mech., 4(5), 489–500.
12. Baithalu R., Mishra S.R., Ali Shah N. (2023). Sensitivity analysis of various factors on the micropolar hybrid nanofluid flow with optimized heat transfer rate using response surface mcthodology: A statistical approach. Phys. Fluids, 35(10).

17 Legendre Wavelet Collocation Approach to Simulate Hybrid Nanofluid Flow with Ohmic Heating and Viscous Dissipation under Magnetic Field

Tanya Gupta

Nomenclature

Ec	Eckert number
B_0	Magnetic field strength (A m^{-1})
Mn	Magnetic parameter
Nu	Nusselt number
K	Porosity parameter
Pr	Prandtl number
Re_x	Reynolds number
C_f	Skin friction coefficient
C_P	Specific heat (J kg^{-1} K^{-1})
S	Suction/injection parameter
T	Temperature (K)
κ	Thermal conductivity (W m^{-1} K^{-1})
Ψ	Legendre wavelet
θ	Non-dimensional temperature
α	Inclination angle
η	Similarity variable

DOI: 10.1201/9781003595786-17

ϕ	Volume fraction of nanoparticles (%)
u, v	Velocity components along the x- and y-axes, respectively ($m \cdot s^{-1}$)
T_w	Wall temperature (K)
ρ	Density ($kg\ m^{-3}$)
Ω_i	Dimensionless constant
μ	Dynamic viscosity ($kg\ m^{-1}\ s^{-1}$)
σ	Electrical conductivity ($\Omega^{-1}\ m^{-1}$)

17.1 INTRODUCTION

Hybrid nanofluids (HNFs), a remarkable fusion of nanotechnology and fluid dynamics, represent a cutting-edge innovation poised to revolutionize a multitude of industries. These extraordinary concoctions are crafted by suspending nanometre-sized particles, often metallic or ceramic, within a base fluid, typically water or oil. The resultant nanofluid exhibits a mesmerizing array of properties that defy conventional fluid behaviour. With their unparalleled ability to enhance thermal conductivity, heat transfer efficiency, and mechanical stability, HNFs are becoming indispensable in applications ranging from advanced electronics cooling to energy-efficient transportation. An experiment was conducted by Madhesh and Kalaiselvam [1] to study the behaviour of the HNF as a coolant. Sah and Ali [2] discussed the application of HNF in solar systems due to its heat transfer characteristics. Additionally, the form (shape) of the nanoparticles is crucial for the transport of mass and heat. Iqbal et al. [3] described the influence of the shape factor on HNFs and also compared different shapes for heat transfer.

In the realm of cutting-edge nanotechnology, one material stands out as a true marvel of science and innovation: graphene. Graphene's role as a nanoparticle is nothing short of impressive, revolutionizing industries and promising a future that once seemed confined to the realm of science fiction. Its extraordinary strength, conductivity, and versatility have ignited a revolution in various fields. From reinforcing materials to revolutionizing electronics, energy, medicine, and environmental solutions, graphene's impact is profound and promising. As we harness its capabilities, the future brims with exciting possibilities where this remarkable nanoparticle reshapes industries and addresses global challenges with innovation. Lee et al. [4] crafted paper sheets from reduced graphene oxide (GO) and then assessed the material's electrical, thermal, and mechanical traits. Their investigation uncovered impressive thermal and electrical conductivity, as well as enhanced tensile strength. Zakeri and Emami [5] conducted an experiment to explore the heat transfer (HT) characteristics of GO nanofluid within a double-pipe heat exchanger. Various neural network models were employed for a numerical investigation of this heat transfer process. MoS_2 nanoparticles, composed of molybdenum and sulphur, are captivating due to their versatility. They excel in reducing friction in machinery, enhancing electronics with high conductivity, catalysing clean energy reactions, and improving energy storage. Additionally, they find applications in targeted drug delivery and precise medical imaging. MoS_2 nanoparticles represent a multifaceted nanotechnology poised to revolutionize numerous industries and technologies. Arif et al. [6] simulated the flow behaviour of GO-MoS_2/engine oil (HNF) in a vertical cylinder (oscillating).

Wavelets play a pivotal role in the computational modelling and analysis of fluid dynamics. Their adaptability, multi-resolution capabilities, and efficiency make them a valuable asset for researchers and engineers striving to solve complex fluid dynamic equations accurately and with reduced computational overhead. Razzagi and Yousefi [7] first defined the operational matrix of integration for the Legendre method. Gupta et al. [8] applied wavelets (Legendre) with the aid of the collocation method to simulate the flow and heat transfer of an HNF which provides the most precise and accurate solution in comparison to other methods. Shiralashetti et al. [9] used the Legendre method to study the radiation effect on the natural convection of a porous vertical plate.

In this study, we employ the Legendre wavelet method to investigate the flow of a HNF (GP-MoS_2/EG-H_2O) over a non-linear porous stretching sheet. Our aim was to analyse the synergistic effect of viscous dissipation and Ohmic heating, a critical consideration in practical applications. The Legendre wavelet method offers a robust approach for this exploration, allowing us to gain valuable insights into the complex dynamics of HNFs under these conditions. The heat transfer enhancement is discussed in a later section. It is observed that skin friction is elevated with the Hartree pressure gradient and Eckert number, while it decreased with porosity and suction.

17.2 MATHEMATICAL MODEL

In this research investigation, we delve into the complex dynamics of two-dimensional unsteady flow over an elongating surface submerged within a porous medium. We adopt the Cartesian coordinates (x,y) to establish a clear framework, designating the horizontal direction as x and the vertical direction as y, as visually depicted in Figure 17.1. The fluid flow under scrutiny exhibits characteristics such as instability, viscosity, electrical conductivity, and incompressibility. The surface of interest is assumed to be aligned parallel to the x-axis, while the y-axis extends perpendicularly from it. Additionally, a magnetic field is applied at an oblique angle α relative to the y direction.

It is worth emphasizing that, in our analysis, we consider the magnetic field (induced) to be significantly smaller in magnitude compared to the applied magnetic field. Furthermore, we incorporate the effects of Ohmic heating, viscous dissipation, and the presence of the porous medium into the governing equations to comprehensively address the system's behaviour.

Based on the assumptions outlined previously, we can formulate the governing equations as follows ([10]):

$$u_x + v_y = 0, \tag{17.1}$$

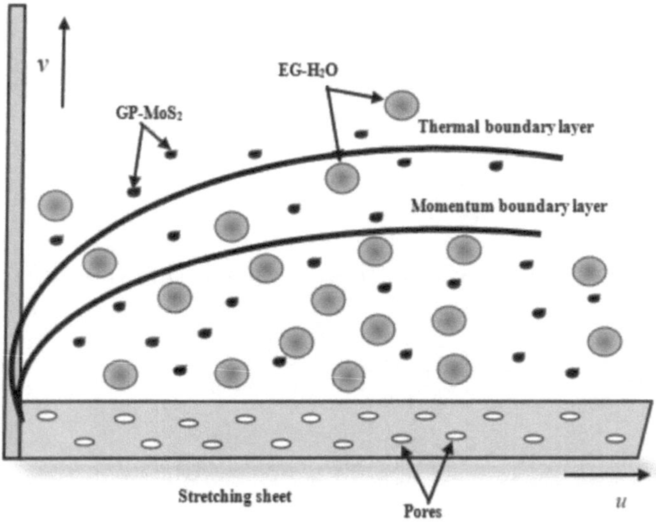

FIGURE 17.1 Geometry of the model.

$$\rho_{hf}\left(u_t+uu_x+vu_y\right)=\frac{\partial}{\partial y}\left(\mu_{hf}u_y\right)-\sigma_{hf}B_0{}^2u\sin^2\alpha-\frac{\mu_{hf}u}{k_p^*}, \tag{17.2}$$

$$\left(\rho C_p\right)_{hf}\left(T_t+uT_x+vT_y\right)=k_{hf}T_{yy}+\mu_{hf}u_y^2+\sigma_{hf}B_0{}^2u^2\sin^2\alpha+\frac{\mu_{hf}u^2}{k_p^*}. \tag{17.3}$$

In the analysis, viscosity is considered as temperature-dependent:

$$\mu_f=\frac{\mu_\infty}{1+\omega^*\left(T-T_\infty\right)}.$$

The boundary constraints are given as

$$\left.\begin{array}{lr} t>0\ :\ u=U_w, v=V_w, T=T_w, & y=0,\\ t\le 0\ :\ u=0,\ v=0, T=T_\infty, & y\ge 0,\\ \quad u\to 0, T\to T_\infty, & y\to\infty. \end{array}\right\}. \tag{17.4}$$

Here U_w, V_w, T_w, B_0 and k_p^* are assumed to be time- and space-dependent, which can be defined as

$$\begin{gathered} U_w=\frac{b}{1-dt}x^m, B_0=\frac{B_1}{\sqrt{1-dt}}x^{\frac{m-1}{2}}, k_p^*=k_1\frac{1-dt}{x^{m-1}},\\ T_w=T_\infty+\frac{T_r}{\left(1-dt\right)^2}\left(\frac{x}{L}\right)^{2m}, V_w=-\frac{V_0}{\sqrt{1-dt}}x^{\frac{m-1}{2}}. \end{gathered} \tag{17.5}$$

The similarity transformation used to transform the equations is

$$\left.\begin{aligned} &u=\frac{bx^m}{1-dt}f'(\eta), v=-\sqrt{\frac{b\nu_f(m+1)}{2(1-dt)}}x^{\frac{m-1}{2}}\left\{\left(\frac{m-1}{m+1}\right)\eta f'(\eta)+f(\eta)\right\}, \\ &\theta=\frac{T-T_\infty}{T_w-T_\infty}, \eta=y\sqrt{\frac{m+1}{2\nu_f x}U_w}. \end{aligned}\right\} \quad (17.6)$$

On applying the similarity transformations, Equations (17.1)–(17.4) are transmuted as follows:

$$\begin{aligned} &\Omega_1\frac{f'''}{(1+\varepsilon\theta)}=\omega\Omega_2(2-\beta)\left(\frac{\eta}{2}f''+f'\right)+\Omega_2\beta f'^2 \\ &-\Omega_2 f f''+\Omega_3\Omega_4(2-\beta)\mathrm{Mn}\, f'\sin^2\gamma+\Omega_1(2-\beta)K f'+\frac{\Omega_1\varepsilon\theta' f''}{(1+\varepsilon\theta)^2}, \end{aligned} \quad (17.7)$$

$$\begin{aligned} &\Omega_5\Omega_6\theta''-\omega Pr\Omega_7(2-\beta)\frac{\eta}{2}\theta'+Pr\Omega_7 f\theta' \\ &+(2-\beta)\Omega_3\Omega_4 Mn\, EcPr\sin^2\gamma f'^2+\Omega_1 K\, EcPr(2-\beta)f'^2=0, \end{aligned} \quad (17.8)$$

with associated boundary conditions

$$\left.\begin{aligned} &f'(0)=1, f(0)=S\sqrt{2-\beta}, f'(\infty)\to 0, \\ &\theta(0)=1, \theta(\infty)\to 0. \end{aligned}\right\}, \quad (17.9)$$

where $\omega=\frac{d}{b}x^{1-n}$, $K=\frac{\nu_f}{bK_1}$, $Mn=\frac{\sigma_f B_1^2}{b\rho_f}$, $\beta=\frac{2m}{m+1}$, $Pr=\frac{\nu_f(\rho C_p)_f}{k_f}$,

$Ec=\frac{b^2L^{2n}}{T_r(C_p)_f}$, $S=\frac{\nu_0}{\sqrt{b\nu_f}}$, and $\varepsilon=\omega^*(T_w-T_\infty)$ represent unsteadiness parameter, porosity, magnetic field, Hartree pressure gradient, Prandtl number, Eckert number, suction, and variable viscosity parameter, respectively. Furthermore,

$$\Omega_1=\frac{\mu_{hf}}{\mu_f},\ \Omega_2=\frac{\rho_{hf}}{\rho_f},\ \Omega_3=\frac{\sigma_{hf}}{\sigma_{nf}},\ \Omega_4=\frac{\sigma_{nf}}{\sigma_f},\ \Omega_5=\frac{k_{hf}}{k_{nf}},\ \Omega_6=\frac{k_{nf}}{k_f},\ \Omega_7=\frac{(\rho C_p)_{hf}}{(\rho C_p)_f}$$

represent dimensionless parameters.

17.2.1 Thermophysical Relations for Hybrid Nanofluid [11]

$$\left.\begin{aligned}
&\frac{\mu_{hf}}{\mu_f}=(1-\phi_{GP}-\phi_{MoS_2})^{-2.5},\frac{\rho_{hf}}{\rho_f}=\phi_{GP}\left(\frac{\rho_{GP}}{\rho_f}\right)+(1-\phi_{GP}-\phi_{MoS_2})+\phi_{MoS_2}\left(\frac{\rho_{MoS_2}}{\rho_f}\right),\\
&\frac{(\rho C_P)_{hf}}{(\rho C_P)_f}=\phi_{GP}\left(\frac{(\rho C_P)_{GP}}{(\rho C_P)_f}\right)+(1-\phi_{GP}-\phi_{MoS_2})+\phi_{MoS_2}\left(\frac{(\rho C_P)_{MoS_2}}{(\rho C_P)_f}\right),\\
&\sigma_{hf}=\frac{\sigma_{MoS_2}+2\sigma_{bf}+2\phi_{MoS_2}(\sigma_{MoS_2}-\sigma_{bf})}{\sigma_{MoS_2}+2\sigma_{bf}+\phi_{MoS_2}(\sigma_{bf}-\sigma_{MoS_2})}\sigma_{bf},\\
&\sigma_{bf}=\frac{\sigma_{GP}+2\sigma_f+2\phi_{GP}(\sigma_{GP}-\sigma_f)}{\sigma_{GP}+2\sigma_f+\phi_{GP}(\sigma_f-\sigma_{GP})}\sigma_f,\\
&\kappa_{hf}=\frac{\kappa_{MoS_2}+2\kappa_{bf}+2\phi_{MoS_2}(\kappa_{MoS_2}-\kappa_{bf})}{\kappa_{MoS_2}+2\kappa_{bf}+\phi_{MoS_2}(\kappa_{bf}-\kappa_{MoS_2})}\kappa_{bf},\\
&\kappa_{bf}=\frac{\kappa_{GP}+2\phi_{GP}(\kappa_{GP}-\kappa_f)+2\kappa_f}{\kappa_{GP}+2\kappa_f-\phi_{GP}(\kappa_{GP}-\kappa_f)}\kappa_f.
\end{aligned}\right\}$$

Here, ϕ stands for volume fraction. Table 17.1 shows the thermal attributes of nanoparticles and the conventional fluid.

17.3 SOLUTION OF THE PROBLEM

The present model is solved using the Legendre wavelet collocation technique, in which a semi-infinite domain is transformed into a finite domain using suitable transformation. This method uses Legendre polynomial and also operational matrix

TABLE 17.1
Thermophysical Properties of the Base Fluid and Nanoparticles ([12–14])

Property	50%W+50%EG	Graphene	MoS_2
Density (kg m^{-3})	1056	2250	5060
Specific heat (J kg^{-1} K^{-1})	3288	2100	397.21
Electrical conductivity (Sm^{-1})	0.00509	1×10^7	2.09×10^4
Thermal conductivity (W m^{-1} K^{-1})	0.425	2500	904.4
Prandtl number	29.86	-	-

of integration. Here, the transformation used is $s = e^{-\eta}$, and Equations (17.1)–(17.4) are transformed into following form:

$$-s^3\frac{d^3 f}{ds^3} + \left[-3 - \frac{\omega\Omega_2}{\Omega_1}(1+\varepsilon\theta)\left(-(2-\beta)\frac{\ln s}{2} + \frac{\Omega_2}{\Omega_1}f + \frac{\varepsilon s}{(1+\varepsilon\theta)^2}\frac{d\theta}{ds}\right)\right]s^2\frac{d^2 f}{ds^2} + \left[\begin{array}{l} -1 + \frac{\omega\Omega_2}{\Omega_1}(1+\varepsilon\theta)(2-\beta)\frac{\ln s}{2} + (1+\varepsilon\theta) + \frac{\Omega_2}{\Omega_1}f \\ + \frac{\Omega_3\Omega_4}{\Omega_1}(2-\beta)Mn\sin^2\alpha + K(2-\beta) + \frac{\varepsilon s}{(1+\varepsilon\theta)^2}\frac{d\theta}{ds} \end{array}\right] s\frac{df}{ds} - \frac{\beta\Omega_2}{\Omega_1}s^2\left(\frac{df}{ds}\right)^2 = 0, \tag{17.10}$$

$$s^2\frac{d^2\theta}{ds^2} + \left[1 + Pr\frac{\omega\Omega_7}{\Omega_5\Omega_6}\frac{(2-\beta)}{2}\ln\frac{1}{s} - \frac{Pr\Omega_7}{\Omega_5\Omega_6}f\right]s\frac{d\theta}{ds} + \frac{(2-\beta)EcPr}{\Omega_5\Omega_6}\left[Mn\Omega_3\Omega_4\sin^2\alpha + K\Omega_1\right]s^2\left(\frac{df}{ds}\right)^2 = 0, \tag{17.11}$$

$$\left.\begin{array}{r} f'(1) = 1, f(1) = S\sqrt{2-\beta}, f'(0) \to 0, \\ \theta(0) = 0, \theta(1) = 1. \end{array}\right\} \tag{17.12}$$

Now introducing Legendre wavelet ([7, 8]),

$$\frac{d^3 f}{ds^3} = J^T\Psi$$

and

$$\frac{d^2\theta}{ds^2} = L^T\Psi, \tag{17.13}$$

where $\Psi(s) = \left[\Psi_{10}, \Psi_{11}, \ldots\Psi_{1(P-1)}, \Psi_{20}, \Psi_{21}, \ldots\Psi_{2(P-1)}, \ldots\Psi_{2^{K-1}0}, \ldots\Psi_{2^{K-1}(P-1)}\right]$.

The residuals are obtained using Equation (17.13) in Equations (17.10) and (17.11) as follows:

$$R_f(s,J,L) = -s^3 J^T\Psi + \left[\begin{array}{l} -3 - \frac{\omega\Omega_2}{\Omega_1}\left(1 + \varepsilon\left(\begin{array}{l} s\left(1 - L^T s^2\Psi(1)\right) \\ + L^T s^2\Psi \end{array}\right)\right) \\ \left(\begin{array}{l} -(2-\beta)\frac{\ln s}{2} + \frac{\Omega_2}{\Omega_1}\left(\begin{array}{l} S\sqrt{2-\beta} - \frac{1}{2}\left(1 - J^T s^2\Psi(1)\right) - J^T s^3\Psi(1) + \\ \frac{s^2}{2}\left(1 - J^T s^2\Psi(1)\right) + J^T s^3\Psi \end{array}\right) \\ + \frac{\varepsilon s}{\left(1 + \varepsilon\left(s\left(1 - L^T s^2\Psi(1)\right) + L^T s^2\Psi\right)\right)^2}\left(1 - L^T s^2\Psi(1) + L^T s\Psi\right) \end{array}\right) \end{array}\right]$$

$$s^2\left(1 - J^T s^2\Psi(1) + J^T s\Psi\right) +$$

$$\begin{bmatrix} -1-\frac{\omega\Omega_2}{\Omega_1}\left(1+\varepsilon\left(s\left(1-L^T s^2\Psi(1)\right)+L^T s^2\Psi\right)\right)(2-\beta)\frac{(-\ln s)}{2} \\ +\left(1+\varepsilon\left(s\left(1-L^T s^2\Psi(1)\right)+L^T s^2\Psi\right)\right)+\frac{\Omega_2}{\Omega_1}f+\frac{\Omega_3\Omega_4}{\Omega_1}(2-\beta)Mn\sin^2\alpha+K(2-\beta)+ \\ \frac{\varepsilon s}{\left[\left(1+\varepsilon\left(s\left(1-L^T s^2\Psi(1)\right)+L^T s^2\Psi\right)\right)\right]^2}\left(1-L^T s^2\Psi(1)+L^T s\Psi\right) \end{bmatrix} \tag{17.14}$$

$$s\left(s\left(1-J^T s^2\Psi(1)\right)+J^T s^2\Psi\right)-\frac{\beta\Omega_2}{\Omega_1}s^2\left(s\left(1-J^T s^2\Psi(1)\right)+J^T s^2\Psi\right)^2,$$

$$R_\theta(s,J,L)$$

$$=s^2\left(L^T\Psi\right)+\begin{bmatrix} 1+Pr\frac{\omega\Omega_7}{\Omega_5\Omega_6}\frac{(2-\beta)}{2}\ln\frac{1}{s} \\ -\frac{Pr\Omega_7}{\Omega_5\Omega_6}\begin{pmatrix} S\sqrt{2-\beta}-\frac{1}{2}\left(1-J^T S^2\Psi(1)\right)-J^T S^3\Psi(1)+ \\ \frac{s^2}{2}\left(1-J^T S^2\Psi(1)\right)+J^T S^3\Psi \end{pmatrix} \end{bmatrix} \tag{17.15}$$

$$s\left(1-L^T S^2\Psi(1)+L^T S\Psi\right)$$

$$+\frac{(2-\beta)EcPr}{\Omega_5\Omega_6}\left[Mn\Omega_3\Omega_4\sin^2\alpha+K\Omega_1\right]s^2\left(s\left(1-J^T S^2\Psi(1)\right)+J^T S^2\Psi\right)^2,$$

and

$$\int_0^s \Psi(t)\,dt=S\Psi(s), \tag{17.16}$$

where S is the operational matrix of integration [7].

Moreover, through the utilization of the collocation method, we derive non-linear algebraic equations that are subsequently resolved using the Newton–Raphson technique. Ultimately, an inverse transformation is employed to yield the desired results.

17.4 QUANTITIES OF ENGINEERING IMPORTANCE [15]

Skin friction (SF) is a quantity of engineering interest that describes the resistance to the motion of a fluid (typically a gas or liquid) along a solid surface. It represents the degree of resistance that the fluid imparts to the surface as it moves across the surface. It can be described as

$$C_f=\frac{\mu_{hf}}{\rho_f U_w^2}\left(u_y\right)_{y=0}, \tag{17.17}$$

The Nusselt number (Nu) is used to assess how effectively heat is transported from a solid surface to a fluid, or vice versa. Engineers use it to design and optimize heat exchangers, cooling systems, and other HT devices, as it provides insights into the efficiency of HT processes.

$$Nu_x = -\frac{x\Omega_6}{\left(T_w - T_\infty\right)}\left(T_y\right)_{y=0} \tag{17.18}$$

In the dimensionless form, these quantities can be derived as

$$C_f\sqrt{\text{Re}_x} = \frac{\Omega_1}{\sqrt{2-\beta}}f''(0), \tag{17.19}$$

$$\frac{Nu_x}{\sqrt{\text{Re}_x}} = -\frac{\Omega_5\Omega_6}{\sqrt{2-\beta}}\theta'(0), \tag{17.20}$$

where $Re_x = \frac{xU_w}{\nu_f}$ is the Reynolds number.

17.5 RESULTS AND DISCUSSION

In this section, we delve into the outcomes of our equation-solving efforts, shedding light on how various key factors influence engineering quantities. The notable parameters we explore here include the magnetic field strength, porosity, Hartree pressure gradient, unsteadiness parameter, Eckert number, and suction parameter. By default, we have assumed specific values for these parameters: $Mn = 0.2$, $K = 1$, $S = 0.05$, Ec = 0.1, $b = 0.4$, and w = 0.05. The current study's validity is established through a meticulous comparison of our findings with those of prior research conducted by Wang [16] and Devi and Devi [17], which is presented in Table 17.2.

This comparative analysis reveals a strong congruence between our work and theirs, thereby substantiating the credibility and reliability of our present investigation.

Heat transfer enhancement (HTE) is portrayed in Figure 17.2 with the increasing values of volume fraction from 0.02 to 0.10. The outcome represents an elevation of 41.43% in HTE, which can be evaluated as follows:

$$\text{HTE (in \%)} = = \frac{Nu\,\text{Re}_x^{-\frac{1}{2}}(\phi \neq 0) - Nu\,\text{Re}_x^{-\frac{1}{2}}(\phi = 0)}{Nu\,\text{Re}_x^{-\frac{1}{2}}(\phi = 0)} * 100\,.$$

We also provide graphical representations that illustrate the trends in SF and Nu along the direction of the magnetic field, allowing for a clearer understanding of how these engineering quantities behave under different conditions. Furthermore, the consequences of SF and Nu are presented in Table 17.3.

TABLE 17.2
Comparison of Numerical Values of $-\theta'(0)$

Pr	Wang [16]	Devi and Devi [17]	Present result ($\phi_1 = 0, \phi_2 = 0$)
2.0	0.9114	0.91135	0.9756
7.0	1.8954	1.89540	1.9303
20.0	3.3539	3.35390	3.3983

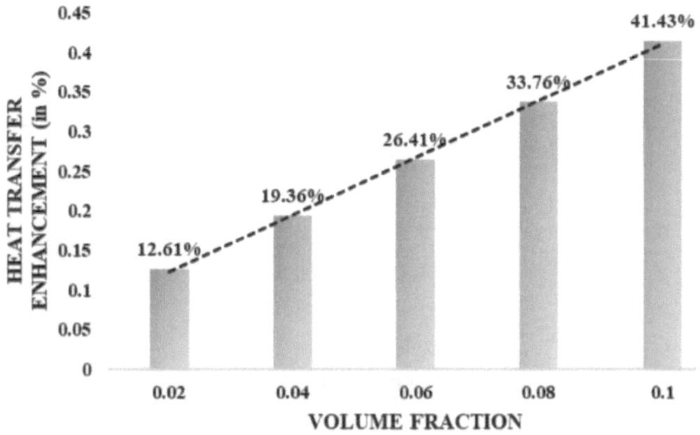

FIGURE 17.2 HTE with volume fraction (ϕ_2).

TABLE 17.3
Values of SF and Nu for Various Dimensionless Parameters

Mn	K	S	ω	Ec	ε	β	Local skin friction	Local Nusselt number
0.2	1	0.05	0.05	0.1	0.05	0.4	−1.43972	2.787520
1.4							−1.69203	2.340625
2							−1.80558	2.139518
	2						−1.83631	2.165455
	3						−2.16234	1.648741
		0.1					−1.47141	3.804692
		0.15					−1.50346	4.913891
			0.15				−1.44357	1.543480
			0.25				−1.44629	0.593195
				0.2			−1.43797	1.952445
				0.3			−1.43613	1.118703
					0.25		−1.64423	2.790007
					0.45		−1.83369	2.790895
						1	−1.43972	2.787520
						1.5	−1.38439	3.034462

17.5.1 Skin Friction Profiles

Skin friction (SF), also known as viscous friction or wall shear stress, is a type of frictional force that acts at the boundary or surface of a fluid (typically a gas or liquid) as it flows over a solid object. This force arises due to the interaction between the fluid particles and the surface of the solid. SF is an important factor in fluid dynamics and is encountered in various engineering and scientific applications. SF profiles are sketched along the magnetic field in Figures 17.3–17.7. Figure 17.3 shows the decrease in SF since higher porosity means more open spaces or voids within the

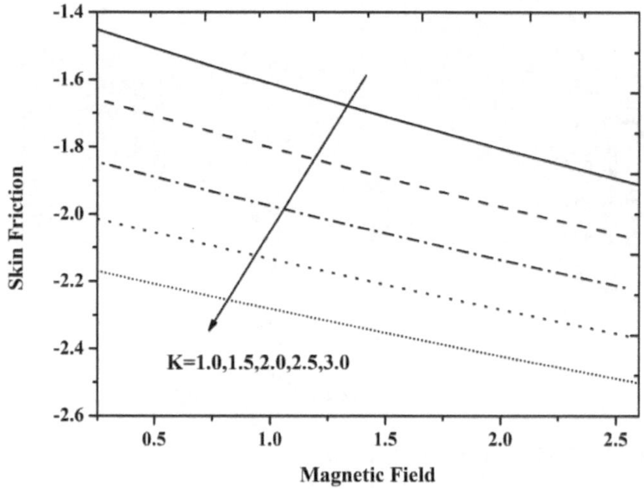

FIGURE 17.3 SF for K along the magnetic field.

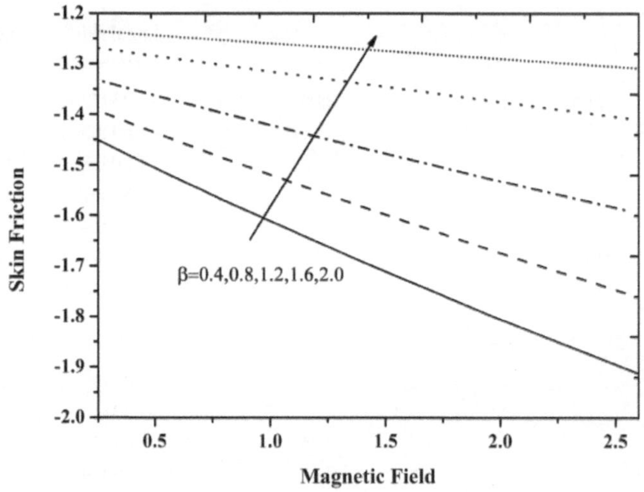

FIGURE 17.4 SF for β along the magnetic field.

porous medium. This increased porosity allows for greater fluid flow and easier passage of the fluid through the medium. Consequently, there is less resistance encountered by the fluid as it flows along the magnetic field lines, resulting in reduced SF. An increasing Hartree pressure gradient acts as a driving force, accelerating fluid flow along the magnetic field direction. This increased acceleration gives rise to heightened interaction between the liquid and the solid substrate, consequently causing a surge in SF as the fluid traverses across the surface, which is delineated in Figure 17.4. The impact of the Eckert number on SF is delineated in Figure 17.5,

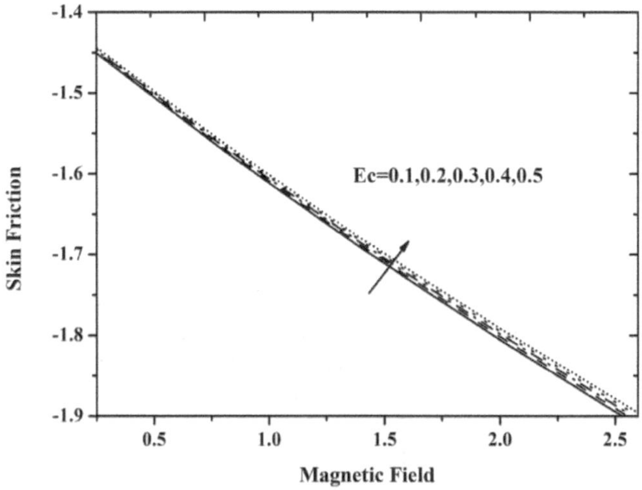

FIGURE 17.5 SF for Ec along the magnetic field.

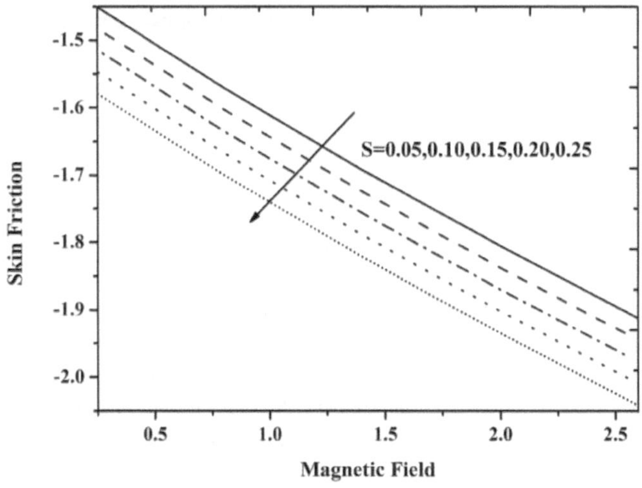

FIGURE 17.6 SF for S along the magnetic field.

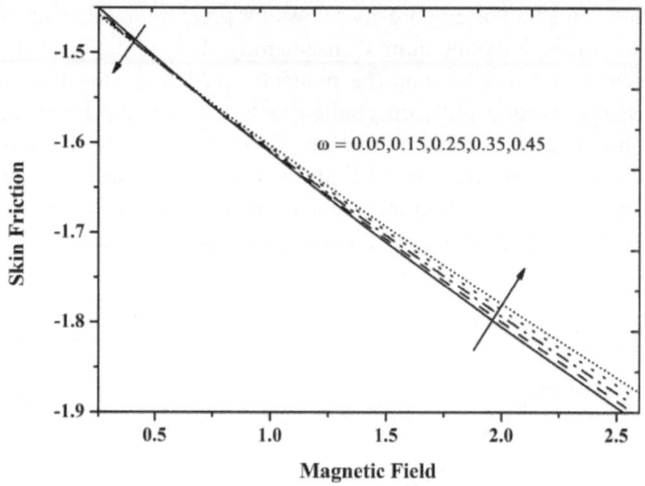

FIGURE 17.7 SF for ω along the magnetic field.

which presents an elevation in profile due to the fact that higher Eckert numbers correspond to greater thermal effects in the fluid. This increased thermal energy causes the fluid near the solid surface to become less viscous, making it flow more easily and creating more friction with the solid surface. Furthermore, it is observed that suction depreciates the SF profile (Figure 17.6). Increasing suction reduces SF along the magnetic field because it pulls fluid away from the surface, reducing the interaction between the liquid and the solid. Less contact means less friction, resulting in a decrease in SF along the magnetic field. When the unsteadiness parameter in fluid flow along a magnetic field is increased, SF exhibits a dual behaviour. Initially, as the unsteadiness parameter is elevated, the flow experiences increased turbulence, resulting in a reduction in SF. However, as the unsteadiness parameter continues to increase beyond a certain critical threshold, the flow becomes even more turbulent and disruptive. This heightened turbulence ultimately leads to an increase in SF. This dual trend is delineated in Figure 17.7.

17.5.2 Nusselt Number Profiles

The Nusselt number, Nu, is a dimensionless parameter in the field of HT. It helps us understand how well heat is transferred through the movement of fluids (convection) compared to HT through direct contact (conduction). Nusselt number profiles are presented in Figures 17.8–17.12. The porosity parameter profile is shown in Figure 17.8. Increasing porosity in a fluid–solid system reduces fluid–solid contact, which can slow down convective HT (lower Nu) along a magnetic field, making it less efficient due to the added impact of the magnetic field on the flow. Increasing the Hartree pressure gradient results in an increase in the Nu along the magnetic field because a higher pressure gradient accelerates the flow of the fluid. This enhanced flow velocity promotes more efficient convective HT, leading to a higher Nu, which signifies improved HT efficiency, which is delineated in Figure 17.9. The effect of the Eckert number on Nu is presented in Figure 17.10. It shows depreciation in the profile due to more significant thermal effects. This causes the fluid near the solid surface to become less effective at convective HT. Figure 17.11 represents the behaviour of suction. Increasing the suction levels along the magnetic field causes an increase in the Nusselt number. This is because increased suction pulls more fluid closer to the solid

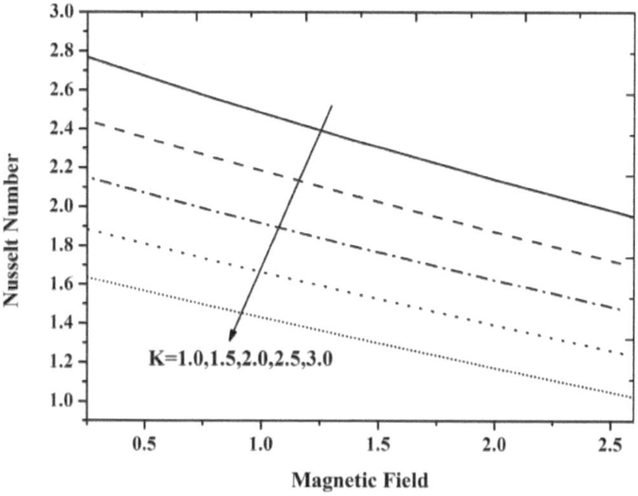

FIGURE 17.8 Nusselt number for K along the magnetic field.

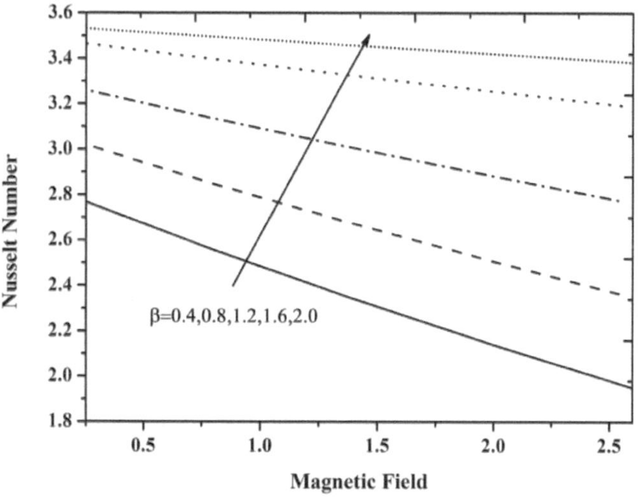

FIGURE 17.9 Nusselt number for β along the magnetic field.

surface, which intensifies the convective HT process. As an outcome, HT becomes more efficient, resulting to a higher Nu. Figure 17.12 delineates the effect of the unsteadiness parameter. Increasing the unsteadiness parameter reduces the Nu along the magnetic field because higher unsteadiness tends to disrupt the orderly flow patterns that aid HT. This disruption leads to less efficient convective HT, resulting in a lower Nu.

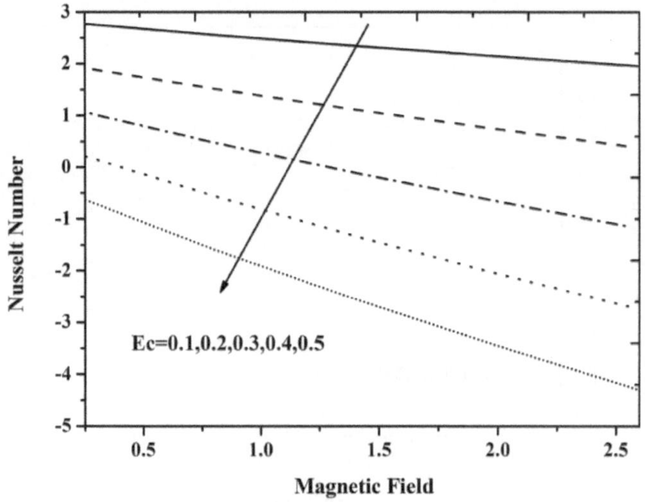

FIGURE 17.10 Nusselt number for Ec along the magnetic field.

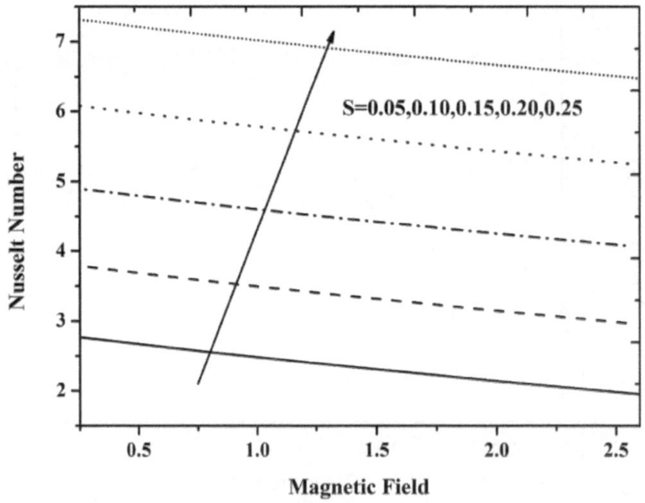

FIGURE 17.11 Nusselt number for ***S*** along the magnetic field.

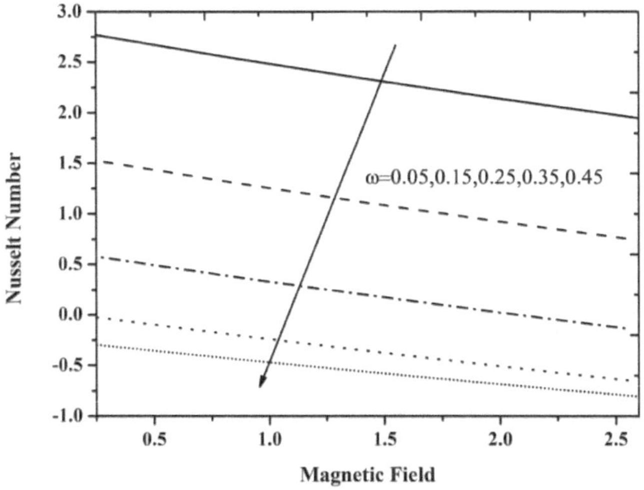

FIGURE 17.12 Nusselt number for ω along the magnetic field.

17.6 CLOSING REMARKS

In this study, the dynamic behaviour of a two-dimensional unsteady flow of a HNF, especially GP-MoS_2/EG-H_2O, over a non-linear stretching sheet was investigated. The work included a thorough investigation of the interactions among Ohmic heating, viscous dissipation, and the presence of a porous medium in this complex fluid flow scenario. The study yielded significant insights and findings:

- Increasing the volume fraction of MoS_2 results in a substantial 41.43% increase in HT performance.
- Dual behaviour is obtained for SF on elevating the unsteadiness parameter along the magnetic field.
- SF profiles are elevated for the Eckert number and Hartree pressure gradient.
- Nusselt number profiles are accelerated for the Hartree pressure gradient and suction parameter.
- SF decreases with suction and porosity, while the Nusselt number decreases the function of the Eckert number, unsteadiness parameter, and porosity.

REFERENCES

1. Madhesh, D., Kalaiselvam, S. (2014). Experimental analysis of hybrid nanofluid as a coolant. Procedia Engineering, 97, 1667–1675.
2. Shah, T. R., Ali, H. M. (2019). Applications of hybrid nanofluids in solar energy, practical limitations and challenges: A critical review. Solar Energy, 183, 173–203.
3. Iqbal, Z., Maraj, E. N., Azhar, E., Mehmood, Z. (2017). A novel development of hybrid (MoS_2–SiO_2/H_2O) nanofluidic curvilinear transport and consequences for effectiveness of shape factors. Journal of the Taiwan Institute of Chemical Engineers, 81, 150–158.

4. Lee, W., Lee, J. U., Jung, B. M., Byun, J. H., Yi, J. W., Lee, S. B., Kim, B. S. (2013). Simultaneous enhancement of mechanical, electrical and thermal properties of graphene oxide paper by embedding dopamine. Carbon, 65, 296–304.
5. Zakeri, F., Emami, M. R. S. (2023). Experimental and numerical investigation of heat transfer and flow of water-based graphene oxide nanofluid in a double pipe heat exchanger using different artificial neural network models. International Communications in Heat and Mass Transfer, 148, 107002.
6. Arif, M., Kumam, P., Khan, D., Watthayu, W. (2021). Thermal performance of GO-MoS_2/engine oil as Maxwell hybrid nanofluid flow with heat transfer in oscillating vertical cylinder. Case Studies in Thermal Engineering, 27, 101290.
7. Razzaghi, M., Yousefi, S. (2001). The Legendre wavelets operational matrix of integration. International Journal of Systems Science, 32(4), 495–502.
8. Gupta, T., Pandey, A. K., Kumar, M. (2023). Numerical study for temperature-dependent viscosity based unsteady flow of GP-MoS_2/$C_2H_6O_2$-H_2O over a porous stretching sheet. Numerical Heat Transfer, Part A: Applications, 1–22.
9. Shiralashetti, S. C., Harishkumar, E., Hanaji, S. (2023). Legendre wavelet operational matrix method for the analysis of thermal radiation effect on natural convection of a vertical plate embedded in a saturated porous medium. International Journal of Ambient Energy, 44(1), 1512–1521.
10. Nandi, S., Kumbhakar, B., Sarkar, S. (2022). MHD stagnation point flow of Fe_3O_4/Cu/Ag-CH_3OH nanofluid along a convectively heated stretching sheet with partial slip and activation energy: Numerical and statistical approach. International Communications in Heat and Mass Transfer, 130, 105791.
11. Nisar, K. S., Khan, U., Zaib, A., Khan, I., Baleanu, D. (2020). Exploration of aluminum and titanium alloys in the stream-wise and secondary flow directions comprising the significant impacts of magnetohydrodynamic and hybrid nanofluid. Crystals, 10(8), 679.
12. Xie, H., Jiang, B., Liu, B., Wang, Q., Xu, J., Pan, F. (2016). An investigation on the tribological performances of the SiO_2/MoS_2 hybrid nanofluids for magnesium alloy-steel contacts. Nanoscale Research Letters, 11(1), 1–17.
13. Acharya, N., Mabood, F. (2021). On the hydrothermal features of radiative Fe_3O_4–graphene hybrid nanofluid flow over a slippery bended surface with heat source/sink. Journal of Thermal Analysis and Calorimetry, 143(2), 1273–1289.
14. Nisar, K. S., Khan, U., Zaib, A., Khan, I., Baleanu, D. (2020). Numerical simulation of mixed convection squeezing flow of a hybrid nanofluid containing magnetized ferroparticles in 50%: 50% of ethylene glycol–water mixture base fluids between two disks with the presence of a non-linear thermal radiation heat flux. Frontiers in Chemistry, 8, 792.
15. Acharya, N., Das, K., Kundu, P. K. (2017). Framing the features of MHD boundary layer flow past an unsteady stretching cylinder in presence of non-uniform heat source. Journal of Molecular Liquids, 225, 418–425.
16. Wang, C. Y. (1989). Free convection on a vertical stretching surface. Journal of Applied Mathematics and Mechanics, 69(11), 418–420.
17. Devi, S. S., Devi, S. P. (2017). Heat transfer enhancement of Cu-Al_2O_3/water hybrid nanofluid flow over a stretching sheet. Journal of Nigerian Mathematical Society, 36(2), 419–433.

18 Numerical Study of Magnetized Hybrid Nanofluid Flow over a Stretching Sheet

Alok Kumar Pandey, Himanshu Upreti, Priya Bartwal, and Ali J. Chamkha

Nomenclature

C_P	Specific heat
C_{fx}	Skin friction coefficient
Ec	Eckert number
f	Non-dimensional velocity field
F_w	Suction/injection parameter
Ha	Magnetic field parameter
K	Porosity parameter
k_{fh}	Permeability of porous medium
Nu_x	Local Nusselt Number
Pr	Prandtl number
Re	Reynolds number
T	Temperature
(u, v)	Velocity in x and y directions
v_w	Mass transfer velocity
(x, y)	Cartesian coordinates

Greek symbols

η	Similarity variable
θ	Non-dimensional temperature
κ	Thermal conductivity
$\lambda_i : 1 \leq i \leq 6$	Constants
μ	Dynamic viscosity
υ	Kinematic viscosity
ρ	Density
σ	Electrical conductivity
ϕ	Volume fraction of nanoparticle

Subscripts

Ag	Silver
Cu	Copper
f	Base fluid
hnf	Hybrid nanofluid
w	Wall
∞	Ambient condition

Superscripts

$'$	Derivative with respect to η

Abbreviations

BCs	Boundary conditions
ODEs	Ordinary differential equations
RKF	Runge–Kutta–Fehlberg

DOI: 10.1201/9781003595786-18

18.1 INTRODUCTION

The influence of magnetic materials and electric currents is known as a magnetic field. There are two parameters that describe the magnetic field at any particular location: its direction and magnitude. In various technical processes, such as the production of paper, glass, and crude oil refinement, as well as in certain geophysical investigations, the effects of the applied magnetic field on fluids are seen. A numerical solution of the problem of magnetohydrodynamic (MHD) flow and mixed convective heat inside a vertical microtube was discussed by Malvandi and Ganji [1]. The effect of a magnetic field on the Couette flow of an incompressible conducting fluid and heat transfer between two concentric vertical pipes using the shooting method was investigated by Makinde and Eegunjobi [2]. Joshi et al. [3] studied the influence of magnetic field on hybrid nanofluid (HNF) flow over a porous stretching surface. Pandey and Upreti [4] used Maxwell Garnetts and Brinkman models for magnetic NFs by addressing the heat and mass mixed convection flow. The numerical spectral analysis of electro-MHD flow of a second-grade NF on a vertical Riga plate with convection was presented by Rasool and Wakif [5] using the spectral local linearization method. Pordanjani and Aghakhani [6] examined the MHDs of water-alumina NF flowing between two inclined concentric cylinders with the radiation effect. Eegunjobi and Makinde [7] developed a mathematical model for studying magnetized two-phase NF flow with buoyancy forces.

An HNF is defined as a mixture of two or more nanoparticles and the base fluid. To enhance heat transmission, an HNF is introduced. HNFs offer a wide range of uses in industries such as manufacturing, microelectronics, medicine, and microfluidics. Hayat and Nadeem [8] explored the heat transfer enhancement on Ag-CuO/water HNF flow over a rotating stretching sheet with chemical reaction using the bvp4c technique. Experimental investigation of natural convection with (Fe_3O_4–CNT/water) HNF in a concentric annulus was inspected by Shahsavar et al. [9]. Dinarvand et al. [10] deliberated the Falkner–Skan problem in a static/moving wedge using a TiO_2-CuO/water HNF. Sheikholeslami et al. [11] scrutinized the impact of base fluid (water) and MWCNT and Fe_3O_4 nanoparticles as HNF in the heat transfer inside a circular cavity. Waini et al. [12] probed the results of assisting and opposing flows on steady mixed convection of a Cu–Al_2O_3/water HNF. Gul et al. [13] numerically compared the magnetic dipole effect on an HNF (Cu-Al_2O_3/H_2O) and NF (Cu/H_2O) in a stretching sheet. Sheikholeslami [14] conveyed the impacts of a (CNT-SiO_2/water) HNF on the solar system connected with a turbulator in a turbulent regime.

After reviewing the works, it is observed that till dvate no study has been made to discuss the heat transfer assessment of magnetized HNF flow over a stretching sheet in the presence of suction, MHD, and viscous dissipation. The impact of acting variables on temperature and velocity is shown by graphs. Moreover, the effect of existing parameters on the heat transfer rate are portrayed by bar diagrams and.elucidated in detail.

18.2 FORMULATION FOR HYBRID NANOFLUID

Assume that the motion of the HNF (Cu-Ag/H_2O-$C_2H_6O_2$) is laminar, steady, and incompressible. The velocity of the sheet is taken as $U_w(x) = cx$, where

$c = streching\ rate > 0$. Moreover, it is supposed that the ambient temperature is T_∞ and the sheet is heated due to a warm HNF with T_∞. An external magnetic field B_0 is used in the y-axis.

The terms MHD and porous medium are included in the momentum equation, and Ohmic–viscous dissipation terms are incorporated in the expression for heat. The model flow configuration is presented in Figure 18.1.

The equations of the model are [3]

$$u_x + v_y = 0, \tag{18.1}$$

$$uu_x + vu_y = \upsilon_{hnf} u_{xx} - \frac{\upsilon_{hnf}}{k_{fh}} u - \frac{\sigma_{hnf} B_0^2}{\rho_{hnf}} u, \tag{18.2}$$

$$\left(\rho C_p\right)_{hnf}\left(uT_x + vT_y\right) = \kappa_{hnf} T_{yy} + \mu_{hnf}\left(u_y\right)^2 + \sigma_{hnf} B_0^2 u^2. \tag{18.3}$$

The boundary conditions (BCs) are

$$\left.\begin{array}{l} u = U_w = cx,\ v = v_w,\ T = T_w\ when\ y = 0, \\ u = 0 = v, T = T_\infty\ when\ y \to \infty. \end{array}\right\} \tag{18.4}$$

Here for suction, v_w should be positive.

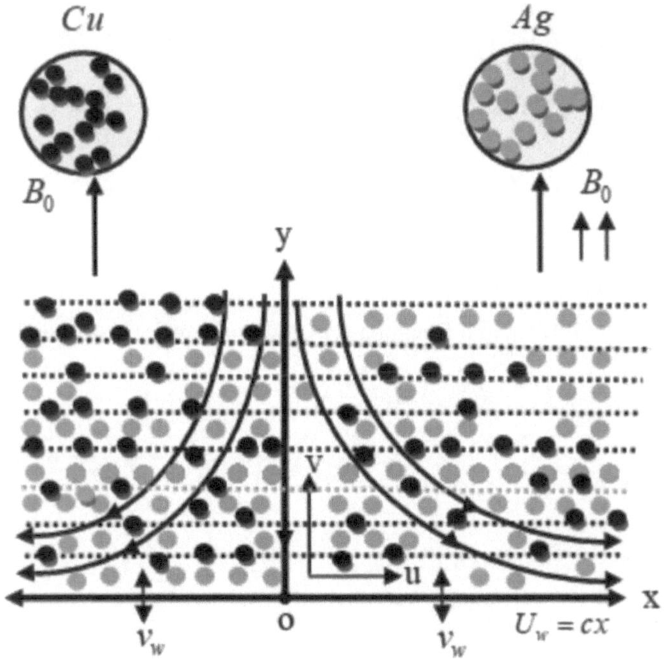

FIGURE 18.1 Geometry of the flow model.

The HNF viscosity is calculated by

$$\mu_{hnf} = \mu_f \left[1 - \left(\phi_{Cu} + \phi_{Ag}\right)\right]^{-2.5}. \tag{18.5}$$

The HNF density is evaluated by

$$\frac{\rho_{hnf}}{\rho_f} = \left(1 - \phi_{Ag} - \phi_{Cu}\right) + \phi_{Cu}\frac{\rho_{Cu}}{\rho_f} + \phi_{Ag}\frac{\rho_{Ag}}{\rho_f}. \tag{18.6}$$

The HNF electrical conductivity is obtained by

$$\frac{\sigma_{hnf}}{\sigma_f} = 1 + \frac{3\sigma_{Cu}\phi_{Cu} + \sigma_{Ag}\phi_{Ag} - 3\sigma_f\left(\phi_{Ag} + \phi_{Cu}\right)}{\sigma_{Cu}\left(1 - \phi_{Cu}\right) + \sigma_{Ag}\left(1 - \phi_{Ag}\right) + \sigma_f\left(2 + \phi_{Ag} + \phi_{Cu}\right)}. \tag{18.7}$$

The HNF heat capacitance is evaluated by

$$\frac{\left(\rho C_p\right)_{hnf}}{(\rho C_p)_f} = \left(1 - \phi_{Ag} - \phi_{Cu}\right) + \phi_{Cu}\frac{(\rho C_p)_{Cu}}{(\rho C_p)_f} + \phi_{Ag}\frac{(\rho C_p)_{Ag}}{(\rho C_p)_f}. \tag{18.8}$$

The HNF thermal conductivity is obtained by

$$\frac{\kappa_{hnf}}{\kappa_f} = \frac{\dfrac{\phi_{Cu}\kappa_{Cu} + \phi_{Ag}\kappa_{Ag}}{\left(\phi_{Cu} + \phi_{Ag}\right)} + 2\kappa_f + 2\left(\phi_{Cu}\kappa_{Cu} + \phi_{Ag}\kappa_{Ag}\right) - 2\left(\phi_{Cu} + \phi_{Ag}\right)\kappa_f}{\dfrac{\phi_{Cu}\kappa_{Cu} + \phi_{Ag}\kappa_{Ag}}{\left(\phi_{Cu} + \phi_{Ag}\right)} + 2\kappa_f - \left(\phi_{Cu}\kappa_{Cu} + \phi_{Ag}\kappa_{Ag}\right) + \left(\phi_{Cu} + \phi_{Ag}\right)\kappa_f}. \tag{18.9}$$

Now, using the following similarity transformations,

$$\begin{gathered} \left(u, v\right) = \left(cxf'(\eta), -\sqrt{cv_f}\, f(\eta)\right), \\ \eta = \sqrt{\frac{c}{v_f}}y, T - T_\infty = \theta(\eta)\left(T_w - T_\infty\right). \end{gathered} \tag{18.10}$$

Equation (18.1) is fulfilled, and Equations (18.2) and (18.3) are changed into

$$f''' - Kf' - \lambda_1 Haf' - \lambda_2\left[\left(f'\right)^2 - ff''\right] = 0, \tag{18.11}$$

$$\lambda_5\theta'' + PrEc\left[\lambda_3\left(f''\right)^2 + \lambda_6 Ha\left(f'\right)^2\right] + \lambda_4 Prf\theta' = 0, \tag{18.12}$$

and the BCs (18.4) are given as

$$\left.\begin{aligned} &f(0) = F_w, f'(0) = \theta(0) = 1, \\ &f'(\infty) = \theta(\infty) = 0 \end{aligned}\right\}, \tag{18.13}$$

where

$$\left.\begin{aligned} &K=\frac{\upsilon_f}{ck_{fh}},Ha=\frac{\sigma_f B_0^2}{\rho_f c},Pr=\frac{\mu_f\left(C_p\right)_f}{\kappa_f},Ec=\frac{U_w^{\ 2}}{\left(C_p\right)_f\left(T_w-T_\infty\right)},\\ &F_w=-\frac{v_w}{\sqrt{c\upsilon_f}},\lambda_1\left(=\frac{\sigma_{hnf}}{\sigma_f}\times\frac{\mu_f}{\mu_{hnf}}\right),\lambda_2\left(=\frac{\upsilon_f}{\upsilon_{hnf}}\right),\\ &\lambda_3\left(=\frac{\mu_{hnf}}{\mu_f}\right),\lambda_4\left(=\frac{\left(\rho C_P\right)_{hnf}}{\left(\rho C_P\right)_f}\right),\lambda_5\left(=\frac{\kappa_{hnf}}{\kappa_f}\right),\text{and }\lambda_6\left(=\frac{\sigma_{hnf}}{\sigma_f}\right) \end{aligned}\right\}. \tag{18.14}$$

The parameters for physical interest as skin friction coefficient C_{fx} and Nusselt number $\mathrm{Nu_x}$ are given as

$$C_{fx}=\frac{\tau_w}{\rho_f U_w^{\ 2}},$$

and

$$Nu_x=\frac{xq}{k_f\left(T_w-T_\infty\right)}, \tag{18.15}$$

where τ_w is the shear stress and q_w is the heat flux, which are given by

$$\tau_w=\mu_{hnf}\frac{\partial u}{\partial y}$$

and

$$q_w=-\kappa_{nf}\frac{\partial T}{\partial y}. \tag{18.16}$$

Now, using the above expressions, Equation (18.16) is converted into

$$C_{fx}\,\mathrm{Re}_x^{1/2}=\lambda_3 f''(0)$$

and

$$Nu_x\,\mathrm{Re}_x^{-1/2}=-\lambda_5\theta'(0). \tag{18.17}$$

18.3 DISCUSSION

In the previous sections, the authors addressed the problem statement and the methodology involved; in this section, the influence of parameters like *K*, F_w, Ec, and Ha on dimensionless velocity, temperature, and local Nusselt number (LNN) is discussed and shown through tables and plots. All the figures are plotted keeping Pr fixed at 6.2 and the volume fraction of dispersed nanoparticles Ag and Cu as 0.1 each. The characteristics of solid particles and hybrid base fluids are presented in Table 18.1.

To validate the accuracy of the present numerical method, a comparison with earlier published data under special case is established. In the present model, the numerical values of LNN for distinct Pr are computed and compared with those reported by Shiaq et al. [15] (see Table 18.2). And, the table reveals that the present results are in good concord.

Figures 18.2–18.10 show the influence of the aforementioned parameters on flow characteristics (velocity and temperature); these plots also show a comparison between HNF and mono-nanofluid ($Ag/H_2O\text{-}C_2H_6O_2$ and $Cu/H_2O\text{-}C_2H_6O_2$) flow characteristics. Figures 18.2 and 18.6 show the effect of increasing porosity parameter (K) on the dimensionless velocity and temperature profiles. It is perceived from these plots that velocity and the associated layer thickness decrease with escalating K; on the contrary, an opposite trend is recorded for the temperature profile. Furthermore, the variation among the profiles of distinct K is very less (see Figure 18.6), but the thermal boundary layer is wider for HNFs compared to mono-nanofluids (see Figure 18.6). The influence of increasing Hartman number (Ha) on the velocity and temperature profiles is shown in Figures 18.3 and 18.7. The Hartman number is associated to the magnetic field; the presence of a magnetic field results in the generation of a retarding force. As a result, this force retards the motion of the fluid (see Figure 18.3) and elevates the temperature of the fluid. Furthermore, Figure 18.7 shows that thermal boundary layer is wider for HNFs. Figures 18.4 and 18.9 illustrate the influence of increasing suction parameter on the velocity and temperature profiles. It is observed from the plots that both velocity and temperature of the fluid decline with the acceleration in the suction parameter. The impact of nanoparticle volume fraction $\left(\varphi_{Ag} \text{ and } \varphi_{Cu}\right)$ on the velocity and temperature profiles

TABLE 18.1
Thermophysical Properties of Hybrid Base Fluids and Nanoparticles

Physical properties	$\rho(kg/m^3)$	$\sigma(s/m)$	$\kappa(W/mK)$	$C_p(J/kgK)$
Silver (Ag)	10500	6.30×10^7	429	235
Copper (Cu)	8933	59.6×10^6	400	385
$H_2O\text{-}C_2H_6O_2$	1063.8	9.75×10^{-4}	0.387	3630

TABLE 18.2
Comparison of Data of Nusselt Number

Prandtl number	Shiaq et al. [15]	Present results
0.70	0.4539	0.4539
2.00	0.9114	0.9114
7.00	1.8954	1.8954

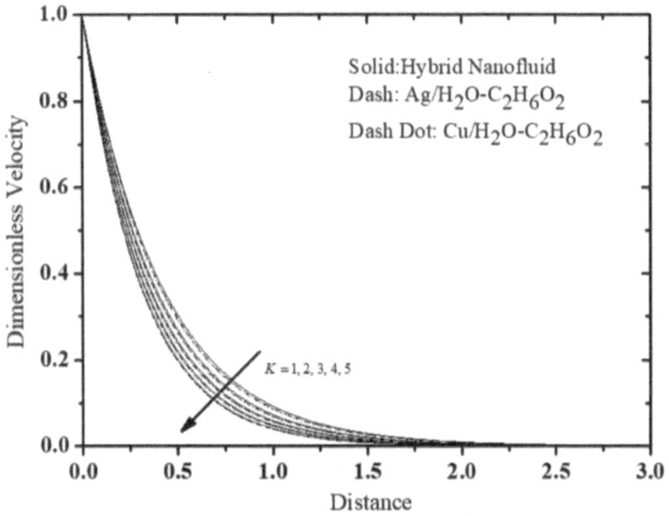

FIGURE 18.2 Effect of *K* on dimensionless velocity.

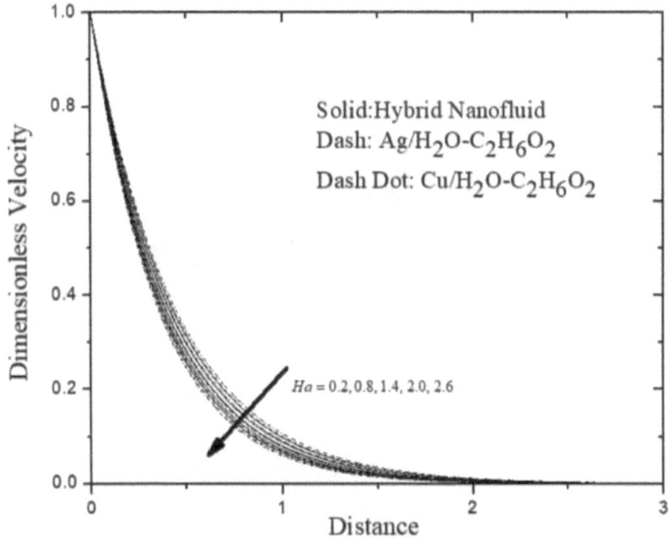

FIGURE 18.3 Effect of Ha on dimensionless velocity.

of HNFs and mono-nanofluids is shown in Figures 18.5 and 18.10. Figure 18.5 shows that increasing the concentration of either nanoparticle (Ag and Cu) in a HNF results an increase in the velocity profile, while the opposite trend was observed for mono-nanofluids (Ag/H_2O-$C_2H_6O_2$ and Cu/H_2O-$C_2H_6O_2$). However, no such contradictory pattern was observed for the temperature profile; for both HNFs and mono-nanofluids, temperature of the fluid increases with an increase in the nanoparticle

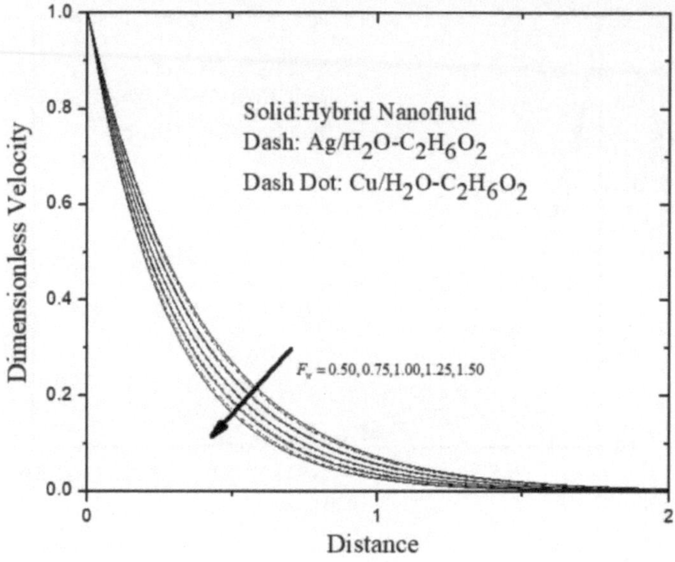

FIGURE 18.4 Effect of F_w on dimensionless velocity.

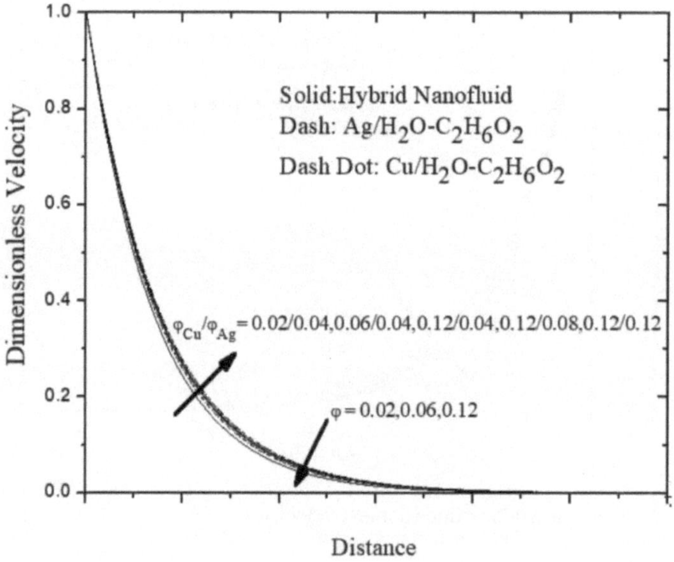

FIGURE 18.5 Effect of φ_{Ag} and φ_{Cu} on dimensionless velocity.

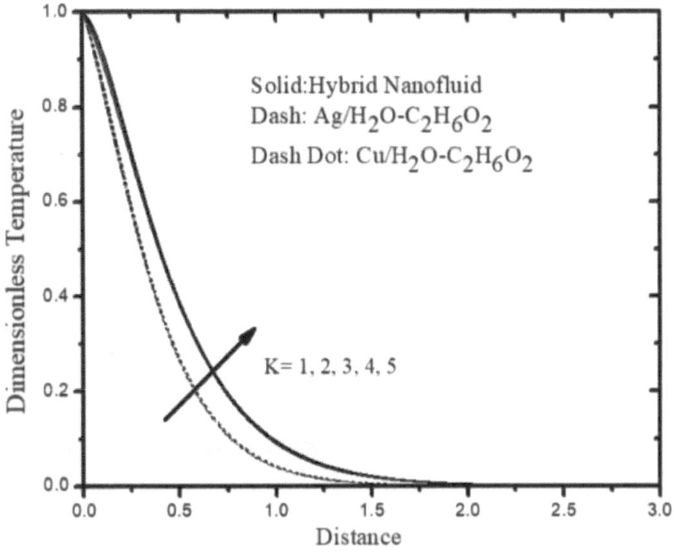

FIGURE 18.6 Effect of K on dimensionless temperature.

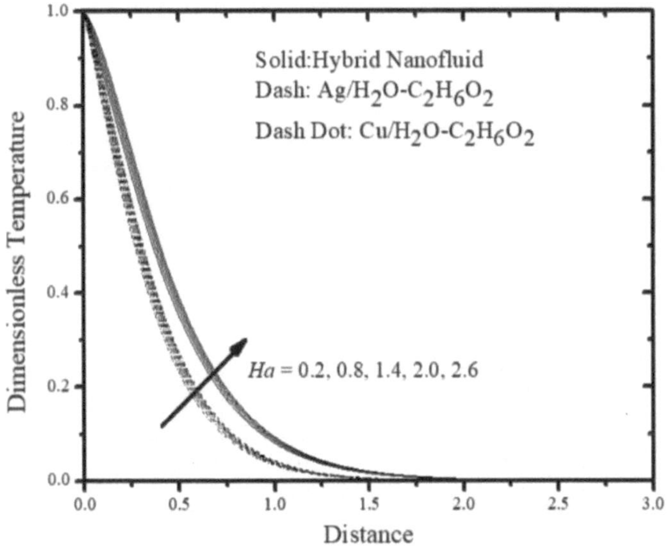

FIGURE 18.7 Effect of Ha on dimensionless temperature.

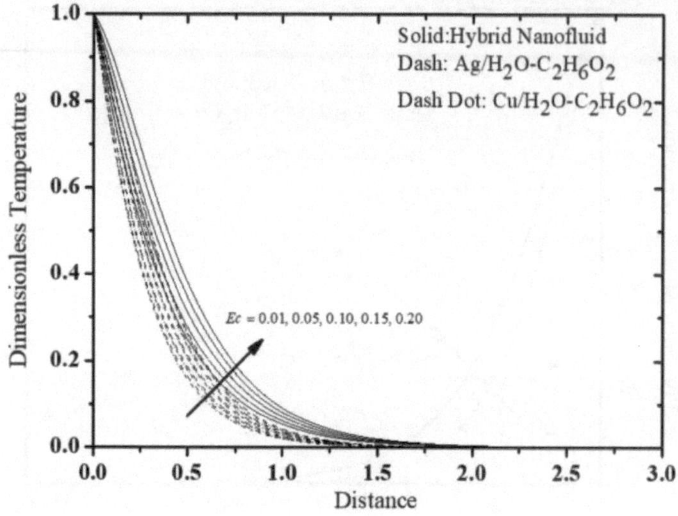

FIGURE 18.8 Effect of Ec on dimensionless temperature.

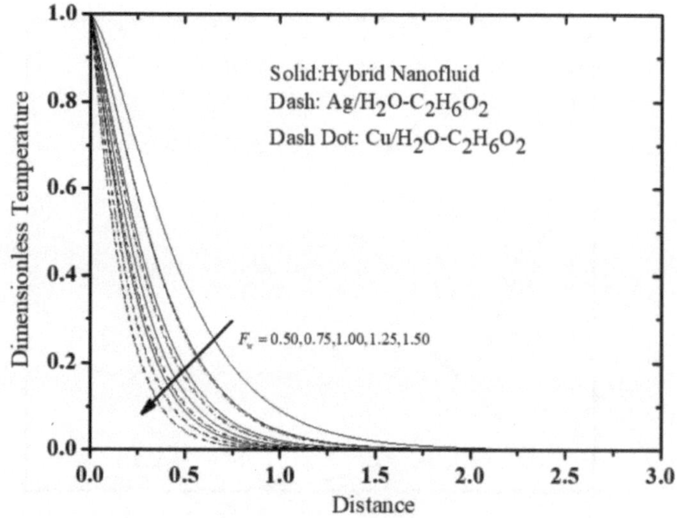

FIGURE 18.9 Effect of F_w on dimensionless temperature.

concentration in the base fluid. Figure 19.8 shows the effect of the dissipation parameter (Ec) on the temperature profile. Increasing the dissipation means more heat to the system due to which the temperature of the fluid also increases. Thus, increasing Ec results in an increase in temperature and associated layer thickness. Here, the HNF has a higher temperature compared to mono-nanofluids.

Figures 18.11(a–c) are plotted to examine the variation in LNN against Ha for distinct values of *K*, Ec, and F_w. It is noted from Figures 18.11(a, c) that LNN decreases with escalating *K* and Ec, respectively. And, LNN increases with increases in F_w for fixed Ha, but for fixed F_w it decreases with increasing Ha (see Figure 18.11(b)).

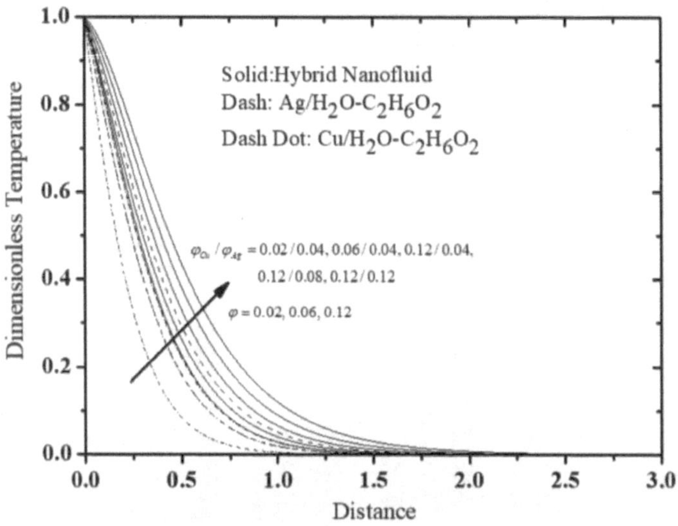

FIGURE 18.10 Effect of φ_{Ag} and φ_{Cu} on dimensionless temperature.

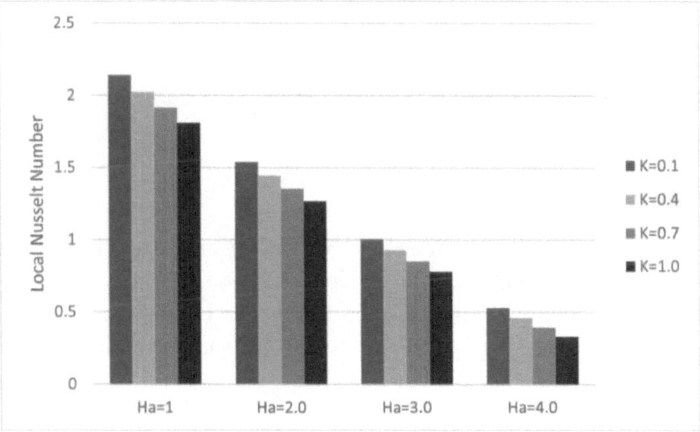

FIGURE 18.11(a) Effect of *K* and Ha on the LNN.

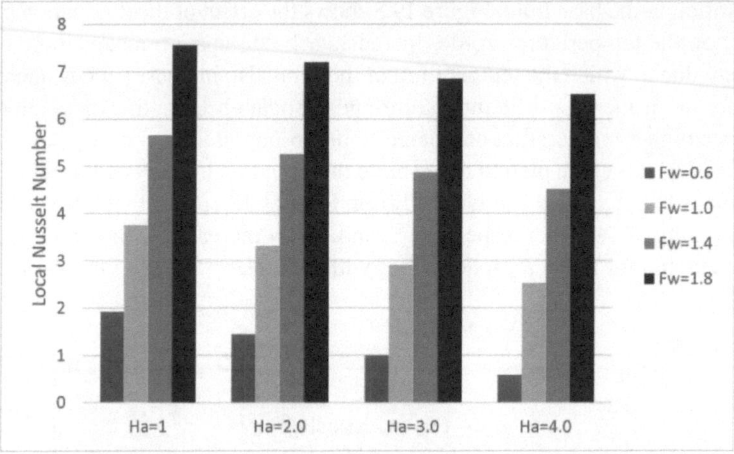

FIGURE 18.11(b) Effect of F_w and Ha on the LNN.

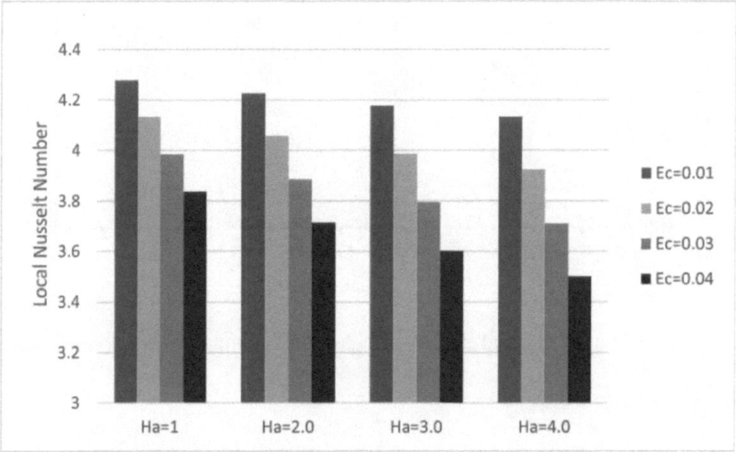

FIGURE 18.11(c) Effect of Ec and Ha on the LNN.s

18.4 CONCLUSIONS

The present work dealt with magnetized HNF flow over a stretching surface in the presence of a porous medium, viscous dissipation, and suction. Moreover, the comparative heat transfer analysis is also discussed between HNFs (Cu-Ag/H_2O-$C_2H_6O_2$) and mono-NFs (Cu/H_2O-$C_2H_6O_2$ and Ag/H_2O-$C_2H_6O_2$). The main outcomes of this work are as follows:

- Heat transfer rate declined with an increase in the Eckert number and porosity parameter.

- An increase in the suction parameter resulted in a decrease in both temperature and velocity outlines.
- The temperature profile constantly reduces with an increase in the values of K and Ha.
- Due to an increase in the values of solid volume fractions, the dual behaviour in the velocity profiles is noticed.

REFERENCES

1. Malvandi, A., & Ganji, D. D. (2014). Magnetohydrodynamic mixed convective flow of Al_2O_3–water nanofluid inside a vertical microtube. Journal of Magnetism and Magnetic Materials, 369, 132–141.
2. Makinde, O. D., & Eegunjobi, A. S. (2020). Entropy analysis of a variable viscosity MHD Couette flow between two concentric pipes with convective cooling. Engineering Transactions, 68(4), 317–334.
3. Joshi, N., Pandey, A. K., Upreti, H., & Kumar, M. (2021). Mixed convection flow of magnetic hybrid nanofluid over a bidirectional porous surface with internal heat generation and a higher-rder chemical reaction. Heat Transfer, 50(4), 3661–3682.
4. Pandey, A. K., & Upreti, H. (2021). Mixed convective flow of Ag–H_2O magnetic nanofluid over a curved surface with volumetric heat generation and temperature-dependent viscosity. Heat Transfer, 50(7), 7251–7270.
5. Rasool, G., & Wakif, A. (2021). Numerical spectral examination of EMHD mixed convective flow of second-grade nanofluid towards a vertical Riga plate using an advanced version of the revised Buongiorno's nanofluid model. Journal of Thermal Analysis and Calorimetry, 143, 2379–2393.
6. Pordanjani, A. H., & Aghakhani, S. (2022). Numerical investigation of natural convection and irreversibilities between two inclined concentric cylinders in presence of uniform magnetic field and radiation. Heat Transfer Engineering, 43(11), 937–957.
7. Eegunjobi, A. S., & Makinde, O. D. (2023). Second law analysis of MHD convection of a radiating nanofluid within the gap between two inclined concentric pipes. International Journal of Modern Physics B, 37(16), 2350153.
8. Hayat, T., & Nadeem, S. (2017). Heat transfer enhancement with Ag–CuO/water hybrid nanofluid. Results in Physics, 7, 2317–2324.
9. Shahsavar, A., Sardari, P. T., & Toghraie, D. (2018). Free convection heat transfer and entropy generation analysis of water-Fe_3O_4/CNT hybrid nanofluid in a concentric annulus. International Journal of Numerical Methods for Heat & Fluid Flow, 29(3), 915–934.
10. Dinarvand, S., Rostami, M. N., & Pop, I. (2019). A novel hybridity model for TiO_2-CuO/water hybrid nanofluid flow over a static/moving wedge or corner. Scientific Reports, 9(1), 16290.
11. Sheikholeslami, M., Mehryan, S. A. M., Shafee, A., & Sheremet, M. A. (2019). Variable magnetic forces impact on magnetizable hybrid nanofluid heat transfer through a circular cavity. Journal of Molecular Liquids, 277, 388–396.
12. Waini, I., Ishak, A., Groşan, T., & Pop, I. (2020). Mixed convection of a hybrid nanofluid flow along a vertical surface embedded in a porous medium. International Communications in Heat and Mass Transfer, 114, 104565.
13. Gul, T., Khan, A., Bilal, M., Alreshidi, N. A., Mukhtar, S., Shah, Z., & Kumam, P. (2020). Magnetic dipole impact on the hybrid nanofluid flow over an extending surface. Scientific Reports, 10(1), 8474.

14. Sheikholeslami, M. (2022). Numerical investigation of solar system equipped with innovative turbulator and hybrid nanofluid. Solar Energy Materials and Solar Cells, 243, 111786.
15. Shaiq, S., Maraj, E. N., & Iqbal, Z. (2019). Remarkable role of $C_3H_8O_2$ on transportation of MoS_2-SiO_2 hybrid nanoparticles influenced by thermal deposition and internal heat generation. Journal of Physics and Chemistry of Solids, 126, 294–303.

Index